clinical epidemiology and biostatistics

The National Medical Series for Independent Study

clinical epidemiology and biostatistics

Rebecca G. Knapp, Ph.D.

Associate Professor
Department of Biostatistics, Epidemiology and Systems Science
Medical University of South Carolina
Charleston, South Carolina

M. Clinton Miller III, Ph.D.

Professor and Chairman
Department of Biostatistics, Epidemiology and Systems Science
Medical University of South Carolina
Charleston, South Carolina

National Medical Series from Williams & Wilkins
Baltimore, Hong Kong, London, Sydney

Harwal Publishing Company, Malvern, Pennsylvania

Williams
& Wilkins

Library of Congress Cataloging-in-Publication Data

Knapp, Rebecca Grant.
Clinical epidemiology and biostatistics / Rebecca Grant Knapp, M. Clinton Miller, III.
p. cm. — (The National medical series for independent study)
ISBN 0-683-06206-9
1. Biometry—Outlines, syllabi, etc. 2. Biotmetry—Examinations, questions, etc. I. Miller, M. Clinton. II. Title. III. Series.
[DNLM: 1. Biometry—examination questions. 2. Biometry—outlines.
WA 18 K67c]
QH323.5.K62 1992
574'.01'5195—dc20
DNLM/DLC
for Library of Congress 90-12218
CIP
r91

ISBN 0-683-06206-9

10 9 8 7 6 5 4 3 2 1

Contents

Preface ix
To the Reader xi

1 Describing Clinical Data 1

I. The "variable" 1
II. Describing variation in clinical data 3
III. Describing grouped clinical data 5
IV. Choosing appropriate descriptive statistics 10
Problems 12
Study questions 14
Answers and explanations 16

2 Probabilities in Clinical Medicine 17

I. Definition and properties of probability 17
II. The 2 × 2 table 18
III. Rules of probability 19
IV. Summary of formulas for calculating probability 23
Problems 25
Study questions 27
Answers and explanations 29

3 Describing the Performance of a Diagnostic Test 31

I. Distinguishing normal from abnormal 31
II. Describing the performance of a new diagnostic test 34
III. Revising estimates of test performance 39
IV. Evaluating a new diagnostic test 42
V. Multiple testing 43
Problems 46
Study questions 48
Answers and explanations 50

4 Defining Normality Using the Predictive Value Method 53

I. The predictive value method for selecting a positivity criterion 53
II. The receiver operator characteristic (ROC) curve 55
Problems 57

Study questions 59
Answers and explanations 60

5 Clinical Decision Analysis 61
I. Decision analysis: a strategy for solving complex problems 61
II. General steps in decision analysis 61
III. Case example: choosing an optimal diagnostic strategy 63
Problems 75
Study questions 76
Answers and explanations 79

6 Probabilities in Genetic Counseling 81
I. Genetic counseling: a diagnostic issue 81
II. Using the binomial formula to assess heritability 82
III. Using Bayes' rule to assess heritability 84
Problems 88
Study questions 89
Answers and explanations 90

7 Frequency of Disease 93
I. Definition and calculation of measures of frequency 93
II. Evaluating frequency measures 100
III. Uses of frequency measures 102
Problems 104
Study questions 106
Answers and explanations 108

8 Risk and Causality 109
I. Associations in clinical medicine 109
II. Risk 109
III. Causality 120
IV. The direct method for calculating age-adjusted rates 122
Problems 124
Study questions 126
Answers and explanations 129

9 Comparing Therapies: The Randomized Controlled Clinical Trial 131
I. The experimental approach to evaluating treatments 131
II. Conducting a randomized clinical trial 132
III. Generalizing from the study results to the target population 138
Problems 139
Study questions 140
Answers and explanations 143

10 Statistical Building Blocks 145
I. Introduction and basic definitions 145

II. The normal (gaussian) distribution 147
III. The sampling distribution and the central limit theorem 152
Problems 160
Study questions 161
Answers and explanations 163

11 Confidence Intervals 167
I. Confidence intervals: a method for estimating population values 167
II. Applications of confidence interval method 169
III. Mathematical derivation of confidence intervals 176
Problems 178
Study questions 180
Answers and explanations 183

12 Hypothesis Testing: The Basis of Statistical Reasoning 187
I. Hypothesis testing using inferential statistics 187
II. The statistical reasoning process: a simple example 187
III. The statistical reasoning process: a clinical example 189
IV. General steps in statistical hypothesis testing 192
V. The quantification of uncertainty in hypothesis testing 192
VI. Design and interpretation of clinical research studies 198
Problems 203
Study questions 204
Answers and explanations 207

13 Tests of Statistical Significance: Chi-square Procedures 209
I. Selecting an appropriate statistical test: a decision tree approach 209
II. Analyzing associations involving frequency (count) data 212
Problems 226
Study questions 228
Answers and explanations 230

14 Quantifying Risk 233
I. Defining the magnitude of associations 233
II. Measures of association (comparative risk) 234
III. Measures of potential impact 241
IV. Derivation of the odds ratio (OR) in case–control studies 244
Problems 246
Study questions 248
Answers and explanations 252

15 Tests of Statistical Significance: Regression and Correlation 255
I. Analyzing associations that involve interval/ratio (i.e., continuous) data 255

II. Corrrelation analysis 259
III. Regression analysis 263
Problems 269
Study questions 271
Answers and explanations 274

16 Tests of Statistical Significance: Paired and Pooled *t* Tests 275
I. Analyzing differences among population means 275
II. Testing the equality of two means: independent samples 275
III. Testing the equality of two means: matched samples 278
IV. Selecting an appropriate sample size 281
Problems 285
Study questions 287
Answers and explanations 291

17 Tests of Statistical Significance: Analysis of Variance 293
I. Analyzing differences among multiple comparison groups 293
II. The completely random design (CRD) 295
III. The randomized complete block (RCB) design 298
IV. Multiple comparison procedures 302
V. Computational formulas and calculations 304
Problems 308
Study questions 310
Answers and explanations 312

Comprehensive Examination 313
Questions 323
Answers and explanations 344

Solutions 359

Appendixes 409

Index 425

Preface

Answers to important clinical questions for an individual patient are often based on observation of groups of similar patients. It is the integrated science of clinical epidemiology and biostatistics that provides the tools for collecting, analyzing, and interpreting information from groups of patients. For the clinical researcher, the need for a strong foundation in clinical epidemiology and biostatistics is unquestioned; for the clinician, an understanding of the basic principles of this science is prerequisite to understanding the reports of research done by others.

NMS *Clinical Epidemiology and Biostatistics* integrates biostatistics and epidemiology with clinical perspectives and examples. The highly successful NMS outline format allows *Clinical Epidemiology and Biostatistics* to accomplish this long-needed integration in a brief, yet well-organized, volume.

The material is organized to guide the student through difficult concepts, presenting basic information first and then gradually increasing the complexity and direct application of the ideas. The discussions and problems focus on clinical problem-solving and clinical relevance rather than on the biostatistical or epidemiological theory.

While *Clinical Epidemiology and Biostatistics* is intended for medical students who need to prepare for NBME exams, FLEX, or FMGEMS, or who seek a concise self-study text, the book will also be useful to those housestaff, practicing physicians, and other health professionals who have never had a formal course in biostatistics and epidemiology but who are called upon to be critical consumers of clinical research. With the incorporation of advanced material in the appendixes, this text can also serve as a review or self-study guide for clinical researchers planning to actively collect, analyze, and interpret clinical data.

The book includes Board-type study questions at the end of each chapter, answer keys and full explanations to accompany those questions, and a comprehensive exam at the end of the book. There are also problem sets with solutions and numerous illustrations to enhance the statistical discussions.

It is our hope that students will not be intimidated by the subject matter of this text but rather become "turned on" to the study and practice of epidemiology and biostatistics.

The authors

To the Reader

Since 1984, the *National Medical Series for Independent Study* has been helping medical students meet the challenge of education and clinical training. In today's climate of burgeoning information and complex clinical issues, a medical career is more demanding than ever. Increasingly, medical training must prepare physicians to seek and synthesize necessary information and to apply that information successfully.

The *National Medical Series* is designed to provide a logical framework for organizing, learning, reviewing, and applying the conceptual and factual information covered in basic and clinical sciences. Each book includes a comprehensive outline of the essential content of a discipline, with up to 500 study questions. The combination of an outlined text and tools for self-evaluation allows easy retrieval of salient information.

All study questions are accompanied by the correct answer, a paragraph-length explanation, and specific reference to the text where the topic is discussed. Study questions that follow each chapter use the current National Board format to reinforce the chapter content. Study questions appearing at the end of the text in the Comprehensive Exam vary in format depending on the book. Wherever possible, Comprehensive Exam questions are presented as clinical cases or scenarios intended to stimulate real-life application of medical knowledge. The goal of this exam is to challenge the student to draw from information presented throughout the book.

All of the books in the *National Medical Series* are constantly being updated and revised. The authors and editors devote considerable time and effort to ensure that the information required by all medical school curricula is included. Strict editorial attention is given to accuracy, organization, and consistency. Further shaping of the series occurs in response to biannual discussions held with a panel of medical student advisors drawn from schools throughout the United States. At these meetings, the editorial staff considers the needs of medical students to learn how the *National Medical Series* can better serve them. In this regard, the Harwal staff welcomes all comments and suggestions.

1 Describing Clinical Data

I. THE "VARIABLE." When reduced to lowest terms, all medical research is simply the study of relationships among **variables**. Most often, medical investigators are interested in studying either associations or differences among variables. A variable is any quality, characteristic, or constituent of a person or thing that can be measured. By definition, a variable is subject to change.

A. Scales used to measure variables. The four basic types of measurement scales listed below represent an increasing refinement of the measurement process.

1. Nominal scale

a. Definition. A nominal scale uses names, numbers, or other symbols to assign each measurement to one of a limited number of **categories that cannot be ordered** one above the other.

(1) The categories of a nominal scale must be exhaustive and mutually exclusive; each measurement must fall into only one category. Within any category, the members are assumed to be equivalent with respect to the characteristic being scaled.

(2) The names or symbols designating the categories can be interchanged without altering the essential information conveyed by the scale.

b. Examples

(1) Measurement of the variable **blood type** results in classification of a person's blood as type A, type B, type O, or type AB.

(2) For the variable **psychiatric diagnosis,** the measurement assigned to each patient is a number that corresponds to a specific diagnosis listed in the *Diagnostic and Statistical Manual of Mental Disorders* (*DSM-III-R*). For example, the number 295. corresponds to the diagnosis of schizophrenia.

(3) Other examples of variables measured on a nominal scale are **sex, race,** and **eye color**.

2. Ordinal scale

a. Definition. An ordinal scale assigns each measurement to one of a limited number of **categories that are ranked** in terms of a graded order.

(1) Differences among the categories are not necessarily equal and often are not even measurable. For example, the amount of the variable represented by a change from category 1 to category 2 is not necessarily the same as the amount represented by a change from category 3 to category 4.

(2) The symbols assigned to represent the categories are not important as long as the ranking system is preserved.

b. Examples

(1) Patient status or condition may be classified as unimproved, stable, or improved. (Note that, although it is known that a patient classified as unimproved is more ill than a patient classified as stable, it is not known how much more ill the first patient is. Also, the difference in illness status from unimproved to stable is not necessarily the same as the difference in status from stable to improved.)

(2) Cancer staging typically uses an ordinal scale to classify disease according to the degree and nature of involvement of body tissues. For example, the staging of Hodgkin's disease is based on the degree of lymph node involvement, with stage I limited to a single lymph node region or single extralymphatic site and stage IV characterized by diffuse or disseminated involvement in extralymphatic tissues.

3. **Interval scale***
   a. **Definition.** An interval scale assigns each measurement to one of an unlimited number of **categories that are equally spaced**. The scale has no true zero point (i.e., the zero point on the scale does not represent the true or theoretical absence of the variable being measured). With an interval scale, it is possible to determine exactly how much more (or how much less) of the variable being measured is represented by each category.
   b. **Example.** Temperature expressed in degrees Fahrenheit or Celsius is a variable measured on an interval scale. (Note that 0° C is the point at which water freezes; it does not represent absence of temperature.)

4. **Ratio scale***
   a. **Definition.** On a ratio scale, measurement begins at a **true zero point** and the **scale has equal intervals**.
   b. **Examples.** Variables measured on a ratio scale are length, time, mass, volume, and temperature in degrees Kelvin.

B. **Types of variables.** Variables can be broadly classified as qualitative or quantitative.

1. **Qualitative variables**
   a. **Definition.** Qualitative variables are variables that are measured at the **nominal** level.
   b. **Example.** A diagnostic test for pregnancy gives a result of either "positive" or "negative." The diagnostic test variable is a qualitative variable.

2. **Quantitative variables**
   a. **Definition.** Quantitative variables are variables that are measured on an **ordinal** or **interval/ratio** scale.
   b. **Examples**
      (1) A measurement of serum sodium concentration (e.g., 140 mEq/L) expresses the **exact amount of sodium** in the serum.
      (2) Serum cholesterol level, systolic blood pressure, and blood urea nitrogen (BUN) level are other quantitative clinical variables.

C. **Basis for variation.** Fluctuation among clinical measurements reflects the combined effects of several phenomena (Figure 1-1). The interpretation of clinical observations depends on the physician's ability to recognize these sources of variation and to account for them in the diagnostic and therapeutic processes.

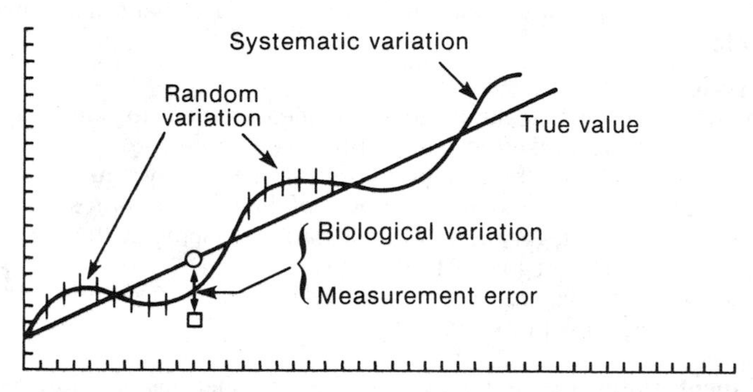

Figure 1-1. Phenomena that cause variation in clinical measurements. The *curved line* represents temporary systematic changes in values of the diagnostic variable, the *hatched lines* represent true random variability inherent in all biologic systems, and the *slanted line* represents the true underlying steady state value of a given variable for a particular person. (Because the true value can shift with the health of the person, the line is slanted rather than horizontal.) The deviation of the actual measurement of the variable (*square*) from the true value of the variable (*circle*) is due to biologic variation (random and systematic), temporary systematic changes, and measurement variation (random and systematic).

*The terms "interval data" and "interval/ratio scale" are used here and throughout the book to designate data based on either an interval or a ratio scale.

1. **True biologic variation**
   a. **Definition.** True biologic variation in clinical measurements is the sum of many unknown factors, each of which contributes a small **random** effect. Random effects, based on the laws of probability are as likely to be positive (causing the measurement to exceed the true value) as they are to be negative (causing the measurement to be less than the true value).
   b. **Example.** A series of consecutive systolic blood pressure measurements on the same patient under theoretically identical conditions will not be exactly equal because of true biologic (random) variability inherent in these measurements within a given patient.

2. **Variation associated with making observations under different conditions**
   a. **Definition.** Variation in clinical measurements occurs when the conditions under which the measurements are made are known to affect the values obtained. This type of variation is **systematic** rather than random, because its effect is predictable and not based on the laws of chance.
   b. **Example.** A patient's systolic blood pressure varies according to the time of day (temporal variation) and the position of the person (postural variation) when the measurement is taken. The pattern associated with these fluctuations is somewhat regular and predictable within a given person.

3. **Measurement variation**
   a. **Definition.** Measurement variation (sometimes called **measurement error**) is variation among clinical observations that is attributed to the measurement process. Measurement error may have both a random component and a systematic component.
      (1) **Random measurement error** is governed by the laws of chance and results in a measurement that is either above or below the true value with equal probability. A series of measurements affected only by random variation will center on the true value of the variable being measured.
      (2) **Systematic measurement error** occurs when, as a result of a flaw in the measurement process, the measurements no longer center around the true value but around a value that is systematically higher or lower than the true value.
   b. **Example.** For a given sample of urine, a series of measurements of pH made with the same meter by the same analyst under theoretically identical conditions will not be exactly equal. Random variation in these measurements is the sum of many components, including instrument precision. Systematic variation results if the instrument is out of calibration.

D. Within-patient variation

1. Variation in the value of a clinical variable within a given patient may result from true biologic (random) variation, varying conditions under which the measurement is made, or measurement error, or it may be due to a pathologic change in the biologic state of the patient.
2. The physician is concerned with identifying the sources of variation within a given patient, because intervention should occur only if there is a true pathologic change.

E. Variation among patients

1. Variation from patient to patient in the value of a clinical variable is attributable to inherent biologic differences among patients, systematic differences in conditions of measurement, and measurement error.
2. The physician is concerned with variation among patients, because most answers to clinical questions for an individual patient are derived from information obtained from groups of patients with similar conditions. For example, a physician would study variation among patients to answer questions such as:
   a. "How unusual is Mr. Smith's creatine kinase (CK) value relative to a group of persons known to be free of myocardial infarction?"
   b. "Do groups of patients who receive antihypertensive agent A have lower diastolic blood pressure measurements, on the average, than those who receive agent B?"

II. DESCRIBING VARIATION IN CLINICAL DATA. Statisticians and clinicians have developed several interrelated concepts to characterize variation in clinical measurements, including bias, accuracy, and precision.

A. Bias. In this chapter, the discussion of bias is limited to cases in which multiple measurements are obtained from one patient. Other types of bias (e.g., selection bias, confounding bias) arise when comparisons are made among groups of patients (e.g., to determine the most effective treatment of a disease); these sources of bias are discussed in Chapter 8 II C.

1. **Definition.** Bias is the systematic component of both biologic variation and measurement variation. Unlike random variation, bias results in measurements that are systematically higher or lower than the true underlying value of a diagnostic variable. Bias may result from a flaw in the measurement process or from sampling error.

2. **Bias due to sampling error**
   a. Sampling error arises when statements about reality (for an individual or group) are made on the basis of incomplete information obtained from a sample.
   b. **Example.** A biopsy of the right side of the liver will give biased results, or results that are not representative of the truth (i.e., the presence of a tumor), if a tumor is located on the left side of the liver.

3. **Bias due to a flaw in the measurement process**
   a. A faulty measuring instrument may yield values that are consistently higher or lower than the true value of the variable being measured.
   b. **Example.** A physician with a hearing impairment may obtain diastolic blood pressure measurements that are consistently above the true diastolic pressure.

4. **The relationship between bias and chance** in the measurement process is illustrated in Figure 1-2, using a series of diastolic blood pressure measurements obtained from a patient by two methods: an intra-arterial cannula (assumed to be an exact and unbiased method) and a sphygmomanometer.
   a. The sphygmomanometer values are systematically shifted (i.e., biased) to the right of the true value, possibly owing to improper cuff size or a hearing deficit of the person making the observation.
   b. The sphygmomanometer values also vary according to chance. These random variations are distributed equally to the left and to the right of the prevailing value.

B. Accuracy

1. **Definition.** When the measurement process yields values that are equal, on the average, to the true underlying value for the diagnostic variable being measured, the measurement (or measurement instrument or process) is accurate, or **unbiased**.

2. **Example.** The blood pressure measurements made using the intra-arterial cannula (*A*) in Figure 1-2 are accurate because they center around the patient's true blood pressure.

3. **Evaluating accuracy.** The accuracy of a set of clinical measurements is determined by comparing the average (also called the **mean;** see III B 1 a) of a set of readings of a given variable with the true underlying value of the variable (if that value is known).

C. Precision

1. **Definition.** The degree to which a series of measurements fluctuates around a central

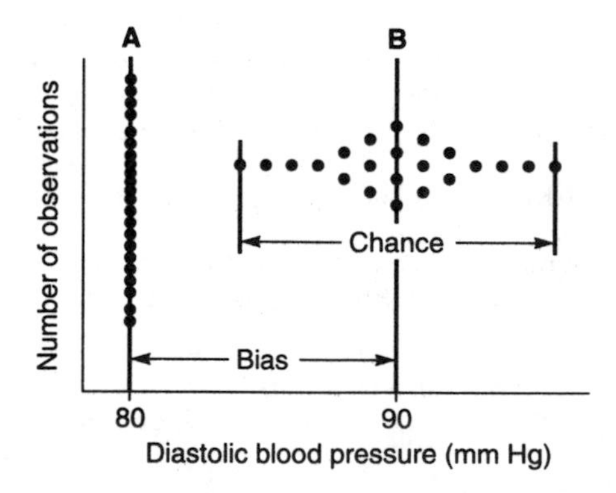

Figure 1-2. Relationship between bias and chance, as illustrated by differences in diastolic blood pressure measurements obtained by (*A*) an intra-arterial cannula (an exact and unbiased method that gives a true value) and by (*B*) a sphygmomanometer. (Reprinted from Fletcher R, Fletcher S, Wagner E: *Clinical Epidemiology: The Essentials.* Baltimore, Williams & Wilkins, 1987, p 10.)

measurement is the precision, or **reproducibility,** of the measurement (or measurement instrument or process). The central value may or may not be the true value of the variable.

2. Precision is independent of accuracy. Measurements may vary from inaccurate (biased) and imprecise, to inaccurate but precise (i.e., systematically shifted from the true value but varying little from a central value), to accurate (unbiased) but imprecise (i.e., averaging around the true value but varying widely from it), to the ideal—accurate and precise.

3. Examples

a. The sphygmomanometer in Figure 1-2B not only is inaccurate since it systematically reads to the right of the true value of 80 mm Hg, but it also is imprecise because it yields readings that vary greatly from those obtained by the intra-arterial cannula.

b. The intra-arterial cannula in Figure 1-2A yields both accurate and precise readings.

4. Evaluating and achieving precision

a. The precision of a set of clinical values is evaluated by observing the **frequency distribution** of the measurements (see III A 1) and by calculating the **standard deviation** of the measurements (see III B 2 b).

b. Perfect precision rarely, if ever, is achieved when measuring biologic phenomena. However, the degree of imprecision (variability) may be reduced by making and recording measurements with great care and by following carefully designed protocols.

III. DESCRIBING GROUPED CLINICAL DATA.*

Clinical data can be described either pictorially (i.e., in **tables** or **graphs**) or quantitatively (i.e., by means of **numerical summaries**).

A. Pictorial descriptions

1. Frequency table

a. Definition. A frequency table consists of a series of intervals that divide a set of clinical data [e.g., the red blood cell (RBC) cholinesterase values shown in Table 1-1] and the frequency of particular data values within each interval. A frequency table describes the **frequency distribution** of a set of clinical measurements, that is, how the values of the clinical variable are distributed among the specified intervals (e.g., the frequency distribution of values in Table 1-1 is shown in Table 1-2). Grouping data into intervals causes the exact value of an individual datum to be lost but highlights features of the data that can be used readily.

b. Class intervals

(1) Definition. Class intervals are a set of nonoverlapping, contiguous intervals encompassing the range of values assumed by the clinical variable being described. The intervals are chosen such that a given data value can be placed in only one interval.

(a) The **number of class intervals** used to group data is arbitrary, although 5–12 intervals usually suffice; too few obscure detail and lead to loss of information, and too many defeat the purpose of grouping.

(b) The **endpoints of the intervals** can be specified in terms of true class limits, which are specified to one decimal place more than the recorded data and reflect the precision of the measuring process.

Table 1-1. Ordered RBC Cholinesterase Values (μmol/min/ml) Obtained from 35 Workers Exposed to Pesticides

7.7	9.8	10.6	11.4	12.3	12.6
8.6	9.9	10.9	11.6	12.3	13.4
8.7	9.9	10.9	11.6	12.4	15.2
9.0	10.1	11.0	11.7	12.5	15.3
9.2	10.2	11.0	11.8	12.5	16.7
9.4	10.2	11.3	12.2	12.6	

Reprinted from Duncan RC, Knapp RG, Miller MC III: *Introductory Biostatistics for the Health Sciences,* 2nd ed. Albany, Delmar, 1983, p 9.

*Sections III and IV are adapted from Duncan RC, Knapp RG, Miller MC III: *Introductory Biostatistics for the Health Sciences,* 2nd ed. Albany, Delmar, 1983, pp 4–19.

Table 1-2. Grouped Frequency Distribution for Data in Table 1-1

RBC Cholinesterase (μmol/min/ml)	Frequency	Relative Frequency	Relative Frequency (%)	Cumulative Frequency (%)
5.95–7.95	1	.029	2.9	2.9
7.95–9.95	8	.229	22.9	25.8
9.95–11.95	14	.400	40.0	65.8
11.95–13.95	9	.257	25.7	91.5
13.95–15.95	2	.057	5.7	97.2
15.95–17.95	1	.029	2.9	100.1
Total	35	1.001	100.1	

Reprinted from Duncan RC, Knapp RG, Miller MC III: *Introductory Biostatistics for the Health Sciences,* 2nd ed. Albany, Delmar, 1983, p 10.

(2) **Example.** The data in Table 1-2 are displayed in six intervals of two-unit width. The true class limits of these intervals are shown in the first column of the table. The first interval includes all observations equal to or greater than 6.0 but less than 8.0 (the sample data were recorded to one decimal place); thus, the true interval limits are 5.95 and 7.95.

c. **Expressions of frequency**

(1) **Frequency** refers to the number of observations that fall into each class interval (see second column in Table 1-2). To determine frequency quickly, raw data should first be placed in ascending (or descending) order (as in Table 1-1).

(2) **Relative frequency** is the number of observations falling into each interval as a fraction or percentage of the total number of observations (see third and fourth columns, Table 1-2). The relative frequencies total approximately 1.0, or 100.0%.

(3) **Cumulative frequency** is the fraction (or percentage) of observations that are less than the upper limit of each interval. Cumulative frequency is obtained by adding the relative frequencies for the intervals in question (see fifth column Table 1-2). For example, the cumulative frequency of RBC cholinesterase values less than 12 μmol/min/ml is 65.8%, which is obtained by adding the relative frequencies 2.9%, 22.9%, and 40.0%.

2. **Histogram**

a. **Definition.** A histogram is a graphic representation of a frequency distribution consisting of a series of rectangles whose widths represent class intervals and whose areas are proportional to the corresponding frequencies of particular data values within each interval. Figure 1-3 is the histogram for the data in Table 1-2.

b. **Shape.** A histogram reveals the shape of a frequency distribution—that is, whether it is **symmetric** or **skewed**.

(1) A **symmetric distribution** is one in which frequencies are equal, or nearly equal, at points on either side of the central value (Figure 1-4A).

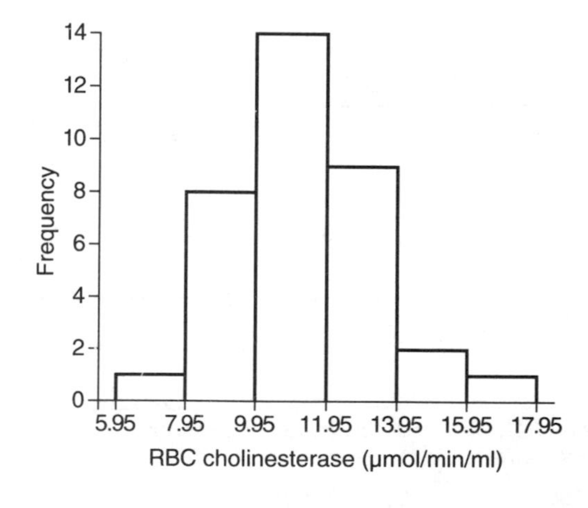

Figure 1-3. Histogram of the frequency distribution of RBC cholinesterase values in Table 1-2. (Reprinted from Duncan RC, Knapp RG, Miller MC III: *Introductory Biostatistics for the Health Sciences,* 2nd ed. Albany, Delmar, 1983, p 6.)

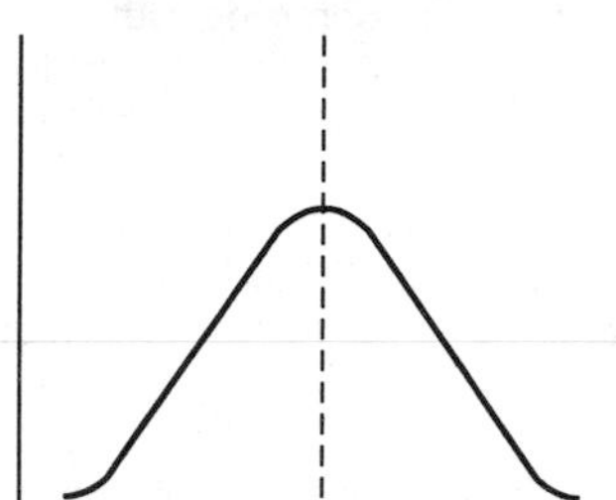

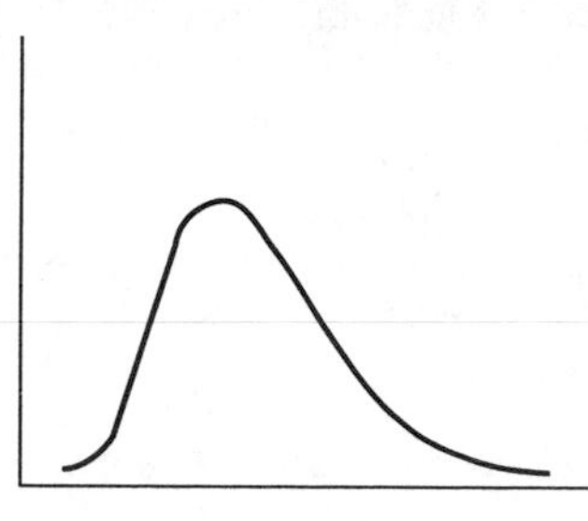

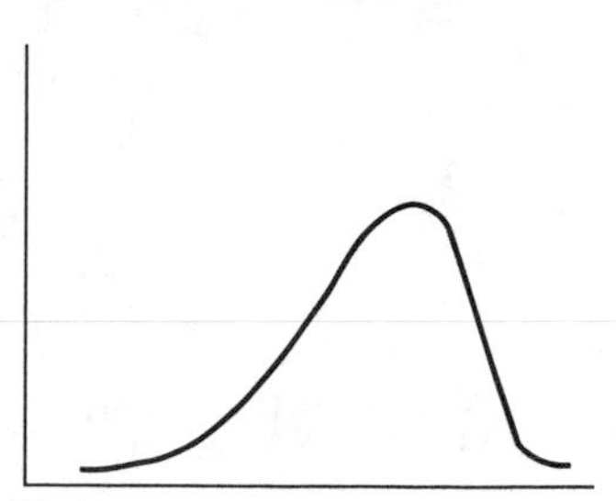

Figure 1-4. Symmetric and skewed frequency distributions.

(2) A **skewed distribution** is one that is not symmetric. It is either skewed left (i.e., has a "tail" to the right; Figure 1-4B) or skewed right (i.e., has a "tail" to the left; Figure 1-4C).

B. Numerical summaries. Quantitative descriptions of sets of clinical values usually involve measures of central tendency and measures of spread.

1. Measures of central tendency. Common measures of central tendency, or **location,** are the mean, the median, and the mode.

a. Mean

(1) Definition. The mean is the arithmetic average of a set of values. It represents the central value of the data that can be determined from the size of the values.

(2) Calculation. The mean of a sample is defined as

$$\bar{Y} = \sum_{i=1}^{n} \frac{Y_i}{n}$$

where n = the number of observations in the data set (sample); Y_i = the sample values; and Σ = the summation operation (see Appendix 1).

(3) Example. To determine the mean length of hospital stay for patients undergoing a particular surgical procedure, investigators observe a sample of five patients. The length of stay (in days) for each patient is as follows: 1, 3, 2, 4, and 5. The mean of this set of observations is

$$\bar{Y} = \sum_{i=1}^{5} \frac{Y_i}{5} = \frac{(1+3+2+4+5)}{5}$$
$$= \frac{15}{5} = 3$$

(4) Properties

(a) If the mean is subtracted from all sample values, the sum of the differences is zero.

(b) When the number of observations in the data set is small, the mean is very sensitive to extreme values. For this reason, the mean is misleading when applied to skewed data.

(i) For example, if the hospital stay of the fifth patient was 50 days instead of 5 days, the mean would change from 3 to 12. If the stay was 500 days, the mean would be 102.

(ii) As the number of observations in the data set increases, however, the influence of a few extreme values on the mean decreases.

(c) The mean is not necessarily equal to one of the sample values.

b. Median

(1) Definition. When a sample of observations is arranged in order of magnitude, the median is the middle value (for an odd number of observations) or the average of the two middle values (for an even number of observations).

(2) **Examples**

(a) The length of hospital stay (in days) for nine patients undergoing different surgical procedures is observed. The values, arranged in order of magnitude, are 1, 1, 3, 4, 8, 9, 12, 13, and 15. Since n is odd, the median is the middle value 8.

(b) The length of hospital stay (in days) for six other patients is observed. The values, arranged in order of magnitude, are 1, 6, 7, 10, 11, and 15. Since n is even, the median is the average of the middle values 7 and 10: $(7 + 10)/2 = 8.5$.

(3) **Properties**

(a) The median is useful for summarizing skewed data because it is insensitive to extreme values. For example, in each group of patients described in the examples above [III B 1 b (2)], the longest hospital stay was 15 days; even if that stay were 150 days, the median for each sample would not change.

(b) For symmetric distributions, the median and mean have the same value.

(c) The median is not necessarily equal to one of the sample values.

c. Mode

(1) **Definition.** The mode is the value in a frequency distribution (or the group in a grouped frequency distribution) that occurs most often.

(2) **Examples**

(a) For the set of values 1, 4, 3, 1, 2, and 5, the value 1 occurs most often. Hence, the mode is 1.

(b) For the set of values 2, 4, 2, 3, 1, 5, and 1, the values 1 and 2 occur most often. Thus, the distribution is **bimodal** (i.e., it has two modes).

(c) For the data in Table 1-2, the modal group is the interval 9.95–11.95 μmol/min/ml.

2. Measures of spread. The amount of spread, or **dispersion,** in a set of data usually is measured by the range of observations or by the standard deviation.

a. Range

(1) **Definition.** The range is the difference between the largest and the smallest values in a set of data.

(2) **Properties**

(a) Because it is easily computed, the range is useful as a simple and quick measure of variability.

(b) Because only two numbers in a data set are used to calculate it, the range may give a misleading impression of the true variability of the data, especially if either or both of the numbers are extreme values. For example, if the sample value of 7.7 in Table 1-1 were changed to 2.7, the range would change from 9 to 14.

b. Variance and standard deviation are more representative measures of variability than is the range because they use all of the measurements in the set and their individual distances (deviations) from the mean of the distribution.

(1) **Variance.** If the average deviation of the observations from the mean were computed, the result would be zero, because the positive and negative differences around the mean would cancel each other. Therefore, the deviations are squared, and the average of the squared deviations, called the **variance,** is taken.

(a) **Calculation.** The variance (s^2) of a set of clinical measurements can be calculated from the formula

$$s^2 = \frac{\sum (Y_i - \bar{Y})^2}{n - 1}$$

Note that the denominator for the variance of data from a *sample* of subjects is $n - 1$ rather than n. Therefore, the sample variance is not a true average of the squared distances of each observation from the mean. The theoretical variance of a finite *population* (rather than a sample) of measurements uses the population size, N, as the denominator.

Table 1-3. Length of Hospital Stay for Five Patients: Deviations from the Mean

Patient	Length of Stay in Days (Y_i)	Deviations ($Y_i - \bar{Y}$)	$(Y_i - \bar{Y})^2$	Y_i^2
1	3	$(3 - 3) = 0$	0	9
2	5	$(5 - 3) = 2$	4	25
3	2	$(2 - 3) = -1$	1	4
4	3	$(3 - 3) = 0$	0	9
5	2	$(2 - 3) = -1$	1	4
	$\sum Y_i = 15$	$\sum (Y_i - \bar{Y}) = 0$	$\sum (Y_i - \bar{Y})^2 = 6$	$\sum Y_i^2 = 51$

(b) Example. The variance for the data in Table 1-3 is calculated as

$$s^2 = \frac{\sum (Y_i - \bar{Y})^2}{n - 1} = \frac{6}{4} = 1.5$$

The value 6 was obtained from the sum of the fourth column in Table 1-3.

(2) Standard deviation. Since the variance uses the square of the deviations, the units of variance also are squared. In the above example, the variance is 1.5 days^2. To avoid this, the positive square root of the variance, known as the **standard deviation,** is taken.

(a) Calculation. The standard deviation (s) is

$$s = \sqrt{s^2} = \sqrt{\frac{\sum (Y_i - \bar{Y})^2}{n - 1}}$$

(b) Example. Standard deviation is illustrated by a graph and a table of hospital stay data collected from five patients undergoing the same operative procedure. The data (in days) are 3, 5, 2, 3, and 2.

(i) Figure 1-5 shows the five measurements (represented by *dots*) in relation to their mean value (i.e., 3 days). The larger the deviation, the more dispersed the measurement is from the center (i.e., the mean).

(ii) Table 1-3 presents the information from Figure 1-5 in tabular form. Using the data from Table 1-3, the standard deviation (s) is calculated as

$$s = \sqrt{\frac{\sum (Y_i - \bar{Y})^2}{n - 1}} = \sqrt{\frac{6}{4}} = 1.22$$

where the numerator of s is the sum of the fourth column in Table 1-3, and n is the number of observations, or sample size (here $n = 5$).

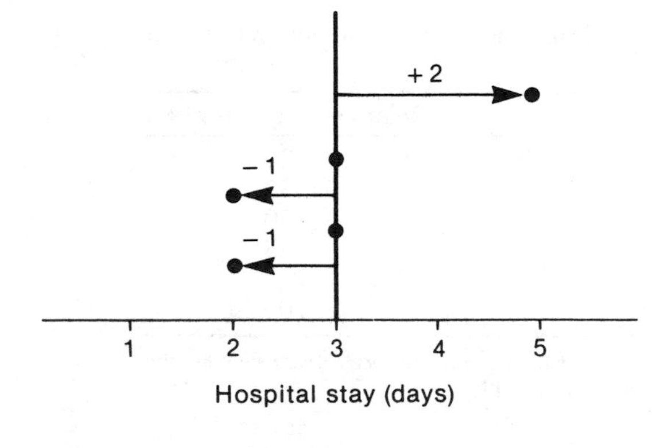

Figure 1-5. Graphic depiction of standard deviation using a set of hospital stay data. Each measurement in the set is represented by a *dot* located above its corresponding value. The numbers above the *arrows* are the deviations of each data point from the mean (3 days).

(c) **Computational formula.** For ease of computation, the following formula is used for standard deviation (s)

$$s = \sqrt{\frac{\sum Y_i^2 - \frac{\left(\sum Y_i\right)^2}{n}}{n-1}}$$

Applying this formula for the sample data in the last column of Table 1-3 gives

$$s = \sqrt{\frac{51 - \frac{(15)^2}{5}}{4}}$$

$$= \sqrt{1.5} = 1.22$$

IV. CHOOSING APPROPRIATE DESCRIPTIVE STATISTICS.

Important in the selection of an appropriate descriptive measure is the level of measurement achieved by the data.

A. Description of nominal data

1. **Statistical methods.** Numerical operations are not performed on nominal scale variables, even if numerical codes are used to define categories. The appropriate descriptive statistics are the **frequency of observation** in the various categories (i.e., frequency table) and the **mode**.
2. **Example.** A group of 100 patients who visited a mental health clinic is divided into groups on the basis of a psychiatric diagnosis (a nominal variable).
   a. Table 1-4 presents information on the frequency of each of four psychiatric diagnoses in these patients.
   b. Of the four psychiatric diagnoses in Table 1-4, psychosis is the most common and, thus, represents the modal diagnosis.

B. Description of ordinal data

1. **Statistical methods**
   a. The most appropriate descriptive statistics are **frequency counts,** the **median,** and the **mode**.
   b. The usefulness of the mean of ordinal data is questionable. Since the intervals between units on the ordinal scale are not necessarily equal in size, arithmetic operations are not valid and, hence, the arithmetic average of ordinal data generally is not considered a useful or valid descriptive measure. However, the mean for ordinal data commonly is reported. In such instances, it has been assumed that the intervals of the scale are equal or, at least, approximately equal.
2. **Example.** A group of 20 geriatric patients is asked to rate the quality of nursing home care on the following scale: 1 = very poor, 2 = poor, 3 = fair, 4 = good, and 5 = very good.
   a. The frequency count for these data is shown in Table 1-5.

Table 1-4. Frequency of Psychiatric Diagnosis in 100 Patients Visiting a Mental Health Clinic

Diagnosis	Frequency	Relative Frequency (%)
Organic brain syndrome	20	20
Psychosis	35	35
Mental retardation	20	20
Personality disorder	25	25
Total	100	100

Reprinted from Duncan RC, Knapp RG, Miller MC III: *Introductory Biostatistics for the Health Sciences*, 2nd ed. Albany, Delmar, 1983, p 17.

Table 1-5. Frequency of Quality of Care Ratings by 20 Geriatric Nursing Home Patients

Rating	Frequency	Relative Frequency (%)
Very poor	3	15
Poor	3	15
Fair	4	20
Good	7	35
Very good	3	15
Total	20	100

Reprinted from Duncan RC, Knapp RG, Miller MC III: *Introductory Biostatistics for the Health Sciences*, 2nd ed. Albany, Delmar, 1983, p 18.

b. The median response is a rating of 3.5. (Usually with data of this type it is more meaningful to specify that the median response is between "fair" and "good," rather than to define it numerically.)

c. The modal response for the data is the rating "good."

C. Description of interval/ratio data

1. Statistical methods

a. An interval scale and a ratio scale both allow the use of ordinary arithmetic operations. Thus, **all of the descriptive statistical methods** presented so far may be used.

b. The **choice of an appropriate measure of central tendency** is influenced by the shape of the frequency distribution and the intended use of the chosen summary.

(1) For **symmetric or nearly symmetric distributions,** the mean and median are equal or nearly equal.

(2) For **skewed data,** the mean and median are not equal. For data whose frequency distribution is skewed to the left (see Figure 1-4B), the mean is smaller than the median. For data whose frequency distribution is skewed to the right (see Figure 1-4C), the mean is larger than the median.

2. Example. The survival time (in months) of five patients receiving treatment for cancer of the pancreas is measured and reported as follows: 3, 5, 4, 4, and 29.

a. The mean (9 months), as a measure of the center of the set of data, gives a misleading impression of the efficacy of the treatment because it is influenced by the extreme value of 29 months.

b. Conversely, the median (4 months) is fairly representative of the central tendency of the data because it divides the distribution by count and therefore is not affected by the numerical value of the extreme observation.

PROBLEMS*

1-1. In a study to determine the effect of cigarette smoking on phenacetin metabolism, phenacetin is administered to 12 smokers and 14 nonsmokers. After two hours, plasma phenacetin level ($\mu g/ml$) is measured in all study participants. Phenacetin level, smoking status, and age for each participant are recorded in the following table.

Plasma Phenacetin Levels in Smoking versus Nonsmoking Study Participants

Study Participant	Age	Smoking Status	Plasma Phenacetin Level ($\mu g/ml$)
1	25	−	.45
2	22	+	.03
3	38	−	.75
4	45	−	1.40
5	25	+	.01
6	28	+	.01
7	28	−	1.28
8	38	+	.02
9	48	−	1.83
10	32	+	.61
11	32	+	.52
12	35	−	1.69
13	32	−	1.91
14	69	−	2.15
15	72	−	2.30
16	43	+	1.12
17	58	−	2.48
18	49	−	2.75
19	55	+	1.01
20	42	−	2.81
21	25	+	1.52
22	26	+	2.55
23	55	−	3.28
24	65	−	3.55
25	28	+	.01
26	71	+	3.80

+ = smoker; − = nonsmoker

a. Compute the mean, median, and mode of the phenacetin levels for each group of participants. Do these measures suggest anything about the shape of the frequency distribution for each group?

b. Compute the range, variance, and standard deviation of the phenacetin levels for each group of participants.

c. Construct a frequency table and a histogram of phenacetin levels for each group of participants. What is the shape of each distribution?

d. What do the statistical results of this study suggest about the qualitative effect of smoking on phenacetin concentration?

e. Is the mean phenacetin level for the smokers an accurate indication of the central value in the group? Why?

f. Suppose it is known that phenacetin metabolism increases with age. How would this factor affect the conclusions made in problem 1-1d?

---

*The problems in this section are adapted from Duncan RC, Knapp RG, Miller MC III: *Introductory Biostatistics for the Health Sciences*, 2nd ed. Albany, Delmar, 1983, pp 21–22.

1-2. In a clinical laboratory, three new instruments for measuring a certain blood component are tested using solutions that contain a known concentration (10 mg/ml) of the substance to be measured. Ten measurements are made with each instrument, and the results are shown in the following table.

Measurements (mg/ml) Obtained by Three Different Instruments on Ten Trials Using a Known Concentration (10 mg/ml)

Instrument #1	**Instrument #2**	**Instrument #3**
5	10	10
10	9	11
7	10	9
15	9	10
16	11	10
12	8	9
4	9	11
8	7	12
10	8	8
13	9	10
$\sum Y = 100$	$\sum Y = 90$	$\sum Y = 100$
$\sum Y^2 = 1148$	$\sum Y^2 = 822$	$\sum Y^2 = 1012$

Reprinted from Duncan RC, Knapp RG, Miller MC III: *Introductory Biostatistics for the Health Sciences*, 2nd ed. Albany, Delmar, 1983, p 22.

a. Determine the mean and standard deviation for each group of measurements.

b. How do the three instruments compare in terms of accuracy and precision?

Solutions on p 359.

STUDY QUESTIONS

Directions: Each of the numbered items or incomplete statements in this section is followed by answers or by completions of the statement. Select the **one** lettered answer or completion that is **best** in each case.

1. The association between exposure to lead-based paint and mental retardation in children is studied, and the following information is recorded for each child in the study: race (black or white), age (in years), and amount of exposure (rated from 0 for no exposure to 3 for more than three years' exposure). Which of the following choices lists these variables in the order of nominal, ordinal, and interval/ratio?

(A) Amount of exposure, race, age
(B) Age, race, amount of exposure
(C) Race, age, amount of exposure
(D) Race, amount of exposure, age
(E) None of the above

2. A series of measurements that center around the true underlying value of the variable being measured, with a large amount of spread around the center point, is said to be

(A) accurate and precise
(B) accurate but imprecise
(C) inaccurate and precise
(D) inaccurate and imprecise
(E) precise but biased

3. The following are survival times (in months) recorded for six tumor-bearing rats being observed after radiation therapy: 1.4, 1.7, 2.3, 2.5, 3.2, and 3.8. If the observed value of 3.8 months is mistakenly recorded as 38 months, what will be the effect on statistics for this study?

(A) An increase in the median
(B) An increase in the mode
(C) An increase in the mean
(D) An increase in both the median and the mean
(E) An increase in the median, the mode, and the mean

4. The following data represent length of hospitalization (in weeks) for five patients with total hip replacement who receive no physical therapy during their stay: 4, 4, 3, 5, 4, and 20. The best measure of central tendency for this set of data is the

(A) mean
(B) range
(C) mode
(D) median
(E) standard deviation

Directions: Each item below contains four suggested answers of which **one or more** is correct. Choose the answer

A if **1, 2, and 3** are correct
B if **1 and 3** are correct
C if **2 and 4** are correct
D if **4** is correct
E if **1, 2, 3, and 4** are correct

5. How are numerical summaries of clinical measurements affected by a frequency distribution that is skewed to the right?

(1) The mean and the median are equal
(2) The mean is larger than the median
(3) The median is larger than the mean
(4) The mode is unaffected

6. In a series of clinical measurements, the degree of spread of the values around the center can be expressed in terms of the

(1) median
(2) range
(3) mean
(4) standard deviation

1-D
2-B
3-C
4-D
5-C
6-C

Directions: Each group of items in this section consists of lettered options followed by a set of numbered items. For each item, select the **one** lettered option that is most closely associated with it. Each lettered option may be selected once, more than once, or not at all.

Questions 7–10

The phrases below describe terms used by clinicians and statisticians to characterize variation in clinical measurement. Match each phrase to the most appropriate term.

(A) Validity
(B) Chance
(C) Bias
(D) Precision
(E) Accuracy

7. Refers to systematic variation in a series of clinical measurements

8. Refers to a series of clinical measurements that, on the average, equal the underlying value of the variable being measured

9. Describes the degree to which a series of measurements fluctuate around a central value

10. Describes the type of variation that, in general, results in values above or below the true value with equal probability

---

7-C
8-E
9-D
10-B

ANSWERS AND EXPLANATIONS

1. The answer is D *[I A 1–4]*
Race is a nominal variable, since no ordered relationship exists between the potential values of the variable. Age in years has an underlying interval/ratio scale (it is a true arithmetic scale). In this study, the amount of exposure is an ordinal variable; the values 0 through 3 represent a ranking of exposure rather than a true quantitative measure of the extent of lead paint exposure.

2. The answer is B *[II B–C]*
A measurement process that produces a series of measurements whose mean is equal to the true underlying value of the variable being measured is described as accurate. If repeated use of this measurement process produces measurements that fluctuate widely from the central value (i.e., the mean), the process is described as imprecise (i.e., not reliable). A biased measurement process produces a series of measurements that are systematically higher or lower than the true underlying value of the variable being measured.

3. The answer is C *[III B 1 a (4) (b)]*
The mean of a set of observed values represents the arithmetic average of those values as determined by size. The mean is sensitive to the presence of extreme values (sometimes referred to as outliers). This is especially true when the number of observations in a data set is small. The median divides a distribution of observed values (arranged from smallest to largest value) into two equal parts by count. The median would not be affected by an extreme value. The mode is the most frequently occurring value; changing 3.8 to 38 does not change the mode.

4. The answer is D *[III B 1 b (3) (a)]*
When a data set contains either a very large or very small value (i.e., when a distribution is skewed), the mean (average of all values in the set) is pulled in the direction of the extreme value and may not adequately represent the center of the frequency distribution. For highly skewed distributions, the median (midpoint of a set of values arranged in order of magnitude) is a more representative measure of the central tendency of the frequency distribution. In this example, the median (4 weeks) is more representative of the center than is the mean (6.67 weeks). Note that only one value (20) falls above the mean, whereas half fall above and below the median.

5. The answer is C (2, 4) *[III B 1 a (4) (b), b (3) (a), c (1)]*
The mean of a set of clinical measurements is "pulled" in the direction of an extreme value (in this case, toward the right of the frequency distribution); therefore, the mean in this case would be larger than the median. If the distribution is symmetric (i.e., not skewed to the right or left), the mean and the median are equal. The mode is the most frequently occurring value; it is not affected by an extreme measurement.

6. The answer is C (2, 4) *[III B 2]*
The degree of spread of a set of measurements around a central value usually is measured by the range of the observations or by the standard deviation. Whereas the range uses only the smallest and largest values in a data set, the standard deviation uses all of the values in a data set and, therefore, is a more representative measure of the degree of variability. The degree of central tendency of the set of measurements usually is measured by the mean, the median, or the mode.

7–10. The answers are: 7-C *[II A]*, **8-E** *[II B]*, **9-D** *[II C]*, **10-B** *[I C 3 a (1)]*
When a measurement process produces measurements that are systematically higher or lower than the true value of the variable being measured, the process is said to be biased. A measuring instrument that is "out of calibration" will produce biased measurements.

A measurement process is accurate if the measurements correspond to the true state of the variable being measured (i.e., the values are equal, on the average, to the true value of the underlying variable). A measurement process is accurate only if the resultant series of measurements are unbiased.

The precision of a measurement process is the extent to which a series of measurements of a relatively stable phenomenon are reproducible (i.e., fall closely around a central value). Reliability and reproducibility are synonyms for precision.

Chance (random) variation is variation that is governed by the laws of probability. Random variation results in values that are as likely to be above the true value as below it.

2
Probabilities in Clinical Medicine

I. DEFINITION AND PROPERTIES OF PROBABILITY. Because medicine is an inexact science, physicians seldom can predict an outcome with absolute certainty. For example, to formulate a diagnosis, a physician must rely on available diagnostic information about a patient (e.g., history and physical examination, laboratory studies, x-ray findings, ECGs). Although no test result is absolutely accurate, it does affect the probability of the presence (or absence) of a disease. Hence, an understanding of probability is fundamental to the clinical decision-making process. Clinicians rely on probability theory for quantifying the uncertainty that is inherent in the decision-making process. Probability theory also allows clinicians to draw conclusions about a **population** of patients based on known information about a **sample** of patients drawn from that population.

A. Definition. To facilitate communication and avoid misinterpretation of clinical information among physicians and between physicians and patients, it is best to define **probability** in terms of relative frequency (i.e., proportion). The probability P that an event E will occur, written $P(\text{E})$, is estimated by

$$P(\text{E}) = \frac{\text{number of times E occurs}}{\text{number of times E can occur}}$$

B. Properties

1. **A probability value must lie between 0 and 1.**
   - **a.** A value of 0 means the event cannot occur.
   - **b.** A value of 1 means the event definitely will occur.
   - **c.** A value of .5 means that the probability that the event will occur is the same as the probability that it will not occur.
2. **The sum of the probabilities** (or relative frequencies) **of all the events that can occur in the sample must be 1** (or 100%).

C. Example. Table 2-1 is a frequency table showing serum cholesterol levels obtained from a sample of 1047 nondiseased men 40–59 years of age. The event (outcome) of interest is the specific interval of cholesterol levels into which a man's cholesterol level falls.

D. Sample problems (use Table 2-1)

1. **Problem #1.** What is the probability that the serum cholesterol level of a man randomly selected from the sample is between 160 and 179 mg/dl?
2. **Solution.** The problem can be solved by using the proportion (i.e., relative frequency) of men falling in the interval 160–179 to estimate the probability that a man selected from this group will have a cholesterol value in the given range. In Table 2-1, 37 of the 1047 men have serum cholesterol levels between 160 and 179 mg/dl. Thus,

$$P(\text{E}) = P(\text{cholesterol level between 160 and 179})$$
$$= 37/1047 = .035$$

3. **Problem #2.** What is the probability that a nondiseased, 50-year old man randomly selected from the same population from which the sample was drawn will have a serum cholesterol level less than 200 mg/dl?

Table 2-1. Grouped Frequency Distribution of Serum Cholesterol Levels in Disease-Free Men Aged 40–59 Years

Serum Cholesterol (mg/dl)	Frequency	Relative Frequency (%)	Cumulative Frequency (%)
120–139	10	1.0	1.0
140–159	21	2.0	3.0
160–179	37	3.5	6.5
180–199	97	9.3	15.8
200–219	152	14.5	30.3
220–239	206	19.7	50.0
240–259	195	18.6	68.6
260–279	131	12.5	81.1
280–299	96	9.2	90.3
300–319	47	4.5	94.8
320–339	30	2.9	97.7
340–359	13	1.2	98.9
360–379	6	0.6	99.5
380–399	4	0.4	99.9
400–419	...	0.0	99.9
420–439	1	0.1	100.0
440–459	...	0.0	100.0
460–479	1	0.1	100.1
Total	1047	100.1	

Adapted from Lewis LA, Olmsted F, Page IH, et al: Serum lipid levels in normal persons: findings of a cooperative study of lipoproteins and atherosclerosis. *Circulation* 16:277, 1957.

4. **Solution.** If the 1047-man sample adequately represents the total population from which it was drawn, this question can be answered using Table 2-1.
   a. According to the cumulative frequency column, the chance of selecting a man with a serum cholesterol level less than 200 mg/dl is 15.8%.
   b. In terms of probability, this is expressed as

$$P(\text{E}) = P(\text{cholesterol level less than 200})$$
$$= \frac{10 + 21 + 37 + 97}{1047}$$
$$= \frac{165}{1047} = .158$$

II. THE 2 × 2 TABLE is introduced here and then used in III to illustrate rules of probability.

A. Definition. The 2 × 2 table (also called **two-way table**) is a simple device for summarizing test results and estimating probabilities. It consists of two rows and two columns, thus creating four **cells**. The column and row totals are called **marginal totals**. The sum of row or column totals is called the **grand total**.

B. Example: determining the accuracy of a diagnostic test. A simple way to evaluate a diagnostic test's performance is by summarizing test results in a 2 × 2 table (for a complete discussion of how to evaluate a diagnostic test, see Ch 3 II). Table 2-2 summarizes results obtained in a study to evaluate an experimental test's ability to detect a certain disease. The investigators randomly selected 100 persons for the study, which they conducted as follows.

1. **Defining the diseased population**
   a. First, the researchers selected a standard diagnostic test, called the **gold standard,** to define those persons who have the disease. (The gold standard may be any diagnostic procedure believed to identify diseased persons with certainty, such as biopsy for prostate cancer.)

Table 2-2. Results of Standard and Experimental Diagnostic Tests

	D+	D−	Totals
T+	7	4	11
T−	3	86	89
Totals	10	90	100

D+ = diseased per standard test; D− = disease-free per standard test; T+ = negative result on experimental test; T− = positive result on experimental test.

b. The study participants then were classified as **diseased (D+)** or **disease-free (D−)** based on results of the gold standard.

2. **Classifying results of the experimental test**
   a. Next, the researchers administered the experimental test to all study participants and recorded positive (T+) or negative (T−) results.
      (1) A result is positive when it is above an arbitrarily chosen **cutoff value** (also called critical value or cutoff point).
      (2) A result is negative when it is below the cutoff value. (Cutoff values are discussed further in Ch 3 I A 2.)
   b. This classification of study participants is independent of the classification for disease using the gold standard diagnostic procedure.

3. **Comparing test results** (see Table 2-2)
   a. Of the 100 persons selected for the study, 10 were defined as diseased (D+) by the gold standard and 90 were defined as disease-free (D−).
   b. In the D− group, 86 persons had negative test results (T−) and 4 had positive test results (T+). The diseased (D+) group included 3 people who were labeled T− and 7 who were labeled T+ using the experimental test.

C. **Sample problems** (use Table 2-2)

1. **Problem #1.** What is the probability that a person selected at random from the 100 study participants has the disease (as determined by the gold standard)?

2. **Solution.** The marginal total for diseased study participants (D+ by the gold standard) is 10. Thus, the probability of selecting a diseased person from the total study group is

$$P(\text{disease}) = P(\text{D}+) = 10/100 = .10$$

Note. If the 100 study participants were selected randomly from the population, $P(\text{D}+)$ is an estimate of the probability that a person selected from this population has the disease.

3. **Problem #2.** What is the probability that a person selected at random has a positive result using the experimental test?

4. **Solution.** From Table 2-2, 11 of the 100 people in the study had a positive result (T+) on the experimental test. Thus,

$$P(\text{positive result}) = P(\text{T}+) = 11/100 = .11$$

III. RULES OF PROBABILITY

A. Combining probabilities

1. **Joint probabilities**
   a. **Definition.** The joint probability of two or more clinical events is the probability that the events occur simultaneously.
   b. **Notation.** The joint probability that both event A and event B occur is written $P(\text{A and B})$.

c. Sample problem (use Table 2-2)

(1) Problem. What is the probability that a study participant who is disease-free (D−) also tests negative (T−)?

(2) Solution. The intersection of column D− and row T− in Table 2-2 indicates that 86 of the 100 study participants satisfy conditions D− and T− simultaneously. Thus,

$$P(\text{D}- \text{ and } \text{T}-) = 86/100 = .86$$

2. Conditional probabilities

a. Definition. Conditional probability is the probability of an event occurring given that another event has already occurred.

b. Notation. The probability that an event B will occur when an event A has already occurred is written $P(\text{B}|\text{A})$.

c. Note. The condition (i.e., the event to the right of the vertical bar) may be controlled by the physician (e.g., treatment) or it may not (e.g., disease).

d. Sample problems

(1) Problem #1 (use Table 2-1). If a man randomly selected from the sample is known to have a cholesterol level less than 240 mg/dl (event A), what is the probability that his level is between 120 and 139 mg/dl (event B)?

(2) Solution. From the table it is found that 523 men in the sample (206 + 152 + 97 + 37 + 21 + 10 = 523) have cholesterol levels below 240 mg/dl, 10 of whom have levels between 120 and 139 mg/dl. Thus, the probability that this randomly selected man's cholesterol level is between 120 and 139 mg/dl is

$$P(\text{B}|\text{A}) = P(\text{level between 120 and 139}|\text{level} < 240)$$
$$= 10/523 = .019$$

(3) Problem #2 (use Table 2-2). What is the probability that a person is diseased, given that his or her experimental test result is positive?*

(4) Solution. Here, the problem is to find the proportion of diseased persons (as diagnosed by the gold standard) who exist in the population of persons labeled T+ by the experimental test. The population of interest (i.e., those persons labeled T+) is only 11 of the 100 total study participants. Of the 11, the number of persons with disease is 7. The probability that a person is diseased given that he or she is among those with a positive test is

$$P(\text{D}+|\text{T}+) = 7/11 = .64$$

Thus, of the group with a positive test, 64% of patients have the disease (as defined by the gold standard).

e. Conditional versus unconditional probability of disease

(1) Unconditional probability. The probability

$$P(\text{disease}) = P(\text{D}+) = 10/100 = .10$$

is an unconditional probability because it assumes no prior knowledge of the results of the diagnostic test (i.e., its value is not based on, or "conditioned on," knowing that the diagnostic test is positive or negative). For this reason, the unconditional probability of disease is also referred to as the **prior,** or **pretest, probability** of disease ("prior to" additional diagnostic information).

(2) Conditional probability. The probability

$$P(\text{disease}|\text{positive test}) = P(\text{D}+|\text{T}+) = 7/11 = .64$$

is the conditional probability that the person has the disease after it is known that the diagnostic test result is positive. Therefore, the denominator in the probability calculation must be reduced to include only those people with a positive test. $P(\text{D}+|\text{T}+)$ is called the **posterior,** or **posttest, probability** of disease.

---

*Positive test results do not guarantee the presence of disease.

B. Formulas used to calculate probabilities

1. Calculating conditional probability in terms of joint probability

a. Formula. Conditional probability, $P(A|B)$, can be defined in terms of the joint probability, $P(A \text{ and } B)$, using the formula

$$P(A|B) = \frac{P(A \text{ and } B)}{P(B)} \tag{2.1}$$

b. Sample problem (use Table 2-2)

(1) Problem. What is the probability that a person has the disease given a positive test result?

(2) Solution. Using formula 2.1 and the data from Table 2-2,

$$P(D+|T+) = \frac{P(D+ \text{ and } T+)}{P(T+)}$$

$$= \frac{7/100}{11/100} = \frac{7}{11} = .64$$

2. Calculating joint probability in terms of conditional probability.

Formula 2.1 can be rearranged to define joint probability in terms of conditional probability. This is called the **multiplication rule of probability**.

a. General multiplication rule. The joint probability of events A and B, $P(A \text{ and } B)$, generally is calculated using the formula

$$P(A \text{ and } B) = P(A|B)P(B) \tag{2.2}$$

Note. In the expression $P(A \text{ and } B)$, the "and" (mathematicians say the "logical and") does not depend on the order of events. In other words,

$$P(A \text{ and } B) = P(B \text{ and } A)$$

Therefore, the multiplication rule can be written

$$P(A \text{ and } B) = P(B \text{ and } A)$$
$$= P(B|A)P(A)$$
$$= P(A|B)P(B)$$

b. Multiplication rule for independent events

(1) Definition. If the occurrence of event B is in no way affected by the occurrence of event A, then the two events A and B are said to be **independent**. Symbolically, if $P(A|B) = P(A)$, then A and B are independent.

(2) Formula. If A and B are independent, the multiplication rule for calculating joint probability (formula 2.2) simplifies to

$$P(A \text{ and } B) = P(A|B)P(B) = P(A)P(B) \tag{2.3}$$

That is, the joint probability that independent events occur together is the **product of their individual probabilities**.

c. Sample problems

(1) Problem #1 (use Table 2-2). What is the probability that a person is disease-free and has a negative test result?

(2) Solution. Using formula 2.2 and the data in Table 2-2,

$$P(D- \text{ and } T-) = P(D-|T-)P(T-)$$
$$= (86/89)(89/100)$$
$$= 86/100 = .86$$

(3) Problem #2 (use Table 2-1). Two men are selected at random from the sample of 1047 disease-free men in Table 2-1. What is the probability that both men have serum cholesterol levels less than 240 mg/dl?

(4) Solution

(a) Let A be the event that the first man has a cholesterol level less than 240 mg/dl and B be the event that the second man has a cholesterol level less than 240 mg/dl.

(b) The selection of the second man is not influenced by the selection of the first man, so the events are independent. Therefore, $P(\text{A and B}) = P(\text{A})\ P(\text{B})$.

(c) From Table 2-2,

$$P(\text{A}) = 523/1047 = .500$$

$$P(\text{B}) = 523/1047 = .500$$

Hence,

$$\begin{aligned} P(\text{A and B}) &= (.500)(.500) \\ &= .250 \end{aligned}$$

(5) **Problem #3.** What is the probability that a couple's first two children are boys?

(6) **Solution.** For this problem, assume that $P(\text{boy}) = P(\text{girl}) = .5$. (The chance of having a girl actually is slightly more than .5.)

(a) The probability that the couple's first two children are boys is

$$P(\text{child \#1 boy and child \#2 boy}) = P(\text{child \#2 boy}|\text{child \#1 boy})P(\text{child \#1 boy})$$

(b) However, the probability of the second child being a certain sex is independent of the sex of the first child. So,

$$P(\text{child \#2 boy}|\text{child \#1 boy}) = P(\text{child \#2 boy})$$

(c) Thus,

$$\begin{aligned} P(\text{child \#1 boy and child \#2 boy}) &= P(\text{child \#2 boy})P(\text{child \#1 boy}) \\ &= (.5)(.5) \\ &= .25 \end{aligned}$$

3. Calculating joint probability using the addition rule

a. General addition rule. The probability that event A or event B will occur generally is calculated using the formula

$$P(\text{A or B}) = P(\text{A}) + P(\text{B}) - P(\text{A and B}) \tag{2.4}$$

b. Sample problem (use Table 2-2)

(1) **Problem.** What is the probability that a patient selected at random from the population represented by the data in Table 2-2 is disease-free or has a negative test result?

(2) **Solution**

(a) From Table 2-2 and formula 2.4,

$$\begin{aligned} P(\text{D}- \text{ or T}-) &= P(\text{D}-) + P(\text{T}-) - P(\text{D}- \text{ and T}-) \\ &= 90/100 + 89/100 - 86/100 \\ &= 93/100 = .93 \end{aligned}$$

(b) **Note.** There are 90 people who are disease-free ($86 + 4 = 90$) and 89 people who test negative ($86 + 3 = 89$). Since the 86 people who satisfy both conditions simultaneously have been counted twice, they must be subtracted once. The correct number of persons who satisfy one or the other condition, therefore, is $90 + 89 - 86 = 93$.

c. Addition rule for mutually exclusive events

(1) **Definition.** Mutually exclusive events are events that cannot occur together. For example, the events D+ and D− are mutually exclusive, since a person cannot simultaneously have disease (D+) and be disease-free (D−).

(2) **Formula.** If A and B are mutually exclusive, $P(\text{A and B}) = 0$ and formula 2.4 is simplified to

$$P(\text{A or B}) = P(\text{A}) + P(\text{B}) \tag{2.5}$$

(3) **Examples**

(a) In Table 2-1, a person's cholesterol level may fall into only 1 of the 18 possible intervals. For example, a person cannot have a cholesterol value that is simultaneously below 160 mg/dl and above 340 mg/dl.

(b) Each serum cholesterol interval in Table 2-1 represents a mutually exclusive outcome of cholesterol measurement. If A represents the event that a person drawn at random has a cholesterol level below 160 mg/dl and B represents the event that the person has a cholesterol level of 340 or more, the probability that A or B will occur is

$$\begin{aligned} P(\text{A or B}) &= P(\text{A}) + P(\text{B}) \\ &= 31/1047 + 25/1047 \\ &= .030 + .024 \\ &= .054 \end{aligned}$$

4. Summation principle for joint probabilities. According to Table 2-2, a diseased (D+) person will test either positive (T+) or negative (T−). Since T+ and T− are mutually exclusive, the probability of a person having the disease is the probability that the person is diseased and tests positive plus the probability the person is diseased and tests negative. This is written as

$$P(\text{D+}) = P(\text{D+ and T+}) + P(\text{D+ and T−})$$

a. Formula. In reality, there are usually more than two causes for an event. Therefore, the summation principle for joint probabilities may be generalized to

$$P(\text{A}) = P(\text{A and B}_1) + P(\text{A and B}_2) \cdots + P(\text{A and B}_n) \qquad (2.6)$$

where $B_1, \ldots, B_n$ are mutually exclusive events.

b. Example. Applying formula 2.6, $P(\text{D+}) = 10/100$ (as read directly from Table 2-2) is verified as

$$\begin{aligned} P(\text{D+}) &= P(\text{D+ and T+}) + P(\text{D+ and T−}) \\ &= 7/100 + 3/100 \\ &= 10/100 = .10 \end{aligned}$$

5. Summation principle for conditional probabilities. Converting the joint probabilities to their conditional probability equivalents, the probability of a person having the disease is written as

$$P(\text{D+}) = P(\text{D+}|\text{T+})P(\text{T+}) + P(\text{D+}|\text{T−})P(\text{T−})$$

a. Formula. The summation principle for conditional probabilities may be generalized to

$$P(\text{A}) = P(\text{A}|\text{B}_1)P(\text{B}_1) + P(\text{A}|\text{B}_2)P(\text{B}_2) \cdots + P(\text{A}|\text{B}_n)P(\text{B}_n) \qquad (2.7)$$

where $B_1, \ldots, B_n$ are mutually exclusive events.

b. Example. Applying formula 2.7, $P(\text{T+}) = 11/100$ (as read directly from Table 2-2) is verified as

$$\begin{aligned} P(\text{T+}) &= P(\text{T+}|\text{D+})P(\text{D+}) = P(\text{T+}|\text{D−})P(\text{D−}) \\ &= (7/10)(10/100) + (4/90)(90/100) \\ &= 7/100 + 4/100 \\ &= 11/100 = .11 \end{aligned}$$

IV. SUMMARY OF FORMULAS FOR CALCULATING PROBABILITY

A. Probability of an event

$$P(\text{E}) = \frac{\text{number of times E occurs}}{\text{number of times E can occur}}$$

B. Multiplication rule of probabilities

1. Nonindependent events

$$P(\text{A and B}) = P(\text{A}|\text{B})P(\text{B})$$

2. Independent events

$$P(\text{A and B}) = P(\text{A})P(\text{B})$$

C. Addition rule of probabilities

1. Mutually exclusive events

$$P(\text{A or B}) = P(\text{A}) + P(\text{B})$$

2. Non-mutually exclusive events

$$P(\text{A or B}) = P(\text{A}) + P(\text{B}) - P(\text{A and B})$$

D. Conditional probability

$$P(\text{A}|\text{B}) = \frac{P(\text{A and B})}{P(\text{B})}$$

E. Summation principle for joint probabilities

$$P(\text{A}) = P(\text{A and B}_1) + P(\text{A and B}_2) \cdots + P(\text{A and B}_n)$$

F. Summation principle for conditional probabilities

$$P(\text{A}) = P(\text{A}|\text{B}_1)P(\text{B}_1) + P(\text{A}|\text{B}_2)P(\text{B}_2) \cdots + P(\text{A}|\text{B}_n)P(\text{B}_n)$$

PROBLEMS

2-1. The table below summarizes results of a study to evaluate the gonodectin (Gd) test as a diagnostic test for gonorrhea in men. The study involved 240 men with symptoms of exudative urethritis who were seen at a medical facility for the diagnosis and treatment of sexually transmitted diseases. Urethral discharge specimens obtained from each of the men were evaluated both by the Gd test and by culture (gold standard). In problems 2-1a through 2-1i, use the results obtained from the 240-man sample to answer the questions about the population of men visiting the clinic.

Culture and Gonodectin (Gd) Test Results for 240 Urethral Discharge Specimens

	Culture Results		
Gd Test Result	**Gonorrhea (D +)**	**No gonorrhea (D −)**	**Totals**
Positive (T+)	175	9	184
Negative (T−)	8	48	56
Totals	183	57	240

Adapted from Janda WM, Jackson T: Evaluation of gonodectin for the presumptive diagnosis of gonococcal urethritis in men. *J Clin Microbiol* 21:143–145, 1985.

a. What is the probability that a man has gonorrhea, or *P*(D+)?

b. What is the probability that a man has a positive Gd test, or *P*(T+)?

c. What is the probability that a man has a positive Gd test and gonorrhea, or *P*(T+ and D+)? Verify the answer obtained directly from the table above by calculating *P*(T+ and D+) using formula 2.2.

d. What is the probability that a man has a negative Gd test and does not have gonorrhea, or *P*(T− and D−). Verify the answer using formula 2.2.

e. What is the probability that a man with gonorrhea has a positive Gd test, or *P*(T+|D+)? Verify the answer using formula 2.1.

f. What is the probability that a man who does not have gonorrhea has a negative Gd test, or *P*(T−|D−)? Verify the answer using formula 2.1.

g. What is the probability that a man who does not have gonorrhea has a positive Gd test, or *P*(T+|D−)? Verify the answer using formula 2.1.

h. What is the probability that a man with gonorrhea has a negative Gd test, or *P*(T−|D+)? Verify the answer using formula 2.1.

i. What is the probability that a man with a positive Gd test has gonorrhea, or *P*(D+|T+). Verify the answer using formula 2.1.

j. At a clinic similar to the one that conducted the Gd test study, a chief resident examines a man with symptoms of exudative urethritis and decides to order the Gd test, which comes back negative. What is the likelihood that this patient has gonorrhea? What is the likelihood that he does not have it?

k. Consider the probabilities *P*(D+|T+) and *P*(T+|D+). Which would be more informative to a physician who is interested in establishing a diagnosis of gonorrhea in a male patient?

l. Based on data in the above table, what is the probability that a man who visits the clinic has a positive Gd test or gonorrhea, or *P*(T+ or D+)?

m. Using data in the above table, confirm the summation principle for joint probabilities:

$$P(D+) = P(D+ \text{ and } T+) + P(D+ \text{ and } T-).$$

2-2. An outbreak of food poisoning occurs in a group of students who attended a back-to-school party. The following table summarizes data obtained from 200 students who were at the party.

Data Obtained From 200 Students Who Attended the Party

	Ill	Not Ill	Totals
Ate Barbecue	90	30	120
Did Not Eat Barbecue	20	60	80
Totals	110	90	200

a. What is the probability that a student becomes ill after eating barbecue?

b. What is the probability that a student becomes ill if no barbecue is eaten?

c. What is the probability that a student does not become ill after eating barbecue?

d. What is the probability that a student who attended the party becomes ill?

e. What is the probability that a student with food poisoning ate barbecue?

f. What is the probability that a student who attended the party did not eat barbecue?

g. What is the ratio of probabilities obtained in problems 2-2a and 2-2b? Interpret its meaning.

2-3. The table below is a frequency table showing red blood cell (RBC) cholinesterase values of 35 agricultural workers exposed to pesticides. In the following problems, it can be assumed that the data are representative of the population of agricultural workers exposed to pesticides.

Frequency Distribution of RBC Cholinesterase Values

RBC Cholinesterase (μmol/min/ml)	Frequency	Relative Frequency	Cumulative Frequency (%)
5.95–7.95	1	.029	2.9
7.95–9.95	8	.229	25.8
9.95–11.95	14	.400	65.8
11.95–13.95	9	.257	91.5
13.95–15.95	2	.057	97.2
15.95–17.95	1	.029	100.1
Total	35	1.001	

Reprinted with permission from Duncan RC, Knapp RG, Miller MC III: *Introductory Biostatistics for the Health Sciences*, 2nd ed. Albany, Delmar Publishers, 1983, p 10.

a. What is the probability that an agricultural worker who is exposed to pesticides has an RBC cholinesterase value between 7.95 and 11.95 μmol/min/ml? Greater than 15.95 μmol/min/ml? Less than 9.95 μmol/min/ml?

b. If a pesticide-exposed worker is known to have an RBC cholinesterase value greater than 11.95 μmol/min/ml, what is the probability that the value is greater than 15.95 μmol/min/ml?

2-4. Assuming that the probability of the birth of a girl is .5 (it actually is slightly higher), what is the probability of the successive birth of four girls? What is the probability that the fourth child is a boy given that the first three children are girls?

2-5. The failure rate for a cardiac arrest alarm in an intensive care unit is .001. For improved safety, a duplicate alarm is installed. What is the probability that a cardiac arrest will not be signaled?*

Solutions on p 361.

*Reprinted from Duncan RC, Knapp RG, Miller MC III: *Introductory Biostatistics for the Health Sciences*, 2nd ed. Albany, Delmar, 1983, p 50.

STUDY QUESTIONS

Directions: Each of the numbered items or incomplete statements in this section is followed by answers or by completions of the statement. Select the **one** lettered answer or completion that is **best** in each case.

1. The joint probability P(A and B) can be expressed in all of the following ways EXCEPT

(A) $P(A) + P(B)$
(B) $P(A|B)P(B)$
(C) $P(B|A)P(A)$
(D) $P(A)P(B)$, if A and B are independent

2. All of the following statements about the conditional probability, $P(A|B)$, are true EXCEPT

(A) $P(A|B) = P(\text{A and B})P(B)$
(B) $P(A|B) =$ the posterior probability that A will occur
(C) $P(A|B) = P(\text{A and B})/P(B)$
(D) $P(A|B) = P(A)$, if A and B are independent
(E) $P(A|B) =$ the joint probability that A and B will occur simultaneously divided by the marginal probability that B will occur

Questions 3–10

The table below summarizes results of a study to evaluate a new screening test for a certain disease. In the study, 1000 people were randomly selected from the population and given the new diagnostic test, and a positive or negative result (T+ or T−) was recorded. Each study participant also was given the gold standard diagnostic procedure to determine the true disease state of the individual (D+ or D−).

	True Disease State		
Test Result	**D+**	**D−**	**Totals**
T+	80	50	130
T−	20	850	870
Totals	100	900	1000

3. Assuming that the sample is representative of the population from which it was drawn, the probability that an individual in this population tests positive and has the disease is estimated by

(A) 80/100 = .80
(B) 80/130 = .62
(C) 80/1000 = .08
(D) 100/1000 = .10
(E) 100/130 = .77

4. The probability calculated in question 3 is written as

(A) $P(D+ \text{ and } T+)$
(B) $P(D+|T+)$
(C) $P(T+|D+)$
(D) $P(D+)$
(E) $P(D+)P(T+)$

5. Assuming that the sample is representative of the population from which it was drawn, the probability that an individual in this population has the disease is estimated by

(A) 80/100 = .80
(B) 80/130 = .62
(C) 80/1000 = .08
(D) 100/1000 = .10
(E) 130/1000 = .13

6. The probability calculated in question 5 is written as

(A) $P(D+ \text{ and } T+)$
(B) $P(D+|T+)$
(C) $P(T+|D+)$
(D) $P(D+)$
(E) $P(D+)P(T+)$

1-A 4-A
2-A 5-D
3-C 6-D

7. If an individual from this population tests positive on the diagnostic test, the probability that the individual truly has the disease is

(A) 80/100 = .80
(B) 80/130 = .62
(C) 80/1000 = .08
(D) 130/1000 = .13
(E) 100/130 = .77

8. The probability calculated in question 7 is written as

(A) P(D+ and T+)
(B) P(D+|T+)
(C) P(T+|D+)
(D) P(D+)
(E) P(D+)P(T+)

9. If an individual from this population is known to be disease-free (as verified by the gold standard), the probability that this individual tests positive on the screening test is

(A) 50/900 = .056
(B) 850/900 = .944
(C) 130/1000 = .130
(D) 50/130 = .385
(E) 130/900 = .144

10. The probability calculated in question 9 is written as

(A) P(T+|D−)
(B) P(D−|T+)
(C) P(D− and T+)/P(T+)
(D) P(D−)P(T+)
(E) P(D− and T+)

Questions 11–12

The table below lists the results of a blood type evaluation of 50 men and 50 women.

	Gender		
Blood Type	**Male**	**Female**	**Totals**
O	20	20	40
A	17	18	35
B	8	7	15
AB	5	5	10
Totals	50	50	100

11. What is the probability that an individual picked at random from this group has type A or type B blood?

(A) .05
(B) .50
(C) .35
(D) .20
(E) None of the above

12. What is the probability that an individual selected at random from this group has type O blood or is a man?

(A) (40/100) (50/100)
(B) (40/100) (50/100) − (20/100)
(C) 20/100
(D) (40/100) + (50/100) − (20/100)
(E) None of the above

7-B 10-A
8-B 11-B
9-A 12-D

ANSWERS AND EXPLANATIONS

1. The answer is A *[III A 1 a, B 2, 3]*
By definition, the joint probability of two or more events is the probability that the events occur simultaneously. For event A and event B, this is given by the multiplication rule of probability:

$$P(\text{A and B}) = P(\text{A}|\text{B})P(\text{B}) = P(\text{B}|\text{A})P(\text{A})$$

If A and B are independent, then

$$P(\text{A}|\text{B}) = P(\text{A}) \text{ and } P(\text{A and B}) = P(\text{A})P(\text{B})$$

The probability that A or B will occur is given by the addition rule of probability

$$P(\text{A or B}) = P(\text{A}) + P(\text{B}) - P(\text{A and B})$$

If A and B are mutually exclusive, then

$$P(\text{A and B}) = 0 \text{ and } P(\text{A or B}) = P(\text{A}) + P(\text{B})$$

2. The answer is A *[III A 2 e (2), B 1 a]*
The conditional probability, $P(\text{A}|\text{B})$, is the posterior probability that event A occurs given that event B is known to have occurred; $P(\text{A})$ is the unconditional probability that event A occurs. In terms of joint probability, the conditional probability, $P(\text{A}|\text{B})$, is stated as $P(\text{A and B})/P(\text{B})$, where $P(\text{A and B})$ is the joint probability of A and B, and $P(\text{B})$ is the marginal, or unconditional, probability of B. If A and B are independent, then $P(\text{A}|\text{B}) = P(\text{A})$; that is, the probability that A occurs given that B is known to have occurred is the same as the unconditional probability that A occurs.

3–10. The answers are: 3-C *[III A 1]*, **4-A** *[III A 1 b]*, **5-D** *[I A, II C]*, **6-D** *[II C 2]*, **7-B** *[III A 2]*, **8-B** *[III A 2 b]*, **9-A** *[III A 2]*, **10-A** *[III A 2 b]*
Among the study group, 80 of 1000 individuals had both a positive test and the disease. Thus, assuming the sample is representative of the population, the probability that an individual in this population tests positive and has the disease is estimated by 80/1000 = .08. The probability of the simultaneous (joint) occurrence of a positive test (T+) and disease (D+) is written as $P(\text{D+ and T+})$. $P(\text{D+}|\text{T+})$ is the conditional probability that an individual has the disease given a positive result on the experimental test. $P(\text{T+}|\text{D+})$ is the conditional probability that an individual tests positive given that the individual is known to be diseased (i.e., it is the probability of a positive test result in the population of patients with the disease). The product, $P(\text{D+})$, is the unconditional—or prior—probability of disease (i.e., the probability that the disease is present prior to knowledge of test results). The product, $P(\text{D+})P(\text{T+})$, equals the joint probability of a positive test and the disease, or $P(\text{D+ and T+})$, only if the occurrence of disease (D+) and a positive test (T+) are independent.

The unconditional probability that an individual in the population has the disease is estimated by the relative frequency of disease among the 1000 study participants if the number of diseased individuals and the number of disease-free individuals are random (i.e., not fixed or determined by the investigator). Among the 1000 study participants, 100 have the disease as determined by the gold standard. Thus, assuming the sample is representative of the population, the probability that an individual in the population has the disease is 100/1000 = .10. This probability is written as $P(\text{D+})$. The product, $P(\text{D+})$, is the unconditional (prior) probability of disease.

The conditional probability that an individual in the population has the disease given a positive test result is estimated by the relative frequency of disease among the study participants who have positive test results. Among the 130 individuals who test positive, 80 have the disease as determined by the gold standard. Thus, the conditional probability of disease given a positive test result is 80/130 = .62. This probability is written as $P(\text{D+}|\text{T+})$.

The conditional probability that an individual in the population who is known to be disease-free will test positive is estimated by the relative frequency of positive test results among the disease-free study participants. Among the 900 individuals who are disease-free as determined by the gold standard, 50 have positive results. Thus, the conditional probability of a positive test result given the absence of disease is 50/900 = .056. This probability is written as $P(\text{T+}|\text{D−})$. Using the definitional relationship between joint and conditional probabilities, this can be written as

$$P(\text{T+}|\text{D−}) = \frac{P(\text{T+ and D−})}{P(\text{D−})}$$

11–12. The answers are: 11-B *[III B 3 c]*, **12-D** *[III B 3 a]*
An individual can have only one type of blood: A, B, AB, or O. Using the data in the table, the probability that an individual selected at random from the group will have type A or type B blood is determined by applying the addition rule of probabilities:

$$\begin{aligned} P(\text{A or B}) &= P(\text{A}) + P(\text{B}) - P(\text{A and B}) \\ &= P(\text{type A}) + P(\text{type B}) - P(\text{type A and type B}) \\ &= 35/100 + 15/100 - 0 \\ &= .50 \end{aligned}$$

Note: $P(\text{A and B}) = 0$, because type A blood and type B blood are mutually exclusive.

Since gender and blood type are not mutually exclusive, this joint probability problem can be solved using the general addition rule. The probability that an individual selected at random from the group has type O blood or is a man is

$$\begin{aligned} P(\text{type O or male}) &= P(\text{type O}) + P(\text{male}) - P(\text{type O and male}) \\ &= (40/100) + (50/100) - (20/100) \\ &= (70/100) = .7 \end{aligned}$$

3
Describing the Performance of a Diagnostic Test

I. DISTINGUISHING NORMAL FROM ABNORMAL. Diagnostic testing, a term that encompasses physical examination, history, imaging techniques (e.g., x-rays, CT scans), and procedures (e.g., ECG), as well as laboratory tests, provides the framework for clinical decision-making. A given diagnostic test is based on the assumption that diseased and healthy individuals can be accurately and reproducibly differentiated by the test.

A. Types of diagnostic tests

1. **Qualitative diagnostic tests** classify patients as diseased or disease-free according to the presence or absence of a clinical sign or symptom. For example, an x-ray might confirm or disprove the existence of a fracture.

2. **Quantitative diagnostic tests** classify patients as diseased or disease-free on the basis of whether they fall above or below a preselected cutoff value known as the **positivity criterion**. This cutoff value is also referred to as the **critical value** or **referent value**.

B. Selection of a positivity criterion. A positivity criterion for a particular diagnostic test is selected in one of six ways.

1. **Gaussian distribution method.** This method is based on the assumption that the frequency distribution of values of the test follows a **normal distribution**.

   a. **Properties** (Figure 3-1). The curve representing the normal distribution is smooth, bell-shaped, and symmetric about its mean (denoted symbolically: μ).

   (1) Fifty percent of the observations lie above the mean and 50% lie below the mean.

   (2) Approximately 68% of the observations lie within 1 standard deviation of the mean.

   (3) Approximately 95% of the observations lie within 1.96 standard deviations of the mean (often rounded to 2 standard deviations).

   (4) Approximately 99% of the observations lie within 2.58 standard deviations of the mean (often rounded to 3 standard deviations).

   b. **Procedure**

   (1) Obtain values for the diagnostic test variable for a large number of individuals in a population believed to be disease-free. This may be done separately for different age-groups, genders, races, or other subgroups.*

   (2) Draw a histogram showing the frequency distribution of these values. Verify that the histogram conforms to the gaussian distribution by visual inspection or a statistical test of "goodness-of-fit."

   (3) Compute the mean and standard deviation.

   (4) To define the normal range, calculate the upper and lower limits (L) using the formula

$$L = \text{mean} \pm 2 \text{ standard deviations}^{\dagger}$$

   (5) In a disease-free population, 95% of the values of the diagnostic variable fall within

---

*The mean and standard deviation and, thus, the shape of the frequency distribution for the test variable may be different for different subgroups within the disease-free population (e.g., men may differ from women). Therefore, normal ranges for each subgroup are often defined separately.

†Additional precision can be obtained by multiplying the standard deviation by 1.96.

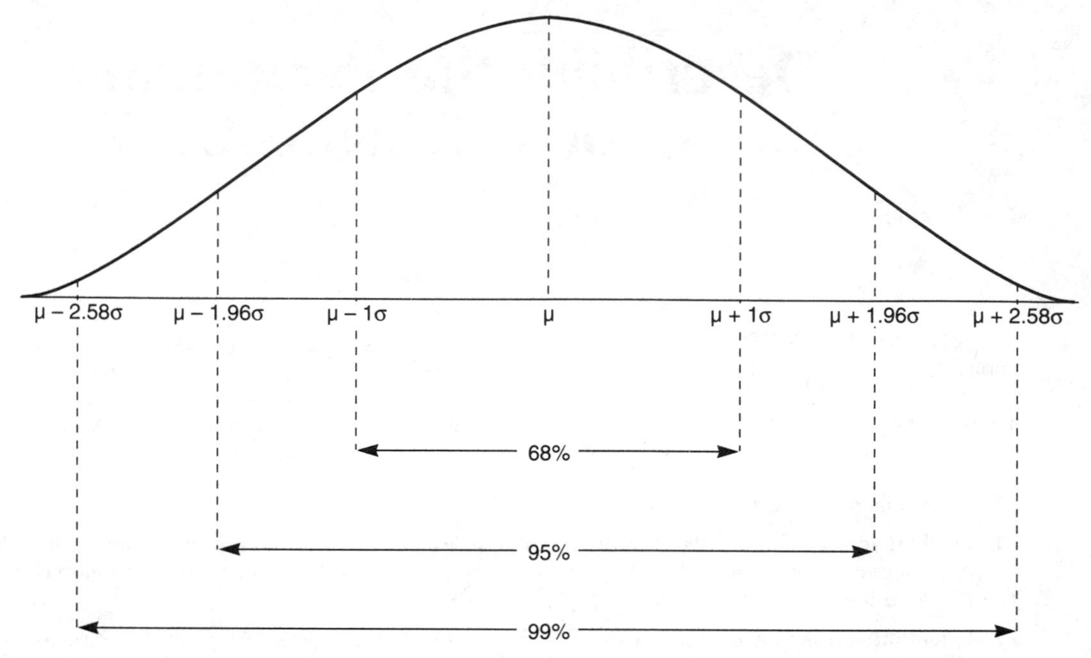

Figure 3-1. A gaussian distribution showing the percentage of values found in ranges about the mean. (Reprinted from Duncan RC, Knapp RG, Miller MC III: *Introductory Biostatistics for the Health Sciences*, 2nd ed. Albany, Delmar, p 59.)

this interval. Values outside the interval are defined as abnormal.

c. Advantages. The advantage of the gaussian method is its simplicity.

d. Disadvantages

(1) Values of a diagnostic test variable are more likely to follow a skewed or bimodal distribution, rather than a gaussian distribution. If the diagnostic test results do not fit a gaussian distribution, the normal range does not necessarily contain 95% of the values.

(2) If the highest 5% of test results are classified as abnormal (or highest and lowest 2.5%), all diseases have the same expected frequency—a medically absurd conclusion.

(3) Multiple testing increases the likelihood that a given patient will be outside the range of normal on at least one test [specifically, the chance is $1 - (.95)^n$, where n is the number of independent tests administered], a consequence leading to the aphorism, "The only 'normal' patient is one not yet sufficiently worked up."

(4) The definition of disease has no biologic basis.

(5) Because the normal range is defined in the disease-free population, a patient classified as abnormal is unusual compared to the normal population but not necessarily diseased. Conversely, an individual whose test results fall within the normal range is not automatically disease-free.

(6) The patient may belong to a subgroup having a different normal range than the one used for comparison.

(7) Changes in the diagnostic test variable over time may be pathologic, even if they remain within the normal range. For example, while the normal range for blood sugar is 40–120 mg/dl, an increase in blood sugar from 45 mg/dl to 110 mg/dl may be abnormal. Rapidly fluctuating values may be particularly significant.

2. Percentile method

a. Procedure

(1) Obtain values for the diagnostic test variable for a large number of individuals in a population believed to be disease-free.

(2) Define the lower 95% of test results as normal and the upper 5% as diseased. If both upper and lower limits are desired, define the highest 2.5% and lowest 2.5% of values as diseased.

b. Advantages of the percentile method are:

(1) Simplicity

(2) Applicability to all possible distributions of the test variable

c. Disadvantages of the percentile method are the same as those of the gaussian method (see I B 1 d).

3. "Culturally desirable" method

a. Procedure. Define as normal those values of the diagnostic test variable considered socially desirable. For example, the normal range for body weight in women would be defined by the popular concept of thinness.

b. Advantages/disadvantages. This method has no advantages and confuses medicine with mores.

4. Therapeutic method

a. Procedure. Define the positivity criterion as that value of the diagnostic test variable above or below which a patient will be treated. As new data about the benefits of therapy become available, the positivity criterion is adjusted accordingly. For example, the definition of hypertensive has shifted to progressively lower values of diastolic blood pressure over the last 3 decades in parallel with research data demonstrating the advantages of treating successively lower blood pressures.

b. Advantages. This method classifies only individuals who will actually be treated as diseased.

c. Disadvantages. This method requires the physician to stay abreast of all therapeutic advances.

5. Risk factor method

a. Definition. A behavioral, genetic, environmental, demographic, or physiologic factor bearing a causal or statistical relationship to the occurrence of a particular disease is called a **risk factor**.

b. Procedure

(1) Assume that the association between the risk factor and the disease indicates that the risk factor is a predictor of the disease.

(2) Define as normal those values of the diagnostic test value that confer no additional risk of mortality or morbidity.

c. Advantages

(1) Risk factors usually are easy to measure.

(2) Associating a risk factor with a specific diagnosis facilitates the design of prevention programs.

d. Disadvantages

(1) For many risk factors, the risk of disease increases steadily over the observed range of values, complicating the definition of the positivity criterion.

(2) The predictive value of the risk factor may be low. Table 3-1 represents a hypothetical study in which the prevalence of lung cancer in the population is assumed to be 100/100,000 and 33.3% of the population are smokers. Because all cases of lung cancer occur in smokers, it may be argued that smoking is a risk factor for lung cancer. However, 99.7% of smokers do not develop cancer, suggesting that smoking status is a poor diagnostic test for lung cancer.

6. Diagnostic or predictive value method. This method is the most clinically sound of the six and is discussed in detail in Chapter 4.

Table 3-1. Hypothetical Data Illustrating Poor Predictive Ability of a Risk Factor (Smoking)

	Lung Cancer (D+)	**No Lung Cancer (D−)**	**Totals**
Smoker (T+)	100	33,233	33,333
Nonsmoker (T−)	0	66,667	66,667
Totals	100	99,900	100,000

Reprinted from Galen R, Gambino R: *Beyond Normality: The Predictive Value and Efficiency of Medical Diagnosis.* New York, John Wiley, 1975, p 165.

II. DESCRIBING THE PERFORMANCE OF A NEW DIAGNOSTIC TEST. Physicians are often faced with the task of evaluating the merit of a new diagnostic test. An adequate critical appraisal of a new test requires a working knowledge of the properties of diagnostic tests and the mathematical relationships between them. This knowledge is based on the principles of probability discussed in Chapter 2.

A. The gold standard. Assessing a new diagnostic test begins with the identification of a group of patients known to have the disorder of interest, using an accepted reference test known as the **gold standard** (see Ch 2 II B 1 a).

1. Limitations

a. The gold standard is often the most risky, technically difficult, expensive, or impractical of available diagnostic options (e.g., postmortem brain biopsy, the gold standard for the diagnosis of Alzheimer's disease).

b. For some conditions (e.g., angina pectoris), no gold standard is available.

2. Comparisons with an imperfect gold standard may lead to the erroneous conclusion that the new test is worse when in fact it is better.

a. If the new test detects diseased individuals more accurately than the gold standard, these patients will be mistakenly labeled **false positives** (see II B 2 b).

b. If the new test is negative in more disease-free individuals, these patients will be mistakenly labeled **false negatives** (see II B 2 d).

B. Graphic and tabular representation of diagnostic test results

1. Example. A medical research team identified a novel peptide growth factor, breast carcinoma promoting factor (BCPF) in explant cultures of breast tumors. A pilot study

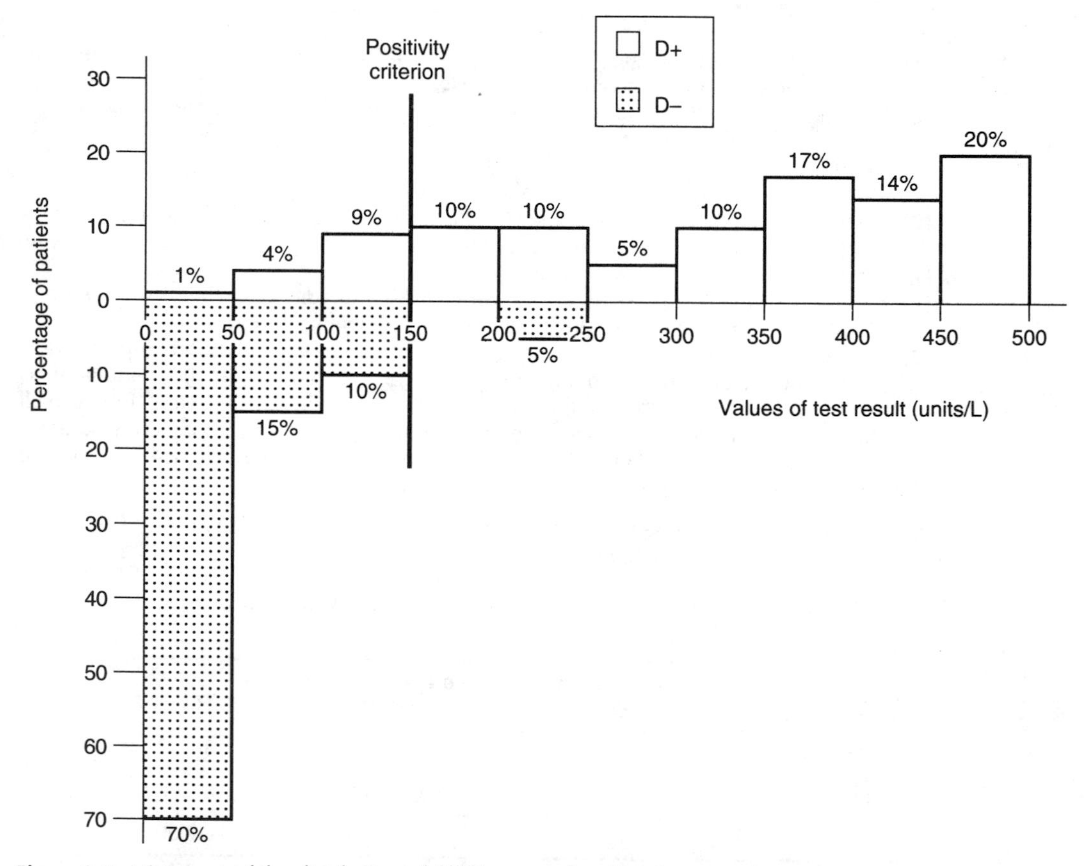

Figure 3-2. Histogram of the distribution of BCPF test result values for 600 patients with breast cancer (*D+*) and 1000 cancer-free patients (*D–*).

revealed that plasma levels of BCPF were elevated in confirmed breast cancer patients. The investigators conducted a clinical trial to determine whether the measurement of plasma levels of BCPF could be used to diagnose breast cancer, selecting a positivity criterion of 150 units/L. Of the 1600 patients available for study, 600 were demonstrated by breast biopsy (the gold standard) to have breast cancer (D+) and 1000 were found to be disease-free (D−).

a. The frequency distributions of the plasma levels of BCPF in the 600 cancer patients (D+) and in the 1000 patients free of cancer (D−) are illustrated in Figure 3-2.

b. The frequency distribution of test values illustrated in Figure 3-3 can be generalized to the diseased and disease-free populations. The frequency distribution for these two populations is a smooth curve.

2. Definitions

a. True positives (TP) are individuals with breast cancer who were correctly identified as diseased by the new test (see Figure 3-3; *nonshaded curve*).

b. False positives (FP) are cancer-free individuals falsely labeled as diseased by the new test (see Figure 3-3; *shaded curve*).

c. True negatives (TN) are cancer-free individuals who are correctly identified as disease-free by the new test (see Figure 3-3; *shaded curve*).

d. False negatives (FN) are individuals with breast cancer falsely labeled as disease-free by the new test (see Figure 3-3; *nonshaded curve*).

3. Summary of test results in a 2 × 2 table. As previously discussed (see Ch 2 II), a 2 × 2 table may be used to summarize the results of the clinical trial of the BCPF test for breast cancer diagnosis (Table 3-2).

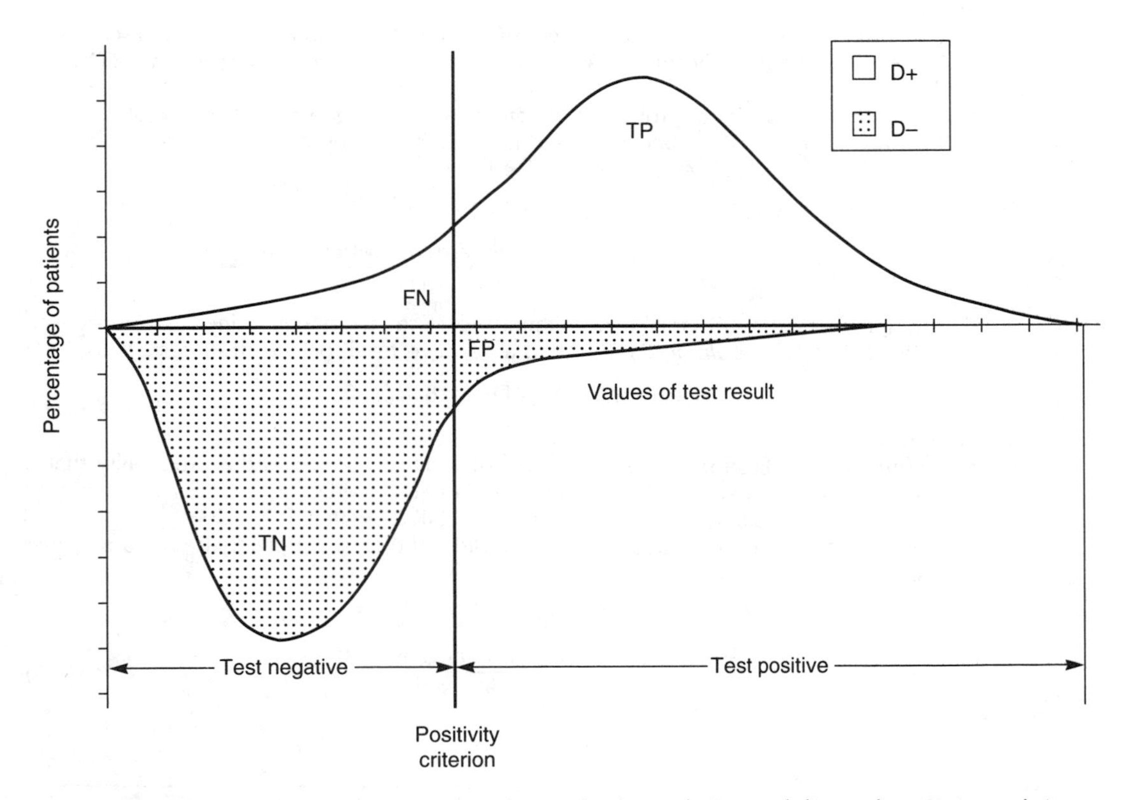

Figure 3-3. Frequency distribution of BCPF test values in the diseased (*D*+) and disease-free (*D*−) populations. *TP* = true positive; *TN* = true negative; *FP* = false positive; *FN* = false negative.

Table 3-2. A 2 × 2 Table Summarizing Results of the BCPF Test

	Breast Cancer (D+)			No Breast Cancer (D−)	Totals
BCPF Test (T+)	570	TP	FP	150	720
BCPF Test (T−)	30	FN	TN	850	880
Totals	600			1000	1600

TP = true positive; FP = false positive; FN = false negative; TN = true negative.

C. Test performance characteristics

1. Sensitivity*

a. Definition. The sensitivity of a diagnostic test is the probability that a diseased individual will have a positive test result. Sensitivity is the **true positive rate (TPR)** of the test.

b. Notation. In conditional probability notation, sensitivity is written $P(\text{T}+|\text{D}+)$.

c. Calculating sensitivity. Sensitivity is calculated as the proportion of diseased individuals with a positive test result, using the formula

$$\text{sensitivity} = P(\text{T}+|\text{D}+) = \text{TPR} = \frac{\text{diseased with positive test}}{\text{all diseased}} \tag{3.1}$$

d. Example. Of the 600 individuals with breast cancer as determined by biopsy (see Table 3-2; *total of column 1*), 570 had a positive result on the BCPF test. Thus,

$$\text{sensitivity} = P(\text{T}+|\text{D}+) = 570/600 = .95$$

2. Specificity*

a. Definition. The specificity of a diagnostic test is the probability that a disease-free individual will have a negative test result. Specificity is the **true negative rate (TNR)** of the test.

b. Notation. In conditional probability notation, specificity is written $P(\text{T}-|\text{D}-)$.

c. Calculating specificity. Specificity is calculated as the proportion of disease-free individuals with a negative test result, using the formula

$$\text{specificity} = P(\text{T}-|\text{D}-) = \text{TNR} = \frac{\text{disease-free with negative test}}{\text{all disease-free}} \tag{3.2}$$

d. Example. Of the 1000 individuals without breast cancer as determined by biopsy (see Table 3-2; *total of column 2*), 850 had a negative result on the BCPF test. Thus,

$$\text{specificity} = P(\text{T}-|\text{D}-) = 850/1000 = .85$$

3. False negative rate

a. Definition. The **false negative rate (FNR)** of a diagnostic test is the probability that a diseased individual will have a negative test result.

b. Notation. In conditional probability notation, FNR is written $P(T-|D+)$.

c. Calculation. FNR is calculated as the proportion of diseased individuals with a negative test result, using the formula

$$\text{FNR} = P(\text{T}-|\text{D}+) = \frac{\text{diseased with negative test}}{\text{all diseased}} \tag{3.3}$$

*Because the calculation of sensitivity is based on the diseased population and the calculation of specificity on the disease-free population, sensitivity and specificity are *mutually exclusive* measures of diagnostic test performance.

d. Example. Of the 600 individuals with breast cancer, 30 had a negative result on the BCPF test (see Table 3-2). Thus,

$$\text{FNR} = P(\text{T}-|\text{D}+) = 30/600 = .05$$

4. False positive rate

a. Definition. The **false positive rate (FPR)** of a diagnostic test is the probability that a disease-free individual will have a positive test result.

b. Notation. In conditional probability notation, FPR is written $P(\text{T}+|\text{D}-)$.

c. Calculation. FPR is calculated as the proportion of disease-free individuals with a positive test result, using the formula

$$\text{FPR} = P(\text{T}+|\text{D}-) = \frac{\text{disease-free with positive test}}{\text{all disease-free}} \tag{3.4}$$

d. Example. Of the 1000 individuals who did not have breast cancer, 150 had a positive result on the BCPF test. Thus,

$$\text{FPR} = P(\text{T}+|\text{D}-) = 150/1000 = .15$$

e. Alternative definitions. FPR may also be defined as $P(\text{T}+ \text{ and } \text{D}-)$ [i.e., the number of false positives divided by the total number of individuals screened], or as $P(\text{D}-|\text{T}+)$ [i.e., the number of false positives divided by the total number with a positive test]. Similar alternative definitions for FNR are preferred by some clinicians.

5. Prevalence

a. Definition. The prevalence of a disease is the proportion of individuals in a population who have the disease. Prevalence is also known as the **prior probability,** or **pretest probability,** of disease, because it is "prior to" (i.e., not based on) diagnostic test results.

b. Notation. In probability notation, prevalence is written $P(\text{D}+)$.

c. Calculation

(1) Prevalence is calculated as the proportion of diseased individuals in a random sample drawn from the population of interest, using the formula

$$\text{prevalence} = P(\text{D}+) = \frac{\text{number with disease}}{\text{total number of individuals in study}} \tag{3.5}$$

(2) Prevalence can be accurately determined only when the number of diseased individuals is not fixed by the investigator and a random sample is selected from the population to which the results will be applied (i.e., a particular practice). In reality, clinical studies designed to evaluate a diagnostic test typically are carried out in hospital settings on clearly diseased individuals. As a result, the number of diseased individuals is often set by the investigator and the study sample is not representative of the population. In this case, an independent estimate of prevalence must be derived from an external source, such as the medical literature, rather than the study data.

6. Predictive value positive

a. Definition. The **predictive value positive (PVP)** of a diagnostic test is the probability that an individual with a positive test result has the disease. It is also the proportion of diseased individuals in the population of individuals with a positive test result. PVP is also known as the **posterior probability,** or **posttest probability,** of disease.

b. Notation. In probability notation, prevalence is written $P(\text{D}+)$.

c. Calculation. PVP is calculated as the proportion of individuals with a positive test result who are actually diseased, using the formula

$$\text{PVP} = P(\text{D}+|\text{T}+) = \frac{\text{diseased with positive test}}{\text{all with positive test}} \tag{3.6}$$

d. Example. In the BCPF study, 720 individuals had a positive test result (see Table 3-2; *total of row 1*); 570 of these individuals were shown by breast biopsy to indeed have breast cancer. Thus,

$$\text{PVP} = P(\text{D}+|\text{T}+) = 570/720 = .79$$

In other words, the probability that a patient with a positive BCPF test result has breast cancer is .79.

7. Predictive value negative

a. Definition. The **predictive value negative (PVN)** of a diagnostic test is the probability that an individual with a negative test result does not have the disease. It is also the proportion of disease-free individuals in the population of individuals with a negative test result.

b. Notation. In conditional probability notation, PVN is written $P(\text{D}-|\text{T}-)$.

c. Calculation. PVN is calculated as the proportion of individuals with a negative test result who truly are disease-free, using the formula

$$\begin{aligned} \text{PVN} &= P(\text{D}-|\text{T}-) \\ &= \frac{\text{disease-free with negative test}}{\text{all with negative test}} \end{aligned} \tag{3.7}$$

d. Example. In the BCPF study, 880 individuals had a negative test result (see Table 3-2; *total of row 2*); 850 of these did not have breast cancer as determined by breast biopsy. Thus,

$$\text{PVN} = P(\text{D}-|\text{T}-) = 850/880 = .97$$

In other words, the probability that a patient with a negative BCPF test result does not have breast cancer is .97.

D. Relationships between test performance characteristics

1. Sensitivity and FNR

a. Calculating sensitivity in terms of FNR

(1) Sensitivity can be defined in terms of FNR, using the formula

$$\begin{aligned} \text{sensitivity} = P(\text{T}+|\text{D}+) &= 1 - P(\text{T}-|\text{D}+) \\ &= 1 - \text{FNR} \end{aligned} \tag{3.8}$$

(2) A **sensitive** test is one that is highly effective at detecting disease (i.e., one that has a **low FNR**). If the test is **perfectly sensitive** (sensitivity = 1.0), all diseased individuals will have a positive result and no cases of the disease will be missed (FNR = 0.0).

b. Calculating FNR in terms of sensitivity

(1) Because T+ and T− are exhaustive and mutually exclusive events, sensitivity $[P(\text{T}+|\text{D}+)]$ and FNR $[P(\text{T}-|\text{D}+)]$ are complements and their sum is 1. Thus,

$$\begin{aligned} P(\text{T}-|\text{D}+) &= 1 - P(\text{T}+|\text{D}+) \\ \text{FNR} &= 1 - \text{sensitivity} \end{aligned} \tag{3.9}$$

(2) Example. Using formula 3.9 and the data in Table 3-2,

$$\begin{aligned} P(\text{T}-|\text{D}+) &= 1 - P(\text{T}+|\text{D}+) \\ \text{FNR} &= 1 - \text{sensitivity} \\ .05 &= 1 - .95 \end{aligned}$$

2. Specificity and FPR

a. Calculating specificity in terms of FPR

(1) Specificity can be defined in terms of FPR, using the formula

$$\begin{aligned} \text{specificity} = P(\text{T}-|\text{D}-) &= 1 - P(\text{T}+|\text{D}-) \\ &= 1 - \text{FPR} \end{aligned} \tag{3.10}$$

(2) A **specific** test is one that is rarely positive in a disease-free individual (i.e., one that has a **low FPR**).

b. Calculating FPR in terms of specificity

(1) Specificity, $P(T-|D-)$, and FPR $P(T+|D-)$, are complements and their sum is 1. Thus,

$$P(T+|D-) = 1 - P(T-|D-)$$
$$\text{FPR} = 1 - \text{specificity} \quad (3.11)$$

(2) Example. Using formula 3.11 and the data in Table 3-2,

$$P(T+|D-) = 1 - P(T-|D-)$$
$$\text{FPR} = 1 - \text{specificity}$$
$$.15 = 1 - .85$$

III. REVISING ESTIMATES OF TEST PERFORMANCE

A. Rationale. Published reports of the performance of a diagnostic test typically contain values for sensitivity and specificity, as well as an estimate of the test's predictive value for the particular study setting. The physician who wishes to use the test in a different practice setting must take into account the effect of any differences between the study population and the practice population on the predictive ability of the test. It is especially critical to assess the impact of any alteration in disease prevalence.

1. An **estimate of prevalence** (see Ch 7 I A) in a particular practice setting may be obtained from:

   a. A systematic review of patient records (provided the test in question has not been used for diagnosis)

   b. Published studies performed in similar patient populations

2. If prevalence can only be expressed as a range of plausible values, rather than a specific estimate, a **sensitivity analysis** (i.e., an evaluation of the predictive value of the test over this range) is indicated (see Ch 5 II F 1).

B. Back calculation method. The effect of changes in disease prevalence on the predictive ability of a diagnostic test can be investigated by modifying the 2 × 2 table to reflect the new prevalence. The published values for sensitivity and specificity may then be used to back calculate the values in the body of the table.

1. **Example.** The following example demonstrates the back calculation method.

   a. Procedure. A physician wishes to assess the predictive ability of the BCPF test in a population in which the prevalence of breast cancer is 5%. Test sensitivity and specificity are unchanged; that is, they equal .95 and .85, respectively.

   (1) Select an **arbitrary sample size,** for example, 1600 (Table 3-3; *step 1*).

   (2) Based on a prevalence, $P(D+)$, of 5%, calculate the number of diseased individuals: (1600)(.05) = 80 (see Table 3-3; *total of column 1, step 2*).

   (3) Calculate the number of disease-free individuals: 1600 − 80 = 1520 (see Table 3-3; *total of column 2, step 3*).

   (4) Calculate the proportion of diseased individuals with a positive test, based on a sensitivity of .95: (.95)(80) = 76 (see Table 3-3; *step 4*).

   (5) Calculate the proportion of disease-free individuals with a negative test, based on a specificity of .85: (.85)(1520) = 1292 (see Table 3-3; *step 5*).

Table 3-3. A 2 × 2 Table Illustrating the Back Calculation Method

	Breast Cancer (D+)	**No Breast Cancer (D−)**	**Totals**
BCPF Test (T+)	76 (step 4)	228 (step 7)	304 (step 8)
BCPF Test (T−)	4 (step 6)	1292 (step 5)	1296 (step 9)
Totals	80 (step 2)	1520 (step 3)	1600 (step 1)

(6) Compute the remaining values for the 2 × 2 table (see Table 3-3; *steps 6–9*).
(7) Calculate the PVP and PVN from the 2 × 2 table. In this example,

$$PVP = P(D+|T+) = 76/304 = .25$$

$$PVN = P(D-|T-) = 1292/1296 = .997$$

b. Summary. A decrease in prevalence from .375 to .05 in the new setting has decreased the PVP from .79 to .25 and increased the PVN from .97 to .997.

2. Relationship between prevalence, sensitivity, and predictive ability of the test

a. A conceptual illustration

(1) A room is filled with individuals diagnosed as "diseased" (D+) by a gold standard procedure. The sensitivity of a new diagnostic procedure is calculated by dividing the number of individuals in the room who have a positive test result by the total number in the room. Thus, this proportion refers only to those in the room and is independent of the frequency at which the disease occurs in the general population (prevalence).

(2) Next, the room is filled with individuals who have a positive result on the new test (T+). The PVP of the new procedure is calculated as the proportion of individuals in the room who are diseased, as defined by the gold standard. This number *is* influenced by the prevalence of the disease in the general population. For rare diseases, the group in the room would be expected to contain few, if any, diseased individuals. If, however, prevalence = 50% (as is the case in many study populations), many more diseased individuals would be found in the room.

(3) As shown in Figure 3-4, sensitivity [TP/(TP + FN)] is the same for both the study (*B*) and general (*A*) populations. PVP [P(D+|T+) = TP/T+ = TP/(TP + FP)], on the other hand, is greater in the study population (*B*), where prevalence = 50%.

b. General rules. Sensitivity and specificity are independent of prevalence.

(1) As prevalence decreases, PVP decreases and PVN increases.
(2) As prevalence increases, PVP increases and PVN decreases.

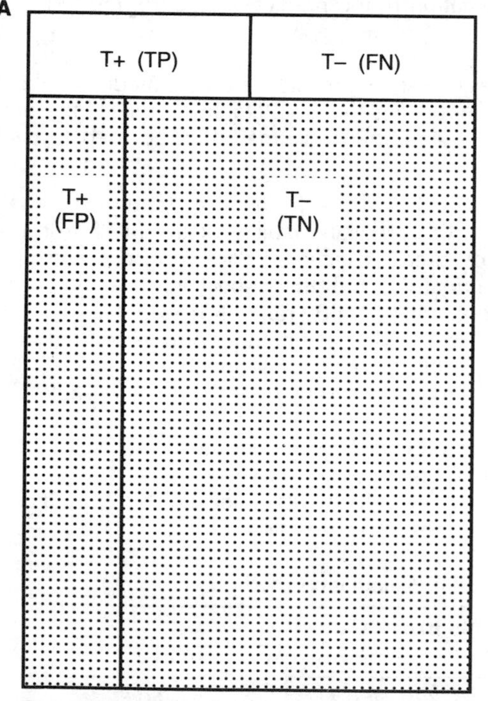

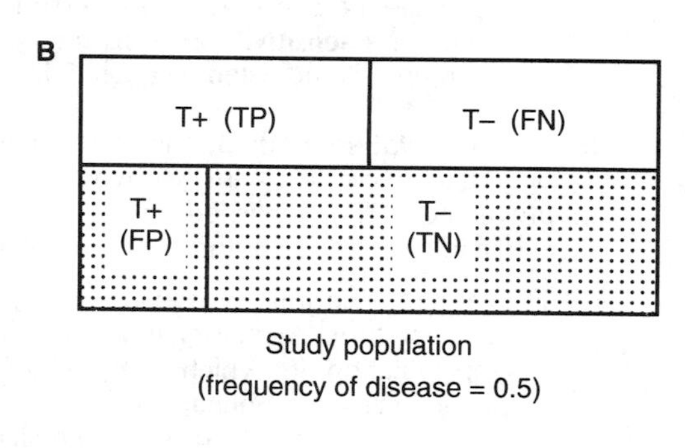

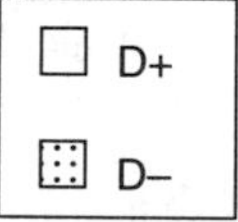

Figure 3-4. Distribution of hypothetical test results in the total and study populations. *D+* = diseased; *D–* = disease-free; *T+* = positive test result; *T–* = negative test result. (Reprinted from Weinstein M, Feinberg H: *Clinical Decision Analysis*. Philadelphia, WB Saunders, 1980, p 87.)

C. Bayes' rule method. Bayes' rule is a mathematical formula that may be used as an alternative to the back calculation method for obtaining unknown conditional probabilities such as PVP or PVN from known conditional probabilities such as sensitivity and specificity.

1. Formula. The general form of Bayes' rule is

$$P(A|B) = \frac{P(B|A)P(A)}{P(B|A)P(A) + P(B|\bar{A})P(\bar{A})} \quad (3.12)$$

where $\bar{A}$ (read "not A") is the complement of A (see Appendix 2 for the derivation of Bayes' rule). This rule can be used to obtain the unknown conditional probability $P(A|B)$ from the known conditional probabilities $P(B|A)$, $P(B|\bar{A})$, and $P(A)$.

2. Calculating PVP using Bayes' rule

a. Formula. Using Bayes' rule, PVP is defined as

$$PVP = P(D+|T+) = \frac{P(T+|D+)P(D+)}{P(T+|D+)P(D+) + P(T+|D-)P(D-)} \quad (3.13)$$

b. Example. The prevalence of breast cancer in a particular practice setting is .05. A patient in this practice is found to have a positive BCPF test result. What is the probability that this patient has breast cancer?

(1) From Table 3-2:

(a) Sensitivity $= P(T+|D+) = .95$

(b) Specificity $= P(T-|D-) = .85$

(c) $P(T+|D-) = 1 - P(T-|D-) = 1 - \text{specificity} = .15$

(d) Prevalence $= P(D+) = .05$

(2) Using formula 3.13,

$$PVP = P(D+|T+) = \frac{P(T+|D+)P(D+)}{P(T+|D+)P(D+) + P(T+|D-)P(D-)}$$

$$= \frac{(.95)(.05)}{(.95)(.05) + (.15)(.95)}$$

$$= .25$$

This is the same result as that obtained using the back calculation method.

3. Calculating PVN using Bayes' rule

a. Formula. Using Bayes' rule, PVN is defined as

$$PVN = P(D-|T-) = \frac{P(T-|D-)P(D-)}{P(T-|D-)P(D-) + P(T-|D+)P(D+)} \quad (3.14)$$

b. Example. A patient in the practice described in III C 2 b has a negative BCPF test result. What is the probability that this individual does not have breast cancer?

(1) From Table 3-2:

(a) Specificity $= P(T-|D-) = .85$

(b) $P(D-) = 1 - P(D+) = 1 - \text{prevalence} = 1 - .05 = .95$

(c) $P(T-|D+) = 1 - P(T+|D+) = 1 - \text{sensitivity} = 1 - .95 = .05$

(2) Using formula 3.14,

$$PVN = P(D-|T-) = \frac{P(T-|D-)P(D-)}{P(T-|D-)P(D-) + P(T-|D+)P(D+)}$$

$$= \frac{(.85)(1 - .05)}{(.85)(1 - .05) + (1 - .95)(.05)}$$

$$= .997$$

Again, this is the same result as that obtained using the back calculation method.

4. Relationship between sensitivity, prevalence, and predictive ability of a test

a. The **formula for PVP** can be rewritten as

$$PVP = \frac{(\text{sensitivity})(\text{prevalence})}{(\text{sensitivity})(\text{prevalence}) + (1 - \text{specificity})(1 - \text{prevalence})}$$

(1) If the test is **perfectly specific** (i.e., specificity = 1.0), the second term in the denominator becomes 0 and PVP = 1. In other words, a perfectly specific test yields no false positive results and a positive test result conclusively indicates the presence of the disease [$P(D+|T+) = PVP = 1$].

(2) If the test is **highly specific,** it will yield few false positives and the TNR, $P(T-|D-)$, is high. If the prevalence is low, however, a positive result is not likely to indicate the presence of disease (PVP decreases as prevalence decreases).

b. The **formula for PVN** may be rewritten as

$$\text{PVN} = \frac{(\text{specificity})(1 - \text{prevalence})}{(\text{specificity})(1 - \text{prevalence}) + (1 - \text{sensitivity})(\text{prevalence})}$$

(1) If the test is **perfectly sensitive** (i.e., sensitivity = 1.0), the second term in the denominator becomes 0 and PVN = 1. In other words, a perfectly sensitive test yields no false negative results and a negative test result conclusively indicates the absence of disease [$P(T-|D-) = PVN = 1$].

(2) Even for a **highly sensitive** test (i.e., one resulting in few false negatives), a negative test result in a population with a high prevalence of the disease may not accurately indicate the true absence of the disease (PVN decreases as prevalence increases).

c. General rules

(1) The more **sensitive** a test, the better its **PVN**.

(2) The more **specific** a test, the better its **PVP**.

IV. EVALUATING A NEW DIAGNOSTIC TEST.

Performance characteristics such as sensitivity and specificity provide the physician with one way of evaluating the merit of a new diagnostic test. The design of a diagnostic clinical trial should also be considered in this analysis. A well-designed study of a new diagnostic test has the following characteristics:

A. Comparison to an acceptable gold standard (see II A)

B. Comparisons that are performed "blind." The investigator carrying out or interpreting the new test should have been "blind" to the gold standard diagnosis. Similarly, the investigator applying the gold standard should have had no knowledge of the diagnostic test result.

C. Results that are not incorporated in the gold standard procedure. In other words, the labeling of patients as diseased or disease-free by the gold standard should be independent of the results of the procedure being evaluated.

D. Reliable and accurate measurements

1. **Reliability (precision).** Imprecision increases the random variation in the measurements and, therefore, increases the spread in the frequency distributions of test values for the diseased and disease-free populations. As a result, the curves representing the two populations overlap more. For a given positivity criterion, increasing the overlap between the two curves increases both the FPR and the FNR and decreases sensitivity and specificity. **Replication and standardization of the test method improve precision.**

2. **Accuracy** affects the location of the two frequency distributions along the horizontal axis (Figure 3-5). Unlike reliability, the accuracy of a test measurement is not improved by replication since it results from systematic rather than random error in measurement. Accuracy is typically accepted "on faith" in a published report of a new diagnostic test.

E. Use of a sample that spans an appropriate spectrum of disease

1. A new diagnostic test that effectively differentiates obviously diseased individuals (e.g., hospitalized patients) from obviously healthy individuals (e.g., healthy, normal volunteers) may be less useful in the real world for distinguishing the disease in question from other diseases with similar signs and symptoms.

2. If values for the diagnostic test variable are correlated with the severity of the disease, the test may readily detect patients with overt or advanced illness (such as the hospitalized patients typically sampled in clinical studies), but be less useful for identifying patients in the early stages of the disease, when therapeutic intervention may be maximally effective.

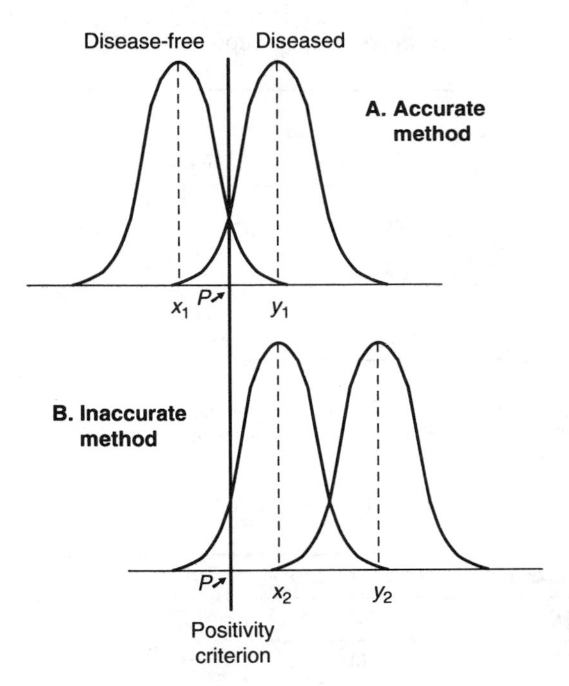

Figure 3-5. Frequency distributions of test results in diseased and disease-free populations using accurate (*A*) and inaccurate (*B*) measurement procedures. In (*A*), the positivity criterion was set at value *P* based on the results of an accurate measurement procedure. In (*B*), the measurement device is "out of calibration," and all test results are systematically higher than the true values of the test variable being measured.

F. Adequate description of the study population. A physician cannot apply the results of a study to her own practice unless she knows how similar her practice is to the study population.

G. Clear and adequate definition of "normality" (see I and Ch 4)

H. New test that offers clear advantages over existing procedures. The new diagnostic test should be easier, less painful, safer, quicker, or cheaper than existing tests. The physician must decide which of these features are most important for his particular practice.

V. MULTIPLE TESTING. Because most diagnostic tests are less than perfect, a single test is frequently insufficient for making an unequivocal diagnosis. Therefore, clinicians often use multiple diagnostic tests, administered either **in parallel** or **serially**. Multiple testing alters the posterior probability of disease (PVP) in a predictable fashion.

A. Parallel testing. In parallel testing, a battery of tests is given **concurrently**. For example, in a clinical chemistry profile, 18 routine chemical determinations are obtained from a single blood sample.

1. Features

a. Uses. Parallel tests should be ordered when rapid assessment is needed (e.g., in emergencies) or for routine physical examinations.

b. Results. A positive result on any one of the group of tests is taken as evidence of the disease.

2. Example. ECG and echocardiography are two procedures used in the diagnosis of ventricular septal defect. When ordered in parallel, a positive result on both tests (T_1+, T_2+), a positive result on ECG and a negative result using echocardiography (T_1+, T_2-), or a positive result using echocardiography and a negative result on the ECG (T_1-, T_2+) all are considered positive results. Table 3-4 describes the joint performance of the two tests in a sample of 200 patients in a large teaching hospital. A 2×2 table (Table 3-5) can be constructed from the data in Table 3-4, from which sensitivity and specificity are found to be:

a. Sensitivity = $P(T+|D+) = 95/100 = .95$

b. Specificity = $P(T-|D-) = 60/100 = .60$

B. Serial testing. In serial testing, a battery of tests is performed **sequentially** until an unequivocal diagnosis can be made.

Table 3-4. Joint Results of Two Tests (T_1 and T_2) for Ventricular Septal Defect (VSD)

Test Results	Type of Testing: Serial*	Type of Testing: Parallel	VSD (D+)	No VSD (D−)	Totals
T_1+, T_2-	Neg	Pos	15	20	35
T_1-, T_2+	Neg	Pos	20	15	35
T_1+, T_2+	Pos	Pos	60	5	65
T_1-, T_2-	Neg	Neg	5	60	65
Totals			100	100	200

T_1 = electrocardiography (ECG); T_2 = echocardiography.

*In actual practice, options T_1-, T_2+ and T_1-, T_2- are not always performed (e.g., if the second test in the series is risky, it will not be ordered if the first test is negative).

Table 3-5. A 2 × 2 Table of Results of Parallel Testing for Ventricular Septal Defect (VSD)

	VSD (D+)	No VSD (D−)	Totals
Positive Test Profile (T+)	95	40	135
Negative Test Profile (T−)	5	60	65
Totals	100	100	200

1. **Features**
   a. **Uses.** Serial tests should be used when rapid assessment is not necessary or when some of the available tests are expensive or risky. These are ordered only if the initial test is positive. If all available tests are equally costly or risky, the most specific test should be used first.
   b. **Results. All tests must have positive results** (T_1+, T_2+) **for the sequence to be considered positive** for the disease.

2. **Example.** Using the example in V A 2, when ECG and echocardiography are ordered serially, both tests must be positive (T_1+, T_2+) to confirm a diagnosis of ventricular septal defect. Table 3-4 describes the joint performance of the two tests in 200 patients. A 2 × 2 table (Table 3-6) can be constructed from the data in Table 3-4, from which sensitivity and specificity are found to be:
   a. Sensitivity = $P(T+|D+) = 60/100 = .60$
   b. Specificity = $P(T-|D-) = 95/100 = .95$

C. Conclusions

1. **Parallel tests** result in:
   a. **Greater sensitivity**
   b. **Increased PVN** (decreased FNR)
   c. **Decreased specificity** (increased FPR)

Table 3-6. A 2 × 2 Table of Results of Serial Testing for Ventricular Septal Defect (VSD)

	VSD (D+)	No VSD (D−)	Totals
Positive Test Profile (T+)	60	5	65
Negative Test Profile (T−)	40	95	135
Totals	100	100	200

2. **Serial tests** result in:
   a. **Lower sensitivity** (higher FNR)
   b. **Increased specificity**
   c. **Increased PVP** (lower FPR)

BIBLIOGRAPHY

Fletcher RH, Fletcher SW, Wagner EH: *Clinical Epidemiology: The Essentials*. Baltimore, Williams & Wilkins, 1982.

Galen RS, Gambino SR: *Beyond Normality: The Predictive Value and Efficiency of Medical Diagnosis*. New York, John Wiley, 1975.

Gehlbach SH: *Interpreting the Medical Literature: A Clinician's Guide*. Lexington, MA, DC Heath, 1982.

Sackett DL, Hayes RB, Tugwell P: *Clinical Epidemiology: A Basic Science for Clinical Medicine*. Boston, Little, Brown, 1985.

Weinstein MC, Feinberg HV: *Clinical Decision Analysis*. Philadelphia, WB Saunders, 1980.

PROBLEMS

3-1. In a study of serum cholesterol levels, 2000 healthy patients in a predominantly white, middle-class, suburban practice are found to have a mean cholesterol level of 150 mg/dl, with a standard deviation of 15 mg/dl. Assuming that cholesterol values in this population follow a gaussian distribution, determine the normal range.

3-2. The frequency distribution of cholesterol values for the population described in problem 3-1 is later discovered to be skewed, rather than gaussian. Using the cumulative frequency distribution of cholesterol values in the following graph, determine the normal range by the percentile method.

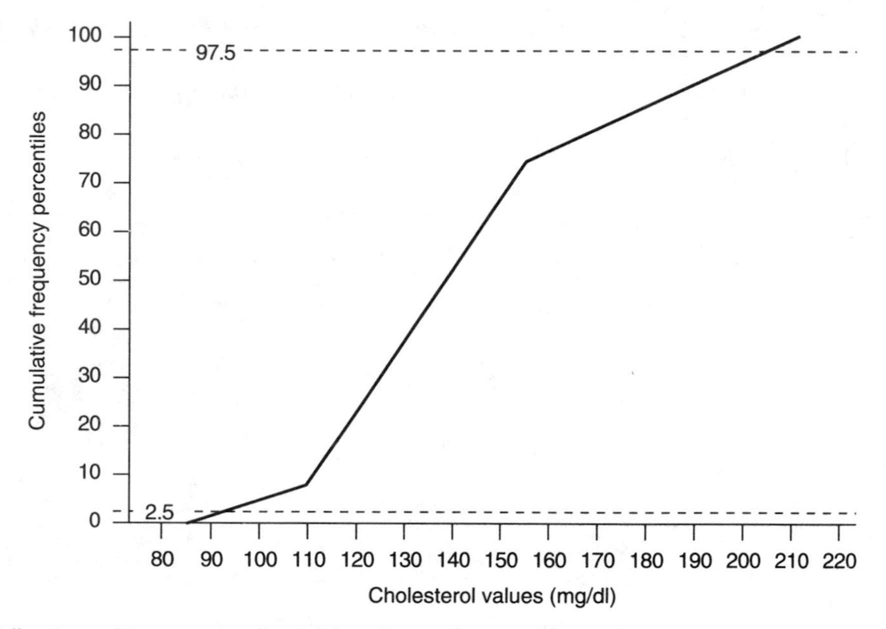

3-3. The following table summarizes results of a study to evaluate the dexamethasone suppression test (DST) as a diagnostic test for major depression. The study compared results on the DST to those obtained using the gold standard procedure (routine psychiatric assessment and structured interview) in 368 psychiatric patients. Use the results obtained in this study to answer problems 3-3a through 3-3d.

A 2 × 2 Table Summarizing Results of the Dexamethasone Suppression Test (DST)

	Depression (D+)	No Depression (D−)	Totals
DST Result (T+)	84	5	89
DST Result (T−)	131	148	279
Totals	215	153	368

Adapted from Carroll BJ: The dexamethasone suppression test for melancholia. *Br J Psychiatry* 140: 292, 1982.

a. What is the prevalence of major depression in the study group?

b. For the DST, determine:
Sensitivity
Specificity
False positive rate (FPR)
False negative rate (FNR)
Predictive value positive (PVP)
Predictive value negative (PVN)

c. What is the probability that a patient who is not depressed will have a negative result using the DST? What is this test characteristic called?

d. What is the probability that a patient with a negative DST result is not depressed? What is this test characteristic called?

3-4. In the study described in problem 3-3, the research team also evaluated DST performance using an alternate positivity criterion. In this instance, test sensitivity was 43% and test specificity was 96%.

a. Using the back calculation method, determine PVP and PVN for the new positivity criterion.

b. Determine PVP and PVN using Bayes' rule.

3-5. A family practice physician estimates the prevalence of major depression in her practice to be approximately 5%. A patient in the practice has a positive result on the DST.

a. Using the sensitivity and specificity reported in problem 3-4, determine the probability that this patient is depressed.

b. A second patient has a negative DST result. What is the probability that this patient is not depressed?

c. A third patient in the practice complains of fatigue, anorexia, and apathy. Based on experience, the physician estimates that the prevalence of depression in patients with these three symptoms is 50%. If this patient has a positive DST result, what is the probability that he is depressed?

3-6. A diagnostic test for polysplenia syndrome has a 7% FPR and a FNR of 18%. The prevalence of this congenital abnormality is .3%.

a. What is the probability that a patient with a positive test result has polysplenia?

b. What is the probability that a patient with a negative test result has polysplenia?

c. What is the probability that a patient with a negative test result does not have polysplenia?

3-7. The prevalence of polysplenia syndrome in a university medical center specializing in the treatment of this defect is 30%. Determine the PVP for the test in this setting.

Solutions on p 364.

STUDY QUESTIONS

Directions: Each of the numbered items or incomplete statements in this section is followed by answers or by completions of the statement. Select the **one** lettered answer or completion that is **best** in each case.

Questions 1–4

The following table summarizes the results of a study to evaluate a clinical "signs and symptoms" test as a diagnostic test for group A β-hemolytic streptococcus infection (strep throat). In the study, pediatric residents used this test to diagnose 150 randomly selected pediatric patients presenting with pharyngitis as either having strep throat (T+) or not (T−). Each study participant was simultaneously diagnosed according to the results of a throat culture (the gold standard procedure; D+ or D−).

A 2 × 2 Table Summarizing Results of the "Signs and Symptoms" (S/S) Test for Strep Throat

	Positive Throat Culture (D+)	Negative Throat Culture (D−)	Totals
S/S Test (T+)	28	30	58
S/S Test (T−)	12	80	92
Totals	40	110	150

1. For the study population, prevalence is estimated by

(A) 58/150 = .39
(B) 28/150 = .19
(C) 40/58 = .69
(D) 40/150 = .27
(E) 28/40 = .70

2. The sensitivity of the "signs and symptoms" test is

(A) 28/58 = .48
(B) 58/150 = .39
(C) 80/110 = .73
(D) 28/150 = .19
(E) 28/40 = .70

3. The probability that a child with strep throat will be diagnosed as disease-free by the "signs and symptoms" test is

(A) 12/92 = .13
(B) 92/150 = .61
(C) 12/40 = .30
(D) 40/150 = .27
(E) 12/150 = .08

4. A 4-year-old child presenting with pharyngitis is diagnosed as having strep throat, based on the "signs and symptoms" test. What is the probability that this child does have strep throat?

(A) 28/40 = .70
(B) 28/150 = .19
(C) 40/150 = .27
(D) 40/58 = .69
(E) 28/58 = .48

1-D 4-E
2-E
3-C

5. If the frequency distribution of values of a diagnostic test variable in the disease-free population is highly skewed, which of the following statements is correct?

(A) The upper and lower limits of the normal range are calculated as the mean ± 2 standard deviations
(B) The gaussian method is an appropriate technique for determining the normal range
(C) Ninety-five percent of the values of this variable fall in the range defined by the mean ± 2 standard deviations
(D) Values falling above the upper 2.5% or below the lower 2.5% of the distribution may be defined as abnormal

6. All of the following are disadvantages of the gaussian method for selecting a positivity criterion EXCEPT

(A) the prevalence of all diseases is at least 5% when this method is used
(B) multiple testing increases the likelihood that a patient will be diagnosed with the disease
(C) the method requires clinicians to keep track of test performance characteristics in their practices
(D) diagnostic test results rarely follow a gaussian distribution
(E) the method does not take into account the frequency distribution of the test variable in the diseased population

7. The PVP of a diagnostic test is

(A) equal to 1 − specificity
(B) a function of only the sensitivity and specificity
(C) equal to $P(T+|D+)$
(D) the probability of disease given a positive test result

Questions 8–9

An experimental screening test for AIDS has a sensitivity of 82% and a specificity of 93%. The prevalence of AIDS in the population to be screened is estimated to be 3%.

8. The probability that an individual with a positive test result does not have AIDS is

(A) .27
(B) .73
(C) .82
(D) .18
(E) none of the above

9. The probability that an individual with a positive test result has AIDS is

(A) .27
(B) .73
(C) .82
(D) .18
(E) none of the above

10. All of the following statements about multiple tests carried out in parallel are true EXCEPT

(A) the sensitivity is greater than for tests carried out in series
(B) the probability of a false negative result is decreased
(C) the probability of a false positive result is increased
(D) the PVP is greater than for tests carried out in series
(E) the PVN is greater than for tests carried out in series

11. All of the following statements about multiple tests carried out in series are true EXCEPT

(A) the specificity is lower than for tests carried out in parallel
(B) the PVP is greater than for tests carried out in parallel
(C) the probability of a false positive result is less
(D) the sensitivity is lower than for tests carried out in parallel
(E) the PVN is less than for tests carried out in parallel

5-D 8-B 11-A
6-C 9-A
7-D 10-D

ANSWERS AND EXPLANATIONS

1–4. The answers are: 1-D *[II C 5]*, **2-E** *[II C 1]*, **3-C** *[II C 3, 4 e]*, **4-E** *[II C 6]*
The prevalence of strep throat, $P(\text{D}+)$, in the study population is estimated by the frequency of strep throat in the 150 study subjects, or $40/150 = .27$. This prevalence is valid only if the numbers of children with and without strep throat (as determined by the gold standard) have not been controlled by the investigators and if the sample of 150 subjects was chosen randomly from the population. In this study, simultaneous diagnosis by the "signs and symptoms" test and throat culture (gold standard) satisfies the first of these criteria.

Sensitivity is the probability of a positive result on the "signs and symptoms" test among children who have strep throat. It is estimated by the relative frequency of a positive test (T+) among individuals known to be diseased (D+). Thus,

$$\text{sensitivity} = P(\text{T}+|\text{D}+) = 28/40 = .70$$

The probability that a child with strep throat will have a negative test result is $P(\text{T}-|\text{D}+)$, the false negative rate (FNR). It is estimated by the frequency of negative test results (T−) in patients known to be diseased (D+). Thus,

$$\text{FNR} = P(\text{T}-|\text{D}+) = 12/40 = .30$$

An alternative definition of FNR is $P(\text{T}- \text{ and } \text{D}+)$, the (unconditional) joint probability of a negative test result and disease. Using this definition, FNR is estimated by the proportion of patients who have both a negative test result and the disease among the total number of patients tested, that is, $12/150 = .08$. A third definition of FNR is $P(\text{D}+|\text{T}-)$, the proportion of individuals with a negative test result who have the disease. With this definition, FNR = 1 − predictive value negative (PVN), that is, $12/92 = .13$.

The probability that a patient with a positive test result (T+) has strep throat (D+) is the predictive value positive (PVP). It is estimated by the proportion of patients with strep throat among all those with a positive test result. Thus,

$$\text{PVP} = P(\text{D}+|\text{T}+) = 28/58 = .48$$

5. The answer is D *[I B 1 b, 2 a]*
The mean ± 2 standard deviations encompasses 95% of values of the test variable (and hence defines the limits of the normal range) only if values of the test variable follow a gaussian distribution. This distribution is symmetric, rather than skewed. The percentile method can be used to define the normal range regardless of the shape of the frequency distribution.

6. The answer is C *[I B 1 d]*
The predictive value method, not the gaussian method, requires a knowledge of test performance characteristics in the particular practice in which the test will be used. When the gaussian method is used to define the normal range, the frequency of all diseases is at least 5%, if both upper and lower limits are employed, or 2.5%, if only an upper or only a lower limit is set (5% of the diseased population that is falsely labeled plus the true proportion of diseased individuals). The normal range defined by the gaussian method is based on the frequency distribution of the test variable in the disease-free population; hence, the frequency distribution of this variable in the diseased population is ignored. When the gaussian method is used, multiple testing increases the probability of an abnormal result $[1 - (.95)^n$, where n is the number of tests].

7. The answer is D *[II C 6]*
The predictive value positive (PVP), $P(\text{D}+|\text{T}+)$, of a diagnostic test is the probability that an individual with a positive test result has the disease. Its complement is $P(\text{D}-|\text{T}+)$. PVP is a function of test sensitivity, specificity, and prevalence.

8–9. The answers are: 8-B *[III C]*, **9-A** *[III C]*
Given sensitivity $P(\text{T}+|\text{D}+)] = .82$, $P(\text{T}-|\text{D}+) = 1 - .82 = .18$. Given specificity $P(\text{T}-|\text{D}-)] = .93$, $P(\text{T}+|\text{D}-) = 1 - .93 = .07$. Given prevalence $P(\text{D}+)] = .03$, $P(\text{D}-) = 1 - .03 = .97$. The probability that a person with a positive test result does not have AIDS, $P(\text{D}-|\text{T}+)$, can therefore be calculated using Bayes' rule as follows:

$$P(D-|T+) = \frac{P(T+|D-)P(D-)}{P(T+|D-)\ P(D-) + P(T+|D+)P(D+)}$$

$$= \frac{(.07)(.97)}{(.07)(.97) + (.82)(.03)}$$

$$= .73$$

The probability that an individual with a positive test result has AIDS is $P(D+|T+)$, the predictive value positive (PVP), which is the complement of the value computed in question 8. Thus, $P(D+|T+) = 1 - .73 = .27$. This result also can be obtained using Bayes' rule as follows:

$$P(D+|T+) = \frac{P(T+|D+)P(D+)}{P(T+|D+)P(D+) + P(T+|D-)P(D-)}$$

$$= \frac{(.82)(.03)}{(.82)(.03) + (.07)(.97)}$$

$$= .27$$

10. The answer is D *[V C]*
Tests carried out in parallel have greater sensitivity and, hence, a lower false negative rate (FNR) than tests carried out in series. In addition, parallel testing lowers specificity [and, hence, increases the false positive rate (FPR)] and increases the predictive value negative (PVN).

11. The answer is A *[V C]*
Tests carried out in series have higher specificity and, hence, a lower false positive rate (FPR) than tests carried out in parallel. In addition, serial testing lowers sensitivity [and, hence, increases the false negative rate (FNR)] and increases the predictive value positive (PVP).

4
Defining Normality Using the Predictive Value Method

I. THE PREDICTIVE VALUE METHOD FOR SELECTING A POSITIVITY CRITERION. None of the six methods of defining the positivity criterion for a diagnostic test (see Ch 3 I B) enjoys universal acceptance. However, because the **predictive value method,** sometimes referred to as the **diagnostic method,** is based directly on the probability of disease as defined by comparison with the gold standard, it is considered the most clinically sound technique for defining normality. With this method, the location of the positivity criterion is chosen such that the diagnostic test will have a given sensitivity, specificity, and predictive value.

A. Tradeoffs between false positive and false negative test results. The choice of values for test sensitivity, specificity, and predictive value and, ultimately, the definition of the positivity criterion depend on a decision whether it is worse to label a patient diseased when he is not (false positive) or to label him disease-free when he is not (false negative). This tradeoff should take into account the needs of the patient as well as the judgment of the physician.

1. **When the consequences of missing a case are potentially grave,** choose a value for the positivity criterion that **minimizes the false negative rate**. For example, in neonatal phenylketonuria (PKU) screening, a false positive result is far less deleterious than a false negative result. While most false positives are identified during follow-up testing, a false negative result may postpone essential dietary intervention until mental retardation is evident.

2. **When a false positive diagnosis may lead to a risky treatment** (e.g., chemotherapy, open heart surgery) **or follow-up procedure** (e.g., cardiac catheterization), choose a value for the positivity criterion that **minimizes the false positive rate**. This rule also applies when a false positive diagnosis may have deleterious effects on a patient's life-style, self-image, or financial situation (e.g., AIDS, mental illness, learning disorders).

B. Procedure. Figure 4-1 illustrates the predictive value method, applied to a hypothetical diagnostic test. The number of times the test yields a correct result is maximized at *x*, the point where the curves representing the diseased and disease-free populations intersect.

1. **To increase sensitivity and the predictive value negative (PVN),** shift the positivity criterion to the **left** (see Figure 4-1B). When the value of the positivity criterion is equal to point *v* on the horizontal axis (not shown), the false negative rate = 0.

2. **To increase specificity and the predictive value positive (PVP),** shift the positivity criterion to the **right** (see Figure 4-1C). When the value of the positivity criterion is equal to point *z* on the horizontal axis (not shown), the false positive rate = 0.

C. Advantages

1. Values for the test variable may assume any distribution.

2. The frequency of the disease is based on actual clinical data and, therefore, is a realistic estimate.

3. The method takes into account the distribution of values of the diagnostic test variable in both the diseased and disease-free populations.

4. The method accounts for the needs of both the physician and the patient, by balancing the potential benefits of a diagnostic test against its clinical costs (missed diagnosis or false labeling).

A. Sensitivity and specificity equal

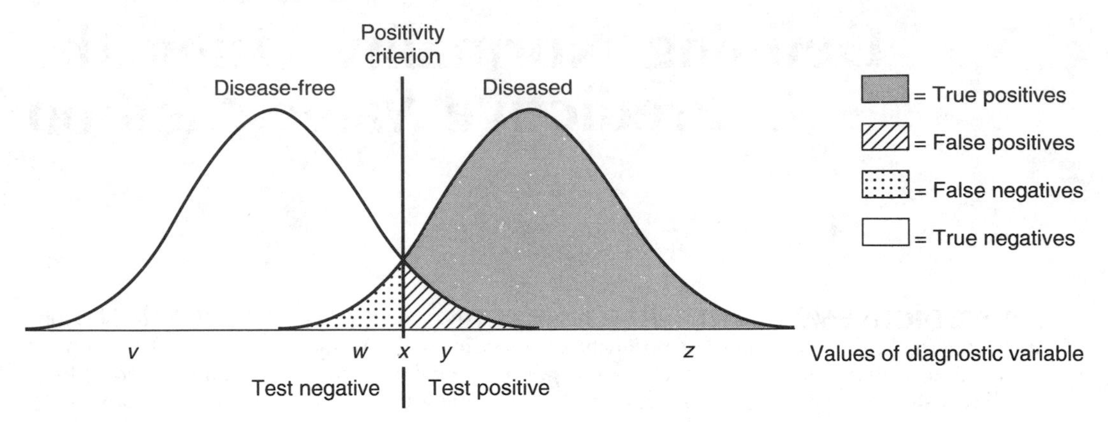

B. Increased sensitivity, decreased specificity

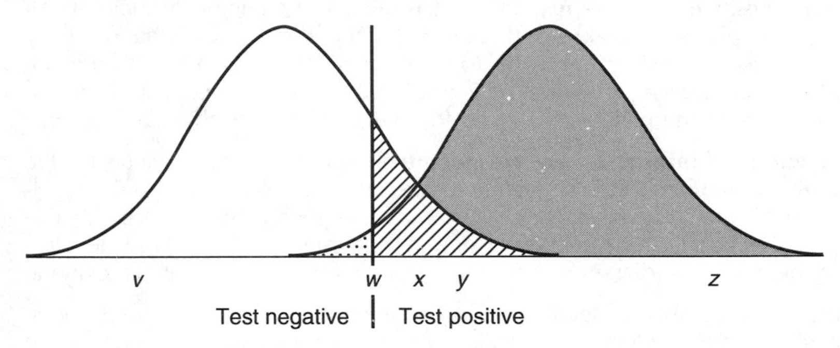

C. Increased specificity, decreased sensitivity

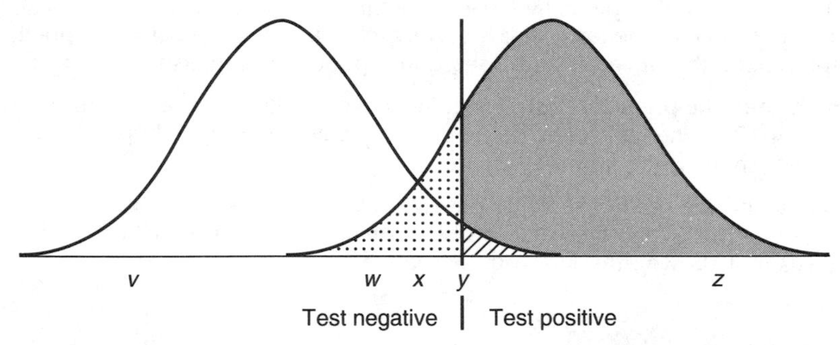

Figure 4-1. Defining normality by the predictive value method; *v*, *w*, *x*, and *y* represent potential choices for the positivity criterion. In (*A*), the positivity criterion is located at *x*, where the sensitivity and specificity of this hypothetical diagnostic test are equal. The number of times the test yields an accurate result is maximal at this point. In (*B*), the false negative rate has been decreased by moving the positivity criterion to the left (to *w*). In (*C*), the false positive rate has been decreased by shifting the positivity criterion to the right (to *y*).

D. Disadvantages. The disadvantage of the predictive value model is that it requires the physician to monitor disease prevalence, PVP, and PVN in her own practice, as well as to determine values for test performance characteristics (e.g., sensitivity, specificity) for several choices of the positivity criterion.

II. THE RECEIVER OPERATOR CHARACTERISTIC (ROC) CURVE. The ROC curve is a graphic representation of the relationship between sensitivity and specificity for a diagnostic test. It provides a simple tool for applying the predictive value method to the choice of a positivity criterion.

A. Drawing the curve. The ROC curve is constructed by plotting the true positive rate (**sensitivity**) against the false positive rate (**1 − specificity**) for several choices of the positivity criterion (Figure 4-2).

B. Using the curve to locate the positivity criterion. The point marked by the dashed circle in Figure 4-2 (*upper left corner*) represents a **perfect diagnostic test**. At this point, both sensitivity and specificity are 100%, that is, all diseased individuals are identified, all healthy individuals are labeled disease-free, and no disease-free individuals are labeled diseased.

1. **When the costs of a false positive and false negative test result are equal,** set the positivity criterion equal to the point on the ROC curve closest to the upper left corner (see Figure 4-2, *x*). At this point, the discriminative ability of the test is maximized and the number of erroneous diagnoses is minimized.

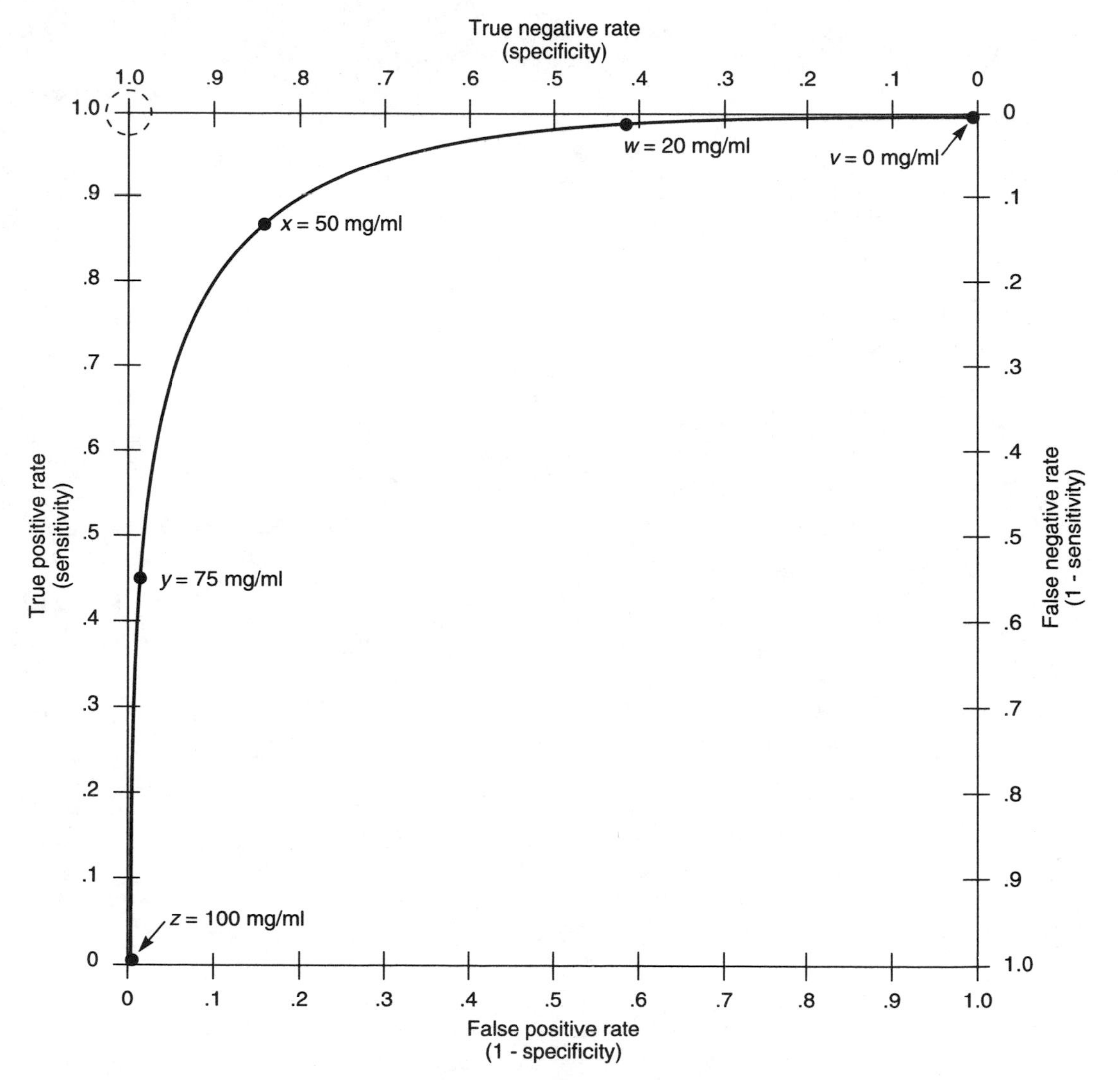

Figure 4-2. A hypothetical receiver operator characteristic (ROC) curve. *v, w, x, y,* and *z* represent the five potential choices for the positivity criterion shown in Figure 4-1, expressed in mg/ml.

2. **When a false positive result is especially undesirable,** set the positivity criterion equal to the point farthest to the left on the ROC curve (see Figure 4-2, *z*).

3. **When a false negative result is especially undesirable,** set the positivity criterion equal to a value toward the right on the ROC curve (see Figure 4-2, *w, v*). Moving to the right in Figure 4-2 is equivalent to shifting the positivity criterion to the left in Figure 4-1. At the point on the ROC curve farthest to the right, all patients with the disease are detected by the diagnostic test.

C. **Using the curve to compare two tests.** ROC curves also can be used to compare two diagnostic tests. The area under the curve represents the overall accuracy of a test; the larger the area, the better the test.

1. The ROC curve for a test that conveys no information falls on the diagonal running from lower left to upper right.

2. In comparing two ROC curves, the one closest to the upper left corner (i.e., the curve with the greatest total area below and to the right of it) has the greater sensitivity and specificity and hence is the more accurate of the two.

PROBLEMS

4-1. The following table provides information on the sensitivity (true positive rate) and specificity (true negative rate) of a hypothetical diagnostic test for eight choices of positivity criterion. This information also is displayed as a ROC curve in the figure below the table. Use the information in the table and ROC curve to answer problems 4-1a through 4-1d.

Test Performance Characteristics for Eight Choices of Positivity Criterion

Value	FPR	TNR	FNR	TPR
202.9	.05	.95	.19	.81
209.2	.025	.975	.29	.71
216.6	.01	.99	.43	.57
221.6	.005	.995	.47	.53
195.6	.10	.90	.11	.89
190.8	.15	.85	.07	.93
186.8	.20	.80	.05	.95
175.2	.40	.60	.01	.99

FPR = false positive rate; TNR = true negative rate; FNR = false negative rate; TPR = true positive rate.

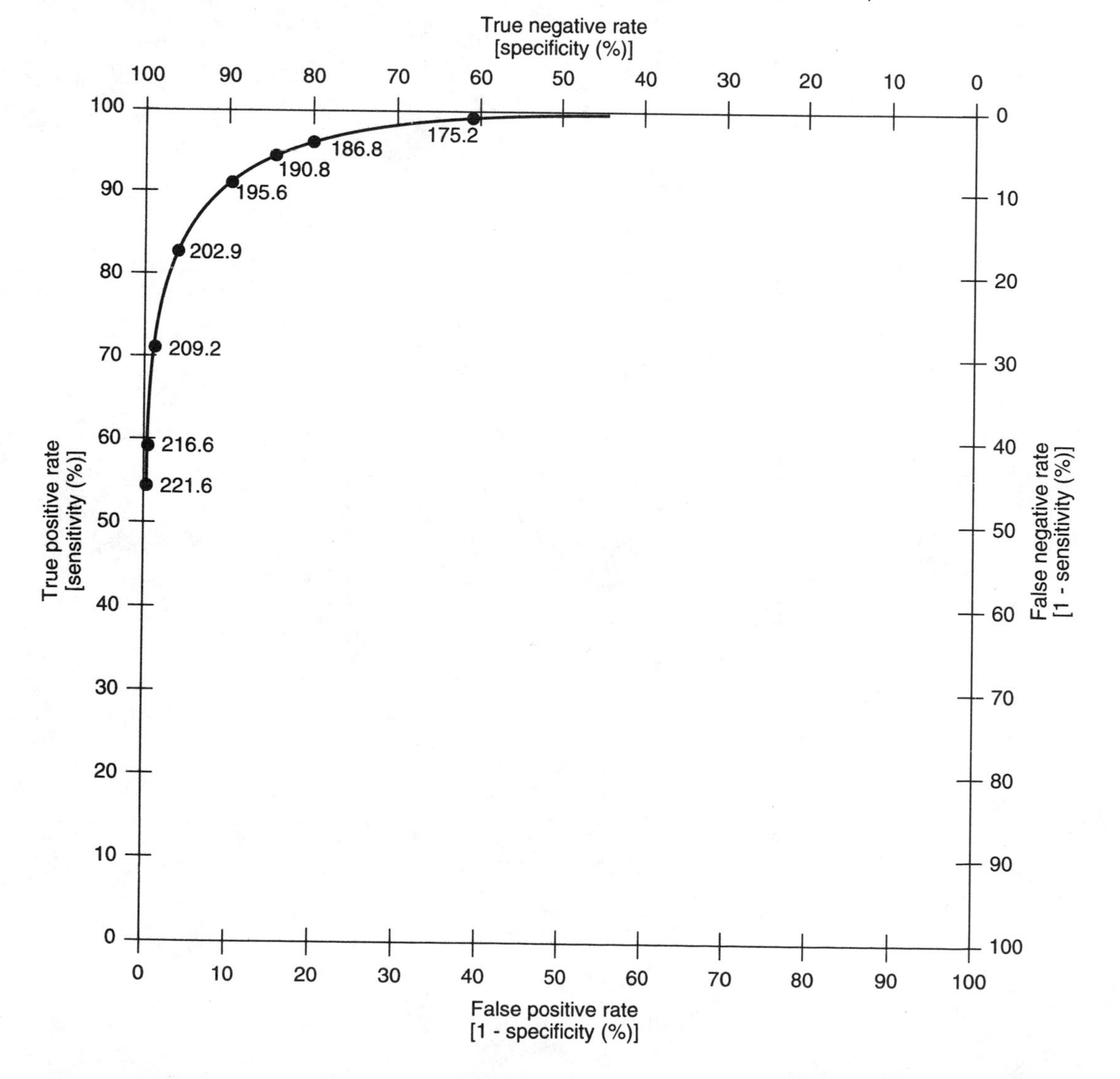

a. Which value of the positivity criterion should be selected if the consequences of a false positive result and a false negative result are equally harmful to the patient?

b. For which value of the positivity criterion is the test maximally accurate?

c. If the consequence of a false positive result is that the patient is subjected to a risky new surgical procedure, while that of a false negative result is additional testing, which value of the positivity criterion should be selected?

d. If a missed diagnosis is potentially fatal and a false positive test result leads to additional testing, which value of the positivity criterion should be selected?

Solutions on p 366.

STUDY QUESTIONS

Directions: Each of the numbered items or incomplete statements in this section is followed by answers or by completions of the statement. Select the **one** lettered answer or completion that is **best** in each case.

1. In the following diagram, which of the following is a consequence of shifting the positivity criterion to the right?

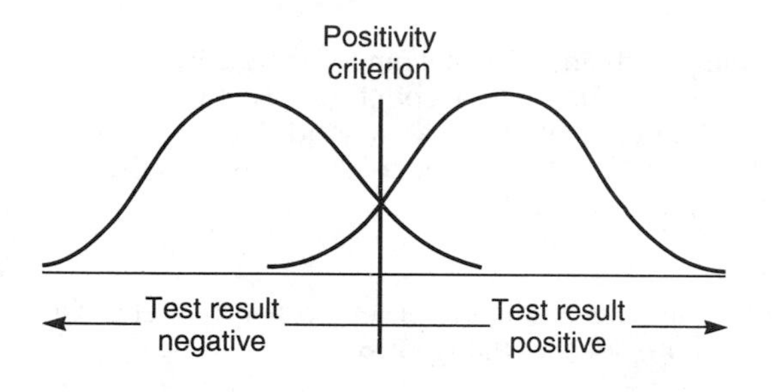

(A) Specificity decreases
(B) Sensitivity decreases
(C) PVP is unaffected
(D) False positive rate approaches 100%

2. Which of the following statements about the predictive value method of defining normality is true?

(A) It requires that values of the diagnostic test variable follow a gaussian distribution
(B) It sets the positivity criterion equal to that value of the test variable above which therapeutic intervention will occur
(C) It sets the positivity criterion such that the test will have a chosen sensitivity and specificity
(D) It does not require information on the frequency distribution of the diagnostic test variable in the diseased population
(E) It is the least clinically sound of the six methods for selecting a positivity criterion

3. The following diagram illustrates the ROC curves for a pair of diagnostic tests, A and B.

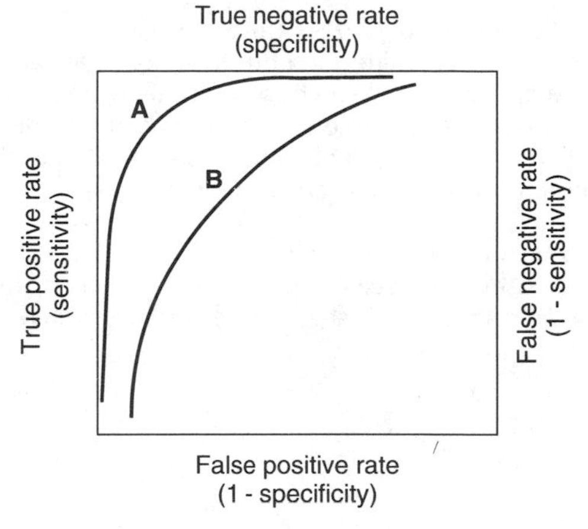

Which of the following statements about the comparative performance of these two tests is true?

(A) Test A performs better than test B
(B) The sensitivity of test B is greater than that of test A
(C) Test A is more risky than test B
(D) Test B has a greater predictive value compared to test A
(E) The specificity of test A is less than that of test B

1-B
2-C
3-A

ANSWERS AND EXPLANATIONS

1. The answer is B *[I B 2]*
As the positivity criterion is shifted to the right, true positive results decrease and false negative results increase; as a result, sensitivity decreases. Similarly, true negative results increase and false positive results decrease, leading to an increase in specificity. Because the false positive rate = 1 − specificity, an increase in specificity will lead to a decrease in the false positive rate. Predictive value positive (PVP) is a function of both sensitivity and specificity and, therefore, is altered by changes in either of these characteristics.

2. The answer is C *[I A; Ch 3 I B 4]*
The predictive value method, the most clinically sound technique for defining abnormality, sets the positivity criterion at a chosen sensitivity and specificity by taking into account the distribution of the diagnostic test variable in both diseased and disease-free populations. The method does not require that values of the test variable fit any particular distribution. The therapeutic method sets the positivity criterion at that value above which therapeutic intervention will occur.

3. The answer is A *[II C]*
The closer the receiver operator characteristic (ROC) curve lies to the upper left corner the better the test performance. Test A is both more sensitive and more specific than test B.

5
Clinical Decision Analysis

I. DECISION ANALYSIS: A STRATEGY FOR SOLVING COMPLEX PROBLEMS

A. Definition. Decision analysis is a mathematical tool designed to facilitate complex clinical decisions in which many variables must be considered simultaneously. Physicians are increasingly challenged to make clinical decisions on the basis of a complex and often conflicting array of diagnostic, therapeutic, financial, and social considerations. This analytical procedure selects among available diagnostic or therapeutic options, based on the probability and predetermined value (utility) of all possible outcomes of those options.

B. Properties. Decision analysis expedites the resolution of complex clinical problems.

1. It provides a systematic framework for organizing all data relevant to the decision.
2. It clearly defines the relationship between possible courses of action and their associated outcomes.
3. It assigns a numerical value to various courses of action, simplifying comparisons between them.

C. Advantages. Decision analysis offers certain advantages over pattern recognition or programmed response approaches to complex cases.

1. Decision analysis allows easier justification of diagnostic or treatment strategies to insurance companies and other third-party carriers, government agencies, and the judicial system.
2. With decision analysis, there is a greater likelihood of informed consent (because patient input can be taken into account in evaluating outcomes).
3. With decision analysis, it is easier to assess and revise assumptions underlying the case.

II. GENERAL STEPS IN DECISION ANALYSIS.

The decision analysis procedure consists of six sequential steps, as described by Pauker and Kassirer (1978); Sackett, Hayes, and Tugwell (1985); and Weinstein and Fineberg (1980). A case example using the steps is described in III.

A. Construct a decision tree. The decision tree is a map of all relevant courses of action and their associated outcomes.

1. **Nodes**
   a. **Decision node.** A branch point representing a diagnostic or therapeutic decision is known as a decision node, which is represented on the tree by a **square**. The first node on the decision tree is always a decision node.
   b. **Chance node.** A branch point representing a chance outcome not directly controlled by the physician is known as a chance node, which is represented on the tree by a **circle**.
   c. **Example.** A physician contemplates ordering a cardiac catheterization for a patient complaining of chest pain. This decision would be represented on a decision tree analyzing the case as a square. The possible outcomes of the procedure (e.g., test results positive or negative, the patient does or does not survive the procedure) are chance outcomes in the sense that they are not directly controlled by the physician. These points would be denoted by circles on the decision tree.

2. **Pruning.** Branches representing sequences of actions and outcomes that are unlikely, unreasonable, or unimportant may be "pruned" from the tree to simplify the analysis.

B. **Determine and assign probabilities.** The probability of occurrence associated with each chance node may be derived from estimates or from literature values.

1. **Estimates** may be based on data from similar patients in the same practice setting or may be obtained from experts (e.g., a leading clinical researcher or an academic medical center specializing in the condition in question).

2. **Literature values.** Probability values are most commonly taken from reports in the medical literature. If the clinical details of the case undergoing analysis differ significantly from the patient population described in the report (e.g., the patient is older or is in a better general state of health), probabilities derived from the report must be adjusted to reflect these differences.

C. **Assign utilities to each potential outcome.** A utility is a numerical estimate of the worth or value of a given outcome.

1. **Single-dimensional measures of utility** are based on an easily-measured quantitative variable and, by definition, provide a numerical estimate of worth.
   a. **Examples** of single-dimensional utility measures include life expectancy, mortality rates, 5-year survival rates, length of hospitalization, and quality-adjusted life expectancy.
   b. Single-dimensional measures of utility are **physician-driven;** that is, they do not take into account patient preferences or estimates of worth.

2. **Multidimensional measures of utility** are based on a trade-off between several components, which may or may not be easily quantified. Multidimensional measures of utility involve both a **quantity dimension** and a **quality dimension**.
   a. **Example.** A numerical estimate of the utility of chemotherapy for advanced colorectal cancer might be based on an assessment of the relative worth of 6 months of life without the additional discomfort imposed by chemotherapy versus 2 years of life complicated by drug-related side effects.
   b. Multidimensional utility measures typically incorporate both **patient and physician input**.
   c. Numerical values for multidimensional measures may be estimated by a simple rank ordering (e.g., the value of four outcomes may be rank ordered and assigned the relative utility values of one to four) or using a more complex method accounting for the relative intensity of the patient's preferences.

D. **Determine the expected utility.** The expected utility of each potential course of action is a function of both the probability of the outcome and its utility. A simple **example** of the concept of expected utility follows (see III B 4 for a more extensive example).

1. In a game of chance in which the participant has a 20% chance of winning $10, a 30% chance of winning $5, and a 50% chance of winning $1, the **utility** of each outcome is its monetary value.

2. The **expected utility** is the average amount of money the participant could expect to win if the game is played many times, or (.2)($10) + (.3)($5) + (.5)($1) = $4.00.

E. **Choose the course of action with the highest expected utility** (see III B 5).

F. **Evaluate the resistance of the chosen course of action to changes in probabilities and utilities.** A clinically valid decision is resistant to credible changes in the assigned probabilities and utilities. If small variations in these values alter the decision, no course of action is clearly preferable to the others and the decision is a toss-up.

1. **Sensitivity analysis** assesses the impact of variations in probabilities and utilities on the final decision. Individual probabilities or utilities are systematically varied over a range of clinically sensible values, and expected utilities are recalculated (see III B 6 a).

2. **Threshold analysis** defines specific pretest probabilities at which decisions should be switched. It can be carried out with any of the input probabilities (see III B 6 b).

III. CASE EXAMPLE: CHOOSING AN OPTIMAL DIAGNOSTIC STRATEGY.

The process of decision analysis is illustrated here, using a case history published by Pauker and Kassirer (1978).* The authors' report reads as follows:

> A 70-year-old woman with "borderline" schizophrenia and chronic heart block controlled with a pacemaker was admitted to the New England Medical Center Hospital with severe biventricular cardiac failure that had developed over the previous month. On the day of admission she developed hemoptysis and on examination she had bilateral pleural friction rubs and premature ventricular beats. Treatment with lidocaine controlled the rhythm disturbance, but the patient developed hypotension and pulsus paradoxus. Echocardiogram gave no evidence of pericardial effusion. A vigorous diuresis was induced with furosemide, and digitalis therapy was begun. Bedside catheterization of the pulmonary artery showed the pulmonary artery pressure to be 60/30 and the pulmonary capillary wedge pressure to be 25 mm Hg.
>
> The diagnosis of pulmonary embolism was considered, and heparin was administered. One week later the patient's condition was less unstable and consideration was given to substantiating this diagnosis. The possibility of carrying out pulmonary arteriography was also considered but the risk of this procedure was thought to be increased. Review of previous records disclosed that the compliance of both the patient and her husband with the advice of physicians was uneven, and for this reason the risk of bleeding from long-term anticoagulation was thought to be substantially increased. (Pauker and Kassirer, 1978, p 326)

A. Clinical summary and relevant considerations

1. **Diagnostic hypothesis.** On the basis of the signs and symptoms outlined above, the medical team caring for this patient arrived at a tentative diagnosis of pulmonary embolism. To confirm this diagnosis, the team must decide whether to order a pulmonary arteriogram, or to forego the arteriogram and continue long-term anticoagulation therapy.
2. **Factors influencing diagnostic strategy.** The factors influencing the team's decision include:
   a. **Increased risk associated with pulmonary arteriography** because of the current health status of the patient
   b. **Increased risk of long-term anticoagulation therapy** because of a history of poor compliance on the part of the patient and her husband

B. Analysis.

The complexity of the case leads the team to use decision analysis in choosing the optimal diagnostic strategy. The steps they follow are outlined here.

1. **Construct a decision tree.** The decision between pulmonary arteriography and long-term anticoagulation therapy is structured as a decision tree, mapping out the relevant courses of action and their respective outcomes. The **clinical starting point** for this analysis is the **decision node D-1**: carry out pulmonary arteriography versus do not order arteriography (Figure 5-1).
   a. **If a decision is made to order the arteriogram,** the **patient may survive or die** from complications of the procedure. Because neither of these outcomes can be directly controlled by the physicians managing the case, they are represented on the tree by **chance node C-1** (see Figure 5-1).
      (1) Given that the patient survives the arteriogram (see Figure 5-1, *lower branch emanating from node C-1*), the **test results may be positive (T+) or negative (T−),** represented by **chance node C-2** (see Figure 5-1).
      (2) Following the test, the medical team must decide whether to **maintain or discontinue anticoagulation therapy**. These decisions are represented on the tree by **decision nodes D-2 and D-3** (Figure 5-2A). However, to discontinue anticoagulation therapy is not a clinically reasonable choice if the pulmonary arteriogram is positive. Similarly, to continue anticoagulation therapy is not a reasonable decision if the arteriogram is negative. Thus, these unnecessary branches may be **pruned** from the tree (see Figure 5-2B and C).
      (3) A positive pulmonary arteriogram may occur in patients who actually have pulmonary emboli (true positives) as well as those who do not (false positives). Similarly, negative test results are obtained in patients with emboli (false negatives) and without emboli (true negatives). These outcomes are represented by **chance nodes C-3 and C-4** (Figure 5-3).

---

*All information from this case study has been reprinted from Pauker SG, Kassirer JP: Clinical decision analysis: a detailed illustration. *Semin Nucl Med* 8:324–330, 1978.

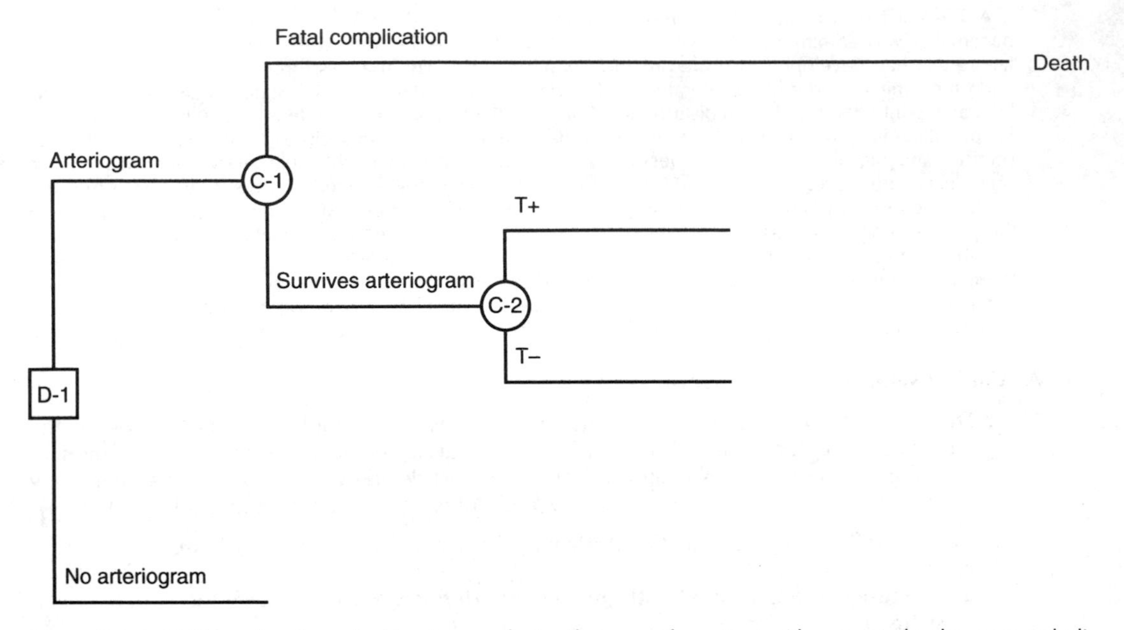

Figure 5-1. Initial branches for a decision tree analyzing the case of a patient with suspected pulmonary emboli. *Squares* indicate decision nodes (labeled from left to right across the tree with the letter D followed by the appropriate number); *circles* indicate chance nodes (labeled from left to right across the tree with the letter C followed by the appropriate number). $T+$ = positive arteriogram; $T-$ = negative arteriogram.

b. If a decision is made to forego the arteriogram (see Figure 5-1, *lower branch emanating from node D-1*), a second decision—whether to **maintain or discontinue anticoagulation**—must be made. This decision is represented on the tree by **decision node D-2** (Figure 5-4).

(1) A patient maintained on anticoagulation medication may or may not actually have pulmonary emboli. In the same fashion, a patient withdrawn from the medication may or may not have emboli. These outcomes are represented by **chance nodes C-5 and C-6** (see Figure 5-4).

(2) The completed decision tree for the case is shown in Figure 5-4.

2. Determine and assign probabilities. The probability of occurrence associated with all outcomes at each chance node is determined next.

a. Test performance characteristics. Pauker and Kassirer report the following test performance characteristics.

> Because data sufficiently specific to be relevant to the precise attributes of the patient were not available, in some instances the estimates of experienced radiologists and cardiologists responsible for the patient's care were used. Based on the clinical setting, it was estimated that the likelihood of recurrent pulmonary emboli was approximately 70%. This probability (0.7) is denoted as the "prior probability."
>
> Selecting a value for the mortality rate of pulmonary arteriography in this patient was difficult. Ordinarily, the risk of death from this study (arteriogram) is only approximately 0.4%. However, because this elderly patient had marked ventricular ectopy, severe cardiac failure, a possible recent myocardial infarction, and persistent pulmonary hypertension, the mortality rate of pulmonary arteriography was taken to be approximately 5%. Furthermore, because 1 week had elapsed since the last suspected embolus, some resolution of emboli might be expected. This factor, combined with observer error was thought to contribute to some increase in the likelihood of a false negative arteriogram. The false negative rate was taken to be 20% (i.e., a sensitivity of 80%) and the false positive rate was taken to be 1% (i.e., a specificity of 99%). (Pauker and Kassirer, 1978, p 327)

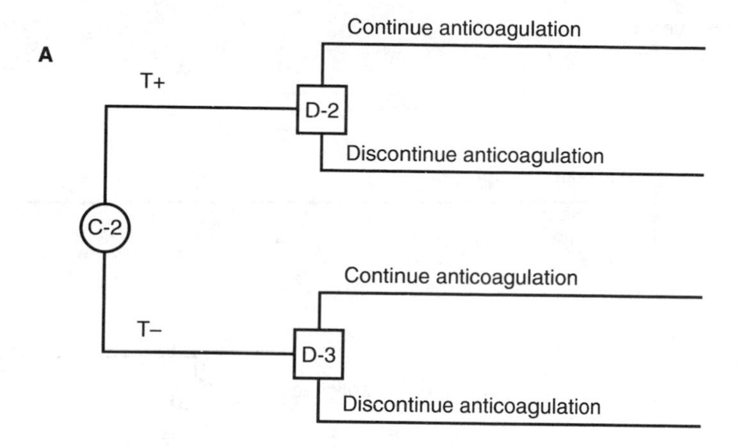

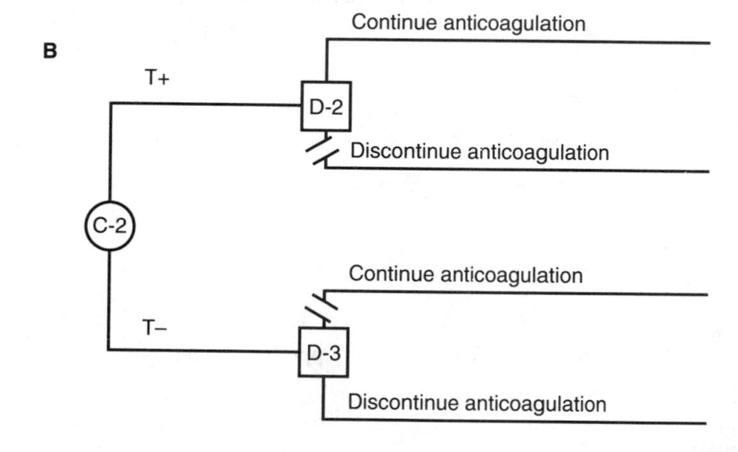

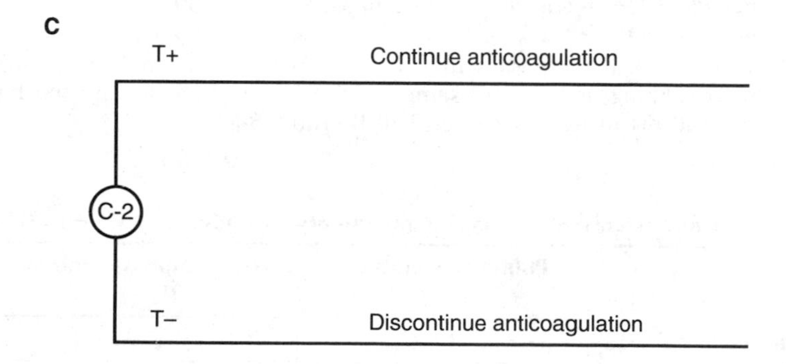

Figure 5-2. Pruning the decision tree. First, a complete decision tree is constructed (*A*). Note that all the branches emanating from node C-2 are intact. Next, branches representing clinically unreasonable decisions are marked to be pruned (*B*). Finally, the marked branches are eliminated, and the pruned tree remains (*C*).

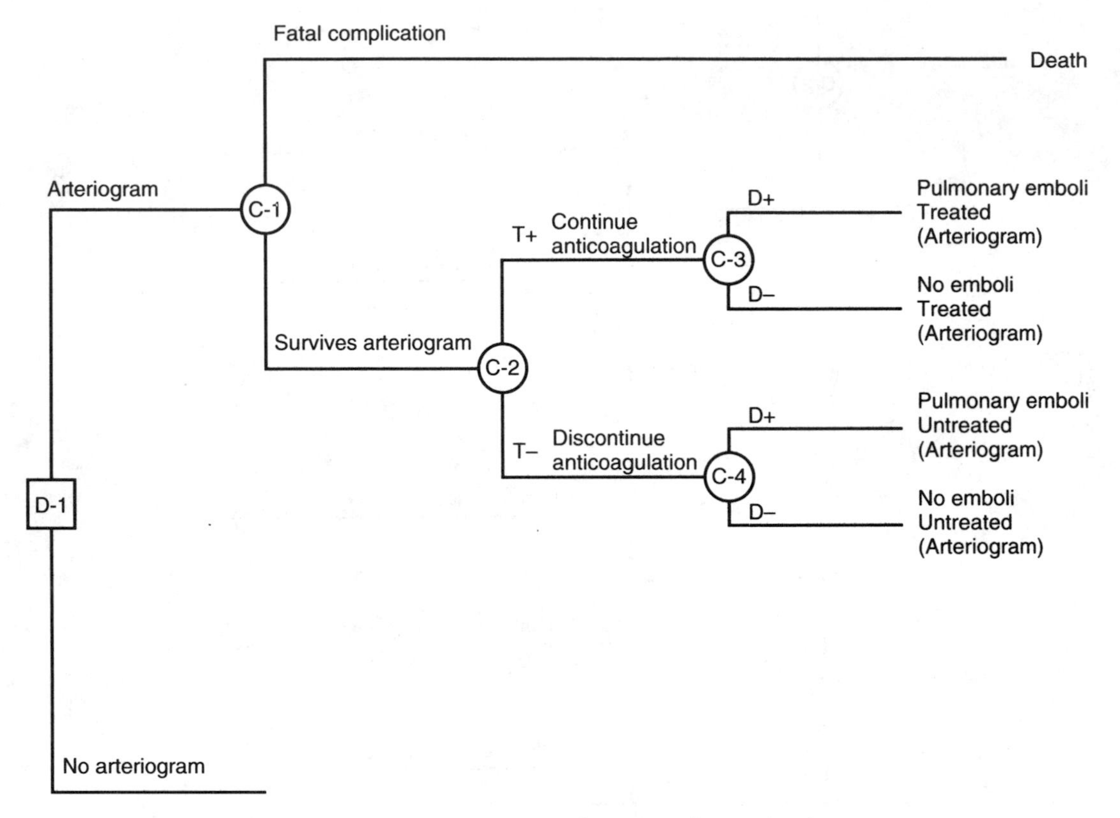

Figure 5-3. Decision tree with upper path completed.

b. Constructing a 2 × 2 table

(1) Summary of test performance characteristics. As stated in the published report:

(a) $P(D+) = P(\text{recurrent emboli}) = .70$

(b) $P(T+|D+) = \text{sensitivity} = \text{true positive rate} = .80$

(c) $P(T-|D+) = 1 - \text{sensitivity} = \text{false negative rate} = .20$

(d) $P(T-|D-) = \text{specificity} = \text{true negative rate} = .99$

(e) $P(T+|D-) = 1 - \text{specificity} = \text{false positive rate} = .01$

(2) These test characteristics may be used to generate a 2 × 2 table, or **decision matrix,** using an arbitrary sample size of 1000 patients and the back calculation method discussed in Chapter 3 III B (Table 5-1).

Table 5-1. A 2 × 2 Table (Decision Matrix) for Pulmonary Arteriography Given $P(D+) = .70$

	Pulmonary emboli (D+)	**No pulmonary emboli (D−)**	**Totals**
Arteriogram positive (T+)	560	3	563
Arteriogram negative (T−)	140	297	437
Totals	700	300	1000

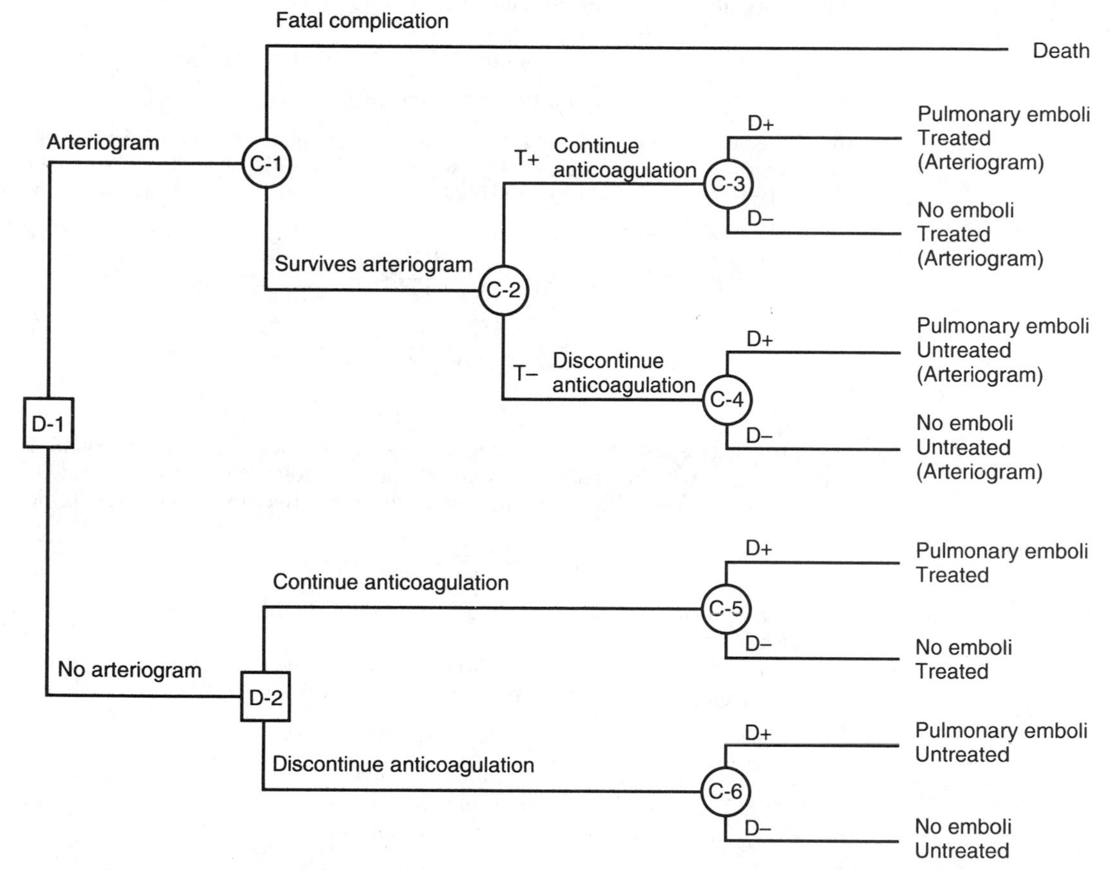

Figure 5-4. Decision tree with both upper and lower paths completed. (Adapted from Pauker SG, Kassirer JP: Clinical decision analysis: a detailed illustration. *Semin Nucl Med* 8:325, 1978.)

c. Assigning probabilities at each chance node

(1) Node C-1. From the published report, the mortality rate for pulmonary arteriography is estimated at 5%. Thus, the probabilities associated with each of the outcomes at chance node C-1 are:

(a) *P*(fatal complication from test) = .05

(b) *P*(survives arteriogram) = .95

(2) Node C-2

(a) The probabilities associated with the chance outcomes at this node may be taken directly from the decision matrix (see Table 5-1). Thus:

(i) $P(T+) = P(\text{positive arteriogram}) = 563/1000 = .563$

(ii) $P(T-) = P(\text{negative arteriogram}) = 1 - P(T+) = .437$

(b) The probabilities at successive nodes (moving from left to right along the tree) actually represent conditional probabilities, given a particular outcome at the preceding node.

(i) For example, the probability of a positive pulmonary arteriogram, $P(T+)$, may actually be more accurately written $P(T+|\text{survives test})$, since having a positive test result depends on having survived the test. Similarly, the probability of a negative test result, $P(T-)$, may be written $P(T-|\text{survives test})$.

(ii) In this example, probabilities to the left of node C-2 and related values in the decision matrix are based on test survivors; hence, conditional probability notation is optional.

(3) **Node C-3**

(a) The probability associated with the upper branch at node C-3 is the probability that the patient has a pulmonary embolism given that the results of the pulmonary arteriogram are positive; that is, $P(D+|T+)$, or the predictive value positive of the test (see Ch 3 II C 6).*

(i) This value can be derived from the decision matrix (see Table 5-1) as

$$P(D+|T+) = 560/563 = .995$$

(ii) Bayes' rule (see Ch 3 III C) also can be used to calculate $P(D+|T+)$ directly from the reported values for sensitivity [$P(T+|D+) = .80$], specificity [$P(T-|D-) = .99$] and prior probability of pulmonary emboli [prevalence, $P(D+) = .70$]. Thus,

$$P(D+|T+) = \frac{P(T+|D+)P(D+)}{P(T+|D+)P(D+) + P(T+|D-)P(D-)}$$

$$= \frac{(.80)(.70)}{(.80)(.70) + (1 - .99)(.30)}$$

$$= .995$$

(b) The probability associated with the lower branch at node C-3 is the probability that a patient with a positive test result has no pulmonary emboli; that is, $P(D-|T+)$. This probability can be derived from the decision matrix (see Table 5-1) as

$$P(D-|T+) = 3/563 = .005$$

(4) **Node C-4.** The probabilities for each of the outcomes at node C-4 can be obtained in a similar fashion to those associated with node C-3.

(a) The probability that pulmonary emboli are present given a negative test result, $P(D+|T-)$, may be derived from the decision matrix (see Table 5-1) as

$$P(D+|T-) = 140/437 = .320$$

(b) The probability that pulmonary emboli are absent given a negative test result, $P(D-|T-)$, is the predictive value negative of the test (see Ch 3 II C 7). This probability can be obtained from the decision matrix (see Table 5-1) as

$$P(D-|T-) = 297/437 = .680$$

or using Bayes' rule [see III B 2 c (3) (a) (ii)].

(5) **Nodes C-5 and C-6.** The probabilities associated with the outcomes at these nodes are the prior probabilities of pulmonary emboli and the absence of emboli in this patient, or .7 and .3, respectively.

(6) Figure 5-5 shows the decision tree with the appropriate probabilities assigned to each outcome at all of the chance nodes.

(7) The probabilities of the final outcomes found in the last column of the tree may be obtained by **multiplying** the probabilities of each branch along the path to that outcome.

(a) For example, the final outcome emanating from the upper branch of node C-3 represents the case in which the patient with pulmonary emboli has survived the pulmonary arteriogram and has had a positive test result.

(b) In this example, the probability of this final outcome can be calculated as

$$P(\text{survives test } \textit{and} \text{ has positive arteriogram } \textit{and} \text{ has emboli})$$

$$= P(T+ \textit{ and } \text{survives test } \textit{and } D+)$$

$$= P(T+ \textit{ and } \text{survives test})P(D+|T+ \textit{ and } \text{survives test})$$

$$= P(\text{survives test})P(T+|\text{survives test})P(D+|T+ \textit{ and } \text{survives test})$$

$$= (.95)(.56)(.995) = .53$$

---

*Because this conditional probability also depends on surviving the arteriography, it is most accurately written $P(D+|T+$ *and* survives procedure).

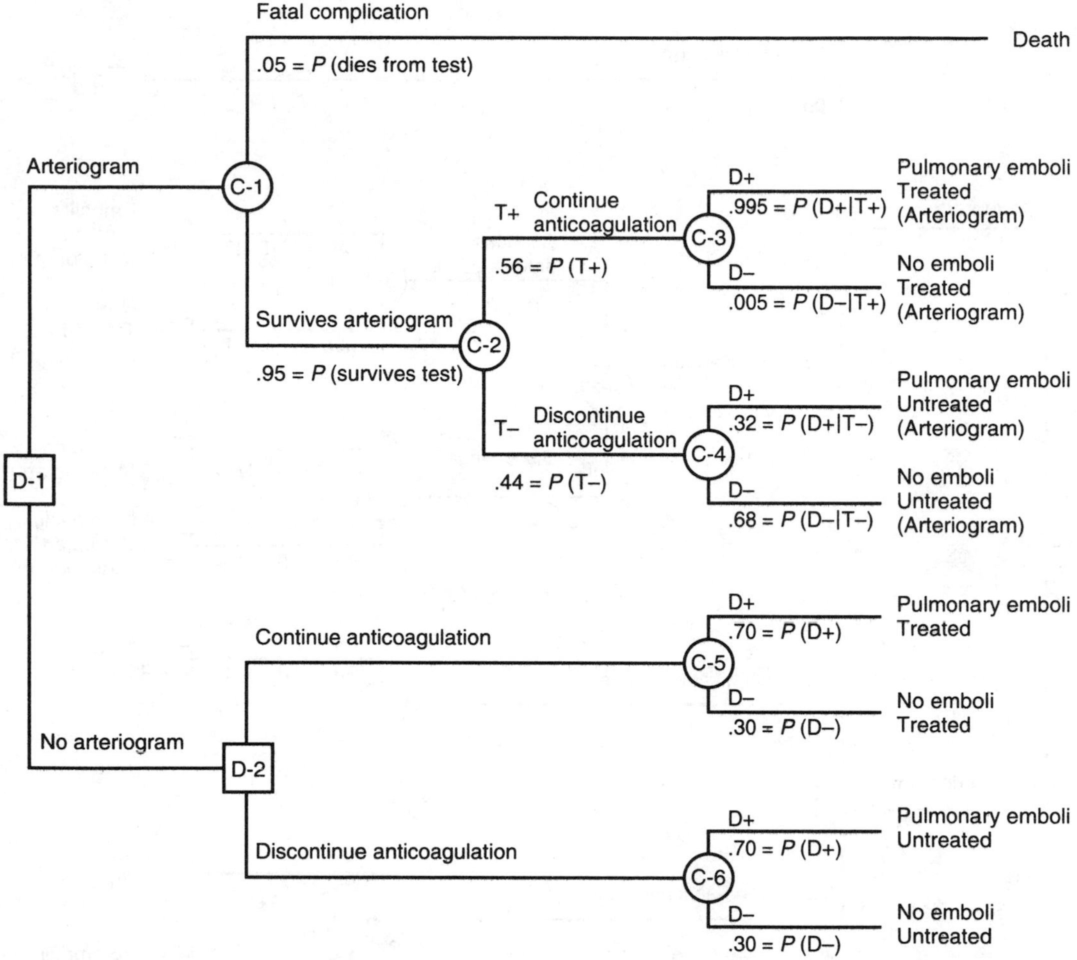

Figure 5-5. Decision tree with assigned probabilities. The probabilities of occurrence for all outcomes have been designated at each chance node.

3. **Assign utilities to each potential outcome.** The medical team selects a single-dimensional measure of utility: expected survival rate (Table 5-2).* Figure 5-6 depicts the decision tree with the relevant utilities assigned to each final outcome.

Table 5-2. Expected Survival Rates for Patients with Treated and Untreated Pulmonary Emboli

Outcomes	Expected Survival (years)
No emboli, untreated	10.0
No emboli, treated	6.7
Emboli, treated	5.0
Emboli, untreated	3.3

Reprinted from Pauker SG, Kassirer JP: Clinical decision analysis: a detailed illustration. *Semin Nucl Med* 8:324–330, 1978.

*Pauker and Kassirer employ a more complex, multidimensional utility structure, accounting for both morbidity and mortality.

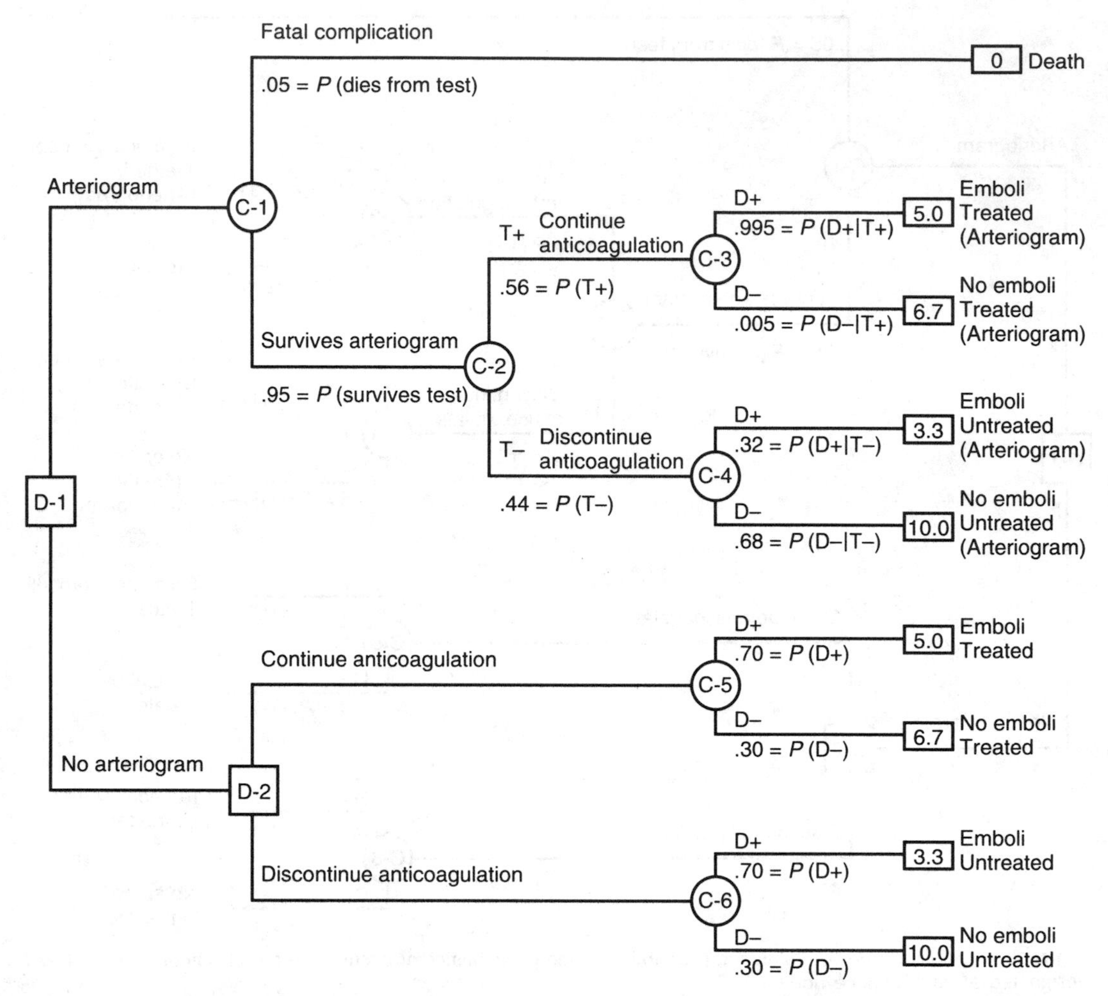

Figure 5-6. Decision tree with assigned utilities. The utility associated with each outcome, expressed as the expected survival rate for that outcome, is shown in the last column (*boxes*). (Adapted from Pauker SG, Kassirer JP: Clinical decision analysis: a detailed illustration. *Semin Nucl Med* 8:328, 1978.)

4. Determine the expected utility for each node.

a. Procedure. The expected utility of an outcome at any given chance node is obtained by multiplying the probability associated with each branch emanating from the node by its corresponding utility and then summing the resulting products over all branches emanating from the node. This process is known as **folding back** or **averaging out**.

(1) Expected utilities are measured in the same units as utilities.

(2) Expected values for each chance node are calculated from left to right on the tree.

b. Example. The expected utility values for this example thus can be determined from Figure 5-6.

(1) The expected values for nodes C-3 through C-6 are:

C-3: (.995)(5.0) + (.005)(6.7) = 5.0085

C-4: (.32)(3.3) + (.68)(10.0) = 7.856

C-5: (.70)(5.0) + (.30)(6.7) = 5.51

C-6: (.70)(3.3) + (.30)(10.0) = 5.31

(2) The expected utility for **node C-2** is the sum of the products obtained by multiplying the expected utility values at nodes C-3 and C-4 by the probabilities associated with the branches emanating directly from C-2, or

$$(.56)(5.0085) + (.44)(7.856) = 6.2614$$

(3) Similarly, the expected utility at **node C-1** is

$$(.05)(0) + (.95)(6.2614) = 5.94833$$

c. Figure 5-7 shows the **completed decision tree,** including the expected utilities associated with each chance node.

5. Choose the course of action with the highest expected utility. The expected utility of the upper branch emanating from decision node D-1, that is, the expected utility of ordering the pulmonary arteriogram, is 5.948. The expected utility of the alternative, foregoing the arteriogram, is 5.51. The option "carry out the test" has the highest expected utility and is therefore the course of action selected by the physicians.

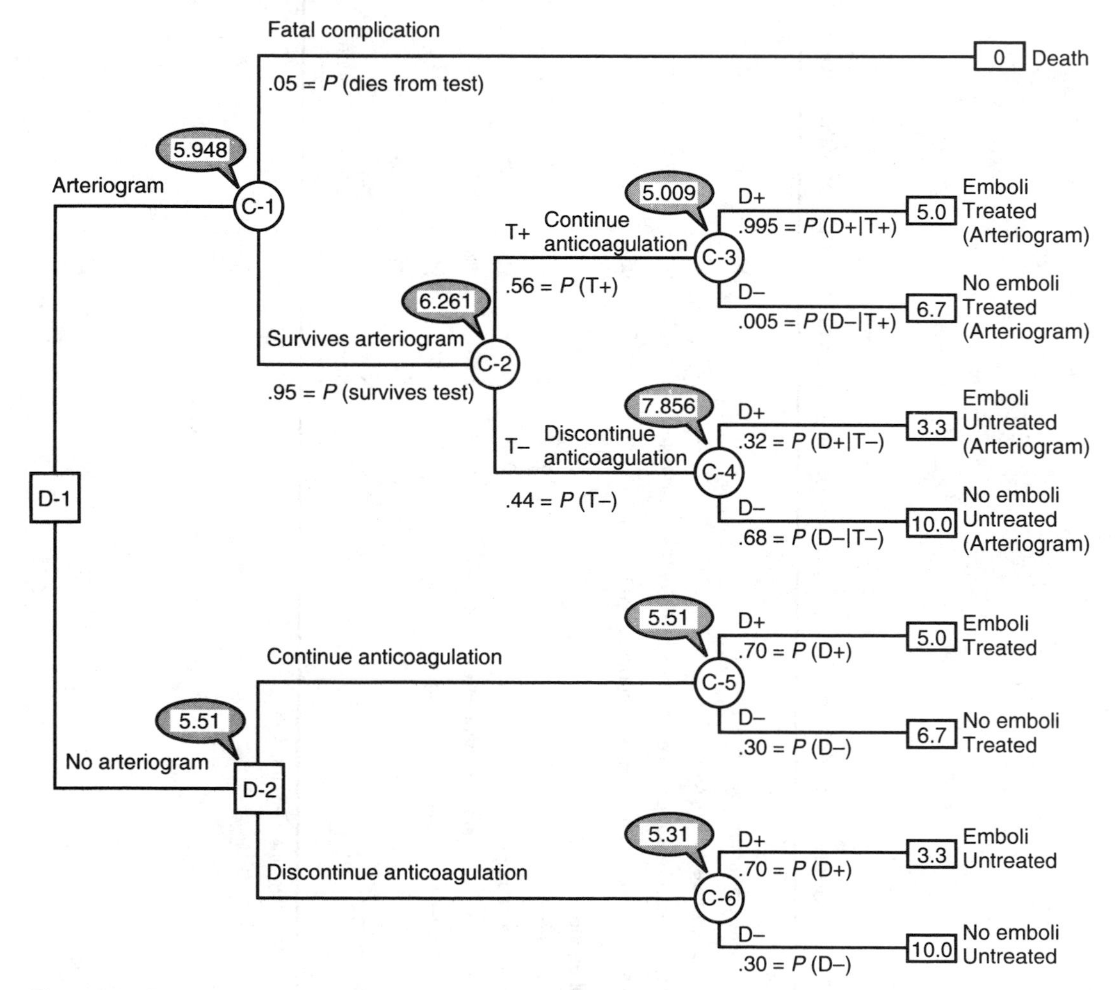

Figure 5-7. Decision tree with expected utilities. The numbers shown in *shaded circles* above each node indicate the expected utility (expressed as survival rate in years) calculated at that node. (Adapted from Pauker SG, Kassirer JP: Clinical decision analysis: a detailed illustration. *Semin Nucl Med* 8:330, 1978.)

Table 5-3. Expected Utilities for Three Courses of Action Over a Range of Values for the Prevalence of Pulmonary Emboli

Probabilities											
Prior probability of pulmonary emboli $P(D+)$	0	.10	.20	.30	.40	.50	.60	.70	.80	.90	1.00
Probability of positive arteriogram $P(T+)$	.01	.09	.17	.25	.33	.41	.48	.56	.64	.72	.80
Probability of emboli given positive arteriogram $P(D+\|T+)$	0	.90	.95	.97	.98	.988	.992	.995	.997	.999	1.00
Probability of emboli given negative arteriogram $P(D+\|T-)$	0	.02	.05	.08	.12	.17	.23	.32	.45	.65	1.00
Expected utilities											
C-1: Order arteriogram	9.47	8.96	8.46	7.96	7.45	6.95	6.44	5.95	5.44	4.93	4.43
D-2: Do not order arteriogram*	10.00	9.33	8.66	7.99	7.32	6.65	5.98	5.51	5.34	5.17	5.00
C-5: Do not order arteriogram; continue treatment	6.70	6.53	6.36	6.19	6.02	5.85	5.68	5.51	5.34	5.17	5.00
C-6: Do not order arteriogram; discontinue treatment	10.00	9.33	8.66	7.99	7.32	6.65	5.98	5.31	4.64	3.97	3.30
Preferred strategy	No arterio-gram	No arterio-gram	No arterio-gram	No arterio-gram	Arterio-gram	Arterio-gram	Arterio-gram	Arterio-gram	Arterio-gram	No arterio-gram	No arterio-gram

*Necessary to determine C-5 and C-6.

6. Evaluate the resistance of the chosen course of action to changes in probabilities and utilities.

a. Sensitivity analysis. To test the validity of the decision to order the pulmonary arteriogram, the physicians analyzing the case calculate the expected utilities for three possible courses of action (order the arteriogram; do not order the arteriogram and continue long-term anticoagulation therapy; and do not order the arteriogram and discontinue anticoagulation therapy) over a range of values for the prior probability (prevalence) of pulmonary emboli (Table 5-3).

(1) Graphic representation (Figure 5-8). The effect of changes in the prior probability of pulmonary emboli on the three courses of action may be depicted graphically by plotting the expected utility of each course of action as a function of prevalence.

(2) Interpretation

(a) At any given prior probability of pulmonary emboli, the optimal course of action is the one having the highest expected utility, that is, the one represented by the uppermost line on the graph (see Figure 5-8). For example, when the absence of pulmonary emboli is guaranteed [*P*(D+) = 0], the best course of action is to forego the arteriogram and discontinue anticoagulation therapy.

(b) The decision to order the pulmonary arteriogram is reached when the value for the prior probability of pulmonary emboli lies between .3 and .8. If the medical team judges that the true prevalence falls within this range, then they may be reasonably assured that they have made the correct decision.

b. Threshold analysis

(1) Graphic representation. The lines on the graph representing the three courses of action **intersect** at three points (see Figure 5-8). These **thresholds** define the prevalence values at which the preferred course of action changes.

(2) Interpretation. In this example, threshold A (see Figure 5-8) occurs at *P*(D+) = .3. For values of *P*(D+) below .3, the optimal course of action is to forego the pulmonary arteriogram and discontinue anticoagulation therapy, while above this value, the best course of action is to order the arteriogram [for values of *P*(D+) up to the next threshold at *P*(D+) = .8]. When *P*(D+) is exactly equal to the threshold value, both strategies are equally valid and the decision is a toss-up.

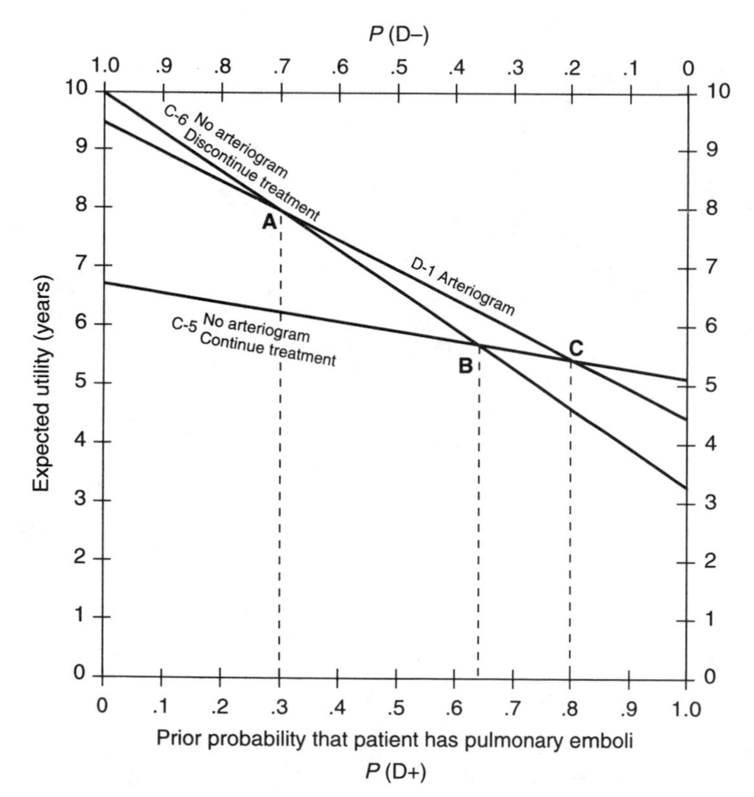

Figure 5-8. Sensitivity analysis assessing the effect of changes in the prior probability of pulmonary emboli (*x-axis*) on the expected utilities (*y-axis*) of three clinical options.

BIBLIOGRAPHY

Hardin G: The tragedy of the commons. *Science* 162:1243, 1968.

Kassirer JP: The principles of clinical decision making: an introduction to decision analysis. *Yale J Biol Med* 49:149–164, 1976.

Lusted LB: General problems in medical decision making with comments on ROC analysis. *Semin Nucl Med* 8:299–305, 1978.

Metz CE: Basic principles of ROC analysis. *Semin Nucl Med* 8:283–298, 1978.

Patton DD: Introduction to clinical decision making. *Semin Nucl Med* 8:272–281, 1978.

Pauker SG: Coronary artery surgery: the use of decision analysis. *Ann Intern Med* 85:8–18, 1976.

Pauker SG, Kassirer JP: Clinical decision analysis: a detailed illustration. *Semin Nucl Med* 8:324–330, 1978.

Pauker SP, Pauker SG: Prenatal diagnosis: a directive approach to genetic counseling using decision analysis. *Yale J Biol Med* 50:275–289, 1977.

Ransohoff DF, Feinstein AR: Is decision analysis useful in clinical medicine? *Yale J Biol Med* 49:165–168, 1976.

Sackett DL, Hayes RB, Tugwell P: *Clinical Epidemiology: A Basic Science for Clinical Medicine.* Boston, Little, Brown, 1985.

Schwartz WB, Gorry GA, Kassirer JP, et al: Decision analysis and clinical judgement. *Am J Med* 55:459–472, 1973.

Weinstein MC, Fineberg HV: *Clinical Decision Analysis.* Philadelphia, WB Saunders, 1980.

PROBLEMS*

5-1. A 20-year-old man is hospitalized complaining of right lower quadrant pain that has steadily increased in severity over the previous 24 hours. The patient also reports diarrhea (unaccompanied by vomiting) of two days' duration. Urinalysis is normal and the patient's white blood cell count is 14,000. Previously published reports in the medical literature yield the following relevant information.

The probability of acute appendicitis in patients with similar signs and symptoms is 25%.
Approximately 75% of patients presenting with this clinical profile have acute gastroenteritis, rather than appendicitis.
Appendectomy is associated with a .5% mortality rate in cases where appendicitis actually exists and a .1% mortality rate in disease-free patients.
The probability of perforation in patients with appendicitis who do not receive immediate surgery is 50%.
If perforation occurs, but is treated appropriately, the resulting mortality rate is approximately 4%.
If perforation does not occur and the patient receives appropriate medical treatment, the chance of recovery is 100%.
If an appendectomy is not performed on a patient who does not have appendicitis, the chance of survival is 100%.

a. Construct a decision tree to determine whether immediate surgery or a 6-hour observation period is the best course of action for this patient. Assign the outcome "death" a utility of 0 and the outcome "recovery" a utility of 1.
b. How can the validity of this decision be tested?
c. What is the relationship between the threshold value for the probability that this patient has acute appendicitis and the preferred treatment strategy?

5-2. Medulloblastoma, a rapidly-growing tumor of the cerebellum, may be treated by surgical resection or craniospinal radiation therapy. However, if exploratory surgery is performed and the tumor proves inoperable, the subsequent effectiveness of radiotherapy may be compromised. An oncologist must decide between exploratory surgery and initiating radiotherapy for a 12-year-old boy with medulloblastoma, based on the following relevant information she has acquired in treating similar cases.

The exploratory surgery itself is associated with no risk to the patient.
The tumor is operable in approximately 50% of patients.
Eighty percent of patients respond to radiotherapy alone.
Operable tumors are removed immediately. The probability of complications following surgical resection is 15%, and patients who experience such complications have a mortality rate of 40%. The survival rate for patients with operable tumors and no post-surgical complications is 95%.
If surgery is performed, and the tumor proves inoperable, only 60% of patients respond to radiotherapy.

Construct a decision tree to determine whether exploratory surgery or radiotherapy alone is the best course of action for this patient. Assume that the utility associated with the outcome "death" is 0 and that associated with the outcome "5-year survival" is 1.

Solutions on p 367.

*The probabilities assumed for clinical events in these problems do not necessarily represent the true likelihood that the events will occur in actual clinical practice.

STUDY QUESTIONS

Directions: Each of the numbered items or incomplete statements in this section is followed by answers or by completions of the statement. Select the **one** lettered answer or completion that is **best** in each case.

Questions 1–4

A particular type of cancer can be treated by surgical resection or radiation therapy. Exploratory surgery is the only method of identifying patients with operable tumors; however, if such surgery is performed and the tumor proves inoperable, postsurgical radiotherapy is less effective. The decision tree below was constructed by an oncologist to weigh the relative merit of exploratory surgery versus radiotherapy for a particular patient.*

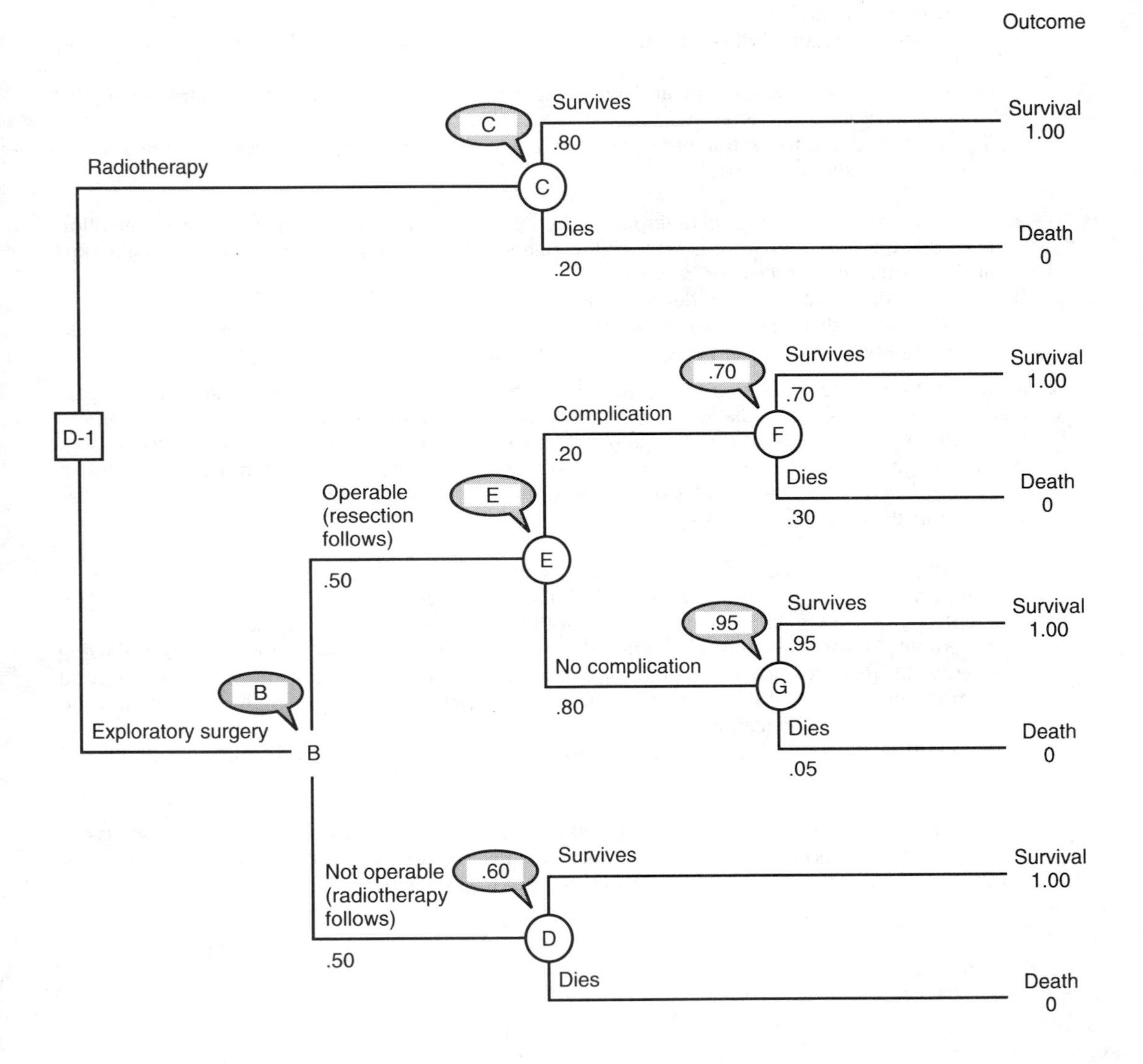

*This example and figure are adapted from Weinstein MC, Fineberg HV: *Clinical Decision Analysis.* Philadelphia, WB Saunders, 1980, p 73.

1. If the oncologist elects to perform exploratory surgery, what is the probability that a patient with an inoperable tumor will die?

(A) .20
(B) .60
(C) .05
(D) .40
(E) The probability cannot be determined from the data given

2. The expected utility associated with chance node E is

(A) .95
(B) .76
(C) .90
(D) .80
(E) not discernible from the data given

3. Branch point B of the decision tree is a

(A) decision node
(B) chance node
(C) threshold value
(D) choice node

4. According to the decision tree, which of the following treatment alternatives represents the best course of action for this patient?

(A) Initiate radiotherapy and forego exploratory surgery
(B) Perform the exploratory surgery first and reserve radiotherapy for inoperable cases
(C) The decision is a toss-up; both radiotherapy and exploratory surgery would be equally acceptable
(D) None of the above

1-D
2-C
3-B
4-A

Questions 5–6

Clinical trials of interferon B demonstrate its effectiveness in treating a rare slow virus infection of the central nervous system, but also identify a number of potentially life-threatening side effects associated with its use. A neurologist caring for a patient with this slow virus infection must decide whether to order a brain biopsy, the only definitive diagnostic procedure, and administer interferon B to those patients with positive test results (strategy 1), or simply to administer the drug to all suspected cases (strategy 2). The risk of mortality associated with brain biopsy is less than .001. The figure below is a sensitivity and threshold analysis showing how variations in the prevalence of the viral infection would affect strategy 1 and strategy 2.

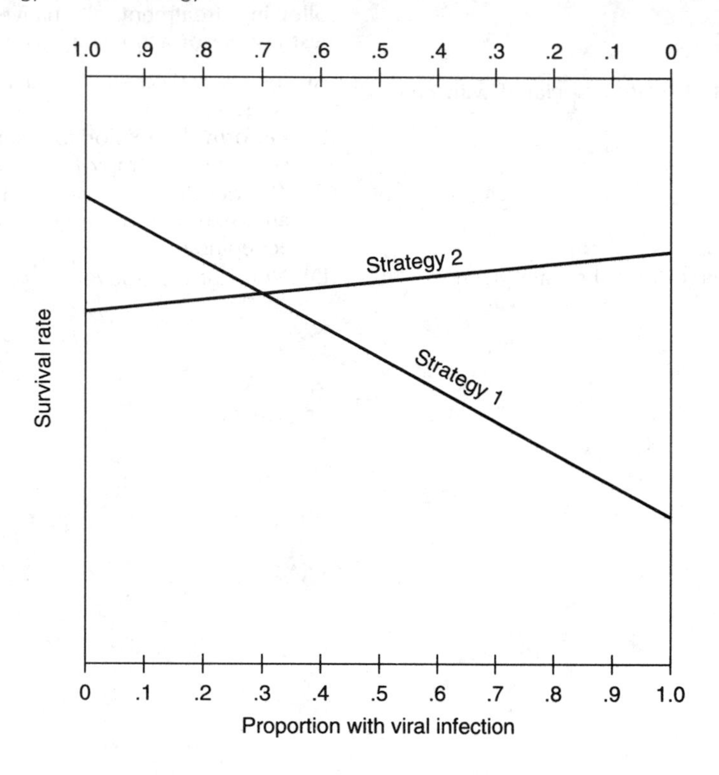

5. Strategy 2 is the best alternative when

(A) the proportion of disease-free patients is greater than .3
(B) the prevalence of the viral infection does not exceed .3
(C) the proportion of diseased patients is greater than .3
(D) the proportion of disease-free patients exceeds 70%
(E) the risk associated with the drug is also taken into account

6. The value "prevalence = .3" is the

(A) sensitivity value
(B) chance of dying from the biopsy
(C) chance of dying from the drug
(D) threshold value

5-C
6-D

ANSWERS AND EXPLANATIONS

1–4. The answers are: 1-D *[III B 4 a]*, **2-C** *[III B 4 a]*, **3-B** *[II A 1 a–b]*, **4-A** *[III B 4–5]*
The expected utility at chance node D, .60, is calculated as

$$\begin{aligned}\text{expected utility} &= (1)P(\text{survives}) + (0)P(\text{dies})\\ &= .60\end{aligned}$$

In other words, the probability that a patient with an inoperable tumor will survive exploratory surgery [i.e., P(survives)] is .60. The probability of death is the complement of the probability of survival: $P(\text{dies}) = 1 - P(\text{survives}) = .40$.

The expected utility at chance node E is the averaged-out value of the branches emanating from this node, or

$$\text{expected utility} = (.20)(.70) + (.80)(.95) = .90$$

Because the branches emanating from node B represent outcomes that are not under the direct control of the physician (i.e., outcomes dictated by nature or chance), this node is a chance node. In contrast, node A is a decision node, since it is the physician, not nature or chance, who determines the course of action—radiotherapy or exploratory surgery.

The expected utility associated with the upper branch of the tree (radiotherapy) is

$$\text{expected utility at node C} = (.80)(1.00) + (.20)(0) = .80$$

The expected utility at node E, calculated in question 2, is .90. The expected utility associated with the lower branch of the tree (exploratory surgery) is obtained by averaging out the probabilities associated with each branch emanating from this node and the expected utilities at nodes D and E. Thus,

$$\text{expected utility at node B} = (.50)(.90) + (.50)(.60) = .75$$

Because the expected utility associated with radiotherapy alone (.80) is higher than the expected utility of exploratory surgery (.75), the preferred course of action for this patient is to forego the exploratory procedure and initiate radiotherapy.

5–6. The answers are: 5-C *[III B 6 a]*, **6-D** *[III B 6 b]*
The preferred strategy is the one associated with the highest survival rate, that is, the strategy corresponding to the uppermost line in the figure for questions 5–6. When the prevalence of the viral infection is greater than .3, strategy 2 is preferred; when the prevalence is less than .3, strategy 1 represents the optimal course of action.

The threshold value for any input variable is the value at which the preferred course of action changes. For prevalence values below the threshold value of .3, strategy 1 is optimal, while above the threshold, strategy 2 is preferred. When prevalence is equal to .3, the expected utility of the two strategies is equal and the decision between them is a toss-up.

6 Probabilities in Genetic Counseling

I. GENETIC COUNSELING: A DIAGNOSTIC ISSUE. Genetic counseling represents a unique type of diagnostic assessment. This increasingly important medical specialty utilizes the rules of probability and a knowledge of the family history of a particular genetic disease to predict future occurrences of the disorder.

A. Using conditional and joint probabilities to assess heritability. The likelihood that a couple will transmit a hereditary disease to their children often can be determined using only the formulas for calculating joint and conditional probability (see Ch 2 III).

B. Example. Sickle-cell anemia (SCA) is a recessive hemoglobin defect occurring almost exclusively in blacks. About 10% of the black population of the United States is heterozygous for the sickle-cell trait. Figure 6-1 is a pedigree for a couple who wish to know the chances that they will have a child with SCA. The man, an adopted child, does not have SCA but knows nothing about the status of his biological relatives. Although the woman's brother has SCA, she herself is unaffected.

1. **Carrier status.** Because neither patient has SCA, the risk to their children depends on whether one or both of them are carriers of the sickle-cell trait.
   a. If **neither patient is a carrier** (i.e., both are homozygotes), none of their children will have SCA, nor will any be carriers.
   b. If **only one patient is heterozygous,** none of their children will be affected, but half will be carriers.
   c. If **both patients are carriers** (i.e., both are heterozygotes), on the average, 25% of their children will not inherit the SCA gene, 50% will inherit a single defective allele and will be asymptomatic carriers, and 25% will have SCA.
2. **Determining probable carrier status for the pedigree**
   a. Because the woman's brother has SCA, both of her parents must be carriers of the sickle cell trait.

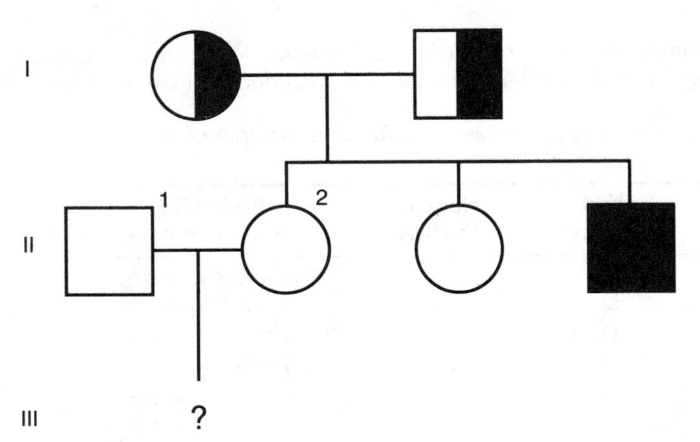

Figure 6-1. A pedigree of sickle-cell anemia. Individuals II-1 and II-2 wish to know their risk of having a child with SCA. *Half-shaded shapes* = carriers of the disease; *completely shaded shapes* = diseased, or affected, individuals.

b. Since the woman herself is unaffected, her chances of being a carrier may be expressed as the conditional probability *P*(woman carrier|woman unaffected), or, in terms of joint probability, given that her parents are known carriers,

$$P(\text{woman carrier}|\text{woman unaffected}) = \frac{P(\text{woman carrier and woman unaffected})}{P(\text{woman unaffected})}$$

$$= \frac{1/2}{3/4} = 2/3 = .67$$

c. Although the man is unaffected, his family history of SCA is unknown. Therefore, the probability that he is a carrier is assumed to be the same as that of the black population in general, or about .10.

d. The probability that both patients are carriers can be calculated as

$$P(\text{woman carrier and man carrier}) = P(\text{woman carrier})P(\text{man carrier})$$

$$= (2/3)(1/10)$$

$$= 1/15 = .067$$

3. Determining the probability of an affected child. The probability that this couple will have a child with SCA, or *P*(child affected), can be calculated as

$$P(\text{child affected and parents carriers}) + P(\text{child affected and parents not carriers})$$

$$= P(\text{child affected}|\text{parents carriers})P(\text{parents carriers}) + 0$$

$$= (1/4)(1/15)$$

$$= 1/60 = .017$$

Thus, *P*(child affected) = .017.

II. USING THE BINOMIAL FORMULA TO ASSESS HERITABILITY.

II. USING THE BINOMIAL FORMULA TO ASSESS HERITABILITY. The binomial theorem is frequently used in genetic counseling. Each conception is considered an independent trial, having the possible outcomes "diseased" or "disease-free."

A. The derivation of the binomial formula relies on the definition of several variables.

1. The proportion of people with a certain hereditary disease in a particular population is defined as ***p***.

a. The probability that an individual selected at random from the population will have the disease, *P*(D+), is *p*.

b. The probability that the individual does not have the disease, *P*(D−), is defined as *q* and is equal to 1 − *p*.

2. Using a sample. For a sample of three individuals selected at random from the population, eight different permutations of diseased and disease-free members can occur (Table 6-1).

Table 6-1. Occurrence of a Genetic Disease in a Random Sample of Three Individuals

First Person	Second Person	Third Person	Number with Disease	Probability of Permutation
D	D	D	3	$p \times p \times p = p^3$
D	D	$\bar{D}$	2	$p \times p \times q = p^2q$
D	$\bar{D}$	D	2	$p \times q \times p = p^2q$
$\bar{D}$	D	D	2	$q \times p \times p = p^2q$
D	$\bar{D}$	$\bar{D}$	1	$p \times q \times q = pq^2$
$\bar{D}$	D	$\bar{D}$	1	$q \times p \times q = pq^2$
$\bar{D}$	$\bar{D}$	D	1	$q \times q \times p = pq^2$
$\bar{D}$	$\bar{D}$	$\bar{D}$	0	$q \times q \times q = q^3$

D = diseased; $\bar{D}$ = not diseased

a. The **number of diseased individuals** in the sample, defined as k, may equal 0, 1, 2, or 3 (Table 6-2).

b. The probability that a sample of three people drawn from this population will contain exactly k people with the disease is given by the formula

$$P = (\text{Number of permutations containing } k \text{ diseased individuals})(p^k)(q^{3-k})$$

3. **Probability of outcomes.** In general, if an event has two possible outcomes, Y (probability of occurrence $= p$) and $\bar{Y}$ (probability of occurrence $= 1 - p = q$), the probability that Y will occur exactly k times in n independent trials is $(C)(p^k)(q^{n-k})$.

a. **The constant C** is defined as

$$C = \binom{n}{k} = \frac{n!}{k!(n-k)!}$$

(1) ***n*!** Read "n factorial" or "factorial n," $n!$ is the product of all integers from 1 to n inclusive. It is denoted symbolically as

$$n! = (1)(2)(3)\ldots(n)$$

(2) For example, $3! = (1)(2)(3) = 6$

(3) By definition, $0! = 1$

b. **The binomial formula.** The probability that Y occurs k times in n independent trials is given by the binomial formula:

$$P(Y = k) = \binom{n}{k} p^k(1-p)^{n-k}$$

where $k =$ the number of trials with outcome Y; $n =$ the total number of trials; and p is the probability that Y will be the outcome on any single trial.

B. **Using the binomial theorem in a genetic counseling example.** Through a neighborhood screening program, a married couple learn that they are both carriers of the sickle cell trait. Although they had planned to have four children, the couple wish to reconsider this decision in light of the knowledge that it is likely that 25% of their children will have SCA. The following sample problems assume that the couple do indeed elect to have four children.

1. **Problem #1.** What is the probability that *none* of the four children will have SCA?

2. **Solution.** In this problem, $k = 0$, $n = 4$, and p, the probability that any child born to the couple, both of whom are known carriers, will have SCA $= .25$. Hence, using the binomial formula,

$$P(Y = 0) = \binom{4}{0}(1/4)^0(3/4)^4$$

$$= \frac{4!}{0!4!}(1)(3/4)^4 = .32$$

3. **Problem #2.** What is the probability that *exactly two* of the four children will have SCA?

4. **Solution.** Here, $k = 2$. Again, using the binomial formula,

$$P(Y = 2) = \binom{4}{2}(1/4)^2(3/4)^2$$

$$= \frac{4!}{2!2!}(1/16)(9/16) = .21$$

Table 6-2. Probability of k Individuals with a Genetic Disease

Number with Disease (k)	Number of Permutations	Probability of k Persons with Disease
0	1	q^3
1	3	$3pq^2$
2	3	$3p^2q$
3	1	p^3

5. **Problem #3.** What is the probability that *at least one* of the four children will have SCA?

6. **Solution.** The condition "at least one" is satisfied if one, two, three, or all four of the children have SCA. Therefore, the probability of at least one affected child can be written $P(Y=1 \text{ or } Y=2 \text{ or } Y=3 \text{ or } Y=4)$. Since these are mutually exclusive events, the addition rule (see Ch 2 III B 3) can be used in conjunction with the binomial formula to solve the problem. Thus,

$$P(Y=1 \text{ or } Y=2 \text{ or } Y=3 \text{ or } Y=4)$$

$$= P(Y=1) + P(Y=2) + P(Y=3) + P(Y=4)$$

$$= \binom{4}{1}(1/4)^1(3/4)^3 + \binom{4}{2}(1/4)^2(3/4)^2 + \binom{4}{3}(1/4)^3(3/4)^1 + \binom{4}{4}(1/4)^4(3/4)^0$$

$$= \frac{4!}{1!3!}(1/4)(27/64) + \frac{4!}{2!2!}(1/16)(9/16) + \frac{4!}{3!1!}(1/64)(3/4) + \frac{4!}{4!0!}(1/256)(1)$$

$$= .42 + .21 + .05 + .004$$

$$= .68$$

7. **Problem #4.** What is the probability that *no more than two* of the four children will have SCA?

8. **Solution.** The condition "no more than two" is satisfied when one or two of the children have SCA, as well as when none of the children is affected. That is, the probability that the couple will have no more than two children with SCA can be written $P(Y=0 \text{ or } Y=1 \text{ or } Y=2)$. Using the addition rule and the binomial formula,

$$P(Y=0 \text{ or } Y=1 \text{ or } Y=2) = P(Y=0) + P(Y=1) + P(Y=2)$$

$$= \binom{4}{0}(1/4)^0(3/4)^4 + \binom{4}{1}(1/4)^1(3/4)^3 + \binom{4}{2}(1/4)^2(3/4)^2$$

$$= \frac{4!}{0!4!}(1)(81/256) + \frac{4!}{1!3!}(1/4)(27/64) + \frac{4!}{2!2!}(1/16)(9/16)$$

$$= .32 + .42 + .21$$

$$= .95$$

III. USING BAYES' RULE TO ASSESS HERITABILITY.

Bayes' Rule is particularly useful in situations where estimates of the likelihood of inheritance are revised on the basis of new knowledge (e.g., in predicting the probable disease status of future siblings following the birth of an affected child).

A. Example. Bruton's agammaglobulinemia is an X-linked inherited immune deficiency disorder (i.e., women are unaffected but may be carriers of the defective gene). A 30-year-old woman contemplating pregnancy has an uncle with this disorder and seeks genetic counseling to determine her risk of having an affected child (her husband is normal).

B. Sample problems

1. **Problem #1.** Following the initial consultation with this patient, the pedigree shown in Figure 6-2 is constructed. Using the information in this pedigree, what is the probability that the patient will have a child with Bruton's agammaglobulinemia?

2. **Solution.** Because Bruton's agammaglobulinemia is an X-linked recessive genetic disorder, transmission of the defective gene depends on the patient's carrier status and on the sex of the child.
   a. **If the patient is a carrier,** on the average, 50% of her daughters will inherit two normal genes (one each from the patient and her husband), while 50% will be asymptomatic carriers, inheriting a defective gene from their mother and a normal gene from their father. Her sons have a 50% chance of inheriting their mother's normal gene (in which case they would be, of course, unaffected) and a 50% chance of inheriting the defective gene and manifesting the disease.

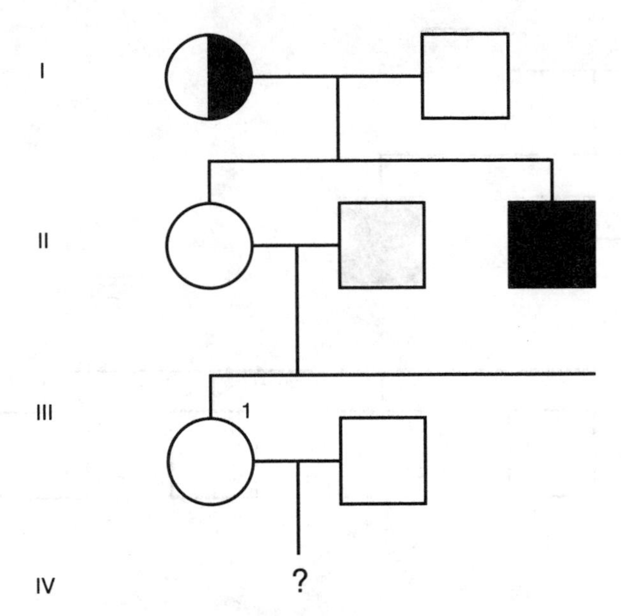

Figure 6-2. A pedigree of Bruton's agammaglobulinemia. Individual III-1 is the patient seeking genetic counseling. *Half-shaded shapes* = carriers of the disease; *completely shaded shapes* = diseased, or affected, individuals.

b. Determining the probability that the patient is a carrier

(1) Because the patient has an uncle with Bruton's agammaglobulinemia, her grandmother is an obligate carrier. Therefore, the probability that the patient's mother is a carrier is 1/2.

(2) If her mother is a carrier, the probability that the patient is also a carrier is 1/2.

(3) Using the summation principle for joint probabilities (see Ch 2 III B 4 a),

$$P(\text{patient carrier}) = P(\text{patient carrier and mother carrier}) + P(\text{patient carrier and mother not carrier})$$

(4) Since the patient **cannot** be a carrier if her mother is not, the last term in this expression = 0. Thus,

$$P(\text{patient carrier}) = P(\text{patient carrier and mother carrier})$$

(5) Using the multiplication rule (see Ch 2 III B 2 a) to convert to conditional probabilities,

$$\begin{aligned} P(\text{patient carrier}) &= P(\text{patient carrier}|\text{mother carrier})P(\text{mother carrier}) \\ &= (1/2)(1/2) \\ &= .25 \end{aligned}$$

c. Determining the probability that the patient will have an affected son

(1) Using the summation principle for joint probabilities, $P(\text{son affected})$ can be calculated as

$$P(\text{son affected and patient carrier}) + P(\text{son affected and patient not carrier})$$

(2) Since the patient cannot have an affected son unless she herself is a carrier, the last term in this expression = 0. Hence,

$$P(\text{son affected}) = P(\text{son affected and patient carrier})$$

(3) Using the multiplication rule,

$$\begin{aligned} P(\text{son affected}) &= P(\text{son affected}|\text{patient carrier})P(\text{patient carrier}) \\ &= (1/2)(1/4) \\ &= .125 \end{aligned}$$

3. Problem #2. In the next counseling session, the physician advising the patient learns that she has three normal brothers (Figure 6-3). Does this new information alter the probability that the patient will have a son with Bruton's agammaglobulinemia?

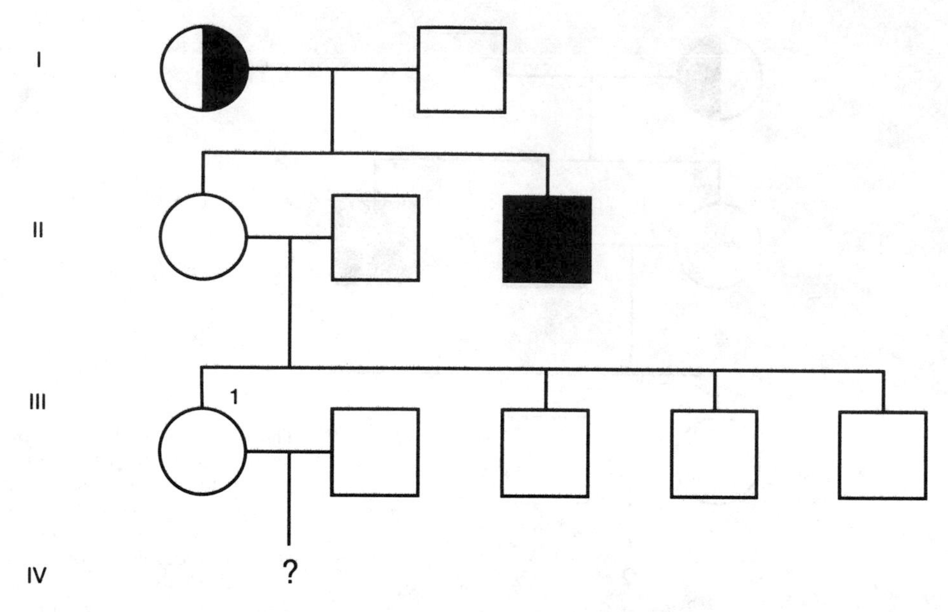

Figure 6-3. Revised pedigree for problem 2 (III B 3).

4. **Solution.** The existence of three unaffected siblings suggests that the patient's mother is more likely to be homozygous for the normal gene than to be a carrier. Bayes' rule (see Ch 3 III C) can be used to revise the probability that the patient's mother is a carrier. Subsequently, estimates of the probability that the patient herself is a carrier, as well as the probability that she will have an affected son, can be revised as well.
   a. **Determining the probability that the patient's mother is a carrier**
      (1) Define M = mother a carrier; $\bar{M}$ = mother *not* a carrier; and s = no affected sons. Then, using Bayes' Rule

$$P(\text{mother carrier}|\text{no affected sons}) = P(M|s)$$
$$= \frac{P(s|M)P(M)}{P(s|M)P(M) + P(s|\bar{M})P(\bar{M})}$$

      (2) $P(M) = P(\bar{M}) = 1/2$ [see III B 2 b (1)]
      (3) **Calculating $P(s|M)$.** If the patient's mother is a carrier, her risk of having an affected son is 1/2. Using the binomial theorem, the probability that the patient's mother will have *no* affected sons (out of a total of three sons) if she is a carrier is

$$P(Y = 0) = \binom{3}{0}(1/2)^0(1/2)^3$$
$$= \frac{3!}{0!3!}(1)(1/8)$$
$$= 1/8 = .125 = P(s|M)$$

      (4) **Calculating $P(s|\bar{M})$.** If the patient's mother is not a carrier, the probability that all three of her sons will be healthy is obviously 1.
      (5) **Calculating $P(M|s)$.** Substituting in Bayes' Rule,

$$P(M|s) = \frac{(1/8)(1/2)}{(1/8)(1/2) + 1(1/2)}$$
$$= 1/9 = .11$$

   b. **Revising the probability that the patient is a carrier.** Substituting the new probability that the patient's mother is a carrier in the expression derived in III B 2 b (5),

$$P(\text{patient carrier}) = P(\text{patient carrier}|\text{mother carrier})P(\text{mother carrier})$$
$$= (1/2)(1/9)$$
$$= (1/18)$$
$$= .056$$

c. Revising the probability that the patient will have an affected son. Following the same line of reasoning used in III B 2 c, P(son affected) could be calculated as

$$P(\text{son affected and patient carrier}) + P(\text{son affected and patient not carrier})$$
$$= P(\text{son affected and patient carrier})$$
$$= P(\text{son affected}|\text{patient carrier})P(\text{patient carrier})$$
$$= (1/2)(1/18)$$
$$= .03$$

PROBLEMS

6-1. Both members of a young couple are identified as carriers of the sickle-cell trait and informed that their risk of having a child with sickle-cell anemia (SCA) is 1/4. The couple plan to have four children.

a. What is the probability that at least two of the four children will have SCA?

b. What is the probability that no more than one of the four children will have SCA?

6-2. A 28-year-old pregnant woman seeks genetic counseling because two maternal uncles had Lesch-Nyhan syndrome, a rare X-linked inherited disorder of purine metabolism characterized by physical and mental retardation, compulsive self-mutilation, choreoathetoid movements, spasticity, and impaired renal function.

a. What is the probability that this patient will have a son with Lesch-Nyhan syndrome?

b. The physician counseling the patient learns that she has four normal brothers. Does this information alter the probability that she will have a son with Lesch-Nyhan syndrome?

6-3. Approximately 10% of the black population in the United States is heterozygous for the sickle-cell trait. The figure below is a pedigree for a couple with a family history of SCA. The couple have one unaffected daughter and plan to have a second child. What is the probability that this child will have SCA?

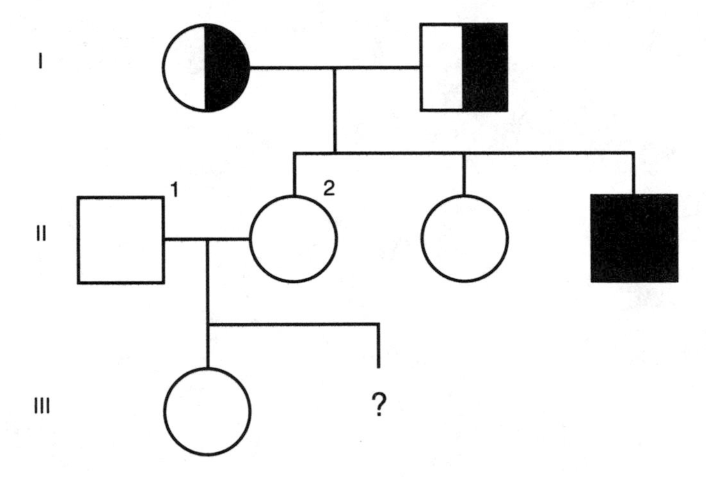

6-4. A husband and wife are both known carriers (i.e., both are heterozygotes) for phenylketonuria (PKU), an autosomal recessive defect in phenylalanine metabolism that results in mental retardation. The couple plan to have three children.

a. What is the probability that all three children will be unaffected?

b. What is the probability that at least one of the three children will have phenylketonuria?

Solutions on p 369.

STUDY QUESTIONS

Directions: Each of the numbered items or incomplete statements in this section is followed by answers or by completions of the statement. Select the **one** lettered answer or completion that is **best** in each case.

1. Duchenne muscular dystrophy is an X-linked hereditary neuromuscular disorder. A 26-year-old woman, who has a brother with Duchenne muscular dystrophy, seeks genetic counseling prior to starting a family. If the patient is a carrier and has four sons, what is the probability that at least one will have Duchenne muscular dystrophy?

(A) 1/2
(B) 15/16
(C) 1/4
(D) 1/16
(E) 5/16

2. The patient in question 1 ultimately has four normal sons and a daughter. What is the probability that the daughter is a carrier of the disorder?

(A) 1/2
(B) 1/16
(C) 15/16
(D) 1/34
(E) 1/4

Questions 3–5

Five percent of those comprising a particular population are known to carry a recessive gene for poliodystrophy, an inherited disorder characterized by the onset of recurrent seizures and dementia in early childhood. A 32-year-old woman who had a brother with this disorder seeks genetic counseling. The patient's husband, an only child, has no knowledge of his family history of the disorder.

3. What is the probability that the patient is a carrier of poliodystrophy?

(A) 2/3
(B) 1/2
(C) 3/4
(D) 1/20
(E) 3/8

4. What is the probability that both the patient and her husband are carriers?

(A) 2/3
(B) 1/20
(C) 1/30
(D) 3/4
(E) 3/8

5. What is the probability that any given child born to this couple will have poliodystrophy?

(A) 1/2
(B) 1/20
(C) 1/30
(D) 1/4
(E) 1/120

1-B
2-D
3-A
4-C
5-E

ANSWERS AND EXPLANATIONS

1. The answer is B *[II B 5–6]*
There is a 50% chance that a son born to a woman who is a carrier of a sex-linked disorder will have the disease and a 50% chance that he will be unaffected. The probability that at least one of four sons born to this patient will have Duchenne muscular dystrophy can be calculated using the binomial formula, based on the fact that the condition "at least one" is satisfied if one, two, three, or all four sons have the disorder. Thus,

$$P(Y \geq 1) = P(Y=1 \text{ or } Y=2 \text{ or } Y=3 \text{ or } Y=4)$$
$$= P(Y=1) + P(Y=2) + P(Y=3) + P(Y=4)$$
$$= \binom{4}{1}(1/2)^1(1/2)^3 + \binom{4}{2}(1/2)^2(1/2)^2 + \binom{4}{3}(1/2)^3(1/2)^1 + \binom{4}{4}(1/2)^4(1/2)^0$$
$$= 4(1/16) + 6(1/16) + 4(1/16) + 1(1/16)$$
$$= 15/16$$

2. The answer is D *[III B 1–4]*
The probability that the daughter is a carrier depends upon the carrier status of the patient herself. Since the patient has a brother with Duchenne muscular dystrophy, her mother is an obligate carrier and the probability that she is also a carrier is estimated, prior to the birth of her children, as 1/2. This probability is modified, however, by the birth of four normal sons. The revised probability is calculated using Bayes' Rule (define C = patient carrier; $\bar{C}$ = patient not a carrier; and s = no affected sons).

$$P(\text{patient carrier}|\text{no affected sons}) = P(C|s)$$
$$= \frac{P(s|C)P(C)}{P(s|C)P(C) + P(s|\bar{C})P(\bar{C})}$$

$P(s|C)$, that is, the probability a woman who is a carrier will have no diseased sons, is found using the binomial formula as follows:

$$P(s|C) = P(Y=0) = \binom{4}{0}(1/2)^0(1/2)^4$$
$$= 1(1/16)$$
$$= 1/16$$

If the patient is not a carrier, the probability that none of her sons will have Duchenne muscular dystrophy, $P(s|\bar{C})$, is 1. Substituting in Bayes' Rule,

$$P(C|s) = \frac{(1/16)(1/2)}{(1/16)(1/2) + 1(1/2)}$$
$$= 1/17$$

The probability that the patient's daughter is a carrier is therefore

$$P(\text{daughter carrier})$$
$$= P(\text{daughter carrier and patient carrier}) + P(\text{daughter carrier and patient not carrier})$$
$$= P(\text{daughter carrier and patient carrier}) + 0$$
$$= (1/2)(1/17)$$
$$= 1/34$$

3–5. The answers are: 3-A *[I B 2 a–b]*, **4-C** *[I B 2 c–d]*, **5-E** *[I B 3]*
Because she has a brother with poliodystrophy, the patient's mother and father must be carriers of the disorder. She herself may be heterozygous or homozygous for the poliodystrophy gene. The probability that a child born to two known carriers will be healthy is 3/4; the probability that such a child is also a carrier is 1/2. Thus,

$$P(\text{patient carrier}|\text{patient healthy}) = \frac{P(\text{patient carrier and patient healthy})}{P(\text{patient healthy})}$$

$$= \frac{1/2}{3/4}$$

$$= 2/3$$

In the absence of information on his family history of poliodystrophy, the probability that the patient's husband is a carrier is assumed to equal that of the population, that is, 1/20. The probability that both the patient and her husband are carriers is therefore

$$P(\text{patient carrier and husband carrier}) = P(\text{patient carrier})P(\text{husband carrier})$$

$$= (2/3)(1/20)$$

$$= 1/30$$

The probability that a child born to this couple will have poliodystrophy, or $P(\text{child affected})$, can be calculated as

$$P(\text{child affected and both parents carriers}) + P(\text{child affected and parents not carriers})$$

$$= P(\text{child affected}|\text{both parents carriers})P(\text{both parents carriers}) + 0$$

$$= (1/4)(1/30)$$

$$= 1/120$$

7
Frequency of Disease

I. DEFINITION AND CALCULATION OF MEASURES OF FREQUENCY.

I. DEFINITION AND CALCULATION OF MEASURES OF FREQUENCY. The frequency with which a disease occurs is critical to the evaluation of diagnostic and treatment options. For example, as noted in Chapters 3 and 5, frequency measures such as prevalence influence the predictive ability of a diagnostic test and provide the underlying assumptions for decision analysis.

A. Prevalence

1. **Definition.** The prevalence (or, more properly, the **point prevalence**) of a disease in a population is **the proportion of that population having the disorder at a given point in time**.
   a. Prevalence is defined in terms of a single point in time even though the actual process of data collection may take place over days, weeks, or years.
   b. Thus, prevalence provides a static measure of disease frequency, analogous to a single frame of a motion picture.

2. **Calculation**
   a. Prevalence is calculated using the formula

$$\text{prevalence} = \frac{\text{total number of diseased individuals at given time}}{\text{total population}}$$

   b. The numerator encompasses both **new** and **ongoing** cases of the disease.

3. **Examples**
   a. Figure 7-1 illustrates the occurrence of hepatitis B in an internal medicine practice for the period January 1, 1991, to December 31, 1991.
      (1) On January 1, 1991, five patients in the practice—cases 1, 4, 6, 8, and 9—have the disorder.
      (2) The prevalence of hepatitis B in this population on January 1, 1991, is 5/100 = .05, or 50 cases per 1000 patients.
   b. Figure 7-2 illustrates the occurrence of coronary heart disease (CHD) in the Charleston Heart Study, an ongoing cohort study begun in 1960 to identify and quantify risk factors for coronary heart disease in blacks and whites.
      (1) In 1960, the initial study sample contained 651 white men. Physical examination demonstrated that 49 members of this group had CHD, while 602 were disease-free.
      (2) The prevalence of CHD in 1960 among white men was 49/651 = .0753, or 75.3 cases per 1000 members of the population.

B. Incidence

B. Incidence of a disease refers to **the number of new cases that develop in a population at risk for the disease over a specified period of time**. Incidence may be measured in one of two ways.

1. **Cumulative incidence (CI)**
   a. **Definition.** CI is the **proportion** of people in a predefined group of fixed size (**fixed**

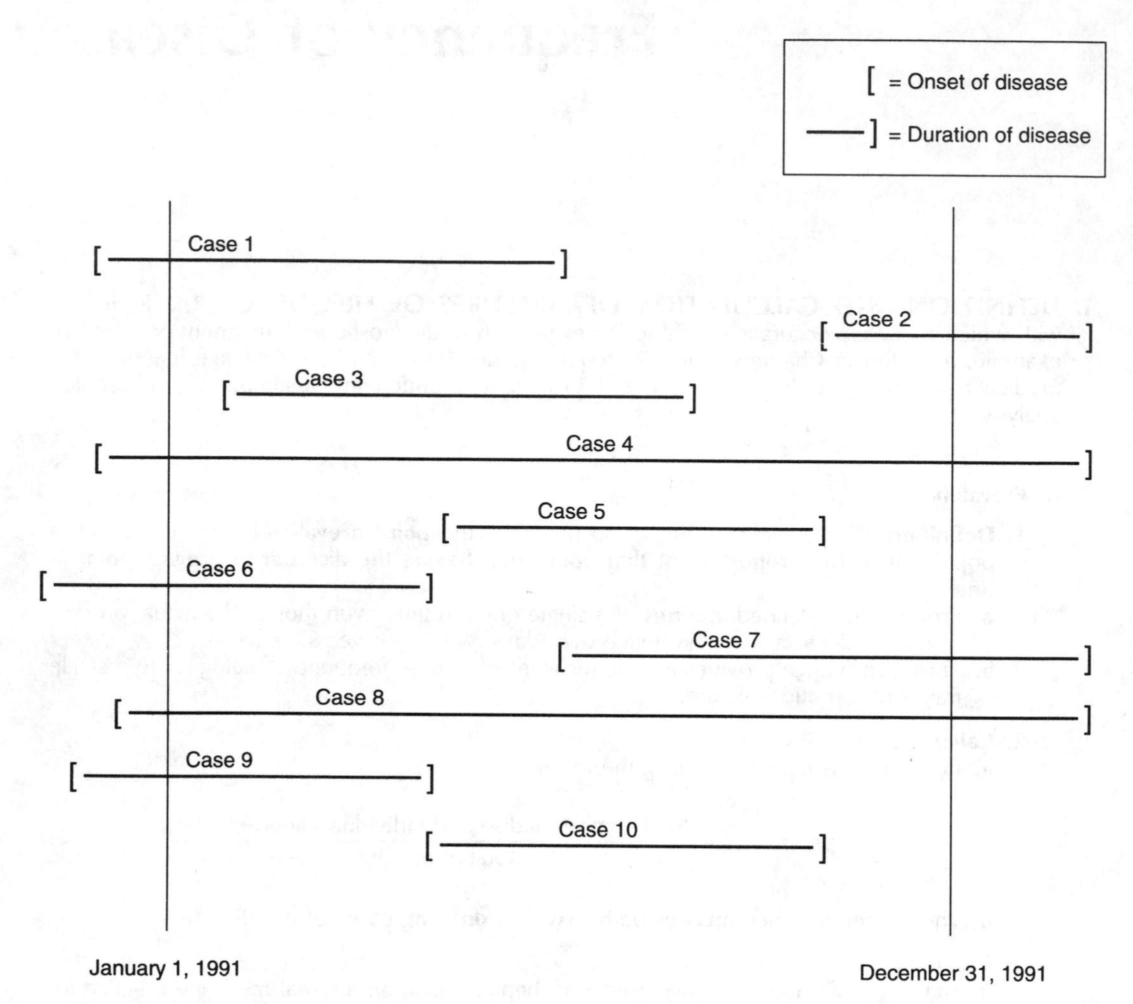

Figure 7-1. Cases of hepatitis B in a 100-patient practice from January 1, 1991, to December 31, 1991.

cohort) who **develop** the disease during a specified time period.* This observation period (known as the **period referent**) must be included in any report of CI.

b. Application

(1) CI can be used to measure **risk**—that is, the probability that a healthy individual will develop the disease in question during the specified period of time.

(2) Inherent in this application is the assumption that *all* members of the population at risk are followed until they develop the disease or the observation period ends.

c. Calculation

(1) CI is calculated using the formula

$$CI = \frac{\text{number of new cases during specified time period}}{\text{total number of people at risk}}$$

(2) Only **new cases** of the disease are included in the numerator; ongoing cases present at the beginning of the period referent are excluded.

d. Examples

(1) During the period January 1, 1991, to December 31, 1991, five new cases of

---

*Cumulative incidence (CI) is often referred to as the "rate" at which new cases of a disease develop over a period of time, although, strictly speaking, a rate is an instantaneous change in one variable per unit change in another (usually time), not a proportion.

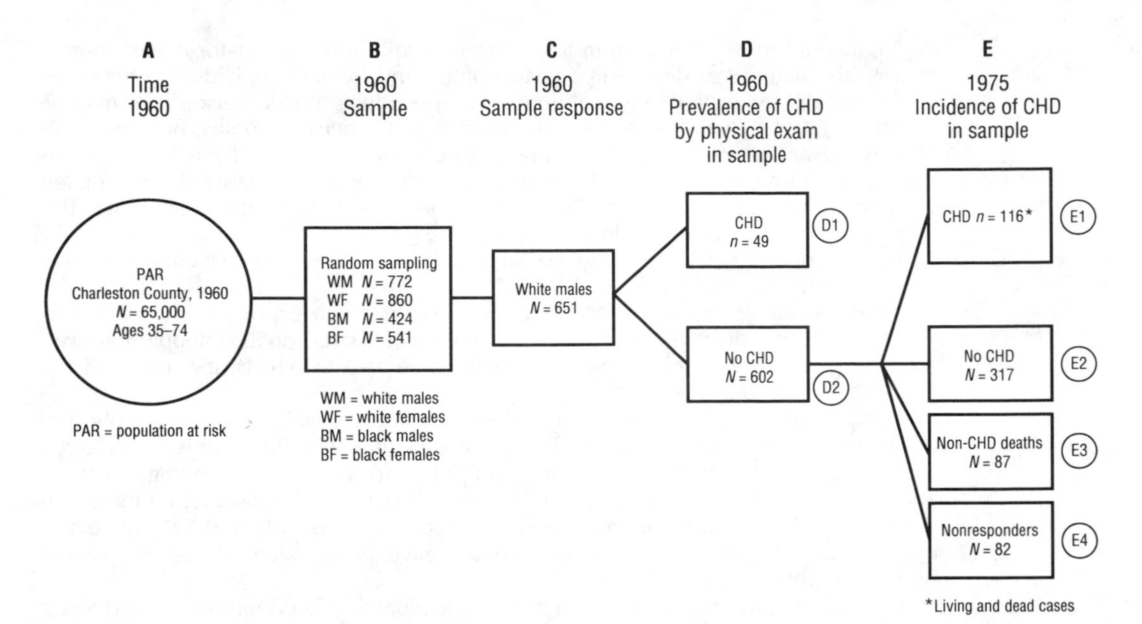

Figure 7-2. Prevalence of coronary heart disease (CHD) in white men, ages 35–74, in 1960 and the 15-year incidence (1960–1975) of CHD among those men, as determined by the Charleston Heart Study. All residents of Charleston in 1960 between the ages of 35 and 74 constituted the population at risk (*A*). There were 772 white men in the random sample selected from the population (*B*), and 651 of those men participated in the study (*C*). Baseline examination revealed that 602 of the 651 participants were free of CHD in 1960 (*D*). Over the 15-year study period, 116 of those 602 men developed the disorder (*E*). (Reprinted from Keil JE, Weinrich MC: *Compendium of Epidemiological Methods.* Student Aid, University of South Carolina School of Public Health and Medical University of South Carolina, 1980, p 4.)

hepatitis B occurred in the same practice described in section I A 3 a (cases 2, 3, 5, 7, and 10; see Figure 7-1).

- **(a)** Because 5 of the 100 patients in the practice already had hepatitis B at the beginning of the observation period (cases 1, 4, 6, 8, and 9), only 95 were at risk for developing the disease in 1991.
- **(b)** The CI of hepatitis B in 1991 for this practice is 5/95 = .053, or 5.3% over the one-year period.

(2) In the Charleston Heart Study, the occurrence of CHD in the cohort of 602 white males identified as disease-free at the start of the study (see Figure 7-2, *D2*) was tabulated for the period 1960–1975.

- **(a)** During the observation period, 116 people developed CHD, including those still living at the end of the period and those known to have died from CHD (see Figure 7-2, *E1*).
- **(b)** The 15-year CI of CHD is 116/602 = .1927; that is, 19.27% of the population developed the disease over the 15-year observation period.

2. Incidence density (ID)

a. Definition. ID (also known as **incidence rate, hazard rate,** and **force of morbidity or mortality**) refers to the rate at which new cases of a disease occur in a population given that the population is both studied and at risk for varying lengths of time.

(1) The variations in number of people studied and the length of time they are observed result from real-life study conditions. For example, patients continuously enrolled in an epidemiological study over a long period of time may either complete the study, drop out for various reasons, or die of other causes.

(2) ID is an **estimate of the average rate** at which a disease develops in a population over a specified time period.

b. Calculation

(1) ID is calculated using the formula

$$\text{ID} = \frac{\text{number of new cases of disease during specified time period}}{\text{person-time at risk for disease}}$$

(2) To account for the variation in follow-up interval, the denominator is **person-time** (i.e., the number of disease-free years contributed by each individual in the study population) rather than the total number of individuals at risk. Person-time may be expressed using various units of time (person-days, person-months, person-years).

(3) A **disadvantage** of using person-time as the denominator for ID is that it lumps together disparate follow-up times. For example, ten individuals followed for ten years and 100 individuals followed for one year would both contribute 100 person-years to the denominator.

(a) If short-term patients differ systematically from long-term patients, bias may result.

(b) For example, if the average length of time between exposure and the occurrence of a disease is four years, a value for ID based on 100 people followed for one year would underestimate the true rate at which new cases of the disease appear.

c. **Example.*** Figure 7-3 illustrates an epidemiological study in which a dynamic cohort of 12 subjects is followed for 5.5 years. The study is initiated with three subjects and three new subjects enter the study at the beginning of years 1, 2, and 3. During the study period, five individuals (subjects 1, 2, 6, 9, and 11) develop the disease. Of the seven individuals who do not, three drop out of the study (subjects 7, 8, and 12), two die of other causes (subjects 3 and 4), and two are alive and healthy at the end of the observation period (subjects 5 and 10).

(1) The person-time at risk is the sum of the individual follow-up times recorded along the right axis (see Figure 7-3):

$$\text{Person-time at risk} = 2.5 + 3.5 + 1.5 + \cdots + 1.5 = 26 \text{ person-years}$$

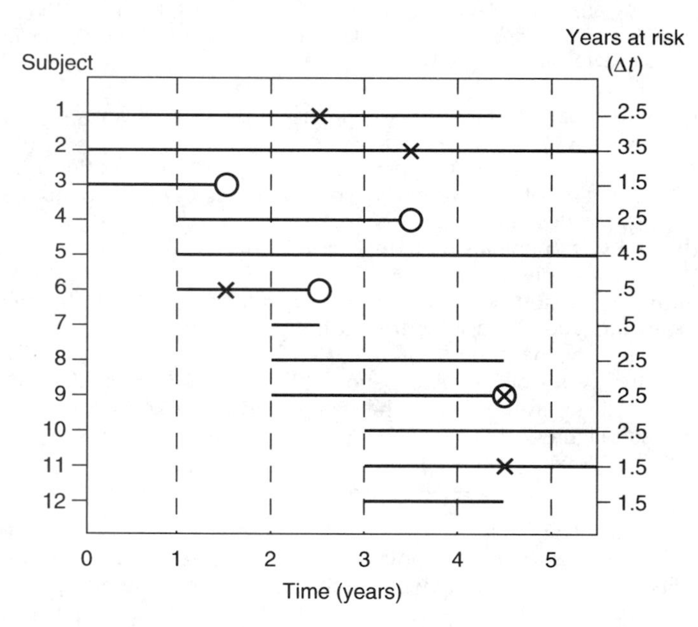

Figure 7-3. Calculation of person-years of follow-up in a hypothetical epidemiological study. A dynamic cohort of 12 initially healthy subjects is followed for 5.5 years. X = first occurrence of the disease; O = death. The number of disease-free person-years contributed by each subject (*years at risk*, Δt) is tabulated in the last column. The total number of person-years contributed by the entire cohort [denominator in the calculation of incidence density (ID)] is the sum of the 12 individual person-year contributions. (Reprinted from Kleinbaum DG, Kupper LL, Morgenstern H: *Epidemiologic Research: Principles and Quantitative Methods.* Belmont, CA, Lifetime Learning Systems, 1982, pp 101–102.)

---

*This example is adapted from Kleinbaum DG, Kupper LL, Morgenstern H: *Epidemiological Research: Principles and Quantitative Methods.* Belmont, CA, Lifetime Learning Systems, 1982, pp 101–102.

(2) Thus,

$$\text{ID} = \frac{5 \text{ cases}}{26 \text{ person-years}}$$
$$= 192 \text{ cases}/1000 \text{ person-years}$$

In other words, new cases of the disease appear at the rate of 192 cases/1000 people/year.

3. Relationship of CI to ID

a. If the observation period is short or the rate of disease occurrence is low, CI is approximately equal to ID multiplied by the length of the observation period (Δt): $CI \approx ID \times \Delta t$.

b. For rare diseases, the one-year CI is approximately equal to the ID.

C. Relationship between incidence and prevalence

1. Expressing prevalence in terms of incidence

a. The prevalence of a disease depends upon its incidence rate and its duration. Assuming that the values for prevalence, incidence rate, and duration are stable over time, this relationship can be written

$$\text{prevalence} \approx \text{incidence} \times \text{average duration of disease}$$

(1) A **high prevalence** may be due to a high incidence, a protracted duration, or both.

(a) Successful treatment of an illness, which prolongs life without effecting a cure, may paradoxically increase its prevalence.

(b) For example, the availability of increasingly effective chemotherapeutic regimens for many types of cancer and the use of insulin in the treatment of diabetes have increased the prevalence of these diseases.

(2) A **low prevalence** may be due to a low incidence, a short duration (i.e., rapid recovery or death), or both.

b. Examples

(1) Figure 7-4 illustrates the occurrence of Crohn's disease, a chronic inflammatory disease of the gastrointestinal tract, for a group of 100 patients in a hypothetical

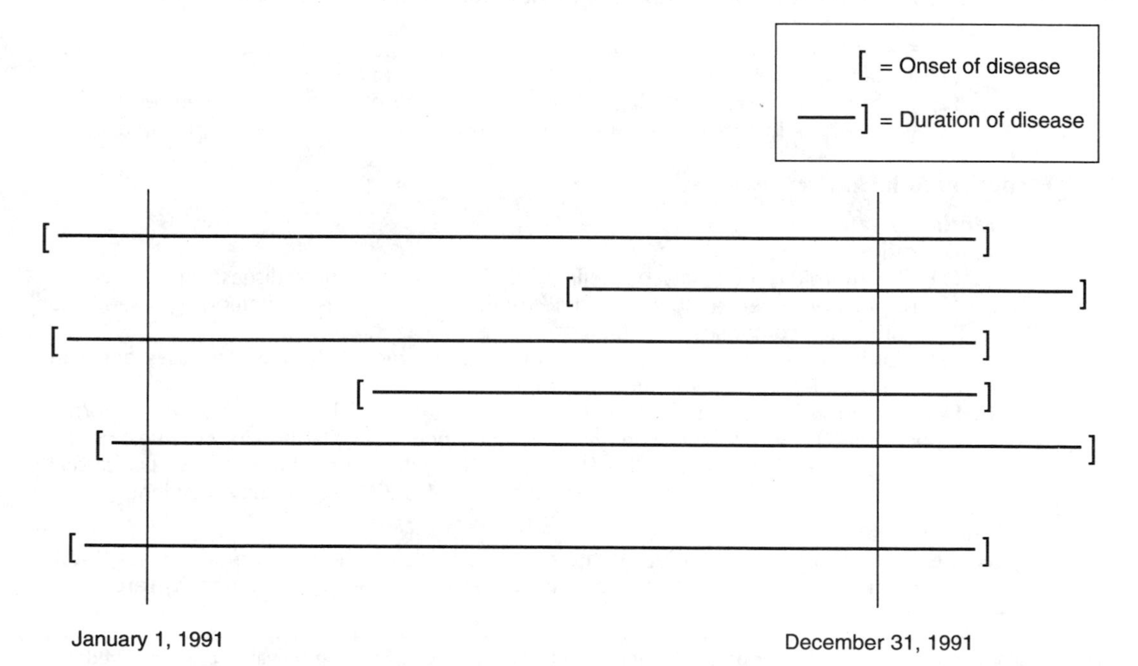

Figure 7-4. Cases of Crohn's disease in a 100-patient practice. Only two new cases occur during the period January 1, 1991, to December 31, 1991. Thus, for this chronic disorder, prevalence is high relative to incidence.

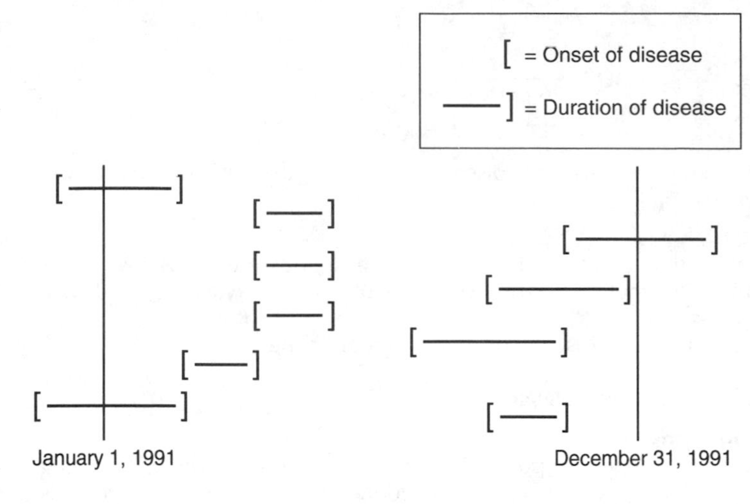

Figure 7-5. Cases of untreated acute myelocytic leukemia in a population of 100 patients. Eight new cases occur during the period January 1, 1991, to December 31, 1991; all but one of these cases are fatal within the observation period. Thus, for this rapidly fatal disease, prevalence is low relative to incidence.

internal medicine practice over the period January 1, 1989, to December 31, 1989. Although only a small number of new cases of Crohn's disease develop during the observation period (CI = 2/96 = .02), the long duration of the disease causes the total number of cases to accumulate. As a result, the prevalence of Crohn's disease is large relative to the incidence (prevalence on December 31, 1989 = 6/100 = .06).

(2) Untreated, acute myelocytic leukemia is rapidly fatal (Figure 7-5). In this instance, where the average duration of the disease is short, prevalence is small relative to incidence.*

2. Variations in incidence and prevalence

a. Because incidence depends on the occurrence of new cases of a disease, a **decrease in incidence** may be due to:

(1) Enhanced resistance to the disease
(2) A change in disease etiology
(3) An effective prevention program that reduces exposure to a known risk factor for the disease

b. A **decrease in prevalence** may be due to:

(1) A decrease in incidence
(2) A shorter duration of the disease due to either improved treatment methods leading to more rapid recovery or an increase in virulence leading to more rapid death

D. Specialized incidence measures

1. Morbidity rate

a. Definitions

(1) The morbidity rate may be defined as the incidence of a disease in a particular population over a specified time period. When this definition is used, the numerator **comprises both fatal and nonfatal cases.**

(2) Alternatively, it may be defined as the rate at which **only nonfatal cases** occur in the population over the time period.

b. Example. In 1978, 130 myocardial infarctions (16 fatal and 114 nonfatal) were reported among the 17,902 white men living in Florence and Darlington counties, South Carolina, (Table 7-1). Hence, the 1978 morbidity rate of nonfatal cases of myocardial infarction is 114/17,902, or 637 cases for every 100,000 white male residents.

2. Mortality rate

a. General definition. The mortality rate enumerates deaths, either from all causes (**total,** or **all-cause, mortality rate**) or from a single disease (**cause-specific mortality rate**), over

---

*Theoretically, prevalence can never be smaller than incidence. However, because prevalence is measured at a single point in time, some cases are necessarily excluded from the calculation when the average duration of a disease is short.

Table 7-1. Population Sizes and Occurrence of Myocardial Infarction (Fatal and Nonfatal) among Residents of Florence and Darlington Counties, South Carolina, 1978

Population Subgroup	Size of Population	Number of Myocardial Infarctions	Deaths Due to Myocardial Infarction
White men	17,902	130	16
White women	20,142	35	7
Black men	8,832	17	3
Black women	11,253	11	1
Total	58,129	193	27

Adapted from Keil JE, et al.: Acute myocardial infarction: period prevalence, case-fatality, and comparison of black and white cases in urban and rural South Carolina. *Am Heart J* 109:776–784, 1985.

a specified time period. Mortality rates may be calculated for the entire population (**crude mortality rate**) or within specific subpopulations (e.g., **age-specific, sex-specific, race-specific mortality rates**).

b. **Cause-specific mortality rate** refers to the rate at which deaths occur from a specific disease during the observation period. For example, the 1978 cause-specific mortality rate for myocardial infarction in Florence and Darlington counties, South Carolina (see Table 7-1), is 27/58,129, or 46 deaths for every 100,000 residents.

c. **Crude mortality rate** is the death rate within an entire population, as opposed to category-specific, age-specific, or age-adjusted rates. Crude mortality rates may be cause-specific or all-cause rates. For example, the 1978 crude death rate from myocardial infarction in Florence and Darlington counties (see Table 7-1) is 27/58,129, or 46 deaths for every 100,000 residents.

d. **Category-specific mortality rate** refers to the rate at which deaths occur in particular subgroups of the population during the observation period. For example, the 1978 race-sex-specific mortality rate for white men residing in Florence and Darlington counties (see Table 7-1) is 16/17,902, or 89 deaths for every 100,000 members of the population.

e. **Age-specific mortality rate** is the rate at which deaths occur within a particular age group (i.e., the numerator is the number of deaths in that age group). For example, the age-specific mortality rate from coronary heart disease for white males between the ages of 55 and 64 (Table 7-2) is 39/129, or 302 deaths for every 1000 members of the population.

f. **Age-adjusted mortality rate** is a statistically computed rate that adjusts death rates across age-specific strata to a common reference age distribution (**standard population**). Because age is an important influence on mortality, age-adjusted rates are useful for comparing mortality rates in populations having different age distributions. The calculation of age-adjusted rates is discussed in more detail in Chapter 8 IV.

Table 7-2. Age-Related Deaths from Coronary Heart Disease, among White Men Participating in the Charleston Heart Study, 1960–1988

Age Range	Size of Population	Deaths from Coronary Heart Disease (CHD)
35–44	243	28
45–54	208	34
55–64	129	39
65–74	55	25
≥75	18	7

Keil JE, principal investigator: Unpublished data, Charleston Heart Study, Medical University of South Carolina.

3. **Case-fatality rate**
   a. **Definition.** The case-fatality rate is the rate at which deaths occur from a particular disease among individuals with the disease (i.e., the proportion of people with the disease who die from it).
   b. **Example.** In 1978, 130 white men living in Florence and Darlington counties suffered a myocardial infarction; 16 of these individuals died as a result. Thus, the 1978 case-fatality rate for myocardial infarction among white men in these counties is 16/130, or 12.3%.

4. **Attack rate**
   a. **Definition.** When the duration of a disease is short (e.g., an acute infectious disease such as measles) and the observation period covers an entire epidemic, the CI of the disease is called the attack rate. An attack rate can be calculated for individuals exposed to a suspected risk factor (see Ch 8 II A) and compared to the rate for unexposed individuals to gain insight into the etiology of acute outbreaks of a disease.
   b. **Example.** Among a group of 140 medical students attending an end-of-school picnic, 90 contracted food poisoning. The attack rate is therefore 90/140, or 64%. By comparing attack rates among students who ate particular foods, it may be possible to pinpoint the cause of the outbreak.

II. EVALUATING FREQUENCY MEASURES

A. **Defining disease occurrences.** The calculation of both prevalence and incidence requires a reliable estimate of the number of individuals with the disease in question. Hence, the method used to define cases affects the accuracy of frequency measures.

1. **Identifying cases by clinical evaluation.** Incidence and prevalence values are influenced by the performance characteristics of the diagnostic procedure and by the rigor of the diagnostic criteria used to label patients as diseased.
   a. The use of a diagnostic test with a high false positive rate (poor specificity) will overestimate incidence and prevalence.
   b. The inclusion of only "classic" cases of the disease, as defined by strict diagnostic criteria, will underestimate incidence and prevalence. In contrast, the inclusion of "probable" cases, according to more lenient diagnostic criteria, may overestimate disease frequency.

2. **Identifying cases by reviewing health records.** The number of cases may be tabulated from existing health records. Estimates of disease frequency based on these data are valid only to the extent that these records are both complete and accurate.

3. **Identifying cases from personal interviews.** Information on disease occurrence obtained by personal interview or from questionnaires may be altered by biased recall. Prevalences computed from such survey data may differ dramatically from prevalences based on data obtained by clinical evaluation (Figure 7-6).

4. **Counting recurrences.** The calculation of incidence may be complicated by recurrences.
   a. **Choice of measures.** The event enumerated in the numerator may be either "the number of individuals developing the disease at least once" or "the number of times the disease occurs."
   b. **Example.** Patients with herpes simplex often experience recurrent outbreaks. In determining the incidence of herpes simplex infections over the course of one year, the event to be counted may be "the number of individuals who develop a herpes simplex eruption" during the observation period or "the number of herpes simplex eruptions," since some patients will develop multiple outbreaks.
      (1) The choice of "number of persons" as the event enumerated in the numerator will provide an estimate of the **probability that any given individual at risk will develop at least one** herpes simplex eruption during the year.
      (2) The choice of "number of eruptions" as the event enumerated will provide an estimate of the **relative frequency of outbreaks** during the observation period.

B. **Defining time of onset.** The calculation of incidence requires an accurate estimate of the time of onset of the disease in question.

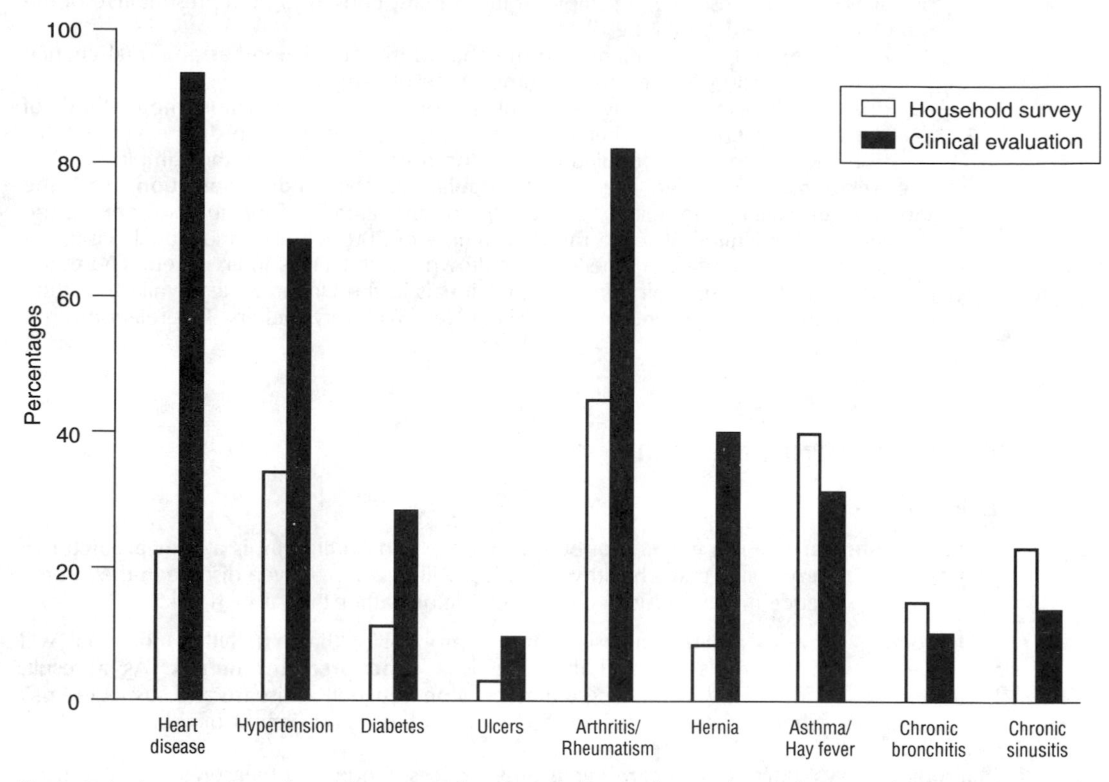

Figure 7-6. A comparison of prevalence rates for selected chronic diseases based on survey data (*open bars*) and rates based on clinical evaluation (*solid bars*). (Reprinted from Sanders BS: Have morbidity surveys been oversold? *Am J Public Health* 52:1648–1659, 1962.)

1. The time of onset is clearly defined in some diseases, for example, acute myocardial infarction, influenza, and appendicitis.
2. In other disorders, such as depression, cancer, and alcoholism, the precise time of onset is difficult, if not impossible, to pinpoint. In these cases, the time of onset may be defined as the date when a definitive diagnosis was reached, the date when symptoms were first noted, or (in the case of most psychiatric disorders) the date of the initial clinic visit.

C. Defining populations

1. Exclusion of risk-free individuals

- **a.** The calculation of the prevalence or incidence of a disease should exclude from the denominator members of the population who are, by definition, not susceptible to the disease. For example, estimates of the frequency of endometrial carcinoma should exclude women who have undergone hysterectomy and men.
- **b.** The calculation of the incidence of an infectious disease, such as measles or chicken pox, should exclude individuals who have been immunized or who have previously had the disease.
- **c.** In practice, the identification of individuals who are no longer at risk for a disease may be difficult. However, if the number of these individuals is small relative to the total population size, failure to totally exclude them will have a minimal impact on the rate determination.

2. Target versus study populations

- **a.** Practical considerations often limit the scope of an epidemiological study. For example, prevalence rates may be computed for the catchment area of a large hospital, rather than for the general population. **The results obtained in the study population are not necessarily generalizable to the larger target population** (see Ch 8 II C 4; Ch 9 III).
- **b. Sampling.** It is often not feasible or even desirable to observe the entire study

population. In this instance, a sample of individuals, chosen to be representative of the population of interest, may be observed.

(1) A **random sample,** in which each member of the population has an equal chance of being selected, is the most common type of sample.

(2) Provided the sample is truly representative of the study population, the methods of statistical inference (see Chapter 12) enable the investigator to draw valid conclusions about the study population from the results obtained in the sample.

c. **The relationship between the target population, the study population, and the sample—an example.** In a study addressing the prevalence of malnutrition among the homeless in the United States, a random sample of 200 homeless individuals visiting a particular randomly selected shelter in midtown Manhattan is interviewed. The target population in this case is all homeless individuals in the United States, while the study population is all homeless individuals visiting New York City shelters. This relationship is illustrated in Figure 7-7.

III. USES OF FREQUENCY MEASURES

A. Estimating risk

1. Because **prevalence** is a function of both incidence and duration, it **is a poor predictor of risk** (i.e., the probability that a healthy individual will develop a given disease in the future). A high prevalence may indicate a chronic condition rather than a high risk.

2. **Incidence** provides a direct measure of the probability that a healthy individual will develop a particular disease and therefore **is a good predictor of risk**. As a result, incidence is widely used in studying the association between a disease and suspected risk factors. (See Chapter 8 for a detailed discussion of the measurement of risk.)

B. Planning and evaluating health care needs and services. Frequency measures are often used to predict health care needs (particularly for chronic diseases) and to evaluate public health programs.

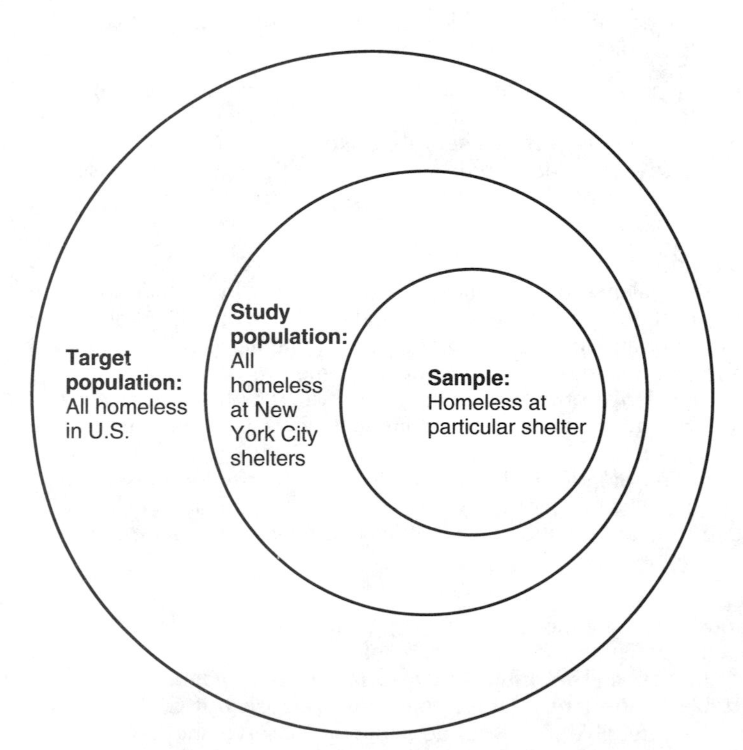

Figure 7-7. The relationship between the target population, the study population, and the sample in a study examining the prevalence of malnutrition among the homeless.

1. **Prevalence is useful in estimating the need for facilities and personnel.** It is also a measure of the "burden" of a particular disease to the health care system.
2. **Incidence provides information on the effectiveness of disease prevention or control programs.**

C. Evaluating the utility of diagnostic or therapeutic options

1. **The predictive value of a diagnostic test depends upon the prevalence** of the disease in question in a given clinical setting (see Ch 3 III A).
2. Thus, the choice of an optimal therapeutic strategy (See Ch 5) requires the following:
   - **a.** An estimate of disease prevalence
   - **b.** Information about the risk (incidence) of the various options under consideration
   - **c.** Knowledge of the posterior probability of disease acquired through diagnostic testing

BIBLIOGRAPHY

Fletcher RH, Fletcher SW, Wagner EH: *Clinical Epidemiology: The Essentials.* Baltimore, Williams and Wilkins, 1982, p 85.

Gruenberg EM: The failures of success. *Milbank Memorial Fund Quarterly* 55:3, 1977.

Keil JE, Weinrich M: *Compendium of Epidemiological Methods.* Student Aid, University of South Carolina School of Public Health and Medical University of South Carolina, 1980.

Kleinbaum DG, Kupper LL, Morgenstern H: *Epidemiological Research: Principles and Quantitative Methods.* Belmont, CA, Lifetime Learning Systems, 1982.

Mausner JS, Kramer S: *Mausner and Bahn's Epidemiology: An Introductory Text,* 2nd ed. Philadelphia, WB Saunders, 1985, p 45.

Sanders BS: Have morbidity surveys been oversold? *Am J Public Health* 52:1648–1659, 1962.

PROBLEMS

7-1. An attack rate is a specialized incidence measure referring to the proportion of individuals in a population who develop a particular disease, among those exposed (or not exposed) to a putative risk factor (see I D 4). The table below summarizes data on a food poisoning outbreak among 200 medical students attending a back-to-school party. In problems 7-1a through 7-1d, use the information in this table to answer the questions about attack rates for students who ate or did not eat various foods at the party.

Cases of Food Poisoning among 200 Medical Students Attending a Back-to-School Party

	Ate Food			Did Not Eat Food		
	Ill	**Not Ill**	**Total**	**Ill**	**Not Ill**	**Total**
Barbecue	90	30	120	20	60	80
Coleslaw	67	33	100	43	57	100

a. What is the attack rate for those students who ate barbecue?

b. What is the attack rate for those who did not eat barbecue?

c. What is the attack rate for those students who ate coleslaw?

d. What is the attack rate for those who did not eat coleslaw?

7-2. A comparison of the attack rates for those students who ate barbecue or coleslaw does not conclusively identify either as the contaminated food. In this instance, the *difference* in attack rates between those who ate and those who did not eat these two foods provides a more informative measure of risk. Calculate and compare the differences in attack rates for barbecue and for coleslaw.

7-3. The table below reclassifies the data on the food poisoning outbreak in terms of the numbers of students who ate just one or both of the suspected foods.

Cases of Food Poisoning among Students Eating Food Combinations

	Ate Barbecue			Did Not Eat Barbecue		
	Ill	**Not Ill**	**Total**	**Ill**	**Not Ill**	**Total**
Ate Coleslaw	65	25	90	2	8	10
Did Not Eat Coleslaw	25	5	30	18	52	70

a. What is the attack rate for those students who ate both coleslaw and barbecue?

b. What is the attack rate for those students who ate coleslaw, but not barbecue?

c. What is the attack rate for those students who ate barbecue but not coleslaw?

d. What is the attack rate for those students who ate neither food?

e. Which food is most likely to have been contaminated?

7-4. What is the ratio of the attack rates calculated in Problems 7-1a and 7-1b? Interpret its meaning.

7-5. A random sample of 1000 black men between the ages of 40 and 60 was selected from a particular community in 1960, as part of an epidemiological study of the frequency of cardiovascular disease. Physical examination revealed 100 cases of coronary heart disease (CHD) among these men. The remaining 900 disease-free men were followed until 1985, when the data depicted on the facing page were obtained.

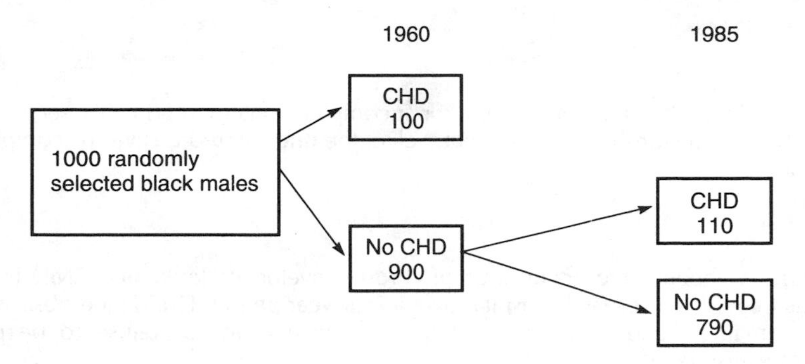

a. What was the prevalence of CHD at the beginning of the study (i.e., in 1960)?

b. What is the CI of CHD in the sample between 1960 and 1985?

7-6. The figure below depicts the occurrence of sinusitis in a population of 10 patients belonging to a health maintenance organization between January 1, 1989, and December 31, 1989.

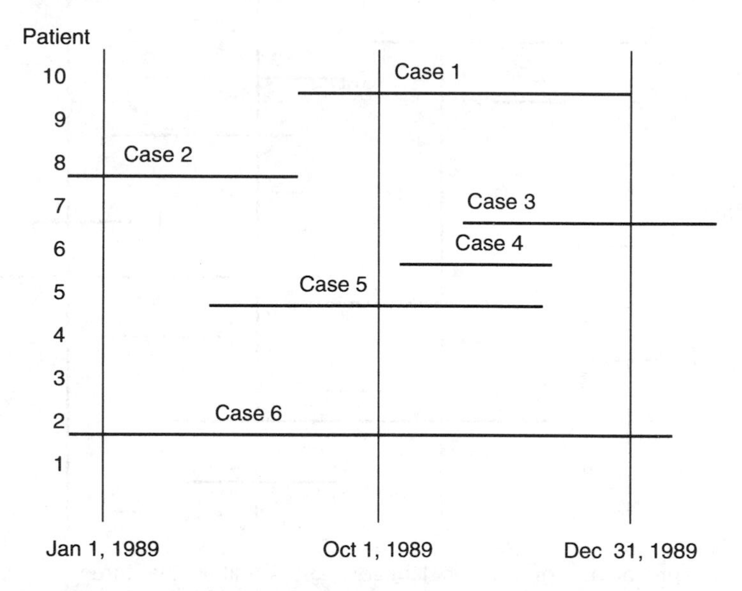

a. What is the prevalence of sinusitis in this population on January 1, 1989?

b. What is the CI of sinusitis during 1989?

7-7. For a given year, the prevalence of Huntington's disease in a certain geographic area is .003 (300 per 100,000) and the incidence is .0003 (30 per 100,000). What is the average duration (in years) of this disorder?

7-8. The following table summarizes data on the incidence and prevalence of chronic obstructive pulmonary disease (COPD) among patients over and under 65 years of age in a particular internal medicine practice. What is the most likely explanation for the observed changes in the incidence and prevalence of COPD after age 65?

Occurrence of COPD among Patients in Two Age Groups

	Under Age 65	**Over Age 65**
Incidence	.003	.01
Prevalence	.05	.01

Solutions on p. 372.

STUDY QUESTIONS

Directions: Each of the numbered items or incomplete statements in this section is followed by answers or by completions of the statement. Select the **one** lettered answer or completion that is **best** in each case.

Questions 1–4

The figure below depicts the occurrence of chronic myelocytic leukemia (CML) in a population of 100 patients treated at a research hospital over a four-year period. During the observation period, the population remains stable (i.e., no members die, move away, or refuse to be examined at the beginning of every year).

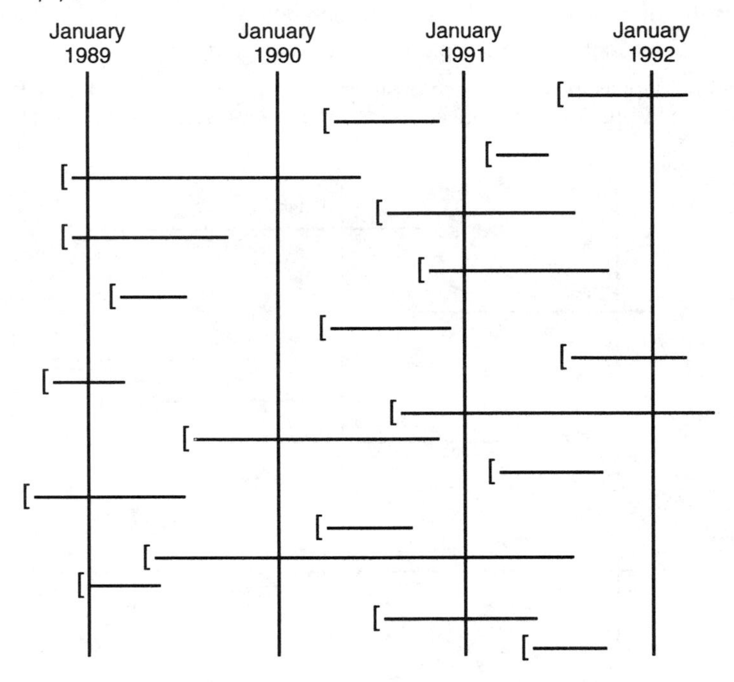

1. What is the prevalence of CML in January 1989?

(A) 5/100
(B) 8/100
(C) 5/95
(D) 8/95
(E) 5/92

2. What is the prevalence of CML in January 1991?

(A) 5/100
(B) 9/100
(C) 5/95
(D) 5/80
(E) 9/80

3. What is the three-year CI of CML for the period 1989–1991?

(A) 20/100
(B) 15/100
(C) 20/95
(D) 15/95
(E) None of the above

4. What is the CI of CML for the year 1991?

(A) 5/100
(B) 5/85
(C) 5/95
(D) 10/85
(E) None of the above

1-A 4-B
2-A
3-D

Questions 5–8

Santa Elena is a seacoast community with a population of 99,000. Its residents can be divided into three age ranges: 25–44, 45–64, and 65 and older—each comprising one third of the population. In 1990, 100 cases of hepatitis A occurred in Santa Elena and were traced to the consumption of contaminated oysters. Of these 100 cases, 20 between the ages of 25 and 44, 10 between the ages of 45 and 64, and five over the age of 64 ultimately proved fatal. Prior to 1990, Santa Elena had never reported a case of hepatitis A.

5. What is the 1990 crude mortality rate for hepatitis A in Santa Elena?

(A) 350 per 1000 residents
(B) 1.01 per 1000 residents
(C) 3.54 per 10,000 residents
(D) 1.06 per 1000 residents
(E) Cannot be determined from the data

6. What was the incidence of hepatitis A in 1990 (assuming that no cases of the disease occurred in Santa Elena prior to 1990)?

(A) .00101
(B) .00035
(C) .35
(D) .04
(E) Cannot be determined from the data

7. What is the age-specific mortality rate for residents over 64 years of age?

(A) 3.03 per 10,000 residents
(B) 4.55 per 10,000 residents
(C) 6.06 per 10,000 residents
(D) 3.54 per 10,000 residents
(E) 1.52 per 10,000 residents

8. What is the case-fatality rate for hepatitis A in Santa Elena?

(A) 3.03 per 10,000 residents
(B) 3.54 per 10,000 residents
(C) 350 per 1000 residents
(D) 1.01 per 1000 residents
(E) Cannot be determined from the data

Questions 9–10

The figure below depicts the occurrence of breast cancer in a four-year follow-up study of 10 women. All relevant events (i.e., disease, death, or withdrawal from the study) occur at the midpoints of time intervals.

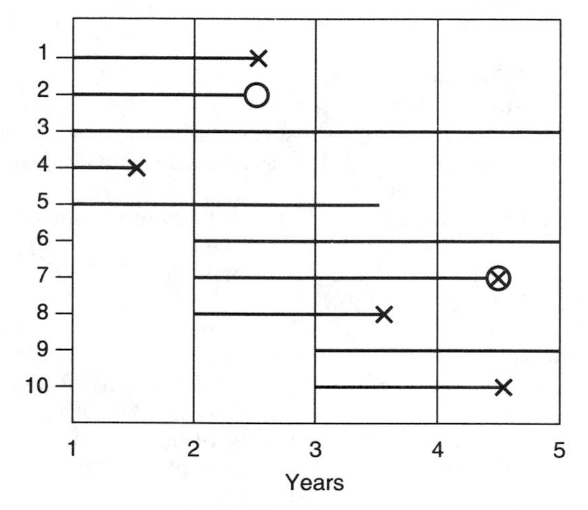

9. The total person-time at risk for this cohort is

(A) 27.5 person-years
(B) 20.5 person-years
(C) 13.0 person-years
(D) 10.0 person-years
(E) Not discernible from the data

10. The incidence density (ID) of breast cancer in this study is

(A) 195 per 1000 person-years
(B) 500 per 1000 person-years
(C) 293 per 1000 person-years
(D) .5 per 1000 person-years
(E) 244 per 1000 person-years

5-C 8-C
6-A 9-B
7-E 10-E

ANSWERS AND EXPLANATIONS

1–4. The answers are: 1-A *[I A]*, **2-A** *[I A]*, **3-D** *[I B 1]*, **4-B** *[I B 1]*
Prevalence is the proportion of the population who have chronic myelocytic leukemia (CML) at the specified point in time. In January 1989, five cases of CML already exist in this population of 100 patients. The prevalence, therefore, is 5/100.

In January 1991, there are five existing cases of CML. Because the population has remained stable since the study began in 1989, the prevalence is still 5/100.

The three-year cumulative incidence (CI) is the proportion of new cases of CML among patients who are initially disease-free. When the study began in 1989, five patients already had CML and 95 were disease-free. Between 1989 and 1991, 15 of the 95 patients at risk developed CML. The three-year CI (i.e., the risk that an initially healthy patient will develop CML during the three-year observation period) is therefore 15/95.

Fifteen cases of CML existed prior to January 1991; therefore, at the beginning of the year, 85 patients were at risk. During 1991, five new cases of CML occurred in the population. Thus, the cumulative incidence of CML for 1991 is 5/85.

5–8. The answers are: 5-C *[I D 2 c]*, **6-A** *[I B 1]*, **7-E** *[I D 2 e]*, **8-C** *[I D 3]*
Because Santa Elena did not report a case of hepatitis A prior to 1990, all 99,000 residents were at risk for dying from the disorder at the beginning of the follow-up period. During 1990, 35 deaths from hepatitis A occurred. The 1990 crude mortality rate is therefore 35/99,000 = .000354, or 3.54 deaths per 10,000 residents.

The population at risk for contracting hepatitis A at the beginning of 1990 (assuming no prior cases) was 99,000. During the one-year observation period, 100 individuals contracted hepatitis A. The cumulative incidence (CI) of hepatitis A in 1990 was therefore 100/99,000 = .00101 (i.e., 10.1 cases per 10,000 residents).

An age-specific mortality rate refers to the death rate within a specified age group. Five deaths from hepatitis A occurred in 1990 among the 33,000 residents over the age of 64. The age-specific mortality rate for this group is therefore 5/33,000 = .000152, or 1.52 per 10,000 residents.

The case-fatality rate is the proportion of cases that ultimately result in death. Thirty-five deaths occurred among the 100 cases of hepatitis A in 1990. Thus, the case-fatality rate is 35/100 = .35, or 350 deaths for every 1000 cases of hepatitis A.

9–10. The answers are: 9-B *[I B 2 c (1)]*, **10-E** *[I B 2]*
The person-time at risk for this cohort is the sum of the individual follow-up times, that is,

$$\begin{aligned}\text{person-time} &= 1.5 + 1.5 + 4 + .5 + 2.5 + 3 + 2.5 + 1.5 + 2 + 1.5 \\ &= 20.5 \text{ person-years}\end{aligned}$$

The incidence density (ID) is the number of new cases of breast cancer that develop over the four-year observation period divided by the person-time at risk (calculated above), or

$$\begin{aligned}\text{ID} &= 5/20.5 \\ &= .244, \text{ or } 244 \text{ per } 1000 \text{ person-years}\end{aligned}$$

8 Risk and Causality

I. ASSOCIATIONS IN CLINICAL MEDICINE. In addition to determining the **frequency of disease occurrence,** clinical epidemiology attempts to define **associations between disease and predisposing or causal factors**. While initial research on a suspected association tries to identify conditions or behaviors that increase the risk of developing a particular disease, ultimately, such studies aim to discover a **cause-and-effect relationship** that will result in effective treatment and prevention strategies.

II. RISK

A. Risk factors. A condition, physical characteristic, or behavior that increases the probability (i.e., risk) that a currently healthy individual will develop a particular disease is termed a risk factor for that disease.

1. Types of risk factors

a. Environmental risk factors are found in the physical environment (e.g., infectious organisms, pollutants, toxins, and drugs).

b. Behavioral, or **life-habit, risk factors** include such behaviors as smoking and failing to observe occupational safety precautions.

c. Social risk factors include divorce, death of a family member, and job loss.

d. Genetic risk factors are inherited factors (e.g., familial hypercholesteremia, which increases the risk of coronary artery disease).

2. Exposure. An individual who has contact with or who manifests the risk factor prior to becoming ill is said to have been exposed to the factor.

a. Exposure may occur at a single point in time. For example, a traffic accident may expose a nearby community to a toxic chemical spill.

b. More commonly, exposure is chronic (e.g., drug abuse, hypertension, sedentary lifestyle). Measures of chronic exposure include current dose, total cumulative dose, years of exposure, and years since first exposure.

3. Relationship of risk factors to disease

a. A risk factor may be a **causal factor** of the disease in question or merely a **marker** for the increased probability of disease.

b. For example, while poor prenatal care and drug use constitute causal factors for neonatal mortality, socioeconomic status would be considered a marker for neonatal mortality.

B. Risk assessment. A number of research designs are used to evaluate the association between a disease and a suspected risk factor.

1. Experimental versus observational studies

a. In **experimental studies,** the investigator controls the assignment of study subjects to experimental and control groups and actively manipulates one variable (called **the independent variable**) while observing the resultant changes in another (**the dependent variable**).

(1) Example. A clinical investigator who wishes to study the effect of caffeine on systolic blood pressure conducts an experimental study in genetically inbred laboratory mice. The mice are randomly assigned to three groups, and each day for six months, one group receives an inert placebo; another, a low dose of caffeine; and the third, a high dose of caffeine (the independent variable). Systolic blood pressure (the dependent variable) is measured daily in all of the animals during the study period.

(2) **Limitations.** Although experimental studies are more scientifically rigorous than observational studies, practical and ethical considerations limit their use in the study of risk. Chapter 9 provides a more detailed discussion of the experimental approach.

b. In **observational studies,** individuals "self-select" into comparison groups. The responses of the groups to differing levels of exposure to the risk factor are observed as they take place "in nature," that is, with no active intervention on the part of the investigator. Observational studies are often the only feasible way to evaluate disease–risk factor associations.

(1) Observational studies may be **descriptive** or **analytic**.

(a) **Descriptive studies** yield estimates of disease frequency. As such, they often call attention to potential disease–risk factor associations.

(b) **Analytic studies** seek to explain patterns in the frequency and distribution of a disease by defining risk factors and, ultimately, cause-and-effect relationships. Analytic studies may also be used to assess the efficacy of treatments or preventive interventions.

(2) **Example.** A second investigator conducts an observational study of the effects of caffeine consumption on blood pressure. Fifty patients from her practice are assigned to one of three groups—control, low-dose, or high-dose—based on personal reports of daily caffeine consumption. Over the course of the next six months, she measures systolic blood pressure monthly in each of the subjects.

(3) **Limitations.** Because the experimenter does not control the assignment of subjects to comparison groups, the experimental environment, or the levels of exposure, observational studies are **far more subject to bias** than experimental studies.

(a) If study groups differ systematically with respect to conditions other than those being monitored, such **confounding variables** (see II C 1) **may bias the results** and account for any observed association between the risk factor and the disease.

(b) **A fundamental challenge in observational studies is the recognition and control of potential sources of bias.** This issue is discussed more fully in II C and II D.

2. The four study designs used in observational studies of risk

a. In a **prospective cohort study** (also known as a **prospective study, cohort study, incidence study, follow-up study,** or **longitudinal study**), an initially healthy **cohort** of individuals, exposed to different levels of a suspected risk factor, is followed **forward in time** to determine the subsequent incidence of the disease in question in each of the comparison groups. This study design is illustrated schematically in Figure 8-1.

(1) The **response,** or **outcome variable** may be any health-related outcome, including remission rate, death rate, or five-year survival rate, as well as incidence.

(2) **Selection of subjects.** Subjects for a prospective cohort study may be either selected from a single population and subsequently divided into those exposed to the suspected risk factor and those not exposed or selected from separate populations.

(a) For example, exposed subjects may be selected from a population with a known exposure to the putative risk factor (such as shipyard workers with a

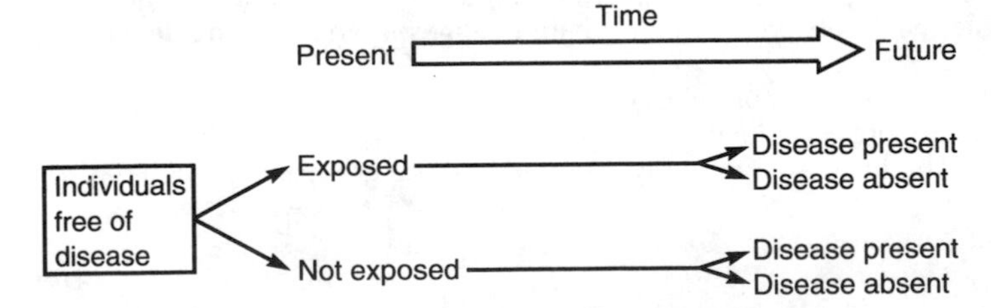

Figure 8-1. Schematic representation of the prospective cohort study design.

Table 8-1. Results of a Hypothetical Prospective Cohort Study Involving a Dichotomous Outcome Variable

	Disease Present (D+)	Disease Absent (D−)	Totals
Exposed (E+)	a	b	$a + b$
Nonexposed (E−)	c	d	$c + d$
Totals	$a + c$	$b + d$	

proven exposure to asbestos). A comparable control group of subjects is chosen from a population with no known exposure to the risk factor.

(b) When comparison groups are selected from different populations, internal validity (see II C) may be compromised if the exposed and nonexposed populations differ with respect to factors other than exposure.

(3) Evaluating outcomes—dichotomous variables

(a) Representation. Data from a prospective cohort study involving a dichotomous outcome variable (e.g., presence versus absence of myocardial infarction) [see II B 2 c (2) (b)] may be presented in a **2 × 2 table** (Table 8-1).

(i) When the comparison groups are selected from separate populations [see II B 2 a (2)], the number of individuals in the exposed and nonexposed categories (see Table 8-1, *row totals*) is **fixed,** that is, determined by the investigator.

(ii) In contrast, the number of individuals who ultimately develop the disease or remain healthy (see Table 8-1, *column totals*) is **random,** that is, determined by nature.

(iii) The incidence of the disease in the exposed group is compared to that in the nonexposed group using an appropriate **statistical test of significance** (see Chapters 13 and 14). A significantly **higher incidence** of disease in the group exposed to the putative risk factor—that is, a difference large enough that it is unlikely to have occurred by random chance—is taken as **evidence of an association** between the risk factor and the disease. Conversely, a significantly lower incidence of the disease in the exposed group suggests that the factor protects individuals from the disease.

(b) When the data are in the form of **counts,** the **strength of the association between the risk factor and the disease** may be expressed as the relative risk or as an odds ratio.*

(i) Relative risk compares the risk of developing the disease when exposed to the risk factor to that in the absence of the risk factor:

$$\text{relative risk} = \frac{\text{risk of disease if exposed to risk factor}}{\text{risk of disease if not exposed to risk factor}}$$

(ii) The **odds ratio** compares the odds† that exposed and nonexposed individuals will have the disease:

$$\text{odds ratio} = \frac{\text{odds that exposed individual will have disease}}{\text{odds that nonexposed individual will have disease}}$$

*In conditional probability notation, relative risk $= \dfrac{P(\text{D+}|\text{E+})}{P(\text{D+}|\text{E−})} = \dfrac{a/(a+b)}{c/(c+d)}$ and the odds ratio $= \dfrac{O_{\text{D+}|\text{E+}}}{O_{\text{D+}|\text{E−}}}$, where $O_{\text{D+}|\text{E+}} = \dfrac{P(\text{D+}|\text{E+})}{P(\text{D−}|\text{E+})} = \dfrac{a}{b}$ and $O_{\text{D+}|\text{E−}} = \dfrac{P(\text{D+}|\text{E−})}{P(\text{D−}|\text{E−})} = \dfrac{c}{d}$.

†The concept of odds is similar to that of probability. The odds of having a disease are equal to the probability of contracting the disease divided by the probability of not contracting the disease, that is,

$$\text{odds of disease} = P(\text{disease})/1 - P(\text{disease})$$
$$= P(\text{disease})/P(\text{no disease})$$

(iii) The calculation of measures of association and measures of impact (**attributable risk**), as well as statistical tests of significance based upon them, are discussed in Chapter 14.

(4) **Evaluating outcomes—continuous variables.** When both the exposure and the response variable are continuous [see II B 2 c (2) (b)] and data are expressed on an **interval/ratio scale,** measures of the strength of the association between exposure and the outcome are provided by correlation and regression analyses (see Ch 15).

(5) **Advantages**

(a) A prospective cohort study is the only type of observational study that yields a **direct estimate of risk,** that is, the probability that a healthy individual will develop the disease over a specified time period.

(b) Of the observational study designs, the prospective cohort study most closely **resembles an experimental study**. As a result, it is the study design least subject to bias.

(c) The **association** between the hypothesized risk factor and several diseases **can be examined simultaneously**.

(6) **Disadvantages**

(a) Incidence rates for most diseases are relatively low. To accurately determine if exposure to a potential risk factor alters the incidence of a rare disease, a large cohort must be observed over a prolonged time period. Such extensive follow-up is both time-consuming and expensive. In the case of potentially lethal risk factors, the protracted duration of a prospective cohort study may be unacceptable.

(b) Prospective cohort studies often suffer from the attrition of study subjects. Differential loss of subjects among comparison groups represents a potential source of bias.

(c) The exposure of individuals to a suspected health hazard may pose serious ethical questions.

b. **Historical cohort study** (also known as a **retrospective cohort study** or **nonconcurrent cohort study**). In this design, the cohort of exposed and nonexposed subjects is assembled from **past records** and followed **forward in time** to determine **present incidence** of the disease. As in the prospective cohort study, comparison groups are defined on the basis of exposure to the suspected risk factor. Figure 8-2 contrasts the prospective and historical cohort study designs.

(1) The **strength of the association** between the risk factor and the disease is determined in the same fashion as for the prospective cohort study [see II B 2 a (3)–(4)].

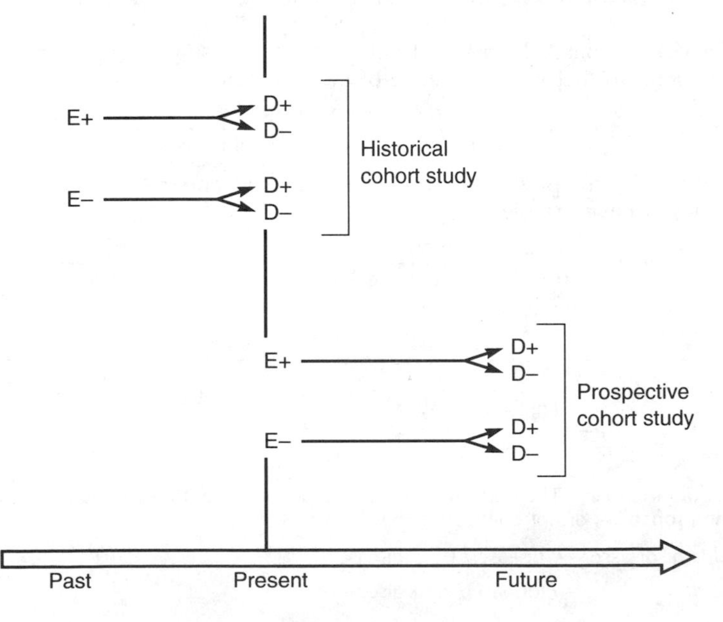

Figure 8-2. Comparison of the historical cohort and prospective cohort study designs. $E+$ = exposed; $E-$ = not exposed; $D+$ = disease present; $D-$ = disease absent.

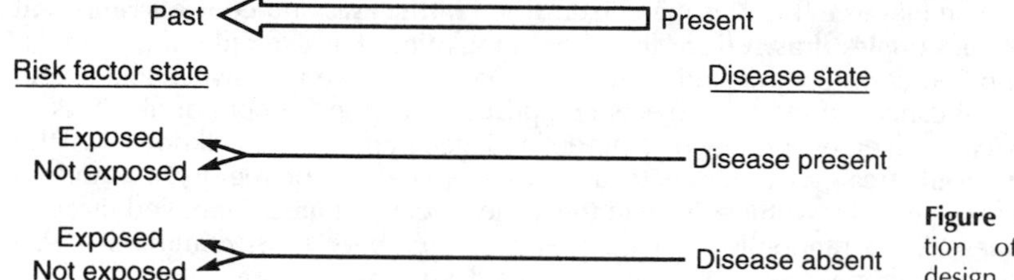

Figure 8-3. Schematic representation of the case–control study design.

(2) **Advantage.** Because it can be carried out with existing data, a historical cohort study can be completed faster and more economically than a prospective cohort study.

(3) **Disadvantage.** The historical cohort study design depends on existing medical records. If these records are incomplete or inaccurate, the conclusions of the study are necessarily compromised.

(4) **Example.** In 1987, Stevens and colleagues initiated a study to investigate the relationship between fat patterning (i.e., the distribution of adipose tissue) and the development of diabetes and coronary artery disease.* Exposure was based on anthropometric measurements, such as abdominal circumference, taken in 1963 on participants in the Charleston Heart Study. Subjects were followed forward to determine 1988 morbidity and mortality rates as a function of 1963 risk factor measurements.

c. In a **case–control study** (also known as a **retrospective,** or **case–referent, study**), subjects are classified as diseased (**cases**) or disease-free (**controls**), then traced **backward in time** to determine exposure to the hypothesized risk factor (Figure 8-3). Figure 8-4 contrasts the case–control and prospective cohort study designs.

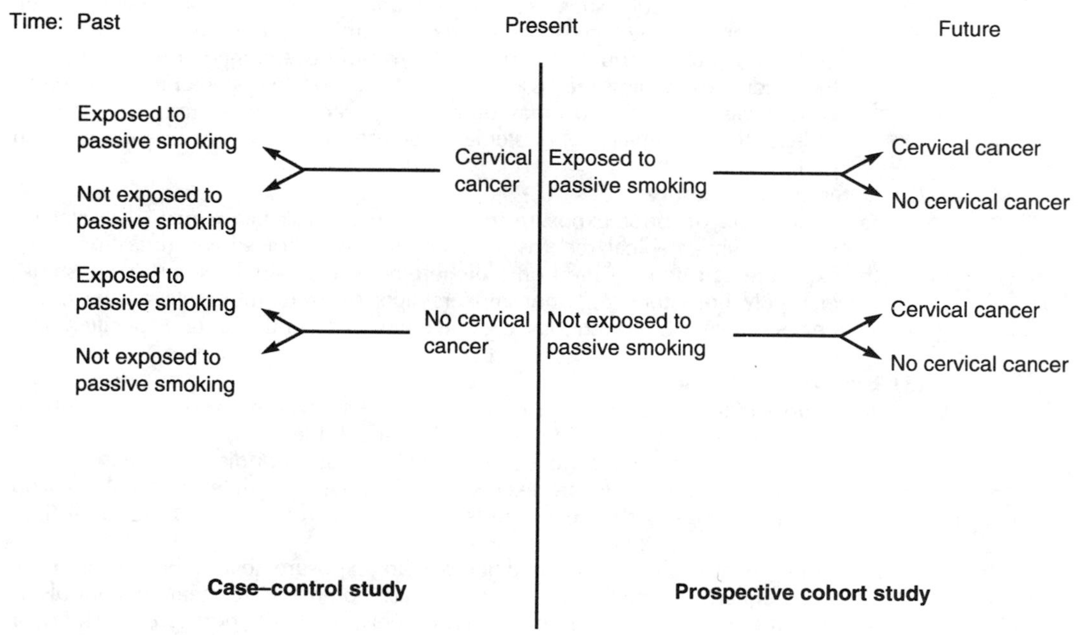

Figure 8-4. Comparison of the case–control and prospective cohort study designs for the evaluation of passive smoking as a risk factor for cervical cancer.

*Stevens J: Adiposity and fat patterning in black Americans. National Heart, Lung, and Blood Institute RO1 HL 42305-01, August 1, 1989–July 31, 1992.

(1) Selection of cases and controls

(a) Selection bias (see II C 2) is minimized when **both cases and controls represent random samples drawn from the same population**. For example, in a population-based case–control study of the association between passive smoking and cervical cancer, diseased subjects comprised a random sample of all cases of cervical cancer occurring in a predefined geographical area (known as the **catchment area**), as reported to a state-mandated tumor registry. A random sample of healthy women living in the same catchment area, identified through phone calls to randomly selected telephone numbers (most community residents had phones), served as controls (see Figure 8-4).

(b) Both **cases and controls should be selected by the same criteria,** especially if the cases are not representative of the general population. For example, in a hospital-based study of the association between passive smoking and cervical cancer, cases were defined as patients admitted for the treatment of cervical cancer to a large teaching hospital, while controls comprised patients admitted to the same hospital for the treatment of other gynecological disorders. Selecting the controls from the same hospitalized population as the cases minimized any confounding effect of hospitalization.

(c) Because of the difficulty of obtaining truly comparable diseased and disease-free comparison groups, case–control studies often employ **multiple control groups**.

(i) If similar results are obtained for all control groups, the likelihood that an observed association is real, rather than due to random chance or bias, is increased.

(ii) For example, in the hospital-based study of passive smoking and cervical cancer described in II B 2 c (1) (b), a population-based control group randomly selected from the community served by the hospital, as well as a control group comprised of healthy relatives, friends, and co-workers of cervical cancer patients, may be observed in addition to the hospitalized control group.

(d) Cases and controls may be **matched** with respect to characteristics related to the response variable, such as age or amount of alcohol consumed weekly (see II D 2). For example, age of first intercourse and number of sexual partners are both known to contribute to a woman's risk of developing cervical cancer. In the studies examining the association between passive smoking and cervical cancer, cases and controls may be matched with respect to these variables to reduce the possibility of systematic differences between the comparison groups.

(2) Determining exposure

(a) Information on prior exposure to the suspected risk factor may be obtained from existing medical records, personal interview, or survey (questionnaire).

(b) Exposure can be measured on a **dichotomous** (e.g., smokers versus nonsmokers), **polychotomous** (e.g., nonsmokers, light smokers, moderate smokers, and heavy smokers) or **continuous** (e.g., nicotine consumption over a specified time period) scale.

(3) Evaluating outcomes

(a) Representation. As with a prospective cohort study, data from a case–control study may be presented in a **2 × 2 table** (see Table 8-1).

(i) The numbers of individuals classified as diseased or disease-free (see Table 8-1, *column totals*) are fixed by the investigator, while the numbers who are exposed or nonexposed (see Table 8-1, *row totals*) are random (i.e., determined by nature).

(ii) A statistically significant difference in exposure (either higher or more frequent) among diseased individuals compared to healthy controls is taken as evidence of an **association** between the hypothesized risk factor and the disease.

(b) Measures of association

(i) Relative risk [see II B 2 a (3) (b) (i)] cannot be directly calculated from the results of a case–control study. Under certain conditions, the odds ratio represents an acceptable estimate of relative risk (see Ch 14).

(ii) When the exposure and outcome variables are dichotomous or polychotomous, the **chi-square statistical test** (see Ch 13) may be used to evaluate the significance of the association between exposure and outcome.

(iii) For continuous variables, **correlation** and **regression** techniques (see Ch 15) are appropriate statistical methods.

(4) Advantages

(a) The case–control study design is the most **practical for studying rare diseases.**

(b) Because case–control studies are conducted with existing data, they can be **completed faster and more economically** than prospective cohort studies. Case–control studies are particularly useful when exposure and disease are separated by a long period of time but the need to identify a cause-and-effect relationship is urgent (e.g., the relationship between other sexually-transmitted diseases and the risk of contracting AIDS).

(5) Disadvantages

(a) Case–control studies are more vulnerable to bias than prospective cohort studies.

(b) Case–control studies do not measure risk directly (see Ch 14 II A 4).

(c) Antecedent–consequence uncertainty. In a case–control design, the correct temporal relationship between the risk factor and the disease may remain ambiguous. For example, a case–control study of the association between depression and alcoholism reveals that individuals undergoing treatment for alcoholism at an outpatient psychiatric clinic are significantly more likely to have been treated for a depressive episode in the preceding five years than other members of the community. However, these results do not resolve the question of whether depression precedes alcoholism and plays a causal role in its development or whether alcohol abuse precipitates depression.

(d) Unlike a prospective cohort study, a case–control study can only examine a single disease.

d. Cross-sectional study (also known as a **prevalence study**). In this study design, individuals are **concurrently** classified as diseased or disease-free and exposed or nonexposed at a single point in time. Prevalence rates are compared between those exposed and those not exposed to the suspected risk factor.

(1) Evaluation of outcomes

(a) The **odds ratio** provides a measure of the strength of the association between dichotomous exposure and outcome variables.

(b) The **chi-square test** (see Ch 13) can be used to evaluate the statistical significance of the association between the suspected risk factor and the disease.

(2) Advantages

(a) The advantages of the case–control study design [see II B 2 c (4)] also apply to cross-sectional studies.

(b) By identifying disease-free individuals, a prevalence study provides the basis for a future prospective cohort study.

(c) A prevalence study may also be used to identify cases and controls for a case–control study.

(3) Disadvantages

(a) Cross-sectional studies suffer from the same limitations as case–control studies, particularly antecedent–consequence uncertainty [see II B 2 c (5)].

(b) Estimates of prevalence may be biased by the exclusion of cases in which death or recovery are rapid (see Ch 7 I C).

(c) Risk factors associated with prevalent cases of a disease may differ from those associated with incident cases.

C. Sources of confounding and bias in observational studies of risk assessment

1. Systematic differences between comparison groups pose a threat to **internal validity**—that is, they invalidate the conclusions of the study with regard to the specific population under investigation. These differences represent potential **extraneous,** or **confounding, variables,** which compete with the hypothesized risk factor as explanations for the observed response.

a. By virtue of their association with both the putative risk factor and the disease in question, confounding variables create an **artificial** relationship between the two.

b. Confounding variables may be introduced by the investigator, in the form of **selection bias** (see II C 2).

c. Alternatively, confounding may result when two or more variables "travel together." For example, because smokers tend to be heavy coffee drinkers and vice versa, smoking and coffee consumption can be regarded as "traveling together," and studies involving one of these variables may be confounded by the other.

d. **Examples**

(1) In a hypothetical prospective cohort study proposed by Michael and colleagues,* the incidence of cancer over a 10-year period is recorded for two random samples, one from a state in which gambling is legal (Nevada) and one from a neighboring state where it is not (Utah). A higher incidence of cancer in the legalized gambling sample may suggest that gambling is a risk factor for cancer. However, the residents of Utah, with its predominantly Mormon population, are more likely to abstain from the use of alcohol and tobacco, while people who participate in legalized gambling are frequently heavy drinkers and smokers. Because excessive use of both alcohol and tobacco are proven cancer risk factors, these variables confound the effect of gambling and represent alternative explanations for the apparent association between gambling and cancer incidence.

(2) In a case–control study of the association between hearing loss and exposure to loud music, patients treated at an outpatient ear, nose, and throat (ENT) clinic for hearing loss are recruited to serve as cases. Control subjects with normal hearing acuity are recruited from the general medicine clinic at the same hospital. All subjects are interviewed to determine the level and duration of their exposure to loud music over the preceding five years. In this example, age represents a potentially confounding variable. If the ENT patients are, in general, older than the controls, they are less likely to have been exposed to loud music. This disparity in age distribution between the cases and controls can mask a true association between loud music and subsequent hearing loss.

2. **Selection bias** occurs when an investigator inadvertently assigns subjects to comparison groups such that they differ with respect to extraneous factors.

a. Selection bias **in prospective cohort studies**

(1) **Examples**

(a) A prospective cohort study was conducted among low-income women to determine the effect of a prenatal nutritional counseling program on infant birth weight. Volunteers for the counseling program were recruited among pregnant women who qualified for financial assistance, while controls consisted of women who chose to forego the program. Such "volunteerism" raises the risk of selection bias, since volunteers for the nutritional counseling program may have been more health conscious than those declining to participate. Thus, additional steps taken by this group to prevent low birth weight (e.g., better prenatal care, giving up smoking), rather than the counseling program, could account for any differences between the comparison groups.

(b) A study of the effects of jogging on coronary heart disease (CHD) compared the incidence of CHD in a sample of joggers and a sample of individuals selected from the general population who did not exercise regularly. Because the joggers were also more likely to engage in other health practices, such as eating a low-cholesterol diet, that reduce the risk of CHD, selection bias may account for any observed difference in the incidence of CHD. This type of selection bias is known as the **healthy worker effect**.

(2) **Migration bias,** another form of selection bias, may occur when individuals drop out of the study or move from one comparison group to the other.

(a) When dropouts occur randomly among the comparison groups and are unrelated to the response variable, migration bias is unlikely to pose a serious threat to internal validity. The risk of bias is significant, however, if the dropouts differ systematically from those remaining in the study.

(b) **Example.** In a prospective cohort study of the effects of nutrition on school performance, a group of high school students from a private suburban school

---

*Michael M III, Boyce WT, Wilcox AJ: *Biomedical Bestiary: An Epidemiological Guide to Flaws and Fallacies in the Medical Literature.* Boston, Little, Brown, p 16.

("good" nutrition) are compared with a group from a public school located in the inner city ("poor" nutrition). If the public school students drop out of school (and hence the study) because of poor school performance at a much higher rate than the private school students, migration bias will invalidate the study results.

b. Examples of selection bias **in case–control studies**

(1) In a case–control study of the association between a new influenza vaccine and viral pneumonia in elderly patients, cases of influenza were selected from the patient records of a family medicine practice. Controls comprised a random sample of the residents of a nearby retirement community. Because cases, on the average, were seen by a physician more often than the controls, they had a greater opportunity to receive the vaccine. This difference in the access to health care resources between the cases and controls suggests that any apparent association between the vaccine and the disease may actually be attributable to selection bias.

(2) In a study of the association between CHD and caffeine consumption, cases are defined as patients undergoing cardiac catheterization at a university medical center. Controls are randomly selected patients hospitalized for conditions other than CHD. The incidence of peptic ulcers is significantly higher in the control group and hence, their overall caffeine consumption is less than that of the cases. As a result, the controls do not accurately represent the disease-free population and the results of the study are undermined by selection bias.

3. Information bias, resulting from systematic differences among the comparison groups in the measurement of the response variable, may also compromise internal validity of a study.

a. Information bias **in prospective cohort studies**

(1) Knowledge of a particular study subject's exposure to the suspected risk factor may influence both the **outcome** and the **rigor** of the diagnostic process. The resultant disparity in surveillance of the comparison groups is known as **surveillance bias** or **diagnostic suspicion bias**.

(2) For example, a prospective cohort study is conducted to determine the association between postmenopausal estrogen use and subsequent incidence of endometrial cancer. If the group receiving estrogen therapy undergoes more frequent and thorough gynecological examinations, more cases of endometrial cancer are likely to be reported in this group, even if the true incidence is similar in both groups.

b. Information bias **in case–control studies**

(1) Knowledge of a subject's disease status may alter the diligence with which exposure information is sought and recorded.

(2) Diseased individuals (i.e., cases) and their families may search their memories more industriously for "causes" (i.e., prior exposure to the suspected risk factor) than controls would. As a result, the cases may not only remember exposure more accurately than controls, but also systematically overestimate or underestimate their exposure, resulting in **recall bias.**

(3) Examples

(a) Medical histories taken on patients undergoing surgical resection for lung cancer (cases) may be more likely to contain information on smoking history than histories taken on patients undergoing other types of surgery (controls).

(b) Mothers of infants born with birth defects (cases) may be more likely than mothers of normal infants (controls) to recall or even overestimate their use of over-the-counter medications during pregnancy.

4. Sampling bias refers to systematic differences between the study population and the target population that pose a threat to **external validity** (i.e., they prohibit generalization from the study results to the target population).

a. Sampling bias **in case–control studies**

(1) Case–control studies based on **prevalent** (i.e., ongoing) rather than **incident** (new) cases are susceptible to sampling bias because they exclude patients who die or recover rapidly, as well as those with mild symptoms whose illness goes undetected. These "survivors" are unlikely to be representative of the total patient population.

(2) Because cases are more easily identified in a hospital setting, case–control studies are often hospital-based rather than population-based. Hospital-based studies fall prey to sampling bias if hospitalized patients are more ill than the general diseased population.

b. **Berkson's bias** (also known as **Berkson's fallacy**)
   (1) This type of sampling bias occurs in hospital- or practice-based case–control and cross-sectional studies in which admission rates differ among the comparison groups.
   (2) **Example.** A hospital-based case–control study is conducted to assess the relationship between asthma and emotional disorders in children. If asthmatic children with emotional disorders are more likely to be hospitalized than those without such disorders, the frequency of emotional disorders among the hospitalized cases will be greater than that among asthmatic children in general. As a result, an artificially inflated association between asthma and emotional disorders (not representative of strength of the association in the general pediatric population) will be observed in the hospitalized study population.

c. Sampling bias due to **nonresponding** or **volunteer subjects**
   (1) Individuals who choose to participate in a study may be motivated to do so by extraneous factors that distinguish the study population from the target population.
   (2) For example, in a prospective cohort study of the effectiveness of a prenatal screening program in reducing mortality and morbidity from cystic fibrosis, the exposed group consisted of pregnant volunteers from an obstetric practice located in a wealthy suburban community. Controls were a random sample of women who were offered the screening test but declined. Clearly, selection bias may occur as a result of systematic differences between the volunteers and those refusing the test. In addition, this study is vulnerable to sampling bias, because the study population, consisting of healthy, well-educated, and highly motivated women, is not representative of the target population of all pregnant women.

d. When **study dropouts differ systematically from those who remain,** the study results are applicable only to the smaller population represented by those completing the study. Even if the initial cohort was representative of the target population, **migration bias** restricts the ability to generalize from the study population to the target population. In general, the more subjects who are lost, the less applicable the results.

D. Controlling for confounding variables and selection bias

1. **Restriction.** The effect of a potentially confounding variable may be minimized by **restricting** study participants to those falling within a narrow range of values for the variable.
   a. **Example.** If race is a risk factor for hypertension and blacks consume more salt than other racial groups, a study of the association between hypertension and salt intake may be confounded by race. By limiting the study to members of a single race, this confounding effect can be controlled.
   b. **Limitations**
      (1) Obviously, the relationship between the restricted factor and the disease cannot be studied.
      (2) Conclusions are applicable only to other individuals falling within the permissible range of the restricted variable.

2. **Matching.** Pairs of subjects may be **matched** with respect to values for a potentially confounding variable (**matching variable**). Matching is often employed in case–control studies.
   a. Subjects can be matched on any number of variables.
   b. Matching variables should be related to both the response variable and the risk factor under study.
   c. **Example.** In a study reported by Gehlbach* on the effect of cigar smoking on baldness, cases were selected from among a group of bald men seen at a hair-loss clinic. Controls comprised patients seen in a family medicine clinic in the same medical center. Because both baldness and cigar smoking are more prevalent in older men, age is a potentially confounding variable in this study. In order to equalize the age distribution

---

*Gehlbach SH: *Interpreting the Medical Literature: A Clinician's Guide.* Lexington, MA, DC Heath, 1982, p 48.

for cases and controls and, hence, eliminate the confounding effect of age, subjects in the "bald" and "hairy" comparison groups were matched with respect to age.

d. Limitations

(1) Only the matching variables are eliminated as a source of confounding.

(2) Identifying matching subjects becomes progressively demanding as the number of matching variables increases.

(3) **Overmatching** may result when subjects are matched on a variable that is related to exposure, artificially equalizing significant differences between the comparison groups. For example, in a case–control study of the association between the sickle-cell trait and malaria, matching cases and controls on the basis of race may negate a significant association and result in a spuriously low estimate of the strength of the association between sickle-cell trait and malaria.

3. Stratification

a. By grouping study subjects into subsets or **strata** with similar characteristics and analyzing data from each subset separately, the effect of a confounding variable can be eliminated.

b. Stratification and control tables

(1) The **control table** method may be used to minimize the effects of potentially confounding variables by ex post facto stratification.

(2) Table 8-2 presents the results of a hypothetical case–control study comparing caffeine intake in individuals with and without CHD. From this table, the occurrence of CHD appears to be associated with caffeine consumption. If, however, smokers consume more caffeine than nonsmokers, and the CHD group contains more smokers than the control group, smoking may confound the apparent effect of caffeine (smoking is a known risk factor for CHD; Table 8-3). By stratifying caffeine intake according to smoking status (Table 8-4), the confounding effect of smoking is eliminated. Consequently, the apparent association between CHD and caffeine consumption proves spurious.

4. Standardization of rates. Three types of morbidity and mortality rates may be obtained from observational studies.

Table 8-2. Mean Daily Caffeine Intake and Coronary Heart Disease (CHD)

	Mean Daily Caffeine Intake (mg)
CHD ($n = 10$)	756.0
No CHD ($n = 10$)	420.0

Table 8-3. Raw Data for Study of Caffeine Intake and Coronary Heart Disease (CHD)

CHD			No CHD		
Subject	**Smoking Status**	**Daily Caffeine Intake (mg)**	**Subject**	**Smoking Status**	**Daily Caffeine Intake (mg)**
1	S	1008	1	S	1008
2	S	1008	2	S	1008
3	S	1008	3	S	1008
4	S	1008	4	NS	168
5	S	1008	5	NS	168
6	S	1008	6	NS	168
7	S	1008	7	NS	168
8	NS	168	8	NS	168
9	NS	168	9	NS	168
10	NS	168	10	NS	168
		Mean = 756.0			Mean = 420.0

S = smoker; NS = nonsmoker.

Table 8-4. Mean Daily Caffeine Intake and Coronary Heart Disease—Effect of Stratification According to Smoking Status

	CHD	No CHD	
Smokers	Mean = 1008 (*n* = 7)	Mean = 1008 (*n* = 3)	Mean = 1008 (*n* = 10)
Nonsmokers	Mean = 168 (*n* = 3)	Mean = 168 (*n* = 7)	Mean = 168 (*n* = 10)
	Mean = 756.0 (*n* = 10)	Mean = 420.0 (*n* = 10)	

a. Unadjusted **crude rates** are based on all data obtained for a particular comparison group.

b. **Specific rates** may be computed for individual strata within a group. For example, age-specific mortality rates may be calculated for the various age strata.

c. Specific rates for a potentially confounding variable can be statistically applied to the overall study population or to a standard population (e.g., the 1980 U.S. population) to obtain **adjusted (standardized) rates**.

(1) See IV for a brief description of a mathematical procedure for adjusting crude rates.

(2) Example. A study is conducted to determine the association between malignant melanoma and sun exposure in two small suburban communities, one located in Florida (high exposure) and the other in Minnesota (low exposure). Because mortality from malignant melanoma is age-related, and the median age in the Florida sample is higher than the Minnesota sample, differences in the crude mortality rates for the two populations may be due to differences in exposure to ultraviolet radiation or to the overall age disparity between the two groups. The unwanted confounding effect of age can be controlled by comparing age-specific mortality rates or by adjusting the crude rates to a common age distribution (age-adjusted rates).

5. **Assuming the worst.** This method is closely related to sensitivity analysis (see Ch 5 II F 1; III B 6 a) and is used when data on potentially confounding variables are weak or unavailable. The confounding effect is estimated by assuming the **"worst possible"** distribution of the factor among the comparison groups.

a. This technique is especially useful in controlling sampling bias due to nonresponders and volunteerism.

b. **Example** (see Figure 7-2). Assuming that all nonresponders in the Charleston Heart Study developed coronary heart disease yields a "worst case" estimate of the incidence of CHD.

6. **Statistical methods. Multivariate regression procedures,** such as logistic regression, Cox's proportional hazards regression, and analysis of covariance, can be used to adjust values of the dependent variable for the influence of one or more confounding independent variables (**covariates**).

III. CAUSALITY

A. Associations that represent cause-and-effect relationships guide prevention, diagnosis, and treatment. Proof of cause, however, is neither simple, straightforward, nor irrefutable.

B. **Establishing causality: Hill's criteria.** The following criteria, proposed by Hill,* are commonly used to evaluate the plausibility of putative cause-and-effect relationships.

1. **Study design.** Study designs may be ranked according to their ability to demonstrate

*Hill AB: The environment and disease: association or causation. *Proceedings of the Royal Society of Medicine, Section on Occupational Medicine,* January 14, 1965, pp 7–12.

causality (Figure 8-5). Because experimental studies assign subjects to comparison groups by randomization, they are less subject to bias than observational study designs and provide the strongest evidence for causality.

2. **Strength of the association.** A strong (i.e., large relative risk or odds ratio) and statistically significant association is more likely to represent a cause-and-effect relationship than a weak association.

3. **Consistency.** Evidence for causality is increased when several investigators, using different study designs at different times, in varying circumstances and locations, and involving dissimilar populations, reach the same conclusions. For example, the case for a causal link between smoking and lung cancer has resulted from the gradual accumulation of consistent evidence from many epidemiological studies.

4. **Correct temporal relationship (i.e., exposure precedes disease).** Common sense dictates that a prospective **cause must precede the effect** in question. However, a correct temporal relationship alone is not sufficient to prove causality. Case–control and cross-sectional studies confuse antecedent and consequence more frequently than experimental or prospective cohort studies [see II B 2 c (5) (c)].

5. **Dose–response relationship between exposure and severity of the outcome**
   a. Evidence for causality is strengthened if risk increases with increasing exposure to the suspected risk factor.
   b. Similarly, if risk decreases following the removal of the risk factor (i.e., the association is **reversible**), a cause-and-effect relationship is more plausible.
   c. Because some exposure variables exhibit a **threshold effect** (i.e., no effect is observed until a critical exposure level has been reached), the lack of a dose–response relationship does not preclude a cause-and-effect relationship.

6. **Plausibility.** A causal relationship should be **plausible,** both from an epidemiological and a biological standpoint. That is, it should be consistent with current knowledge of the distribution and underlying biological mechanisms of the disease.
   a. A biologically implausible association may reflect the limitations of current knowledge, rather than the lack of a causal relationship. For example, one hundred years ago, a viral basis for an infectious disease would have been rejected as biologically implausible.
   b. Plausibility is neither necessary nor sufficient evidence to establish a cause-and-effect relationship.

7. **Specificity.** If a single suspected cause can be linked to a single effect, this **specificity** provides weak evidence in favor of a causal relationship. Specificity is more common in the case of acute infectious diseases or hereditary diseases involving a single-gene mutation, and can be demonstrated far less frequently for chronic degenerative disorders, such as coronary heart disease.

8. **Analogy.** The existence of other cause-and-effect relationships analogous to the one in question supports a causal interpretation, albeit weakly. For example, knowledge that a particular drug is a potent teratogen lends credence to the possibility that other drugs in the same class are also teratogenic.

C. Individually, the criteria listed in III B are, at best, **necessary** but not **sufficient** conditions to establish causality. Thus, demonstrating that an observed association is more likely to

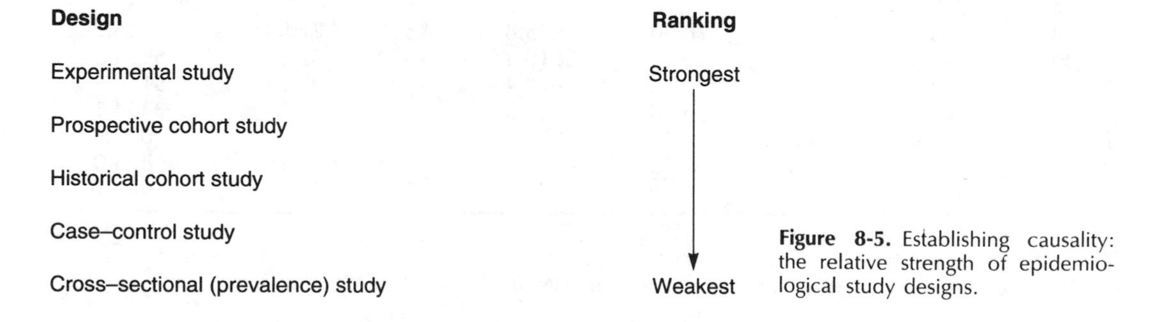

Figure 8-5. Establishing causality: the relative strength of epidemiological study designs.

represent a cause-and-effect relationship than not typically requires accumulation of evidence from several sources.

IV. THE DIRECT METHOD FOR CALCULATING AGE-ADJUSTED RATES

A. Tabulated data. Table 8-5 presents data for cancer mortality in single and married women over a three-year period, grouped into seven age intervals. The **total** number of women in each age interval for a selected standard population is also tabulated (Table 8-5, *column 8*).

1. The standard population may be the age distribution of the overall study population (single and married women), the age distribution of one of the comparison groups (e.g., married women), or it may be any general population (e.g., the 1980 U.S. population). In this example, the standard population is the total 1930 population of American women—single, married, widowed, and divorced.

2. The crude mortality rate in single women = 5845/9472.8 = 61.7 per 100,000 members of the population (Table 8-5, *last line in column 4*).

3. The crude mortality rate in married women = 29,014/21,144.7 = 137.22 per 100,000 members of the population (Table 8-5, *last line in column 7*).

B. Calculation of age-adjusted mortality rates

1. **Single women**
   a. Age-specific rates for each age interval (Table 8-5, *column 4*) are multiplied by the number of women in that interval for the standard population (Table 8-5, *column 8*).
   b. These values are summed and divided by the total number of women in the standard population (Table 8-5, *last line in column 8*) to obtain the age-adjusted mortality rate:

$$\frac{(8847.1)(3.69) + \cdots + (825.0)(1567.63)}{34{,}967.1} = \frac{6{,}152{,}004}{34{,}967.1}$$

$$= 175.94 \text{ per } 100{,}000$$

2. **Married women.** The calculation of the age-adjusted rate for married women is analogous to that for single women:

$$\frac{(8847.1)(5.37) + \cdots + (825.0)(981.19)}{34{,}967.1} = \frac{5{,}277{,}634}{34{,}967.1}$$

$$= 150.93 \text{ per } 100{,}000$$

Table 8-5. Cancer Mortality According to Age for Single and Married Women in the United States, 1929–1931

	Single Women			Married Women			
1	2	3	4	5	6	7	8
Age (yrs)	**Number in Population***	**Cancer Deaths**	**Cancer Death Rate†**	**Number in Population***	**Cancer Deaths**	**Cancer Death Rates†**	**Population Distribution for All Women***
15–24	6129.5	226	3.69	2627.5	141	5.37	8847.1
25–34	1485.4	251	16.90	5923.8	1335	22.54	7725.2
35–44	759.0	612	80.63	5610.3	4630	82.53	6924.3
45–54	522.4	1198	229.33	3876.4	7894	203.64	5189.4
55–64	343.5	1511	439.88	2124.3	8155	383.89	3454.1
65–74	187.9	1340	713.15	828.2	5346	645.50	2002.0
≥75	45.1	707	1567.63	154.2	1513	981.19	825.0
Total	9472.8	5845	61.70	21,144.7	29,014	137.22	34,967.1

*in thousands
†per 100,000 population
Reprinted from Colton T: *Statistics in Medicine.* Boston, Little, Brown, 1974, p 48.

C. The higher crude mortality rate observed in married women is attributable to the greater proportion of older women in this group. Age-adjustment removes the confounding effect of age. As a result, age-adjusted mortality rates indicate a higher rate of cancer mortality among single women.

D. Age-adjusted rates are artificial, rather than actual, rates. They represent the theoretical rates that would have occurred for single and married women, had the age distributions for these groups been equivalent to that of the standard population. Age-adjusted rates provide useful summary rates for comparing groups with disparate age distributions. Age-specific rates should be used instead when describing true rates of disease within a population.

BIBLIOGRAPHY

Esdaile JM, Horwitz RI: Observational studies of cause–effect relationships: an analysis of methodologic problems as illustrated by the conflicting data for the role of oral contraceptives in the etiology of rheumatoid arthritis. *J Chron Dis* 39:841–852, 1986.

Fletcher RH, Fletcher SW, Wagner EH: *Clinical Epidemiology: The Essentials.* 2nd ed. Baltimore, Williams and Wilkins, 1988.

Gehlbach SH: *Interpreting the Medical Literature: A Clinician's Guide*. Lexington, MA, DC Heath, 1982.

Hill AB: The environment and disease: association or causation. *Proceedings of the Royal Society of Medicine, Section on Occupational Medicine,* January 14, 1965, pp 7–12.

Kahn HA, Sempos CT: *Statistical Methods in Epidemiology*. New York, Oxford University Press, 1989.

Kleinbaum DG, Kupper LL, Morgenstern H: *Epidemiologic Research: Principles and Quantitative Methods*. Belmont, CA, Lifetime Learning Publications, 1982.

Linos A, Worthington JW, O'Fallon WM, Kurland LT: Case–control study of rheumatoid arthritis and prior use of oral contraceptives. *Lancet* 1:1299–1300, 1983.

Michael M III, Boyce WT, Wilcox AJ: *Biomedical Bestiary: An Epidemiological Guide to Flaws and Fallacies in the Medical Literature*. Boston, Little, Brown, 1984, p 16.

Murphy EA: *The Logic of Medicine*. Baltimore, Johns Hopkins University Press, 1976.

Royal College of General Practitioners' Oral Contraception Study: Reduction in incidence of rheumatoid arthritis associated with oral contraceptives. *Lancet* 1:569–571, 1978.

Sackett DL: Bias in analytic research. *J Chron Dis* 32:51–63, 1979.

Sackett DL, Haynes RB, Tugwell P: *Clinical Epidemiology: A Basic Science for Clinical Medicine.* Boston, Little, Brown, 1985, p 232.

Stevens J: Adiposity and fat patterning in black Americans. National Heart, Lung, and Blood Institute RO1 HL 42305-01, August 1, 1989–July 31, 1992.

Vandenbroucke JP, Valkenburg HA, Boersma JW, Cats A, Festen JJM, Huber-Bruning O, Rasker JJ: Oral contraceptives and rheumatoid arthritis: further evidence for a protective effect. *Lancet* 2:839–842, 1982.

PROBLEMS

8-1. In a hospital-based case–control study of the relationship between rheumatoid arthritis and oral contraceptive use, the cases consist of a random sample of 100 women with rheumatoid arthritis undergoing treatment at one of five outpatient rheumatology clinics.

a. Suggest an appropriate control group for this study.

b. What is the response (outcome) variable and how might it be measured?

c. List potential sources of bias that may compromise the conclusions of the study.

d. List potentially confounding variables and suggest methods for limiting their influence.

8-2. **a.** Describe how the relationship between rheumatoid arthritis and oral contraceptive use might be examined using a prospective cohort study design. Include details of how study subjects might be selected, and identify the response variable.

b. Identify potential sources of bias and describe their possible effects on the outcome of the study.

8-3. A psychiatrist hypothesizes that men who abuse drugs are more likely to commit rape than men who are not drug abusers. She conducts a case–control study to test this hypothesis, selecting as cases a random sample of men serving sentences for rape in the state penitentiary.

a. Are the cases likely to be representative of the population of all rapists? Why?

b. Identify the response variable. What difficulties might the investigator encounter in measuring this variable?

c. Suggest a suitable control group for this study.

d. List potentially confounding variables and suggest methods to control them.

e. What are the advantages and disadvantages of using a case–control study design to examine the relationship between rape and drug abuse?

8-4. Briefly outline a prospective cohort study design to assess the hypothesis that drug abusers are more likely to commit rape than men who do not abuse drugs.

8-5. A team of health care planners wishes to estimate the prevalence of AIDS in a particular community.

a. What study design is the most appropriate for achieving this objective?

b. Briefly describe how such a study might be conducted.

c. How is the prevalence of AIDS estimated?

d. List uses of this prevalence estimate.

8-6. The health care planners described in problem 8-5 also wish to estimate the incidence of AIDS in the community in question.

a. What is the best study design for accomplishing this objective?

b. Briefly describe how such a study might be conducted.

c. How is the incidence of AIDS estimated?

d. List potential uses of this incidence estimate.

8-7. The following table lists death rates from colorectal cancer for a particular community by race and age (for simplicity, community residents have been divided into only two age groups—"under 65" and "over 65"). Use the data in this table to answer the questions below.

a. Calculate the crude (unadjusted) overall death rate from colorectal cancer for black residents of this community.

b. Calculate the crude overall death rate from colorectal cancer for white residents of the community.

c. The crude overall death rate for whites is higher than that for blacks even though the age-specific death rates for blacks are higher in both age groups. Explain this paradox.

d. Compare the age-specific death rate for blacks under 65 to that for whites under 65. Can the higher rate for blacks be explained by the fact that there is a greater proportion of people under 65 in the black community than there is in the white community?

Death Rates from Colorectal Cancer, Listed by Race and Age

	Proportion of Population Comprising Age Interval	Death Rate (per 1000 people)
Blacks		
Under 65	.8	40.0
Over 65	.2	65.0
Whites		
Under 65	.2	28.0
Over 65	.8	60.0

e. From the data in the table above, calculate the age-adjusted death rates from colorectal cancer for both blacks and whites. Assume that 40% of the individuals comprising the selected standard population are under 65 years of age and 60% are over 65.

8-8. A particular developing country reports a much lower crude death rate than the United States. What is the most likely explanation for this observation?

Solutions on p 374.

STUDY QUESTIONS

Directions: Each of the numbered items or incomplete statements in this section is followed by answers or by completions of the statement. Select the **one** lettered answer or completion that is **best** in each case.

1. All of the following are advantages of the case–control study design EXCEPT

(A) it is easier to identify a sufficient number of diseased subjects for this type of study than a cohort study
(B) this design is more practical for studying rare diseases
(C) this design is subject to fewer ethical concerns than a prospective cohort study
(D) this design is less vulnerable to bias than other observational study designs
(E) this design is less expensive than a cohort study

2. In a hospital-based case–control study of the association between coffee consumption and the occurrence of myocardial infarction, a group of patients hospitalized after suffering a myocardial infarction was compared to a control group hospitalized for other reasons. The patients hospitalized for myocardial infarction (cases) were found to consume significantly more coffee than the controls. All of the following statements represent possible explanations for the observed positive association between coffee consumption and myocardial infarction EXCEPT

(A) heavy coffee drinkers may also be heavy smokers, so smoking rather than coffee consumption is the relevant causal factor
(B) excessive coffee consumption causes myocardial infarction
(C) the hospitalized controls consumed less coffee, on the average, than individuals in the general population with no history of myocardial infarction, resulting in a spurious association between coffee consumption and myocardial infarction
(D) the cases restricted their coffee intake after suffering a myocardial infarction
(E) there is a higher incidence of anxiety disorders among the cases; anxiety increases both coffee consumption and the incidence of myocardial infarction

3. All of the following statements about observational studies are true EXCEPT

(A) events are observed as they occur in nature, with no active intervention on the part of the investigator
(B) the comparison groups may differ with respect to factors related to the response variable
(C) potentially confounding variables are controlled by randomization
(D) subjects may be followed forward in time from exposure to occurrence of outcome, backward in time from outcome to exposure, or evaluated at a single point in time to concurrently assess outcome and exposure
(E) they are especially useful in situations where it is impossible, impractical, or unethical to manipulate exposure to a suspected risk factor

4. A physician conducted a cross-sectional study and observed that the prevalence of respiratory diseases is higher in Arizona than in Pennsylvania. The most appropriate conclusion he can draw from this result is

(A) living in Arizona causes respiratory illness
(B) people with a respiratory illness may be more likely to live in Arizona because of the beneficial effect of its climate
(C) the incidence of respiratory illness is also higher in Arizona
(D) an individual living in Arizona is at greater risk of developing a respiratory illness than an individual living in Pennsylvania
(E) individuals with a respiratory illness should not move to Arizona

---

1-D 4-B
2-D
3-C

Questions 5–7

A pediatrician hypothesizes that children who consume large amounts of a particular food additive are more likely to be hyperactive than children who do not. To test this hypothesis, he interviews the parents of 100 hyperactive children (identified from the medical records of six child psychiatrists) and 100 children undergoing treatment by one of the same six physicians for other behavioral problems to determine the dietary habits of their children over the preceding two years.

5. This study is an example of

(A) a cohort study
(B) a case–control study
(C) a cross-sectional study
(D) an experimental study
(E) a prevalence study

6. The investigator observes that hyperactive children consume significantly more foods containing the additive in question than children in the control group and concludes that the additive causes hyperactivity. This conclusion is

(A) valid for the study population
(B) potentially flawed because recall bias may affect parents' responses
(C) not valid because parents of hyperactive children may have limited their children's consumption of foods containing the additive
(D) generalizable to the target population of all hyperactive children
(E) free of antecedent–consequence uncertainty

7. What is the most appropriate outcome for the investigator to measure and report?

(A) Incidence of hyperactivity in those children who do and in those who do not consume foods containing the additive
(B) Prevalence of hyperactivity among children who do and among those who do not consume foods containing the additive
(C) Rates of exposure to the putative risk factor (i.e., the additive) for hyperactive versus nonhyperactive children
(D) Rates of hyperactivity among those children who eat foods containing the additive versus rates among those who do not

8. Antecedent–consequence uncertainty is defined as which of the following?

(A) A form of bias occurring when the comparison groups differ with respect to exposure to the antecedent factor
(B) A form of bias resulting from a statistically insignificant increase in the occurrence of the outcome (consequence) as a result of exposure (antecedent)
(C) A form of bias resulting from confusion as to whether exposure precedes or follows an outcome
(D) A form of bias occurring when the level of uncertainty associated with the measurement of the outcome (consequence) differs for cases and controls
(E) A form of bias in which the antecedent is more likely to precede the consequence in cases than in controls

9. In a large cross-sectional study, 10 of every 100,000 black men are found to have prostate cancer compared to 20 out of every 100,000 white men. The epidemiologist conducting the study concludes that white men are at greater risk for developing prostate cancer. This conclusion is

(A) valid for the study population
(B) incorrect because of the failure to distinguish between incidence and prevalence
(C) incorrect because only men were studied, therefore the study lacks an appropriate control group
(D) incorrect if there are more white men than black men in the study population
(E) incorrect because a case–control study design should be used to estimate risk

10. A prospective cohort study employs

(A) subjects known at the onset to have the disease in question
(B) subjects known at the onset to be disease-free
(C) subjects whose exposure to a suspected risk factor is comparable to that of the control group at the onset of the study
(D) a cohort with a high incidence of the disease in question
(E) comparison groups of equal size

5-B 6-B 7-C 8-C 9-B 10-B

Questions 11–12

A study is conducted to evaluate the effectiveness of a new influenza vaccine in preventing hospitalization and death due to influenza-related complications in elderly patients. A group of elderly patients who received the vaccine in the six months preceding the observation period were identified from the patient records of participating physicians. A second group of elderly patients who had not received the vaccine was also identified from these records. Both groups of subjects were followed for a two-year period to determine the number of hospitalizations and deaths from all causes. The mortality rate from all causes was found to be significantly higher in the vaccinated group than in the nonvaccinated group.

11. This study is an example of

(A) a randomized controlled clinical trial
(B) a cohort study
(C) a case–control study
(D) a prevalence study
(E) an experimental study

12. The physicians conducting this study concluded that elderly patients should not receive the vaccine. This conclusion is

(A) valid for the study population
(B) invalid because elderly patients who received the vaccine may have been less healthy than those who did not
(C) invalid because knowledge of patients' vaccination status biased measurement of the response variable
(D) invalid because the observed result can be attributed to recall bias

13. All of the following represent true statements about the prospective cohort study design EXCEPT

(A) it resembles a true experimental study more closely than other observational study designs
(B) it is appropriate for estimating risk
(C) it is impractical for studying rare diseases
(D) it is poorly suited for studying diseases with a long latency between exposure and illness when the need for information is urgent
(E) it can be conducted with existing data

Questions 14–17

In a study of the association between exposure to lead-based paint and mental retardation, the parents or guardians of 1000 mentally retarded children and 1000 children with normal IQs were questioned about their child's exposure to lead-based paint. The results revealed that 100 of the mentally retarded children had been exposed to lead paint, compared to only one child of normal IQ.

14. This study is an example of

(A) a cohort study
(B) a case–control study
(C) an experimental study
(D) a cross-sectional study
(E) a prevalence study

15. The absolute risk (incidence) of mental retardation among the children exposed to lead-based paint is

(A) .1
(B) .01
(C) 100 times greater than the risk of those not exposed to lead paint
(D) equal to the prevalence of mental retardation in these children
(E) cannot be calculated directly from the results of this study

16. The investigator conducting this study used matching to control potential sources of confounding. Appropriate matching variables include all of the following EXCEPT

(A) socioeconomic status
(B) length of exposure to lead-based paint
(C) age
(D) parental IQ

17. Suppose the investigator had selected only children of comparable socioeconomic status as study subjects. This method of controlling the potentially confounding effect of socioeconomic status is called

(A) matching
(B) stratification
(C) restriction
(D) randomization
(E) statistical adjustment

11-B 14-B 17-C
12-B 15-E
13-E 16-B

ANSWERS AND EXPLANATIONS

1. The answer is D *[II B 2 c (4)–(5)]*
Case–control studies, by their nature, are more susceptible to bias than cohort studies. Both exposure and outcome have already occurred when the study is initiated, therefore, the investigator is less able to control potential sources of bias. Obtaining a sufficient number of cases is easier in a case–control study than in a prospective cohort study, since cases are identified at the beginning of the study rather than as they occur in initially healthy comparison groups. For this reason, case–control studies require fewer subjects and are more practical for studying rare diseases. This study design is also faster and more economical than prospective cohort studies. It is the most appropriate design in situations where exposing subjects to a putative risk factor may be unethical.

2. The answer is D *[II C 1]*
Excessive coffee consumption alone may indeed cause myocardial infarction. However, a number of alternative explanations for the data cannot be ruled out. For example, smoking is a potential confounding variable in this study; it is associated with the outcome (myocardial infarction) as well as with exposure (coffee consumption). Similarly, anxiety may increase both coffee consumption and the incidence of myocardial infarction. If the hospitalized control group included many individuals with illnesses limiting coffee consumption (e.g., peptic ulcer disease), the association between coffee consumption and myocardial infarction will be erroneously high. In contrast, if patients who have had a myocardial infarction subsequently reduce their coffee consumption, an investigator is less, not more, likely to observe a positive association between these factors.

3. The answer is C *[II B 1 a–b, 2 a, b, c, d]*
Randomization (i.e., the random allocation of study subjects to comparison groups) distinguishes experimental studies from observational studies. In an observational study, subjects "self-select" into comparison groups. While observational studies can be used when it is difficult or unethical to employ an experimental design, they are more subject to potential sources of bias. Cohort studies (in which healthy subjects are followed forward in time after exposure), case–control studies (in which cases and healthy controls are followed backward in time to assess prior exposure), and cross-sectional studies (in which outcome and exposure are determined concurrently) are examples of observational studies.

4. The answer is B *[II B 2 d (1)–(3)]*
A cross-sectional study can provide an estimate of prevalence, but not incidence. Conclusions drawn from the results of a cross-sectional study are subject to antecedent–consequence uncertainty (i.e., did living in Arizona lead to respiratory illness or did a prior respiratory illness cause an individual to move to Arizona?). Causation cannot be inferred solely from the results of this study.

5–7. The answers are: 5-B *[II B 2 c]*, **6-B** *[II C 3 b (2)]*, **7-C** *[II B 2 c (2) (b)]*
This study follows a group of children diagnosed as hyperactive (cases) and a group of nonhyperactive children (controls) backward in time to determine exposure to the questionable food additive; it is therefore a case–control study.

Food additives have been widely blamed for hyperactivity in the popular press. As a result, parents of hyperactive children are more likely to recall or even overestimate their children's consumption of the food additive than the control parents. If parents of hyperactive children restricted their consumption of the additive, the apparent association between the additive and hyperactivity would be expected to appear even stronger than that reported. The conclusions of any case–control study are potentially subject to antecedent–consequence uncertainty and are generalizable to the target population of all hyperactive children only if there are no systematic differences between the study population and the target population.

Neither incidence nor prevalence can be estimated directly in a case–control study. The rate at which hyperactivity occurs among children who consume the additive and those who do not is a measure of incidence and, hence, cannot be determined from the results of this study.

8. The answer is C *[II B 2 c (5) (c)]*
Antecedent–consequence uncertainty, a form of bias common to case–control and cross-sectional studies, results from confusion about whether exposure occurred before or after the disease process began.

9. The answer is B *[II B 2 d (1), (3)]*
Cross-sectional (prevalence) studies may be used to determine prevalence, the proportion of existing cases of a disease, but not incidence, the rate at which new cases occur. Only incidence provides an estimate of risk.

10. The answer is B *[II B 2 a, b; Figure 8-4]*
Cohort studies employ initially healthy subjects who have been or who will be exposed or not exposed to the suspected risk factor. Although the two groups need not be equal in size, the subsequent statistical analysis will be more efficient if they are. A case–control study employs study subjects known at the outset of the study to have the disease in question (cases) or known to be disease-free (controls).

11–12. The answers are: 11-B *[II B 2 a–b]*, **12-B** *[II C 2 a]*
Two groups of patients exposed or not exposed to the putative risk factor (the vaccine) were followed forward in time to determine the outcome (rate of hospitalization and death from all causes); therefore, this is a cohort study.

The comparison groups may have differed systematically with respect to overall health status, a variable likely to affect the response (hospitalization or death from all causes). This selection bias provides an alternative explanation for the observed association. Because both hospitalization and death from all causes represent objective response (outcome) variables, their measurement should be unaffected by the knowledge of vaccination status. Since data were obtained from patient records rather than personal interviews, this study is not subject to recall bias.

13. The answer is E *[II B 2 a (5), (6), b (2), c (4) (b)]*
Only the case–control and historical cohort study designs make use of existing data. Of the observational study designs, the prospective cohort design most closely resembles a true experimental study. While it provides information on the incidence of a disease in individuals exposed or not exposed to a putative risk factor and hence, may be used to estimate risk, a prospective cohort study may require large sample sizes and lengthy follow-up periods. As a result, it is often impractical for studying rare diseases or in situations where the need for information is urgent.

14–17. The answers are: 14-B *[II B 2 c]*, **15-E** *[II B 2 c (5) (b)]*, **16-B** *[II D 2 b]*, **17-C** *[II D 1]*
This is an example of a case–control study, in which cases (mentally retarded children) and controls (children of normal IQ) were followed backward in time to determine exposure to a putative risk factor (lead-based paint).

Incidence (risk) cannot be directly determined from a case–control study. The probability that a mentally retarded child has been exposed to lead paint is $P(E+|D+) = 100/1000 = .1$. The probability that a child of normal IQ has been exposed to lead paint is $P(E+|D-) = 1/1000 = .001$. Cases are thus 100 times more likely to have been exposed to lead paint than controls.

The length of exposure to lead-based paint is an outcome to be measured; hence, it cannot be used as a matching variable. Socioeconomic status, parental IQ, and age are potentially confounding variables that could be controlled by matching. For example, low socioeconomic status and low parental IQ may increase the chance that the child lives in an older home with peeling lead-based paint. A child born to parents with low IQs may be genetically predisposed to mental retardation. The older the child, the greater his or her opportunity of exposure to lead paint.

Restriction limits study subjects to a single homogeneous subgroup. In this instance, study participants are limited to children of the same socioeconomic class (e.g., only middle-class children).

9
Comparing Therapies: The Randomized Controlled Clinical Trial

I. THE EXPERIMENTAL APPROACH TO EVALUATING TREATMENTS. A definitive diagnosis is only the first step in patient care. Next, the physician must choose the safest and most effective treatment. While observational studies are used to assess risk, the experimental approach, specifically, the **randomized controlled clinical trial,** is the tool of choice for comparing therapies. Experimental studies may also compare health care services, patient education programs, and administrative strategies.

A. The experimental study: key features. In contrast to observational studies, experimental studies allow the investigator to retain a high degree of control over the study subjects and conditions. As a result, three features distinguish experimental studies, including most clinical trials.

1. Subjects are **randomly assigned** to comparison groups. In contrast, participants in an observational study "self-select" their group status (see Ch 8 II B 1).

2. The investigator **compares** the subjects studied **with an appropriate control group**.
   a. The inclusion of a separate control group provides a comparison basis for evaluating the worth of a new treatment.
   b. In an **uncontrolled** clinical trial, subjects serve as their own controls (i.e., treatment efficacy is determined by comparing the same subjects before and after therapy). The results of such studies are subject to alternative explanations, including the following.
      (1) **Predictable improvement.** In the case of many diseases, the majority of patients eventually recover without treatment. Therefore, if the study period is sufficiently long, nature, not the treatment, may effect the cure.
      (2) **Fluctuating disease severity.** The course of some diseases, such as ulcerative colitis or multiple sclerosis, is characterized by alternating periods of exacerbation and remission. If the experimental treatment is initiated during an acute period, a treatment "effect" may actually be a spontaneous remission.
      (3) Patients who have **volunteered** for a clinical trial are often **anxious to please** the investigator by getting well. As a result, they may respond, either consciously or unconsciously, in a way that biases the experimental results. This tendency for subjects to alter their behavior simply because of the attention they receive in a study is known as the **Hawthorne effect.***
      (4) **Regression to the mean**
         (a) Subjects chosen for study because they represent extreme values of the response variable typically have lower values on subsequent measurements. This statistical phenomenon, known as **regression to the mean,** may account for apparent treatment effects in an uncontrolled clinical trial.
         (b) **Example.** An uncontrolled trial of a novel antihypertensive drug enrolled 100 patients with blood pressures exceeding 140/100 during a routine physical. Because of the random variation inherent in biologic variables (see Ch 1 I C 1), some of these "hypertensive" subjects actually represented normotensive individuals with chance elevations in blood pressure. Subsequent blood

*The Hawthorne effect was first observed in studies of worker productivity and illumination carried out at the Hawthorne Works of the Western Electric Company in Chicago. The work output of study participants increased regardless of whether illumination intensity was increased, decreased, or remained the same.

pressure readings in such individuals would, on the average, be likely to improve, even if the drug actually is ineffective.

3. The investigator **manipulates** the treatment under investigation, also known as the **independent variable**.

B. Efficacy versus effectiveness

1. An **efficacious** treatment is one that "works" (i.e., is proven significantly better than an appropriate alternative) for those who **receive** it.
2. An **effective** treatment, in contrast, is one that works for those to whom it is **offered**. An ineffective treatment may lack efficacy, patient acceptance, or both.
3. The majority of clinical trials seek to address the question of efficacy rather than that of effectiveness.

II. CONDUCTING A RANDOMIZED CLINICAL TRIAL. An investigator conducting a randomized controlled clinical trial must perform the following five sequential operations (Figure 9-1).

A. Formulate a hypothesis. All analytic studies, including clinical trials, begin with a research question, or **hypothesis** (e.g., Does new drug X improve memory in patients with Alzheimer's disease?).

1. In experimental studies, the hypothesis takes the general form **"if A, then B."**
   a. **"A"** represents the **independent** or **treatment variable,** that is, the factor that is manipulated by the investigator.
   b. **"B"** represents the **dependent (response) variable,** which is measured and analyzed to determine the efficacy or effectiveness of the treatment.
2. The hypothesis must **constrain the research problem** so that the study is manageable

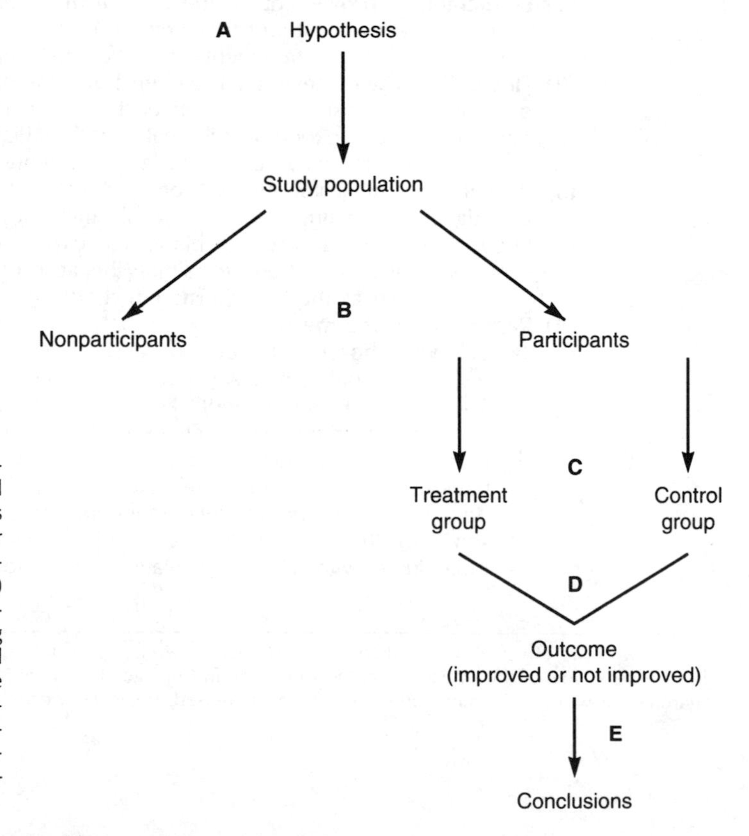

Figure 9-1. The five steps comprising a randomized controlled clinical trial. First, the purpose of the trial is framed as a research question or hypothesis (*A*). Participants are selected from the study population (*B*) and randomly allocated (*C*) to control or treatment groups. Following treatment, the outcome is measured in each group (*D*). Appropriate statistical methods are used to analyze the data (*E*) and draw conclusions regarding the efficacy or effectiveness of the experimental treatment.

within the confines of available resources (such as money, time, personnel, or number of available subjects).

B. Select participants (see Figure 9-1).

1. Selection method

a. Ideally, study participants should be **selected randomly** from the target population. This ensures that the study results will be applicable to the target population.

b. Alternatively, **volunteers** may be recruited from a population of accessible patients (e.g., patients admitted to a university-affiliated referral hospital). In this case, the results may apply only to the study population.

2. Sample size

a. The appropriate sample size is determined by the desired **power** of the statistical test to be used to analyze the data.

(1) Power is the ability of a statistical test to detect differences among comparison groups, given that such differences exist in the study population.

(2) The **larger** the sample size, the **greater** the power of the statistical test. Conversely, reducing sample size decreases power.

b. An experimental treatment may be judged ineffective if no statistically significant differences are observed between comparison groups. However, negative results may merely be evidence of inadequate sample size. Chapter 12 provides a more detailed discussion of the relationship between power and sample size.

3. Controls

a. Selection of an appropriate control group

(1) Possible pitfalls. The use of control subjects drawn from a different time or location than experimental subjects may lead to erroneous conclusions.

(a) Patients who receive the experimental treatment may be compared to a **historical (nonconcurrent) control group** of former patients who did not receive the treatment. Trials based on a historical control group tend to be biased in favor of the experimental treatment due to:

(i) Improvements in overall patient care

(ii) Improvements in diagnostic procedures, leading to the selection of less ill patients

(iii) Altered virulence of the disease

(iv) Psychological effects, such as the Hawthorne effect [see I A 2 b (3)]

(v) Unconscious supportive therapy provided to the experimental group, but not the controls (e.g., experimental patients receiving more extensive postoperative care).

(vi) Differences in confounding variables, such as age, sex, or race

(b) The results of studies in which the subjects receiving the experimental treatment are compared to controls drawn from **different settings** (e.g., a different hospital) may also be misleading. Differences in referral patterns or the skill of health care providers, for example, may account for observed differences between comparison groups.

(2) Proper control selection. A **concurrent control group,** treated at the same time and evaluated in the same setting as experimental subjects, provides the greatest protection against potential sources of bias.

b. Selection of an alternative treatment

(1) Control subjects may receive **no treatment;** rather, they are simply observed over the same time period as subjects receiving the experimental treatment.

(2) More commonly, controls receive a **placebo** treatment, indistinguishable from the experimental treatment but lacking its specifically active component.

(a) In drug trials, the placebo is typically an inert pill identical in appearance and taste to the experimental medication. A trial of a new surgical procedure may employ a "sham" procedure, in which controls receive an identical anesthetic, skin incision, and postoperative care as the experimental subjects.

(b) Placebo effect

(i) Control subjects may improve in response to placebo administration, due to an unknown, nonspecific mechanism of action. This **placebo effect** often skews comparisons involving a placebo group.

(ii) For example, a clinical research team investigated the efficacy of a new appetite suppressant. For successive two-week periods, obese subjects received in random order either the new drug, a placebo, or no medication. Both drug and placebo effectively reduced daily caloric intake, compared to the drug-free period. The investigators concluded that the efficacy of the new appetite suppressant was attributable, in part, to the placebo effect.

(3) The experimental treatment may be compared with a **standard treatment**. When an efficacious treatment already exists, withholding treatment or administering a placebo is often considered unethical.

C. Allocate subjects to comparison groups.

1. Prior to treatment, subjects must be **randomly allocated** to experimental and control groups.
   a. A **random number table** or a computer- or calculator-based **random number generator** may be used for randomization. Subjects are often first randomly assigned to groups that are then in turn randomly assigned to the various treatment categories.
   b. Because all subjects are equally likely to be assigned to any of the comparison groups, random allocation tends to negate the influence of confounding variables and minimize selection bias (see Ch 8 II C 2).
      (1) Although randomization tends to equalize the comparison groups with respect to confounding variables over the long run, it does not guarantee equivalence over the short term.
      (2) **Example.** A fair coin flipped a large number of times (the long run) will result in an equal number of heads and tails. If, however, the coin is flipped only a few times (the short term), the majority of flips may come up heads even if the coin is fair. Similarly, a confounding factor may be distributed unequally among randomly allocated groups just by chance, especially if the sample size is small.
   c. If subjects are systematically, rather than randomly, allocated to comparison groups, selection bias is likely to result.
   d. After allocation, the distribution of potentially confounding factors should be compared among the experimental and control groups. In the unlikely event that systematic differences exist (e.g., controls are, on the average, significantly younger than experimental subjects), subsequent statistical analysis of the data must compensate for these differences.

2. **Study designs**
   a. A **completely random design** imposes no restriction on the random allocation of subjects to study groups (Figure 9-2). For example, 50 individuals volunteer to participate in a study evaluating the efficacy of a new antihypertensive drug. Thirty of these individuals meet the inclusion criteria and are recruited for the study. Fifteen subjects are randomly assigned to receive the new drug and fifteen are randomly allocated to the control (placebo) group.
   b. In the **randomized block design,** subjects are first stratified into **blocks,** based on a potentially confounding variable such as disease severity, age, or baseline value of the dependent variable. Subjects are then randomly assigned to comparison groups within each strata (Figure 9-3). Thus, random allocation takes place only within the defined "blocks." For example, subjects in the antihypertensive study described above may be stratified into four age groups: 18–35 years, 36–55 years, 56–75 years, and over 75 years. Participants within each block are assigned at random to receive either the new drug or a placebo.

3. **Post-randomization changes in comparison groups.** Study participants may drop out of the study, switch treatment groups, or fail to comply with the prescribed treatment regimen.
   a. Differential dropout rates among the comparison groups represent a source of **migration bias** [see Ch 8 II C 2 a (2)], especially if experimental subjects drop out of the study for reasons related to the treatment.
   b. Bias can also result from differential rates of compliance or when patients who do not respond favorably to one treatment are switched to a different treatment group.
   c. **Example.** A randomized controlled clinical trial of a β-adrenergic blocking agent demonstrated that patients recovering from a myocardial infarction who received the

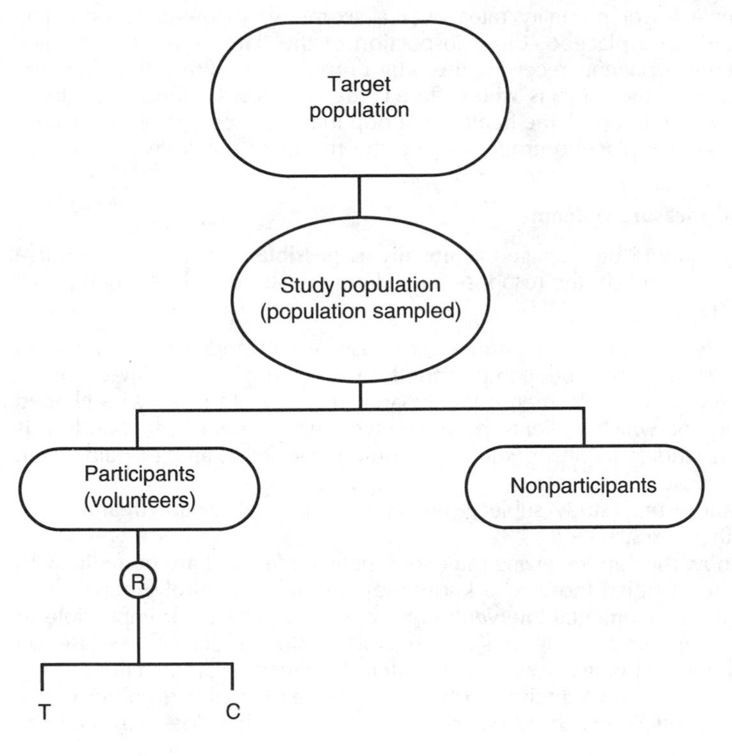

Figure 9-2. Schematic representation of the completely random study design. R = random allocation; T = experimental group; C = control group.

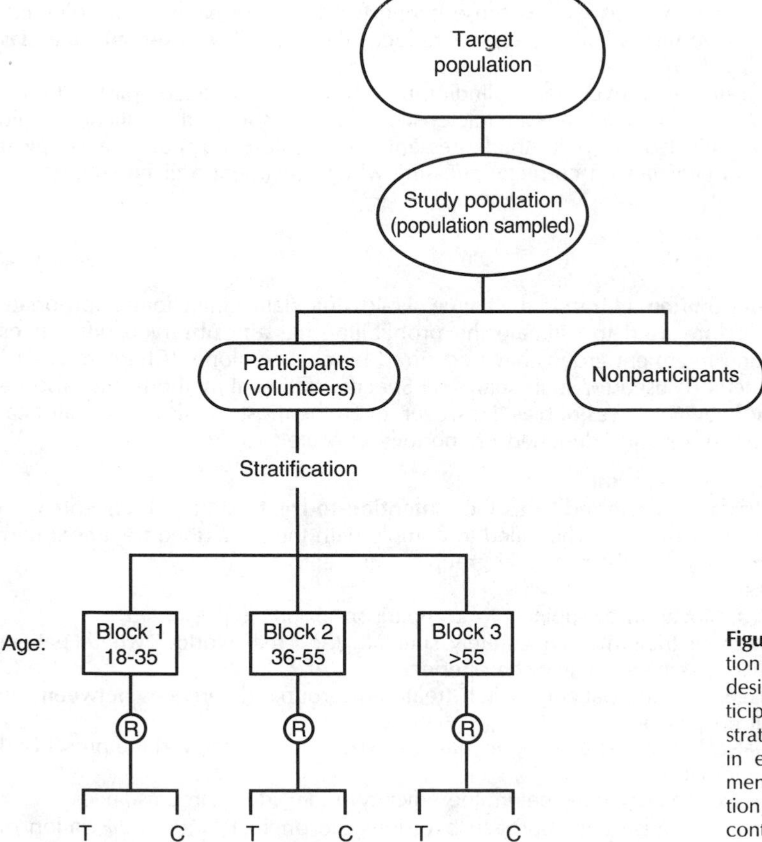

Figure 9-3. Schematic representation of the randomized block study design. In this example, study participants are grouped into three age strata, then randomly assigned within each block to control or treatment groups. R = random allocation; T = experimental group; C = control group.

drug had significantly lower mortality rates over a six-month follow-up period than similar patients receiving a placebo. Close inspection of the data, however, revealed that three times as many patients receiving the drug dropped out of the study because of adverse side effects. If the subjects who suffered side effects were those most likely to die during the follow-up period, the treatment group may have comprised selectively healthier patients than the placebo group, biasing the results of the study.

D. Administer treatment and measure outcome.

1. The comparison groups should be managed **as equally as possible** throughout the course of the study. In each of the groups, the response variable is measured and recorded for all subjects (see Figure 9-1).

2. **Blinding.** During the intervention and measurement phase of the randomized controlled clinical trial, internal validity may be compromised by the subjects' or investigators' awareness of comparison group assignments. Participants in a clinical trial may be **blinded** to keep them unaware of which patients have received which treatment. Blinding is particularly valuable in studies involving subjective outcomes, for example, pain relief, mood elevation, or nausea.
   a. In a **single-blind** study, only study subjects are unaware of the group (treatment or control) to which they are assigned.
      (1) Patients who know they are receiving the experimental treatment are more likely to report positive results than those who know they are in the control group.
      (2) The nature of the experimental intervention may make it difficult or impossible to blind subjects. For example, in a study examining the effect of exercise on emotional well-being in patients who have suffered a myocardial infarction, group assignment (exercise or no exercise) is obvious to the participants. In other cases, characteristic side effects, such as bitter taste, nausea, or hair loss may cue the subjects.
   b. In a **double-blind** study, both subjects and researchers (i.e., the investigators measuring the experimental outcome as well as those caring for the study patients) are blinded.
      (1) Blinding the investigators and caregivers reduces the possibility of **surveillance bias** (see Ch 8 II C 3 a).
      (2) Like subject blinding, investigator blinding may be difficult to accomplish. Physiologic effects, such as heart rate, characteristic side effects (e.g., the anticholinergic side effects typical of tricyclic antidepressants), or obvious cues such as surgical scars may aid clinicians in correctly guessing which treatment a given subject has received.

E. Analyze data.

1. **Estimating the contribution of random chance.** Following data collection, appropriate statistical techniques are used to estimate the probability that any observed differences between control and treatment groups have occurred by chance alone. (Chapters 12–17 review statistical methods used for such analyses.) Specific statistical methods may also be used to control for known sources of bias. However, even the most sophisticated statistical analysis cannot rescue a poorly designed or shoddily executed study.

2. **Controlling for subject migration**
   a. **Management trials** (also referred to as the **"intention-to-treat"** strategy). Patients who switched treatment groups or who failed to comply with the prescribed treatment may be analyzed according to their original group assignment.
      (1) **Advantages**
         (a) Random allocation of subjects to comparison groups is preserved.
         (b) Management trials more accurately simulate the "real world," where patient compliance is typically less than perfect.
      (2) **Disadvantage.** If many patients switch treatment groups, differences between the groups will be obscured.
   b. **Explanatory trials.** With this strategy, only patients who actually received the prescribed treatment are compared.
      (1) **Advantage.** Explanatory trials determine efficacy under ideal circumstances.
      (2) **Disadvantage.** Because this approach overrides the original random allocation of

subjects to comparison groups, it is vulnerable to all the sources of bias that characterize a cohort study (see Ch 8 II C 2 a, 3 a).

3. Procedures that bias statistical analyses. Incorrect applications of statistical methods are among the most common sources of misleading conclusions in any experimental study, including clinical trials.

a. Post-hoc comparisons

(1) The use of study data to answer new questions that arise after completion of the clinical trial is particularly likely to produce invalid results.

(2) Example. A randomized controlled clinical trial is conducted to determine the efficacy of a chemotherapeutic regimen for treating Hodgkin's disease. A significant improvement in the treated group leads the research team conducting the trial to carry out a post-hoc comparison of efficacy in patients treated early in the course of the disease versus patients in the advanced stages. That is, treatment and control groups are **post-stratified** into subgroups of patients with early and advanced Hodgkin's disease. Statistical analyses to determine treatment efficacy are then carried out for each of the newly formed subgroups.

(a) Because the assignment of subjects to the "early" or "advanced" subgroups is not random, subsequent comparisons are subject to the same biases that plague cohort studies (see Ch 8 II C 2 a, 3 a).

(b) Even if the original sample is large enough to satisfy the requirements of the subsequent statistical analysis, the post-stratification subgroups are likely to be too small to provide enough power to detect significant differences (see II B 2 a).

b. Multiple comparisons. Comparisons between a large number of dependent variables or post-stratification subgroups increase the likelihood of spurious significant differences. This issue is discussed in greater detail in Chapter 17.

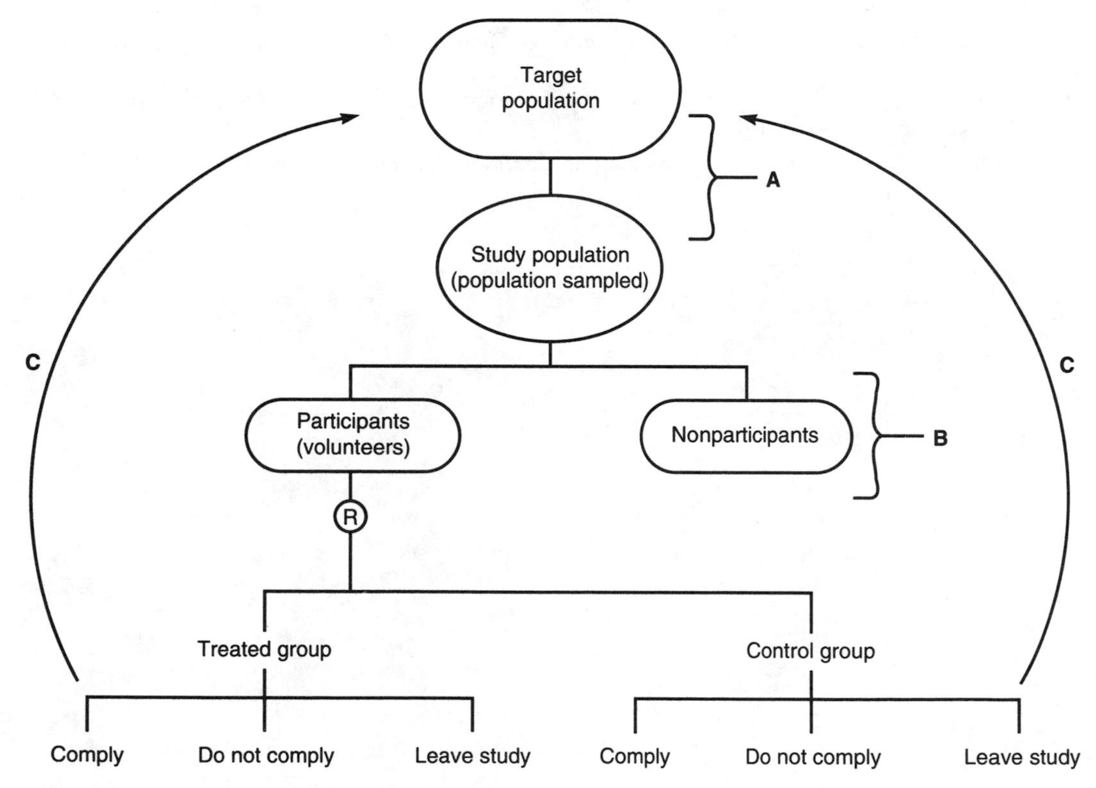

Figure 9-4. Threats to the external validity of a randomized controlled clinical trial. The ability to generalize from study results to the target population may be compromised by systematic differences between the study and target populations (*A*), volunteerism (*B*), or disparities in compliance between study participants and ordinary patients (*C*). *R* = random allocation.

III. GENERALIZING FROM THE STUDY RESULTS TO THE TARGET POPULATION

A. The results of even a well-conducted, properly analyzed clinical trial may apply only to the study population. For study results to be applicable to the target population, the study subjects and environment must resemble those found in the target population as closely as possible.

B. **Factors that can compromise the external validity** of a randomized controlled clinical trial may occur at various stages of the trial (Figure 9-4) and include the following.

1. **Systematic differences between the study and target populations.** For example, study patients cared for at an academic referral center are likely to be more ill than the target population of patients presenting for treatment in a particular family practice setting (see Figure 9-4, *A*).

2. **Volunteerism.** As discussed in Chapter 8 II C 4 c, individuals who volunteer for a study are likely to differ systematically from those who decline to participate. Volunteers, for example, may be healthier, more motivated, or better educated than nonresponders (see Figure 9-4, *B*).

3. **Disparities in patient compliance.** During a randomized controlled clinical trial, heroic measures may be employed to ensure patient compliance (see Figure 9-4, *C*). Because compliance in the target population is likely to be far more haphazard, a treatment proven effective under the ideal conditions of the trial may be less so in the "real world."

BIBLIOGRAPHY

Fletcher RH, Fletcher SW, Wagner EH: *Clinical Epidemiology: The Essentials.* Baltimore, Williams and Wilkins, 1982, p 134.

Gehlbach SH: *Interpreting the Medical Literature: A Clinician's Guide.* Lexington, MA, DC Heath, 1982, p 21.

Mausner JS, Kramer S: *Mausner and Bahn's Epidemiology: An Introductory Text, 2nd ed.* Philadelphia, WB Saunders, 1985, p 197.

Sackett DL, Gent M: Controversy in counting and attributing events in clinical trials. *N Engl J Med* 301:1410–1412, 1979.

PROBLEMS

9-1. Two neurologists hypothesize that patients who receive physical therapy after a cerebrovascular accident (CVA) as soon as their condition stabilizes experience a lesser degree of permanent impairment than those for whom physical therapy is delayed. Outline a randomized controlled clinical trial that these physicians could conduct to test their hypothesis.

9-2. Is an observational or an experimental study more likely to yield a valid conclusion regarding the hypothesis proposed in problem 9-1? Why?

Solutions on p 377.

STUDY QUESTIONS

Directions: Each of the numbered items or incomplete statements in this section is followed by answers or by completions of the statement. Select the **one** lettered answer or completion that is **best** in each case.

1. Which of the following statements explains the principal difference between an experimental study and a prospective cohort study?

(A) In an experimental study, subjects are followed forward in time from intervention to outcome
(B) An experimental study requires larger sample sizes
(C) In an experimental study, subjects are randomly allocated to comparison groups
(D) A control group is required in an experimental study
(E) Participants in an experimental study are randomly selected from the target population

Questions 2–3

In a study to evaluate the efficacy of a new antiviral agent in curing the common cold in young children, 100 children between the ages of 2 and 8, diagnosed with colds by participating pediatricians, were given the new drug. One week later, the investigators conducting the study observed that 90 of the 100 subjects were asymptomatic. They concluded that the antiviral drug was highly effective in curing children's colds.

2. Which of the following statements regarding this conclusion is correct?

(A) The conclusion is valid for the study population
(B) The conclusion is invalid because the investigators measured prevalence rather than incidence
(C) The conclusion is invalid because the frequency measure reported does not have an appropriate denominator
(D) The conclusion is invalid because it is not generalizable to the target population of all children with colds
(E) The conclusion is invalid because the study lacks an appropriate control group

3. Plausible explanations for the observed result include all of the following EXCEPT

(A) recovery may be attributed to the typical course of the common cold in children, rather than to the new drug
(B) subjects may have come from families that were especially health conscious; as a result, they received care measures in addition to the new drug (e.g., over-the-counter cold remedies, special diets, or extra bed rest)
(C) symptomatic relief (the reported outcome) is not necessarily equivalent to a cure (true response variable)
(D) the new drug is effective in alleviating cold symptoms in children
(E) selection bias, rather than the drug, is likely to be responsible for the observed result

4. An occupational health planner who wanted to see if a new safety device would help reduce accident rates among auto workers conducted a study over a two-year trial period involving two randomly selected manufacturing plants. The new device was installed at the first plant (plant A), while the second plant (plant B) continued to operate with only routine safety equipment and procedures. During the observation period, five accidents were recorded at plant A and 20 at plant B. The investigator concluded that the new safety device was efficacious in decreasing work-related accidents. Which of the following statements best describes this conclusion?

(A) It is valid for the study population
(B) It is invalid because prevalence, rather than incidence, was measured
(C) It is invalid because the study lacked a control group
(D) It is invalid because the control group and the treatment group were drawn from different settings

---

1-C 4-D
2-E
3-E

5. The internal validity of a randomized controlled clinical trial would be most seriously compromised by

(A) the failure of randomization to generate comparable comparison groups
(B) systematic differences between the study and target populations
(C) comparison groups of disparate sizes
(D) the use of a statistical test with low power
(E) the use of volunteers, rather than randomly selected individuals, as study subjects

6. In a randomized controlled clinical trial of a new antihypertensive agent, patients with diastolic blood pressure readings greater than 90 mm Hg on their most recent clinic visit are selected as subjects. A patient assigned to the experimental treatment group is normotensive at his first clinic visit after initiating drug therapy. All of the following factors represent plausible explanations for this observation EXCEPT

(A) the Hawthorne effect
(B) random variation in blood pressure measurements
(C) regression to the mean
(D) efficacy of the experimental drug
(E) measurement error

7. The principal reason why a randomized controlled clinical trial is superior to an observational study is

(A) study results are always applicable to the target population
(B) migration bias is eliminated
(C) subjects are assigned to comparison groups, rather than self-selecting their group status
(D) random allocation guarantees that comparison groups are equally vulnerable to potentially confounding variables
(E) both measurement and observer error are eliminated

Questions 8–9

A pharmaceutical company conducted a double-blind randomized controlled clinical trial of a new antipruritic drug. One group of subjects received a lotion containing the experimental drug, a second group received an identical lotion lacking the active ingredient, and a third group received no treatment at all. Two weeks later, the "itching intensity" scores (the lower the score, the less severe the itching) for the three groups were 12 (experimental group), 15 (placebo group), and 65 (no treatment).

8. Which of the following factors best explains the result obtained for subjects who received the lotion lacking the active ingredient?

(A) The Hawthorne effect
(B) Regression to the mean
(C) Selection bias
(D) The placebo effect
(E) Failure to adequately blind subjects and investigators

9. Based on the data, the company would be most correct in making which of the following conclusions?

(A) The experimental drug is equivalent to no treatment at all because its most likely mechanism of action is the placebo effect
(B) The experimental drug is not beneficial because its effect is similar to that of the placebo
(C) The efficacy of the experimental drug cannot be determined because the study lacks a concurrent control group
(D) The experimental drug is beneficial but may not be efficacious

10. Post-randomization bias can occur for all of the following reasons EXCEPT

(A) varying rates of noncompliance among comparison groups
(B) cross-over of subjects from one treatment group to another
(C) differential dropout rates among comparison groups
(D) migration bias
(E) random loss of subjects among comparison groups

5-A 8-D
6-A 9-D
7-C 10-E

Directions: Each group of items in this section consists of lettered options followed by a set of numbered items. For each item, select the **one** lettered option that is most closely associated with it. Each lettered option may be selected once, more than once, or not at all.

Questions 11–13

Match each protocol described below to the relevant study design.

(A) Experimental study: completely random design
(B) Experimental study: randomized block design
(C) Observational study: prospective cohort study
(D) Observational study: historical cohort study

11. Subjects volunteering for a study of atherosclerotic heart disease were divided into three groups according to their degree of vascular plaque accumulation (low, moderate, or high). Within each of the three strata, subjects were assigned one of three diets: very low fat diet (<10% of total calories from fat), low fat diet (<30% of total calories from fat), or a reduced calorie diet with no restrictions on fat intake. At the end of a five-year follow-up period, vascular plaque accumulation was assessed in the three diet groups to determine which diet led to the greatest reduction in plaque accumulation.

12. A study is carried out to compare the effect of two low fat diets on vascular plaque accumulation. The investigator conducting the study obtained a list of individuals who had been continuously enrolled between 1985 and 1990 in one of two commercial weight loss programs based on the two diets. Participants in both diet programs were invited to participate in the study.

13. A group of patients with coronary heart disease volunteered to participate in a study of the relationship between diet and vascular plaque accumulation. The subjects were randomly assigned to one of three diets: very low fat diet (<10% of total calories from fat), low fat diet (<30% of total calories from fat), or a reduced calorie diet with no restrictions on fat intake. At the end of a five-year follow-up period, vascular plaque accumulation in the three diet groups was compared to determine which diet had achieved the greatest reduction in plaque accumulation.

11-B
12-D
13-A

ANSWERS AND EXPLANATIONS

1. The answer is C *[I A 1]*
Randomization (i.e., the investigator controls which subjects receive the experimental intervention by random allocation of subjects to comparison groups) occurs in an experimental study design and distinguishes that study design from an observational design, such as the prospective cohort study. In both an experimental study and a prospective cohort study, patients are followed forward in time from exposure or intervention to outcome. Both designs employ a control group and, ideally, subjects who have been randomly selected from the target population.

2–3. The answers are: 2-E *[I A 2 b]*, **3-E** *[Ch 8 II C 2; I A 2 b (1)–(4)]*
This is an example of an uncontrolled clinical trial, in which subjects serve as their own controls. Without a separate control group to provide a basis for comparison, the investigator cannot be certain that the new drug was responsible for the observed recovery rate (i.e., the study lacks internal validity). The response variable in this study is the cure rate, rather than incidence or prevalence. Its numerator is the number of asymptomatic (presumably "cured") children (90), while its denominator is the total number of study subjects (100). Lack of external validity (i.e., the study results are not applicable to the target population of all children with colds) does not compromise the validity of conclusions regarding the study population.

Selection bias cannot explain the observed results, since only a single group was studied. Predictable recovery from the disease (most cold symptoms resolve within a week) or extra intervention on the part of parents may explain the observed outcome. The response measured (symptomatic relief) does not necessarily equal the hypothesized outcome (cure rate); patients may still harbor the virus in the absence of symptoms. The experimental drug may actually be effective in providing symptomatic relief, although the validity of this alternative cannot be demonstrated on the basis of the study.

4. The answer is D *[II B 3 a (1) (b)]*
Studies that utilize a control group drawn from a different location or time than the experimental group may easily produce erroneous conclusions. In this study, for example, differences in the enforcement of safety regulations, work quotas, or the average age of employees, rather than the new safety device, may have accounted for the observed differences in the occurrence of accidents. A study using two groups of workers employed at the same plant would have been less vulnerable to such potential sources of bias.

5. The answer is A *[II B 1, C 1 a–d]*
The internal validity of a study depends upon an unbiased comparison between groups. In a randomized controlled clinical trial, the purpose of randomization is to equalize the effect of potentially confounding variables and reduce the chance of bias. However, while randomization guarantees that comparison groups will be equal "in the long run" (i.e., for large sample sizes), it may fail to do so "in the short term" (i.e., for small sample sizes). In a similar fashion, while an infinitely large number of tosses of a fair coin will yield an equal number of heads and tails, a small number of tosses (e.g., 10) often results in a disparate proportion of either heads or tails. When randomization fails to generate comparable study groups, a biased comparison will ensue. Systematic differences between the study and target populations threaten the external, but not internal, validity of an experimental study. The use of comparison groups of differing sizes reduces the efficiency of statistical analysis, but does not affect the internal validity of study conclusions. Similarly, while the use of a statistical test of low power may lead to a false negative result (i.e., failure to observe a significant difference between comparison groups when such a difference actually exists), it does not compromise internal validity. By increasing the likelihood that the study population is not representative of the target population, volunteerism jeopardizes external, not internal, validity.

6. The answer is A *[I A 2 b (1)–(4)]*
The Hawthorne effect (i.e., the tendency of subjects to alter their behavior because they are the object of scrutiny), is an unlikely explanation for changes in a physiological variable, such as blood pressure. However, such measurements are vulnerable to random variation. An abnormally high diastolic pressure on one reading may simply represent a random excursion of the measurement from its normal steady state value. A second reading may represent a random excursion below the normal steady state value. The tendency of physiological measurements to ultimately return to this baseline value is known as regression to the mean. Measurement error on one or both readings may account

for the observed change in diastolic pressure. Obviously, the normotensive reading may also result if the new drug is efficacious.

7. The answer is C *[I A 1; II C 1 b]*
Random assignment of subjects to comparison groups eliminates self-selection of group status and reduces bias. Consequently, the results of a randomized controlled clinical trial are more likely to be valid than the results of an observational study. Randomized controlled clinical trials are no more likely to use randomly selected study subjects than an observational study. Therefore, they are equally likely to generate results that are applicable only to the study population. Migration bias, which results when subjects drop out of the study or switch comparison groups for reasons related to the treatment, as well as measurement and observer error, are all potential sources of bias, regardless of study design. Although randomization equalizes the effect of potentially confounding variables in the long run (i.e., for large sample sizes), it may fail to do so in the short run (analogous to the failure of 10 tosses of a fair coin to result in 5 heads and 5 tails). Migration bias may also threaten the original random assignment of subjects to study groups. Thus, although randomization strives to equalize comparison groups with respect to potential confounding variables, it does not guarantee that this goal will be achieved.

8–9. The answers are: 8-D *[II B 3 b (2) (b)]*, **9-D** *[I B 7]*
The placebo effect—the difference in outcome experienced by patients who receive a placebo compared to the outcome experienced by patients who receive no treatment at all—occurs in response to the intervention itself, rather than to its specific mechanism of action. The similar intensity scores of the experimental group and the placebo group suggest that the placebo effect has occurred. The results obtained for both the treatment and the placebo groups would be equally distorted by the Hawthorne effect; hence, it is an unlikely explanation. Regression to the mean is not a plausible explanation because subjects were not selected for study based on an extreme value of the outcome being measured. Failure to adequately blind investigators or subjects would have been expected to result in a value for the placebo group similar to that obtained for untreated controls.

The company may conclude that the drug is beneficial (i.e., it relieves itching), although its primary mechanism of action may be the placebo effect. Most drugs achieve their effects through a combination of a specific (i.e., over and above placebo) and a nonspecific (i.e., placebo) component. The magnitude of the specific component determines the efficacy of the drug, while the total of both components determines whether the drug is beneficial. Many beneficial drugs have a nonspecific mechanism of action. Clearly, the agent under investigation in this study may fall into this category—it represents an improvement over no treatment at all, but its efficacy is uncertain.

10. The answer is E *[II C 3]*
If subjects are randomly, rather than systematically, lost to follow-up, bias does not ensue. However, bias may be introduced when post-randomization changes are related to the treatment received and may alter the outcome. For example, a greater rate of noncompliance among the members of one comparison group is a potential source of bias. Subjects who systematically switch treatment groups, or drop out of the study for reasons related to the treatment, may also introduce post-randomization bias.

11–13. The answers are: 11-B *[II C 2 b]*, **12-D** *[Ch 8 II B 2 b]*, **13-A** *[II C 2 a]*
In the first protocol, the investigators reasoned that the baseline degree of plaque accumulation could alter any reduction resulting from dietary modification (e.g., a large decrease might occur in patients with high initial levels of accumulation, while only small decreases would be observed in patients with lower baseline levels of plaque accumulation). To control for this potentially confounding effect, they chose a randomized block design, in which subjects were first grouped into strata according to their baseline values for plaque accumulation. The random assignment of subjects to the three diets was then carried out within each group. Because subjects were randomly assigned to comparison groups, this is an experimental, rather than an observational, study.

In contrast, subjects in the second study self-select their group status. Thus, this is an observational study design. Because both exposure (to one of the two low fat diets) and the outcome (plaque accumulation) had already occurred when the study was initiated, this is an example of a historical cohort study.

The third protocol follows an experimental study design, since subjects were randomly allocated to the three comparison groups. Unlike the first study, no restrictions were imposed on randomization. Hence, this is an example of a completely random design.

10
Statistical Building Blocks

I. INTRODUCTION AND BASIC DEFINITIONS. Every clinician needs to understand the fundamental principles of statistical reasoning in order to appraise the medical literature intelligently. The basic "building blocks" for this understanding are introduced in this chapter and applied specifically in subsequent chapters, which review individual tests of statistical significance.

A. Parameters and statistics

1. Definitions

a. A **parameter** is a value or characteristic associated with a **population**. Parameters are fixed constants and are typically denoted by Greek letters, such as the **population mean,** denoted by the Greek letter μ, and the **population standard deviation,** represented by the Greek letter σ.

b. A **statistic** is a value or characteristic calculated from a sample. A statistic calculated from a single sample is a fixed quantity, while that based on repeated sampling may be considered an outcome of a random variable. Statistics are typically denoted by Roman letters, for example, the **sample mean** ($\bar{Y}$) and **sample standard deviation** (s).

2. Relationship. Sample statistics are **estimates** of the corresponding population parameters (e.g., $\bar{Y}$ is an estimate of μ).

B. Frequency distributions

1. Definition. A **frequency distribution** is a tabulation, by table or graph, of the frequency at which values occur in a set. It may be empirical or theoretical.

2. Empirical frequency distribution

a. Definition. An empirical frequency distribution is tabulated from data based on a **sample** of people drawn from the population of interest.

b. Example. Table 10-1 and Figure 10-1 are tabular and graphic (i.e., a histogram) frequency distributions calculated directly from the data in Table 1-1.

3. Theoretical frequency distribution

a. Definition. A theoretical frequency distribution is the frequency distribution that could be constructed if data were collected from all possible elements of the **population**.

Table 10-1. Grouped Frequency Distribution for Data in Table 1-1

RBC Cholinesterase (μmol/min/ml)	Frequency	Relative Frequency
5.95–7.95	1	.029
7.95–9.95	8	.229
9.95–11.95	14	.400
11.95–13.95	9	.257
13.95–15.95	2	.057
15.95–17.95	1	.029
Total	35	1.000

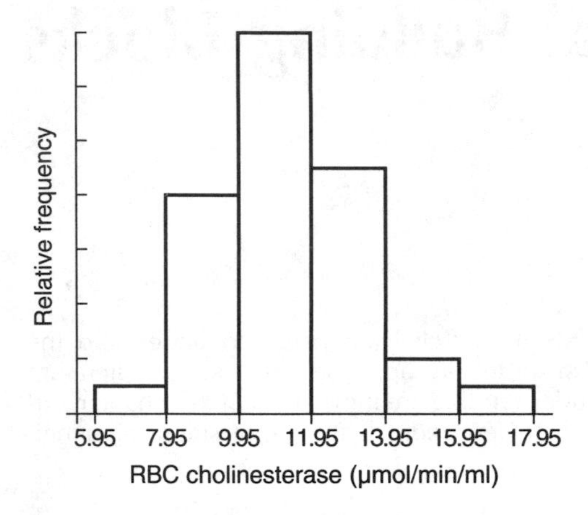

Figure 10-1. Histogram of the frequency distribution of the RBC cholinesterase data presented in Table 10-1. (Reprinted from Duncan RC, Knapp RG, Miller MC III: *Introductory Biostatistics for the Health Sciences,* 2nd ed. Albany, Delmar, 1983, p 6.)

b. Shape. The theoretical frequency distribution of a **continuous random variable** (i.e., a variable measured experimentally using an interval/ratio scale) is a smooth curve.

(1) The shape of a histogram based on a sample (i.e., the empirical frequency distribution) provides an "estimate" of the shape of the corresponding theoretical frequency distribution of the population giving rise to the sample.

(2) The **probability density function** is a mathematical formula that describes the shape of a particular theoretical frequency distribution. The area under the probability density function $f(y)$ between two points, a and b, represents the specific probability that a particular value of the variable will fall within this range.

c. Example. The empirical frequency distribution shown in Figure 10-1 can be used to approximate the theoretical frequency distribution of red blood cell (RBC) cholinesterase values in the entire population.

(1) In Figure 10-1, the area enclosed by each rectangle corresponds to the frequency at which RBC cholinesterase values fall within the given interval. In terms of relative frequency, each rectangle represents the proportion of the total area of the histogram that lies between the specified points (e.g., 2.9% of the total area is enclosed by the rectangle corresponding to RBC cholinesterase values between 5.95 and 7.95).

(2) If 3500, rather than 35, RBC cholinesterase values were tabulated, the width of each interval would be considerably smaller (Figure 10-2). Thus, as the sample size approaches the true population size, the discrete "jumps" in the histogram grow smaller and the shape of the histogram approaches a smooth curve.

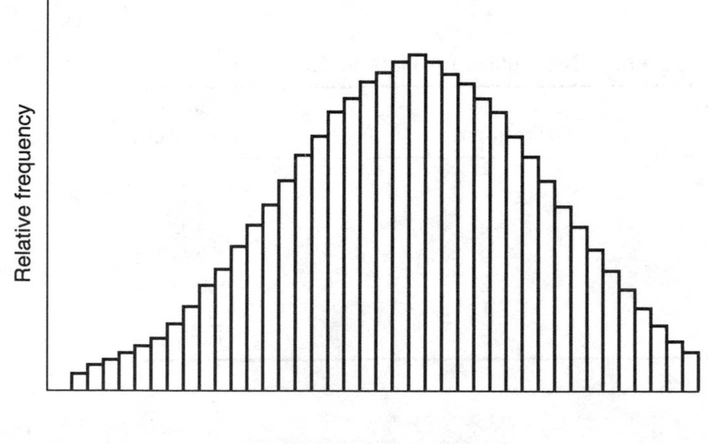

Figure 10-2. Histogram of the frequency distribution of RBC cholinesterase values collected from a large sample. Compare the small intervals in this histogram with the intervals comprising the histogram shown in Figure 10-1.

II. THE NORMAL (GAUSSIAN) DISTRIBUTION

A. Introduction. The most important probability density function used in statistical applications is the **normal,** or **gaussian, distribution**. The distinguishing features of this distribution are summarized in Chapter 3 I B 1 a, in the context of determining normal ranges of clinical values. The most important of these features are reviewed below.

B. Key features of the normal distribution

1. **Shape.** The normal distribution is represented by a **smooth, bell-shaped curve that is symmetric about the population mean μ** (Figure 10-3). The exact shape of a particular normal distribution is governed by:
   a. The **location of its center** (i.e., μ) on the horizontal axis
   b. The **degree of spread** of points around the center, which is the standard deviation, σ

2. **Area under the curve**
   a. The **relative frequencies** of values comprising the normal distribution are represented graphically in Figure 10-4.

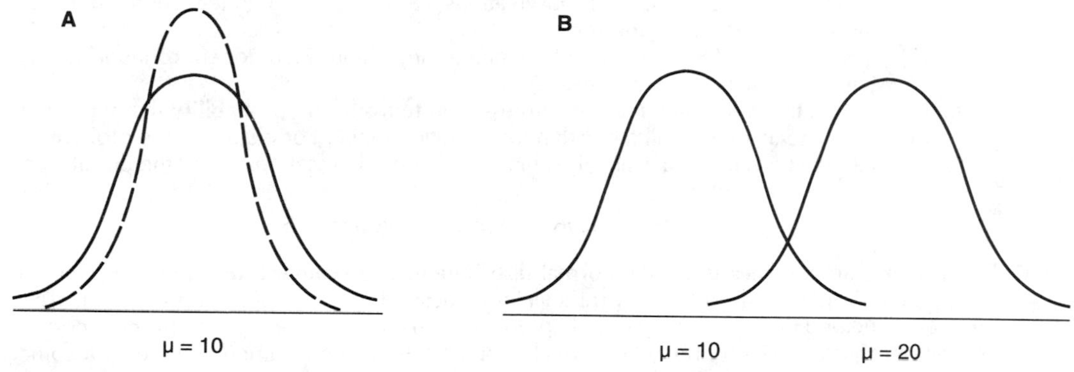

Figure 10-3. A comparison of normal distributions with different standard deviations (σ) and means (μ). The centers of the two normal distribution curves in *A* are located at the same point on the horizontal axis (i.e., the curves have the same mean). The frequency distribution of the *dashed curve* has a smaller spread (i.e., a smaller standard deviation) about the mean than the frequency distribution represented by the *solid curve*. The two normal distribution curves in *B* have different means but the same spread of values about their respective means and, hence, the same standard deviation.

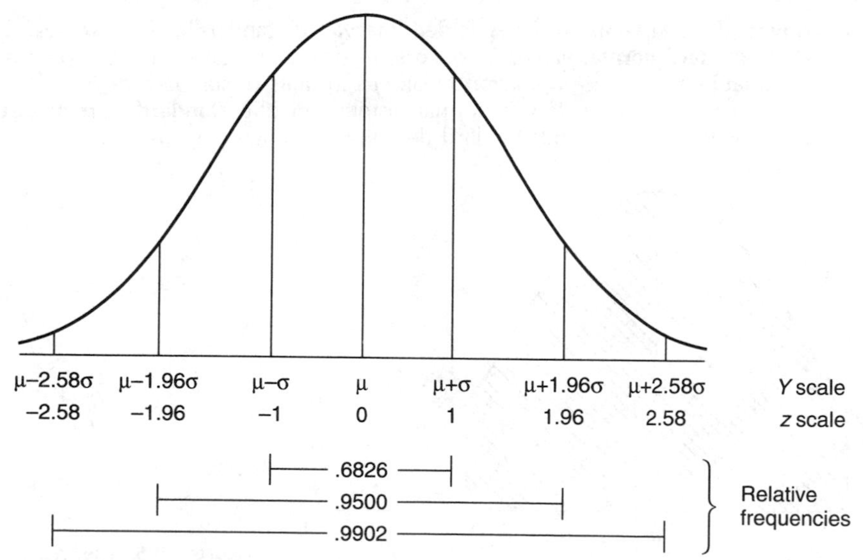

Figure 10-4. A normal distribution, showing the relative frequencies (probabilities) of values falling within specified intervals about the mean. Also shown are the standard deviations (*z scale*) corresponding to these values.

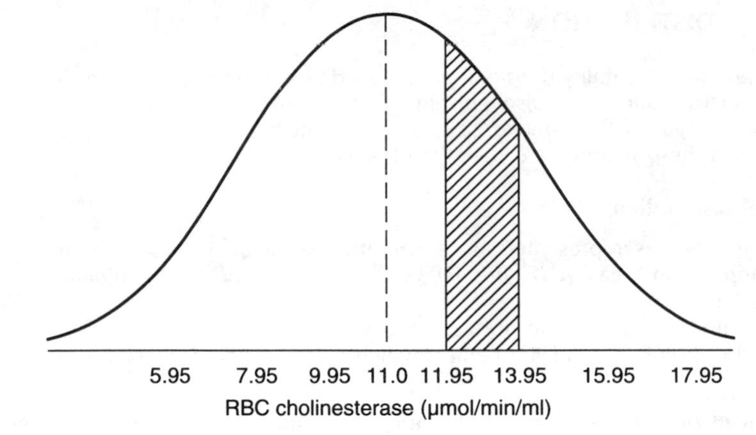

Figure 10-5. Normal curve depicting the interval of interest (i.e., RBC cholinesterase values between 11.95 and 13.95 μmol/min/ml) for problem 1 (II C 1). The *shaded area* under the curve is equivalent to the relative frequency (probability) sought.

(1) Fifty percent of the observations lie above and 50% lie below the mean.

(2) Approximately 95% of the observations lie within 1.96 (often rounded to 2) standard deviations of the mean.

(3) Approximately 99% of the observations lie within 2.58 (often rounded to 3) standard deviations of the mean.

b. A given relative frequency may be expressed in terms of the **probability** that the value of a random variable Y will fall within the specified range. For example, the probability (P) that a given value of Y will fall within 1.96 standard deviations (σ) of the population mean μ may be written

$$P(\mu - 1.96\sigma \leq Y \leq \mu + 1.96\sigma) = .95$$

C. Determining probabilities using the normal distribution. The **standard normal table** (Table B in Appendix 3) can be used to find probabilities associated with any normal distribution. This process is outlined in the following sample problems, which are based on the RBC cholinesterase data in Figure 10-1 and Table 10-1. In all problems, it can be assumed that the data come from a healthy population with a mean of 11 μmol/min/ml and a standard deviation of 2 μmol/min/ml.

1. Problem #1 (Figure 10-5). What is the probability (i.e., the area under the curve) that an individual randomly selected from the population will have an RBC cholinesterase value between 11.95 and 13.95 μmol/min/ml?

2. Solution

a. Convert the endpoints of the specified interval to standardized (z) scores.

(1) A **standard normal value,** or ***z* score,** measures the difference between a random variable Y and the population mean (μ) in units of standard deviation (σ).

(a) The z scores follow a normal distribution (the **standard normal distribution**) with mean = 0 and standard deviation = 1 (Figure 10-6).

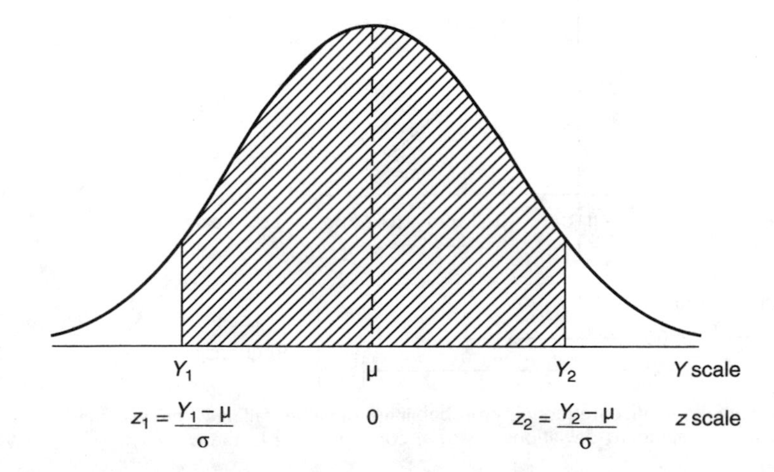

Figure 10-6. The relationship between raw scores (Y scale) and standardized scores (z scale).

(b) Ninety-five percent of z scores lie between -1.96 and $+1.96$ (the signs indicate values to the left and right of the mean, respectively)

(2) Values of Y are converted to z scores using the formula

$$z = \frac{\text{observation} - \text{its population mean}}{\text{its standard deviation}}$$

For this problem, the observations to be converted are RBC cholinesterase values, represented symbolically as Y_i, and the population mean and standard deviation are μ and σ, respectively. Thus,

$$z = \frac{Y_i - \mu}{\sigma}$$

(3) The z scores corresponding to the endpoints of the stated interval (i.e., $11.95 = Y_1$ and $13.95 = Y_2$) are calculated as follows:

$$z_1 = \frac{Y_1 - \mu}{\sigma} = \frac{11.95 - 11.00}{2} = .48$$

$$z_2 = \frac{Y_2 - \mu}{\sigma} = \frac{13.95 - 11.00}{2} = 1.48$$

b. Represent the desired area under the standard normal curve between z_1 and z_2 (i.e., the required probability) **graphically and use the standard normal table** (Table B, Appendix 3) **to compute the area defined by this interval.**

(1) **Graphic representation of z scores** (see Figure 10-6)

(a) Point Y_1 lies z_1 standard deviations below the population mean, μ (therefore, z_1 will have a **negative** sign), while point Y_2 lies z_2 standard deviations above the mean.

(b) $P(Y_1 \leq Y \leq Y_2)$ on the Y scale is equivalent to $P(z_1 \leq z \leq z_2)$ on the z scale.

(2) **Interpretation of the standard normal table**

(a) Column 1 of the standard normal table (Table B, Appendix 3), labeled z, contains values of z to the first decimal place.

(b) The values **above** columns 2–11 represent the second decimal place values associated with a given z value.

(3) **Finding the area below a given z value.** The body of Table B (see Appendix 3) is composed of four-digit numbers, each representing the area under the curve **below** (i.e., to the **left** of) the z score defined by that row–column intersection.

(a) **Example 1.** The area below (to the left of) $z = -2.55$ on the standard normal curve is indicated by the four-digit number at the intersection of the row labeled $z = -2.50$ and the column labeled $-.05$ (i.e., .0054). This area is illustrated in Figure 10-7.

(b) **Example 2.** The area below $z = +1.96$ is given as the four-digit number at the intersection of the row labeled $z = +1.90$ and the column labeled .06 (i.e., .9750). In other words, the probability that a given z score will have a value less than $+1.96$ is .9750. The area corresponding to this probability is depicted in Figure 10-8.

(4) **Finding the area above a given z value.** Because the total area under the normal curve must add up to 1, or 100%, the area **above** (i.e., to the **right** of) a given z value is given by (1 − the area below the value).

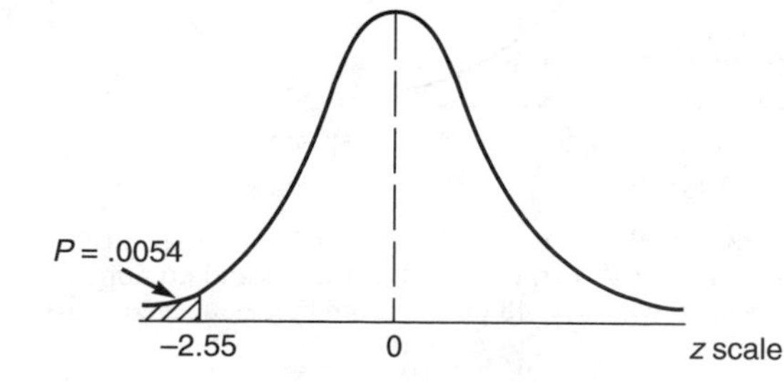

Figure 10-7. Standard normal curve illustrating the area under the curve to the left of (i.e., below) $z = -2.55$ (*shaded area*). This area corresponds to a probability of .0054.

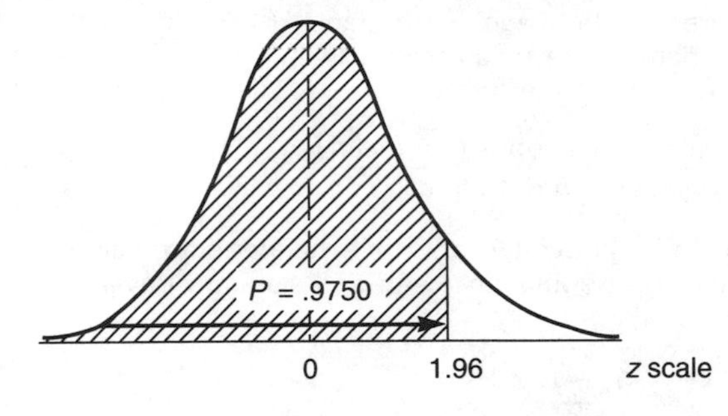

Figure 10-8. Standard normal curve illustrating the area under the curve to the left of (i.e., below) $z = +1.96$ (*shaded area*). This area corresponds to a probability of .9750.

(5) Application to problem 1

(a) Figure 10-9 depicts the specified interval on both the *Y* scale and the standard normal (*z*) scale (*shaded area*).

(b) Probabilities for $z \leq 1.48$ (i.e., .9306) and $z \leq .48$ (i.e., .6844) are obtained from Table B (see Appendix 3). The probability corresponding to the shaded area (i.e., the area under the curve between the defined endpoints) is determined by subtraction:

$$P(.48 \leq z \leq 1.48) = .9306 - .6844 = .2462$$

(c) Thus, the probability that an individual selected at random will have an RBC cholinesterase value between 11.95 and 13.95 μmol/min/ml is .2462. In other words, 24.62% of the population (the **relative frequency**) may be expected to have RBC cholinesterase values falling within this interval.

3. Problem #2. What proportion of people in the population have RBC cholinesterase values greater than 13.95 μmol/min/ml?

4. Solution

a. In problem 1 (II C 1), the area (i.e., proportion, or probability) below $Y = 13.95$ was found to be .9306. The area above $Y = 13.95$, therefore, is calculated as

$$\begin{aligned} P(Y > 13.95) &= P(z > 1.48) \\ &= 1 - .9306 \\ &= .0694 \end{aligned}$$

b. In other words, 6.94% of people in the population can be expected to have RBC cholinesterase values exceeding 13.95 μmol/min/ml.

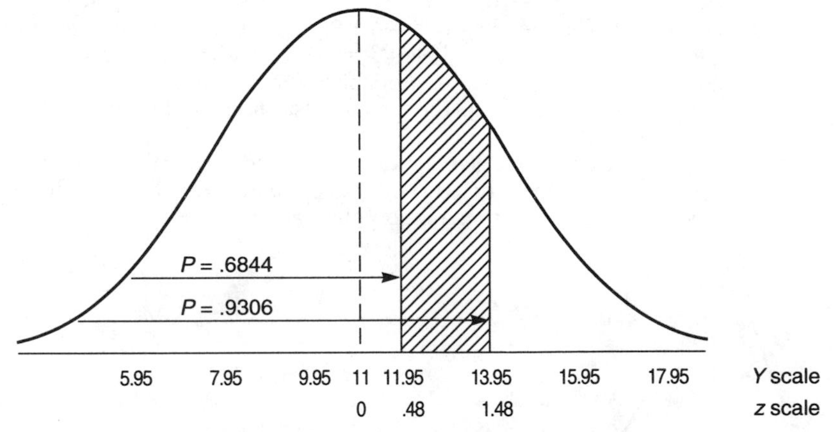

Figure 10-9. Calculation of the probability that a given RBC cholinesterase value will fall between 11.95 μmol/min/ml and 13.95 μmol/min/ml. This probability (*shaded area*) is shown on both the *Y* and standard normal (*z*) scales, and is computed by subtracting the probability associated with $z \leq .48$ (.6844) from that associated with $z \leq 1.48$ (.9306).

5. **Problem #3.** After prolonged contact with a sample of a novel cholinomimetic compound, a 41-year-old pharmaceutical chemist is brought to the emergency room with vomiting, headache, severe hypotension, and bradycardia. Laboratory results reveal an RBC cholinesterase level of 20 μmol/min/ml. How unusual is this RBC cholinesterase value in the healthy population (i.e., how likely is it that this patient's symptoms constitute a toxic reaction to the chemical she handled)?

6. **Solution.** Recall that for any normal distribution, 99% of all values for the given population fall within 2.58 standard deviations of the population mean, while only 1% of values lie beyond 2.58 standard deviations. The specific probability that an RBC value will be equal to or greater than 20 μmol/min/ml is determined as follows.
   a. **Convert the value of interest to a standardized (*z*) score.**
      (1) The *z* score corresponding to an RBC cholinesterase value (Y) of 20 is

$$Y = \frac{Y - \mu}{\sigma} = \frac{20 - 11}{2} = 4.5$$

      (2) Thus, an RBC cholinesterase value of 20 μmol/min/ml lies 4.5 standard deviations above the mean for the healthy population.
   b. **Represent the specified area (*probability*) graphically and use the standard normal table** (Table B, Appendix 3) **to obtain the probability.**
      (1) Figure 10-10 shows the area of interest (*shaded area*) on both the Y and z (standardized) scales.
      (2) The probability corresponding to the shaded area $= P(z \geq 4.5)$. Since the area above the maximum z value (3.89) listed in Table B (see Appendix 3) is .0001, the area above $z = 4.5$ is

$$P(z \geq 4.5) < 1 - .9999 < .0001$$

      (3) In other words, the probability that an individual in the healthy population will have an RBC cholinesterase level greater than or equal to 20 μmol/min/ml is less than .0001. In terms of relative frequency, less than 0.1% of the healthy population will fall in this category. Although the exact mechanism of toxicity is unclear, the chemist's cholinesterase level is clearly abnormal and bears further investigation.

7. **Problem #4.** Between which two RBC cholinesterase values would 95% of the values in the disease-free population be expected to fall?

8. **Solution.** For any normal distribution, 95% of the values for the population lie within 1.96 standard deviations of the population mean. Since, by definition, a z score is the distance (in units of standard deviation) from the mean, the raw score values (Y_1 and Y_2) sought here can be determined by converting from the corresponding z scores ($z_1 = -1.96$ and $z_2 = +1.96$).
   a. Figure 10-11 shows the z values that correspond to the Y values expected in 95% of the healthy population.

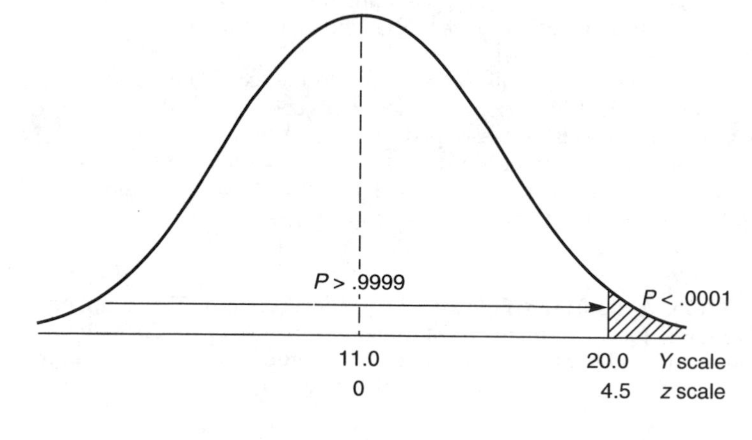

Figure 10-10. Theoretical frequency distribution of RBC cholinesterase values in a healthy population, showing the standardized (z) score for an RBC cholinesterase value of 20 μmol/min/ml. The probability corresponding to the area to the right of (i.e., above) this value (*shaded area*) is calculated to be less than .0001.

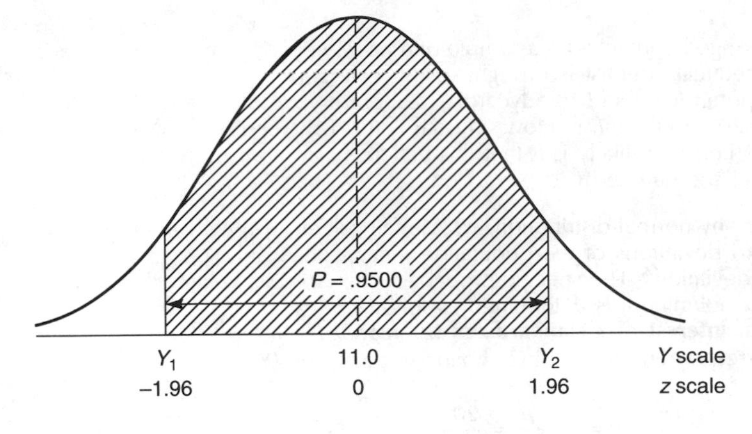

Figure 10-11. Theoretical frequency distribution of RBC cholinesterase values in a healthy population, showing the standardized (z) scores corresponding to the range of Y values (*shaded area*) expected in 95% of the population.

b. The z scores are converted to their raw score equivalents as follows:

$$z_1 = \frac{Y_1 - \mu}{\sigma} \qquad z_2 = \frac{Y_2 - \mu}{\sigma}$$

$$-1.96 = \frac{Y_1 - 11}{2} \qquad +1.96 = \frac{Y_2 - 11}{2}$$

$$Y_1 = 11 - 2(1.96) \qquad Y_2 = 11+2(1.96)$$

$$Y_1 = 7.08 \qquad Y_2 = 14.92$$

c. Thus, 95% of RBC cholinesterase values in the healthy population can be expected to fall between 7.08 and 14.92 μmol/min/ml. This interval constitutes the **normal range** for RBC cholinesterase values, as defined by the gaussian distribution method (see Ch 3 I B 1).

D. Comparing empirical (sample) and theoretical (population) probabilities

1. If the data in a study are based on a sample drawn from a population having a normal frequency distribution, estimates of relative frequency derived from the empirical frequency distribution (i.e., the frequency distribution of the sample) should be similar to estimates of relative frequency obtained using the standard normal table (i.e., the theoretical population frequency distribution). In such a case, the area under the standard normal curve provides a reliable estimate of the true relative frequencies (probabilities) in the population from which the sample was drawn.

2. Example

a. For the RBC cholinesterase data used in problems 1–4 (II C), Table 10-1 indicates that 25.7% of the 35 individuals in the **sample** have RBC cholinesterase values between 11.95 and 13.95 μmol/min/ml.

b. Using the theoretical frequency distribution (i.e., standard normal table; see problem 1, II C 1–2), 24.62% of the **population** have RBC cholinesterase values in this interval.

c. The relative frequency observed in the sample is very close to that calculated using the theoretical frequency distribution. Close concordance between the empirical and theoretical frequencies for all intervals suggests that the population from which the sample was drawn is normally distributed (or at least approximately so). In this case, the standard normal distribution can be used to estimate the true relative frequencies (probabilities) in the population of interest.

III. THE SAMPLING DISTRIBUTION AND THE CENTRAL LIMIT THEOREM. The sampling distribution of means is another name for the frequency distribution of the population of sample means. The central limit theorem, which describes the tendency of this frequency distribution to assume a normal distribution, is a primary tenet of statistical inference.

A. Population of sample means

1. **Definition.** The population of sample means (i.e., of $\bar{Y}$s) is composed of the group of averages calculated for all possible samples of a given size n, selected at random from the population of interest, known as the **parent population**.

2. **Example.** A cardiovascular research team wishes to estimate the average cholesterol level of the population of men who have had an acute myocardial infarction. In this example, the parent population is the population of men who have had an acute myocardial infarction.
   a. The team randomly selects a sample of 100 men from this population and calculates the average cholesterol level of this sample. Because the sample does not constitute the entire population of men who have had an acute myocardial infarction, this sample mean ($\bar{Y}$) is only an **estimate** of the mean cholesterol level in the parent population (μ).
   b. If additional samples of 100 men are selected, a series of different sample means ($\bar{Y}$s) is obtained, each representing a separate estimate of the unknown population mean, μ.
   c. If **all possible samples** of 100 men are selected and the average cholesterol level is calculated for each, a new population—the **population of sample means**—is generated (Figure 10-12).

3. **Properties.** The characteristics of the population of sample means (i.e., its mean, standard deviation, and the shape of its frequency distribution) are central to the process of statistical reasoning. These properties may be obtained empirically for very small finite populations, or they may be derived mathematically.
   a. **Mean.** The **mean of the population of sample means ($\mu_{\bar{y}}$)** equals the mean of the parent population from which the samples were drawn (μ_y):

$$\mu_{\bar{y}} = \mu_y$$

   b. **Standard deviation.** The **standard deviation of the population of sample means** is a measure of how closely the sample means cluster around the population mean μ. Because the averaging process cancels out extreme values, thus reducing variation, the sample means should cluster closer to the true population mean than the individual measurements (Ys) themselves.
      (1) The standard deviation of the population of sample means, $\sigma_{\bar{y}}$, is equal to $\sigma/\sqrt{n}$, where σ is the standard deviation of the individual measurements comprising the parent population and n is the size of the samples constituting the population of sample means (i.e., of $\bar{Y}$s):

$$\sigma_{\bar{y}} = \sigma/\sqrt{n}$$

      (2) The standard deviation of the population of sample means is known as the **standard error of the mean.**
      (3) **Estimating $\sigma_{\bar{y}}$.**
         (a) The **standard deviation (s) of a sample of size n** is an estimate of the population standard deviation, σ.
         (b) The standard error, $\sigma/\sqrt{n}$, is estimated by $s/\sqrt{n}$
         (c) Values for $\bar{Y}$, s, and $s/\sqrt{n}$ can thus be derived from a single sample of n observations. These values provide estimates of the corresponding population parameters, μ, σ, and $\sigma_{\bar{y}}$ (Figure 10-13).
      (4) **Relevance to clinical studies.** Confusion as to whether a particular estimate of the spread about a mean represents the standard deviation of individual observations or the standard error of the mean is rampant in the clinical literature.
         (a) **Standard deviation** (σ) measures the spread of individual measurements (Y_is) about the population mean μ; it is **appropriate for describing the subject-to-subject variation** among measurements of the variable Y.
         (b) **Standard error of the mean** measures the spread of all possible sample means about the population mean μ (i.e., the sample-to-sample variation among sample means). It is **appropriate for describing how precisely the sample mean $\bar{Y}$ estimates the population mean.** That is, the smaller the value of the standard error, the more precise the estimate of μ.
   c. **Frequency distribution**
      (1) If samples are drawn from a normally distributed parent population, the frequency

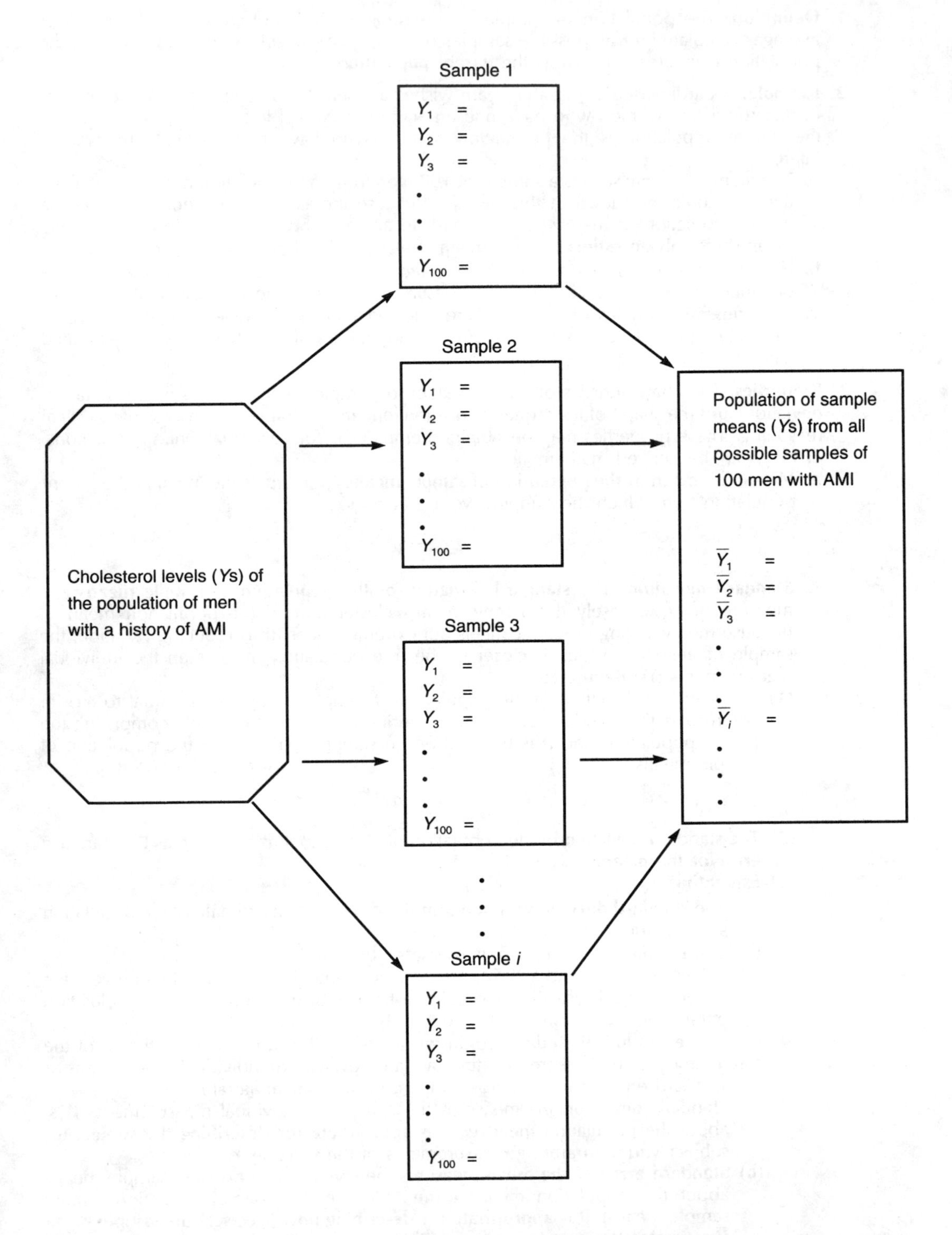

Figure 10-12. The population of sample means. From the parent population of all men with a history of acute myocardial infarction (AMI) are drawn all possible samples of 100 men each. The cholesterol levels of the individual men comprising each sample (Y_1, Y_2, Y_3, and so on) are obtained, and the mean ($\overline{Y}$) for each sample is calculated. The collective means of all possible samples comprise the population of sample means.

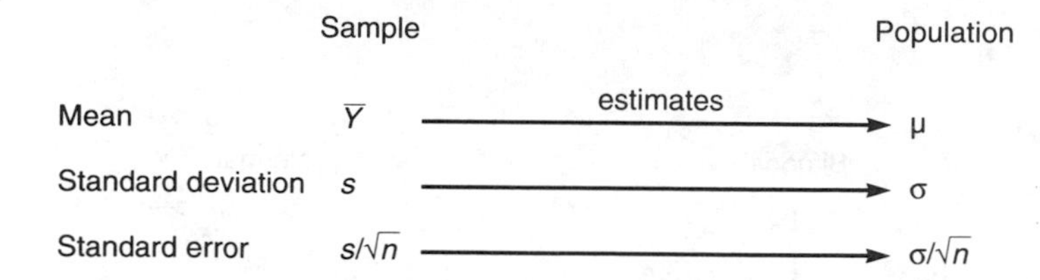

Figure 10-13. Relationship between sample estimates and corresponding population parameters.

distribution of the population of sample means, known as the **sampling distribution,** is also normal. If samples are drawn from a non-normally distributed parent population, the frequency distribution of the population of sample means **approaches the normal distribution as the sample size increases.** This property of the sampling distribution, known as the **central limit theorem,** is the cornerstone of statistical inference.

(2) **The central limit theorem.** Given any parent population with mean μ and standard deviation σ, as the size of the samples (n) drawn from the population increases, the distribution of sample means ($\overline{Y}$s) approaches the normal distribution, with mean μ and standard deviation $\sigma/\sqrt{n}$.

(3) **Examples**

(a) **Normally distributed parent population.** The frequency distribution shown in Figure 10-14A is that of individual cholesterol levels for the population of men who have had an acute myocardial infarction. The sampling distributions shown in Figure 10-14B and Figure 10-14C represent the frequency distributions of sample means for all possible samples of 10 and 100 men, respectively. These frequency distributions, like that of the individual cholesterol levels, follow the normal distribution.

(b) **Non-normally distributed parent population.** Figure 10-15 depicts frequency distributions for two non-normally distributed parent populations (*A*) and for sample means based on all possible samples of two (*B*), five (*C*), and 20 (*D*) individuals that could theoretically be drawn from these parent populations. Note that even for fairly modest sample sizes, the frequency distributions of the sample means begin to assume the characteristic bell shape of the normal distribution.

B. Computing probabilities associated with sample means using the normal distribution. The standard normal table (Table B, Appendix 3) can be used to find probabilities associated with any sample mean, following the same procedure as that outlined in II C.

1. **Problem #1.** If the true mean cholesterol level for the population of men who have had an acute myocardial infarction is 240 mg/dl and the standard deviation is 40, what is the

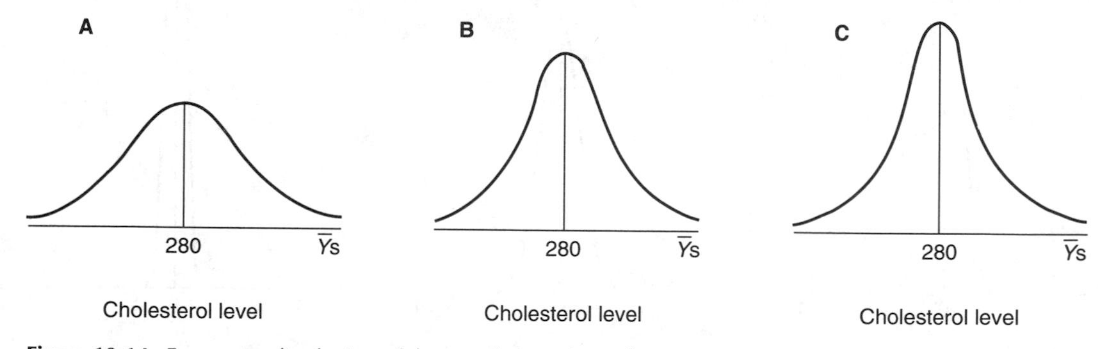

Figure 10-14. Frequency distribution of the population of sample means when the parent population is normally distributed. (*A*) Frequency distribution of the parent population—individual cholesterol levels of all men with a history of acute myocardial infarction (AMI) ($\mu = 280$ mg/dl; $\sigma_{\overline{y}} = 20$). (*B*) Frequency distribution of the population of sample means ($\overline{Y}$s), based on a sample size of 10 men ($\mu = 280$ mg/dl; $\sigma_{\overline{y}} = 20/\sqrt{10}$). (*C*) Frequency distribution of the population of sample means, based on a sample size of 100 men ($\mu = 280$ mg/dl; $\sigma_{\overline{y}} = 20/\sqrt{100}$). Note that the frequency distributions in *B* and *C*, like that of the parent population, are normal distributions. The spread of sample means around the population mean μ is smaller in *C* than in *B*.

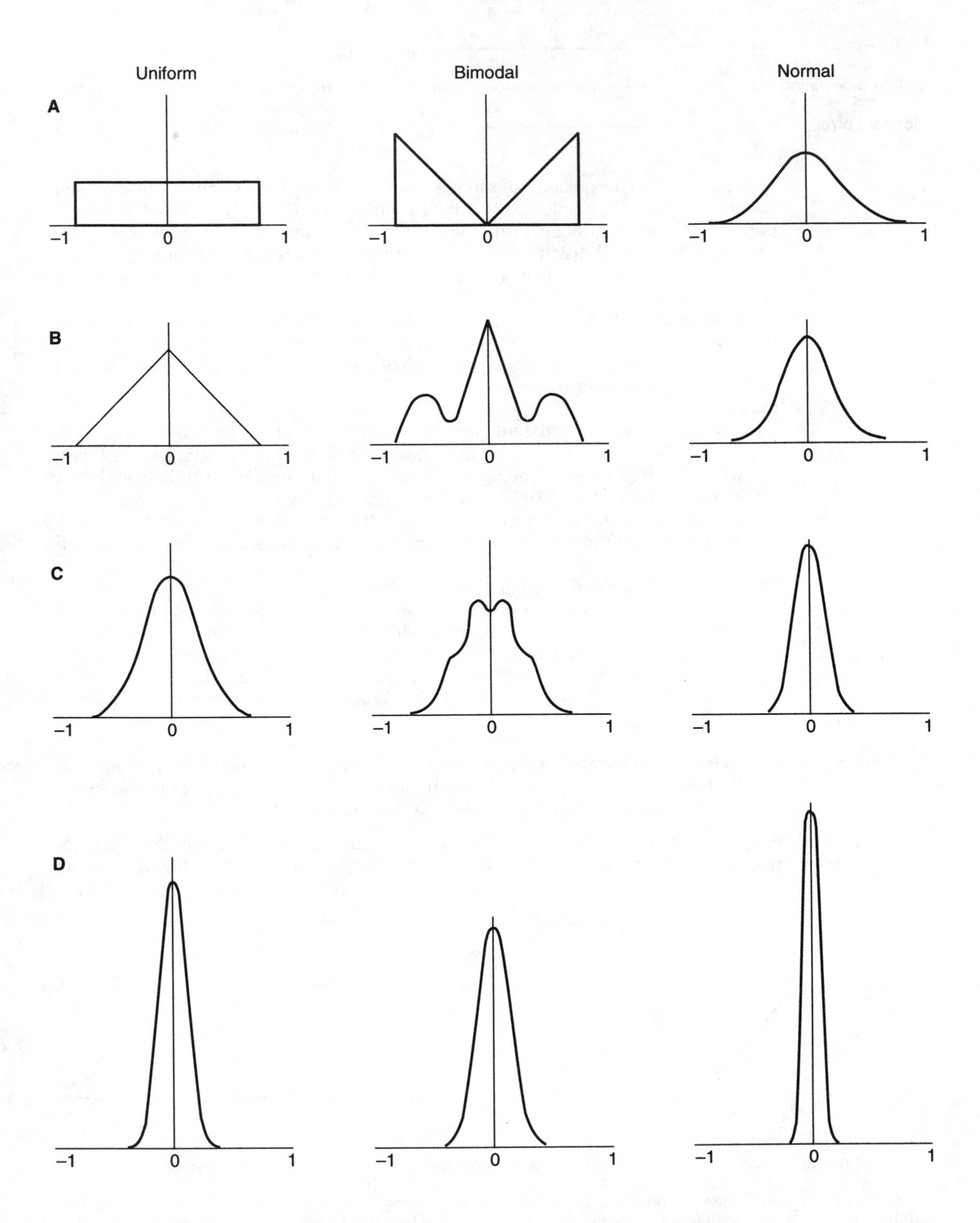

Figure 10-15. Comparison of the frequency distributions of the population of sample means ($\bar{Y}$s) from uniformly, bimodally, and normally distributed parent populations. (*A*) Frequency distributions of individual values comprising the three parent populations. (*B*) Frequency distribution of the population of sample means, based on a sample size of 2. (*C*) Frequency distribution of the population of sample means, based on a sample size of 5. (*D*) Frequency distribution of the population of sample means, based on a sample size of 20.

probability that the mean cholesterol level of a sample of 100 men randomly selected from this population will be equal to or greater than 260 mg/dl?

2. **Solution**
   a. **Convert the sample mean (Y) to a standardized (z) score.**
      (1) The formula used to standardize $\bar{Y}$ is

$$z = \frac{\text{distance between a given } \bar{Y} \text{ and } \mu}{\text{standard deviation of the population of } \bar{Y}\text{s}}$$

$$= \frac{\bar{Y} - \mu}{\sigma_{\bar{Y}}}$$

where σ = the standard deviation of the parent population of individual cholesterol values and $\sigma_{\bar{Y}} = \sigma/\sqrt{n}$

      (2) Using this formula, the z value corresponding to $\bar{Y} = 260$ is calculated as

$$z = \frac{260 - 240}{40/\sqrt{100}} = \frac{20}{4} = 5$$

      (3) In other words, a sample mean of 260 mg/dl lies 5 standard deviations above the population mean of 240 mg/dl.
   b. **Represent the desired probability (area under the curve) graphically and calculate this probability using Table B** (see Appendix 3).
      (1) Figure 10-16 depicts the probability of interest (*shaded area*) on both the $\bar{Y}$ scale and the standardized (z) scale.
      (2) The largest z score tabulated in Table B (see Appendix 3) is $z = +3.89$, for which the associated probability is .9999. The area to the right of (i.e., above) this point is $1 - .9999$, or .0001.
      (3) Therefore, the probability of obtaining a mean cholesterol level in a single sample of 100 men that is equal to or greater than 260 mg/dl from a population with a true mean of 240 mg/dl is much less than .0001, or extremely unlikely.
3. **Problem #2.** A sample of 16 men is selected from the population, rather than a sample of 100. How does this decrease in sample size affect the probability of obtaining a sample mean that is equal to or greater than 260 mg/dl?
4. **Solution**
   a. The variability (standard deviation) among all possible sample means based on samples of 16 men is much greater than that among sample means based on samples of 100 men. That is,

$$\sigma_{\bar{Y}}\ (n = 100) = \sigma/\sqrt{100} = 40/10 = 4$$

$$\sigma_{\bar{Y}}\ (n = 16) = \sigma/\sqrt{16} = 40/4 = 10$$

   b. Hence, a sample mean as extreme as 260 mg/dl is more likely to be obtained with a sample size of 16 than a sample size of 100. The exact probability associated with the

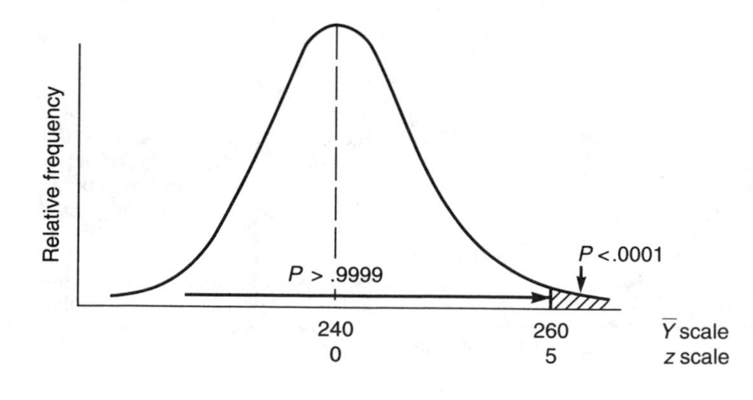

Figure 10-16. Frequency distribution of the population of sample means for all possible samples of 100 men drawn from a population of men with a history of acute myocardial infarction (AMI) (μ = 240 mg/dl; $\sigma_{\bar{Y}} = 40/\sqrt{100}$), showing the standardized (z) score for a sample mean cholesterol level of 260 mg/dl. The probability corresponding to the area to the right of (i.e., above) this z score (*shaded area*) is calculated to be less than .0001.

smaller sample size is

$$z = \frac{\overline{Y} - \mu}{\sigma/\sqrt{n}} = \frac{260 - 240}{40/\sqrt{16}} = \frac{20}{10} = 2$$

$$P(\overline{Y} \geq 260) = P(z \geq 2)$$
$$= 1 - .9772$$
$$= .0228$$

c. Thus, in a population with a true mean cholesterol level of 240 mg/dl, less than .01% of the sample means based on samples of 100 men will be 260 mg/dl or greater, compared to 2.28% of the sample means based on samples of 16 men (Figure 10-17).

5. Problem #3. The frequency distribution of blood ammonia levels in a healthy pediatric population is found to follow a normal distribution, with a mean of 60 μg/dl and a standard deviation of 6.67 μg/dl. What is the probability that a random sample of nine children from this population will have a mean blood ammonia level ≥ 70 μg/dl?

6. Solution

a. Convert the sample mean to a standardized (z) score.

(1) Using the formula in III B 2 a (1), the z score corresponding to $\overline{Y} = 70$ μg/dl is calculated as

$$z = \frac{\text{distance of } \overline{Y} \text{ from the true mean}}{\text{standard deviation of } \overline{Y}\text{s}}$$
$$= \frac{\overline{Y} - \mu}{\sigma/\sqrt{n}}$$
$$= \frac{70 - 60}{6.67/\sqrt{9}}$$
$$= 4.5$$

(2) In other words, the sample mean of 70 μg/dl lies 4.5 standard deviations above the population mean of 60 μg/dl.

b. Represent the desired probability (area under the curve) graphically and calculate this probability from the standard normal table (Appendix 3, Table B).

(1) Figure 10-18 portrays the probability of interest (*shaded area*) on both the $\overline{Y}$ scale and the z scale.

(2) The maximum z score tabulated in Table B (see Appendix 3) is 3.89, for which the corresponding probability value is .9999. The area to the right of (i.e., above) 3.89 is 1 − .9999 = .0001.

(3) Thus, the chance that a random sample of nine children drawn from this population will have an average blood ammonia level equal to or greater than 70 μg/dl is less than .01%.

7. Problem #4. Laboratory results obtained in nine children with Reye's syndrome reveal an average blood ammonia level of 70 μg/dl. Based on the probability calculated in problem

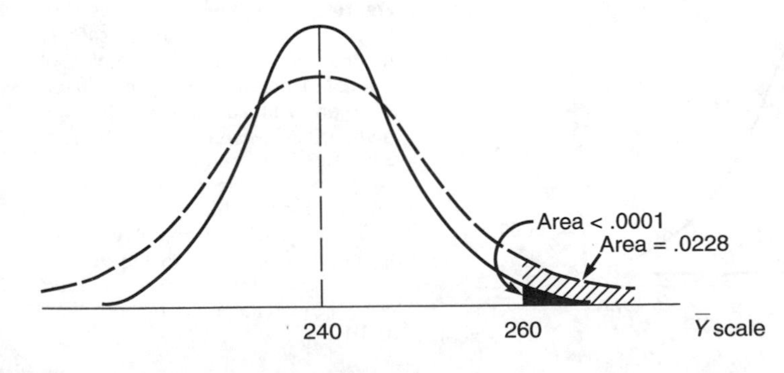

Figure 10-17. Comparison of the frequency distributions of the population of sample means based on all possible samples of 100 men (*solid curve*) and all possible samples of 16 men (*dashed curve*), showing the effect of sample size on the probability of obtaining a sample mean equal to or greater than 260 mg/dl.

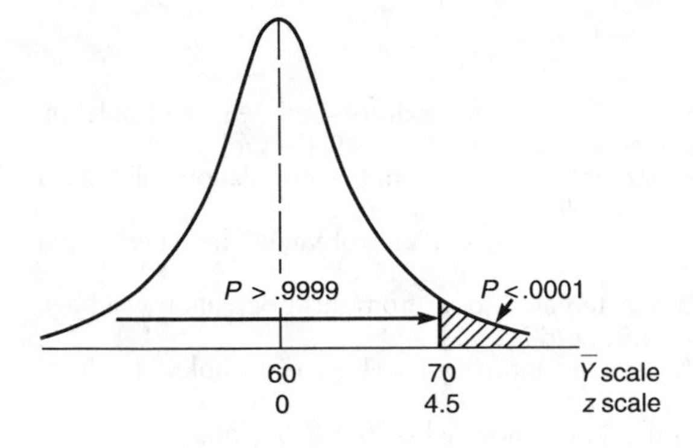

Figure 10-18. Frequency distribution of the population of sample mean blood ammonia levels for all possible samples of nine children drawn from a healthy pediatric population ($\mu = 60$ μg/dl; $\sigma = 6.67$), showing the standardized (*z*) score for a sample mean blood ammonia level of 70 μg/dl. The probability corresponding to the area to the right of (i.e., above) this *z* score is calculated to be less than .0001.

3 (III B 5–6), what is the likelihood that the true mean of the population of pediatric patients with Reye's syndrome is equal to the mean for the healthy pediatric population (i.e., $\mu = 60$ μg/dl)? (For a more detailed discussion of the concepts in this problem, see Ch 12.)

8. Solution. This problem may be restated as "Is it likely that a sample mean equal to or greater than 70 μg/dl could be obtained from a population with a mean $\mu = 60$ μg/dl?"

a. In practice, population means are unknown. The value of a population mean is inferred from the information obtained in a sample drawn from the population.

b. As shown in problem 3 (III B 5–6), the probability (i.e., the likelihood) that a sample mean of 70 μg/dl or greater will be obtained from a population having a true mean of 60 μg/dl is less than .0001, or highly unlikely. It is far more likely that the sample of patients with Reye's syndrome was drawn from a population having a true mean greater than 60 μg/dl. This leads to the following conclusion:

$$\mu_{\text{Reye's syndrome patients}} > 60\ \mu\text{g/dl}$$

$$\mu_{\text{Reye's syndrome patients}} > \mu_{\text{healthy population}}$$

PROBLEMS*

10-1. Serum cholesterol levels in a particular population are normally distributed, with a population mean (μ) = 242.2 mg/dl and a population standard deviation (σ) = 45.4 mg/dl.

a. What is the probability that an individual randomly selected from this population will have a cholesterol level between 219.5 and 259.5 mg/dl?

b. What proportion of individuals in this population have cholesterol values between 139.5 and 219.5 mg/dl?

c. What is the probability that an individual selected at random from this population will have a cholesterol level between 159.5 and 179.5 mg/dl?

d. What is the probability that a randomly selected individual will have a cholesterol level greater than 242.2 mg/dl?

e. What proportion of cholesterol values in this population fall below 219.5 mg/dl?

10-2. Use the standard normal table (Appendix 3, Table B) to find the following values:

a. The area within ± 1.96 standard deviations of the mean

b. The area within ± 2.58 standard deviations of the mean

10-3. The mean systolic blood pressure in a certain population of men between the ages of 20 and 24 is known to be 120 mm Hg, with a standard deviation of 20 mm Hg.

a. A 22-year-old man is selected at random from the population. What is the probability that his systolic blood pressure is less than or equal to 150 mm Hg? Less than or equal to 110 mm Hg?

b. A 20-year-old man selected at random from the population had a systolic blood pressure of 160 mm Hg. How far does this value lie above the population mean, in units of standard deviation? (Note: z values are in units of standard deviation.)

c. Ninety-five percent of all systolic blood pressure readings for the men comprising this population will lie between which two values? Between which two values will 90% of all systolic blood pressure readings lie?

d. What proportion of readings fall between 100 and 140 mm Hg?

e. What proportion of readings are less than 60 mm Hg or greater than 180 mm Hg?

10-4. As part of a hypertension screening project, systolic blood pressures were determined in a random sample of 100 male medical students drawn from the population described in problem 10-3. The mean systolic pressure in this sample was 126 mm Hg.

a. For any sample of 100 men randomly selected from this population, what is the probability of obtaining a sample mean greater than or equal to 126 mm Hg? Recall that the true population mean $\mu = 120$ mm Hg. (Note: The distribution in question is the distribution of mean systolic blood pressure readings, derived from all possible samples of 100 men that could be drawn from this population, rather than the distribution of individual systolic blood pressure readings. According to the central limit theorem, what are the mean and standard deviation of this population of sample means?)

b. How many standard deviations above the true population mean μ does the observed sample mean lie?

c. If samples of 100 men ($n = 100$) were repeatedly drawn from the population, between which two values would 95% of all the sample means lie?

d. Based on the values calculated in problems 10-1a through 10-1c, what conclusions can be drawn regarding the systolic blood pressure of male medical students?

e. If samples of 100 men ($n = 100$) are repeatedly drawn from the population, above which value would 10% of the sample means be expected to fall?

f. What is the probability of selecting a sample with a mean $\bar{Y}$ less than or equal to 115 mm Hg from a population with a true mean systolic blood pressure (μ) = 120 mm Hg?

Solutions on p 378.

*The problems in this section are adapted from Duncan RC, Knapp RG, Miller MC III: *Introductory Biostatistics for the Health Sciences*, 2nd ed. Albany, Delmar, 1983, pp 74–76.

STUDY QUESTIONS

Directions: Each of the numbered items or incomplete statements in this section is followed by answers or by completions of the statement. Select the **one** lettered answer or completion that is **best** in each case.

1. All of the following statements about the normal distribution are true EXCEPT

(A) this distribution is represented by a symmetric, bell-shaped curve
(B) a normally distributed random variable is also continuous
(C) this distribution is a theoretical frequency distribution
(D) this distribution may be skewed
(E) this distribution may be used to calculate the normal range of clinical values for a diagnostic test

2. The central limit theorem stipulates all of the following EXCEPT

(A) when the sample size n is large, the means of repeated samples drawn from a non-normally distributed parent population will be approximately normally distributed
(B) for large sample sizes, the sampling distribution of sample means has a mean μ and a standard deviation σ
(C) the means of repeated samples of n individuals drawn from a normally distributed parent population will also be normally distributed
(D) as the sample size increases, the frequency distribution of the population of sample means more closely approximates a normal distribution, even if the samples are drawn from a parent population that is not normally distributed

Questions 3–5

During a hypertension screening program, a team of community health planners determine that the mean systolic blood pressure for normal men comprising a particular ethnic population is 124 mm Hg, with a standard deviation of 10 mm Hg. Use the standard normal table (Table B, Appendix 3) to answer questions 3–5.

3. The probability that a man selected at random from the specified population will have a systolic blood pressure greater than 146 mm Hg is

(A) .0139
(B) .4861
(C) .9861
(D) 2.2

4. Ninety percent of the systolic blood pressure readings will lie between which two values?

(A) 112.2 to 136.8 mm Hg
(B) 107.6 to 140.5 mm Hg
(C) 104.4 to 143.6 mm Hg
(D) 98.2 to 149.8 mm Hg

5. If repeated samples of 25 men are drawn from this population, 10% of the sample means will fall above which mean value?

(A) 127.29
(B) 127.92
(C) 126.56
(D) 121.44

1-D
2-B
3-A
4-B
5-C

Questions 6–8

In a study of lead poisoning among urban adults, an epidemiologist observed that the mean blood lead level for the general population is 25 μg/dl, with a standard deviation of 15 μg/dl.

6. The probability that an individual selected at random from the urban population will have a blood lead level greater than or equal to 60 μg/dl is

(A) 2.33
(B) .9901
(C) .4901
(D) .0099

7. Ten percent of blood lead levels would be expected to lie above which value?

(A) 25.6 μg/dl
(B) 44.2 μg/dl
(C) 28.8 μg/dl
(D) 40.0 μg/dl

8. The probability that a random sample of nine people drawn from this population will have a mean blood lead level greater than 60 μg/dl is

(A) .0099
(B) .9901
(C) .2206
(D) <.0001

9. The average systolic blood pressure for a random sample of 25 women, drawn from a particular population, is 120 mm Hg. What is the probability that a sample mean less than or equal to this value would be obtained from a population whose true mean μ is 124 mm Hg, assuming the standard deviation is 10 mm Hg?

(A) .3446
(B) .0228
(C) .9772
(D) <.0001

10. All of the following statements about the standard error of the mean are true EXCEPT

(A) the standard error describes the sample-to-sample variation among the set of sample means estimating a given population mean
(B) the standard error decreases as the sample size increases
(C) the standard error is smaller than the subject-to-subject variation in values of the response variable
(D) the standard error decreases as the subject-to-subject variation in the response variable increases

11. The standard error of the mean is used for which of the following functions?

(A) To describe the subject-to-subject variation in values of the response variable
(B) To describe the precision with which the sample mean estimates the unknown population mean
(C) To measure the average distance between a measurement and the population mean μ
(D) To calculate the normal range of clinical values for a diagnostic test

Directions: Each group of items in this section consists of lettered options followed by a set of numbered items. For each item, select the **one** lettered option that is most closely associated with it. Each lettered option may be selected once, more than once, or not at all.

Questions 12–15

For each distinguishing characteristic given below, select the corresponding variable or frequency distribution.

(A) Empirical distribution
(B) Theoretical distribution
(C) Continuous random variable
(D) Discrete random variable

12. A conceptual distribution of all possible population values

13. A frequency distribution tabulated from sample data

14. May assume any value within a given interval

15. Approximates the true frequency distribution of a population

---

6-D 9-B 12-B 15-A
7-B 10-D 13-A
8-D 11-B 14-C

ANSWERS AND EXPLANATIONS

1. The answer is D *[II B 1]*
The normal, or gaussian, distribution is represented by a continuous bell-shaped curve. This distribution is not skewed, but rather is symmetric about a central mean. When a diagnostic variable is known to follow a normal distribution, the interval defined by the mean ± 2 standard deviations encompasses 95% of permissible values for the diagnostic variable and can be used to specify the normal range of clinical values for that variable.

2. The answer is B *[III A 3 c (1)–(2)]*
The central limit theorem states that the shape of the frequency distribution of the population of sample means (based on repeated samples of size n) is normal when the parent population follows a normal distribution. Even when the samples are drawn from a non-normal parent population, the frequency distribution of sample means is approximately normal; as the sample size n increases, this frequency distribution is progressively better estimated by the normal distribution. The mean of the sampling distribution (i.e., the frequency distribution) of sample means is μ, the true population mean. The standard deviation of the sampling distribution is $\sigma/\sqrt{n}$.

3–5. The answers are: 3-A *[II C 1–4]*, **4-B** *[II C 7–8]*, **5-C** *[III A 3 c, B 1–2]*
Assuming that values for systolic blood pressure in this ethnic population are normally distributed, with a true population mean $\mu = 124$ mm Hg and standard deviation $\sigma = 10$ mm Hg, the probabilities to be computed in questions 3–5 can be obtained from the standard normal table (Table B, Appendix 3). In questions 3 and 4, the specified Y values (blood pressure observations) must first be converted to z scores, using the formula

$$z = \frac{\text{observation} - \text{its population mean}}{\text{its standard deviation}}$$

$$= \frac{Y - \mu}{\sigma}$$

The area corresponding to the calculated z score is then derived from the standard normal table.
To find the probability that an individual selected at random from the population will have a systolic pressure exceeding 146 mm Hg, the value $Y = 146$ is first converted to the corresponding z score:

$$z = \frac{146 - 124}{10} = 2.2$$

The probability that a given systolic pressure reading exceeds 146 mm Hg is therefore

$$P(Y > 146) = P(z > 2.2)$$
$$= 1 - .9861$$
$$= .0139$$

In terms of relative frequency, 1.39% of the members of this ethnic population have systolic blood pressures greater than 146 mm Hg.
Ninety percent of the observations comprising a normally distributed population lie within 1.645 standard deviations (i.e., between $z = -1.645$ and $z = +1.645$) of the population mean μ. The corresponding Y values are calculated using the formula given above.

$$1.645 = \frac{Y_1 - 124}{10} \qquad -1.645 = \frac{Y_2 - 124}{10}$$

$$Y_1 = 140.45 \qquad Y_2 = 107.55$$

In other words, 90% of the systolic blood pressures in this population fall between 107.55 and 140.45 mm Hg.
Regardless of whether the values for systolic blood pressure in the parent population (this ethnic population) are normally distributed or not, according to the central limit theorem, samples drawn from the population will produce a frequency distribution that is normal if the sample size is

sufficiently large. Therefore, the area above (i.e., to the right of) a given z score is equal to (1 − the area below) [i.e., to the left of] the score. Thus, the z score with an area of 10% above it is equivalent to the z score with an area of 90% below it. Table B (see Appendix 3) lists the z score with an area of 90% below it as 1.28. This z score is converted to the corresponding $\overline{Y}$ value using the formula

$$z = \frac{\text{observation} - \text{its population mean}}{\text{its standard deviation}}$$

Since here the observation is the population of sample means ($\overline{Y}$), the standard deviation is $\sigma/\sqrt{n}$, and the equation becomes

$$z = \frac{\overline{Y} - \mu}{\sigma/\sqrt{n}}$$

$$1.28 = \frac{\overline{Y} - 124}{10/\sqrt{25}}$$

$$\overline{Y} = 126.56 \text{ mm Hg}$$

6–8. The answers are: 6-D *[II C 1–4]*, **7-B** *[II C 7–8]*, **8-D** *[III A 3 c (1), B 1–2]*
Blood lead levels in the general population are assumed to follow a normal distribution with a mean $\mu = 25$ μg/dl and standard deviation $\sigma = 15$ μg/dl. The specified value $Y = 60$ μg/dl is converted to its corresponding z score using the formula given in the answer to question 3:

$$z = \frac{Y - \mu}{\sigma} = \frac{60 - 25}{15} = 2.33$$

The probability that an individual selected at random from the general population will have a blood lead level greater than or equal to 60 μg/dl can be calculated as

$$P(Y > 60) = P(z > 2.33) = 1 - .9901 = .0099$$

The z score that has an area of 10% above (i.e., to the right of) it is equivalent to the z score that has an area of .9000 below (i.e., to the left of) it. Table B (see Appendix 3) gives this z score as 1.28. Solving the formula given in the answer to question 3 for the corresponding value of Y:

$$1.28 = \frac{Y - 25}{15}$$

$$Y = 44.2 \ \mu\text{g/dl}$$

That is, 10% of the blood lead levels in the urban population can be expected to exceed 44.2 μg/dl.

The specified sample mean is converted to its corresponding z score using the formula given in the answer to question 5:

$$z = \frac{\overline{Y} - \mu}{\sigma/\sqrt{n}} = \frac{60 - 25}{15/\sqrt{9}} = 7$$

The probability that a random sample of 25 people will have a mean blood lead level greater than 60 μg/dl is

$$P(\overline{Y} > 60) = P(z > 7) < 1 - .9999 < .0001$$

9. The answer is B *[III B 1–2]*
The answer to this problem is similar to the answer to question 8. Again, the specified sample mean, $\bar{Y} = 120$ mm Hg, must first be converted to the corresponding z score:

$$z = \frac{\bar{Y} - \mu}{\sigma/\sqrt{n}} = \frac{120 - 124}{10/\sqrt{25}} = -2$$

The required probability is then calculated as

$$P(\bar{Y} \leq 120) = P(z \leq -2) = .0228$$

In other words, the chance that a random sample of 25 women, drawn from a population whose mean systolic blood pressure $\mu = 124$ mm Hg, will have an average systolic pressure less than or equal to 120 mm Hg is 2.28%.

10. The answer is D *[III A 3 b]*
The standard error of the mean is the standard deviation of the population of sample means. It provides a measure of the sample-to-sample variation among possible values of the sample mean. The standard error of the mean is estimated by $s/\sqrt{n}$, where s is the estimate of the subject-to-subject variation among possible values of the response variable. From this formula, it is apparent that the standard error **decreases** as the sample size n increases and **increases** as s, the subject-to-subject variation, increases. The standard error is always smaller than the subject-to-subject variation (s), except when $n = 1$ (in which case, s cannot be determined).

11. The answer is B *[III A 3 b (4) (b)]*
The standard error of the mean measures the sample-to-sample variation among the set of sample means (based on a given sample size) that estimate a population mean μ. Thus, the standard error describes the "closeness" or precision with which a single sample mean estimates the common unknown population mean. The standard deviation of the population of individual measurements (σ) defines the subject-to-subject variation among values of the response variable. It therefore provides a measure of the average distance between a given value of the response variable and the population mean. The standard deviation (or its estimate, the sample standard deviation s) may be used to define the normal range of values for a diagnostic test. This range comprises 95% of possible values for the diagnostic variable and is calculated as $\bar{Y} \pm 1.96s$.

12–15. The answers are: 12-B *[I B 3]*, **13-A** *[I B 2]*, **14-C** *[I B 3 b]*, **15-A** *[I B 3 b (1)]*
The conceptual frequency distribution for an entire population of values is known as a theoretical distribution. Because the frequency distribution of values in a sample drawn from a particular population is generated from the raw data, it is an empirical distribution.

A continuous random variable can assume any value along a continuum. Blood oxygen level, tidal volume, and glomerular filtration rate are examples of continuous random variables. In contrast, a discrete random variable can assume only integer values; hence it is measured in terms of "counts." The number of children per family or the number of heart beats per minute are examples of discrete random variables.

The frequency distribution tabulated from the sample data (i.e., the empirical distribution) is an estimate of the corresponding frequency distribution from which the sample was drawn. The larger the sample size, the closer the empirical distribution approximates the theoretical distribution.

11
Confidence Intervals

I. CONFIDENCE INTERVALS: A METHOD FOR ESTIMATING POPULATION VALUES.

Medical studies employ statistical methods to estimate differences between comparison groups as well as to determine if such differences have occurred by random chance (i.e., if they are statistically **significant**). This chapter describes the use of **confidence intervals** to estimate population parameters.

A. The concept of confidence intervals. In practice, it is usually not possible to calculate population parameters directly. Instead, the value of the parameter is estimated by the corresponding sample statistic (see Ch 10 I A 1–2; Figure 10-13). Because the value of the sample statistic varies from sample to sample (as measured by the standard error), uncertainty is introduced into this estimation process. The confidence interval method of estimating a population parameter takes into account the sample-to-sample variation of the statistic by defining an interval within which the true population parameter is likely to fall.

1. **Example.** A cardiovascular research team determines that the average blood cholesterol level of a sample of 100 men, drawn from a population of men with a history of acute myocardial infarction, is 260 mg/dl. Based on this sample, the team wishes to estimate the mean blood cholesterol level for the population. Individual cholesterol levels for this population are known to follow a normal frequency distribution with a standard deviation (σ) of 20 mg/dl.

   a. The mean blood cholesterol level computed for the sample of 100 men ($\overline{Y} = 260$ mg/dl) provides a **point estimate** of the mean cholesterol level (μ) of the population of men with a history of acute myocardial infarction. For each additional sample of 100 men that could potentially be drawn from this population, a different point estimate (sample mean) could be obtained (Figure 11-1).

   b. As discussed in Chapter 10, 95% of the values in any normal distribution fall within 1.96 standard deviations of the population mean (see Figure 11-1, *shaded area*).

   (1) If a particular sample mean ($\overline{Y}$) lies on the horizontal axis under the shaded area in Figure 11-1, **the true population mean, μ, will be included in the interval $\overline{Y} \pm 1.96\sigma_{\overline{Y}}$**. If, however, $\overline{Y}$ lies outside the shaded area, in the "tails" of the frequency distribution, **this interval will not include μ.**

   (2) When the population mean is unknown, the exact location of $\overline{Y}$ on the horizontal axis is also unknown. However, the **probability** that the sample mean will fall within 1.96 standard deviations of the unknown population mean μ (i.e., that it will lie under the shaded area in Figure 11-1) is known to be 95%. Conversely, there is a 5% probability that $\overline{Y}$ will lie more than 1.96 standard deviations beyond the population mean.

   (3) Thus, before the sample is drawn, the probability that any interval $\overline{Y} \pm 1.96\sigma_{\overline{Y}}$ will contain the unknown population mean μ is 95%. In this example, the research team may be **95% confident** that the mean cholesterol level in the population of men who have had an acute myocardial infarction lies within the interval $260 \pm (1.96)(20)$ mg/dl, that is, between 221 mg/dl and 299 mg/dl.

   c. **Formula.** The **95% confidence interval estimate of the population mean μ** is thus calculated as

$$\overline{Y} \pm 1.96\sigma_{\overline{Y}}$$

where $\overline{Y}$ (the sample mean) is the point estimate of the corresponding population

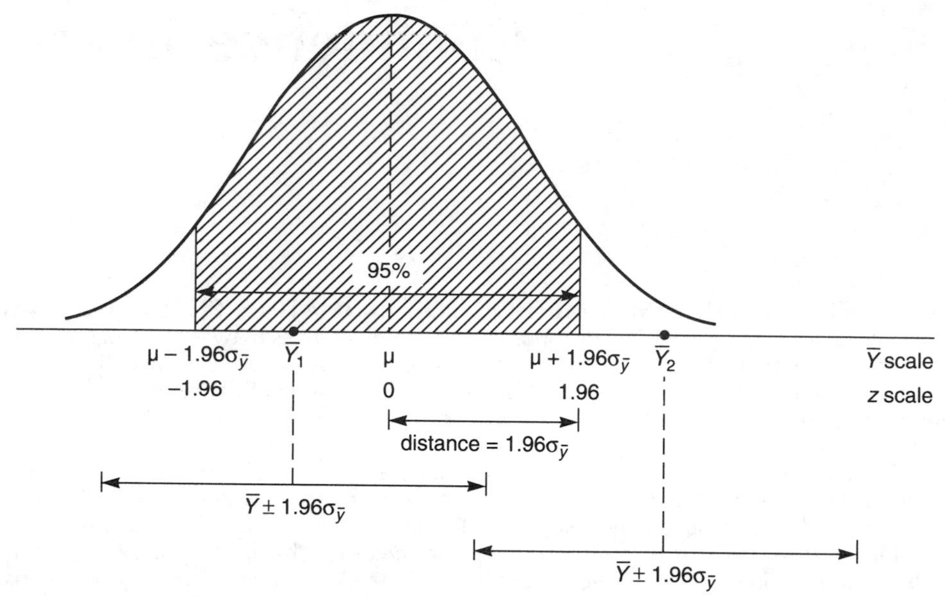

Figure 11-1. Frequency distribution of the population of sample means ($\bar{Y}$s) about the true population mean μ. Ninety-five percent of the values of $\bar{Y}$ (*shaded area*) fall within 1.96 standard deviations of the mean. For any value of $\bar{Y}$ in this area (e.g., $\bar{Y}_1$), the interval $\bar{Y} \pm 1.96\sigma_{\bar{y}}$ contains the unknown population mean μ. For any value of $\bar{Y}$ falling in one of the tails (*unshaded areas*), such as $\bar{Y}_2$, the interval $\bar{Y} \pm 1.96\ \sigma_{\bar{y}}$ does not contain the population mean μ.

parameter (μ) and $\sigma_{\bar{y}}$, the **standard error of the estimate** (here, $\sigma_{\bar{y}} = \sigma/\sqrt{n}$ = standard error of the sample mean $\bar{Y}$), defines the sample-to-sample variation in values of the point estimate. The value 1.96, known as the **confidence coefficient,** represents the **level of confidence** specified for this interval estimate. (See III for the mathematical derivation of this estimate.)

2. **General definition.** In general, the $100(1 - \alpha)\%$ confidence interval defines an area, $(1 - \alpha)$, that lies within the interval and an area, α, that falls outside the interval. The confidence coefficient for a $100(1 - \alpha)\%$ confidence interval is $z_{(1-\alpha/2)}$.
   a. For this example, the 95% confidence interval defines the area within the interval, $1 - \alpha$, as .95 and the area outside the interval, α, as .05. The corresponding 95% confidence coefficient is $z_{(1-.025)} = z_{.975} = 1.96$.
   b. The 95% confidence coefficient is $z_{.975}$, rather than $z_{.95}$, because the values comprising the standard normal table (Appendix 3, Table B) are listed in terms of the areas to the left of (i.e., below) the corresponding z scores on the horizontal axis of the frequency distribution. The two z values that encompass an area of .95 (i.e., $1 - \alpha$) correspond to an area, $1 - \alpha/2$, or .975 to the left of the upper z value (i.e., $z = +1.96$) and an area, $\alpha/2$, to the left of the lower z value (i.e., $z = -1.96$).

B. General form of a confidence interval estimate

1. **Formula.** In general, a $100(1 - \alpha)\%$ confidence interval estimate of a given population parameter can be calculated using the formula

   endpoints of interval = estimate ± (confidence coefficient × standard error of estimate)

   where **estimate** refers to the point estimate computed from the sample, the **confidence coefficient** reflects the desired level of confidence, and the **standard error of the estimate** is the sample-to-sample variation among all possible values of the point estimate.

2. **Interpretation**
   a. If a series of samples, all of a given sample size n, are obtained from a particular population and $100(1 - \alpha)\%$ confidence intervals estimating a population parameter are constructed for each sample, the **relative frequency** with which these intervals actually contain the true population parameter is $100(1 - \alpha)\%$.

b. For a single interval computed for a single sample, a $100(1 - \alpha)\%$ confidence interval signifies that the investigator can be $100(1 - \alpha)\%$ confident that this interval contains the unknown population parameter.

C. Factors governing confidence interval width

1. Confidence coefficient. The greater the degree of confidence that the interval will encompass the unknown population value, **the larger the confidence coefficient and the wider the corresponding confidence interval.** In other words, the "penalty" paid for increased confidence is a wider confidence interval.

2. Standard error of the estimate

a. The value of the standard error (and hence, the degree of sample-to-sample variation in the point estimates) depends on the following:

(1) The amount of variation among **individual** values in the population from which the samples are drawn (i.e., the value of σ): **the more individual variation (i.e., the larger σ), the wider the confidence interval**

(2) The sample size n: **for a fixed standard deviation (σ), the larger the sample size, the narrower the confidence interval**

b. If σ is large, the confidence interval will be excessively wide (i.e., it will provide a poor estimate of the population parameter) unless the sample size n is sufficiently large.

II. APPLICATIONS OF CONFIDENCE INTERVAL METHOD.

The confidence interval method may be used to estimate a number of population parameters.

A. Estimating a single population mean μ when the population standard deviation σ is known

1. Example. A study is conducted to assess the effect of a new formula on weight gain in infants with a particular carbohydrate malabsorption syndrome. Nine infants with the syndrome are fed the new formula for a six-week period. At the end of this period, the mean weight gain for the group is 10.5 lbs. What is the 90% confidence interval estimate for the mean weight gain in the population from which this sample of nine patients was drawn (i.e., the population of all infants with the malabsorption syndrome who receive the new formula)? Weight gain values in this population are known to be normally distributed; the variation in weight gain among the infants (i.e., σ) is 12 lbs.

2. Parameters and sample values

a. The parameter to be estimated is μ, the average weight gain for the population of infants suffering from the malabsorption syndrome who are fed the new formula.

b. The **point estimate** of the population mean is **$\bar{Y}$, the sample mean,** which is 10.5 lbs.

c. The **standard error** of the estimate is [see Ch 10 III A 3 b (2)]

$$\sigma_{\bar{Y}} = \sigma/\sqrt{n} = 12/\sqrt{9} = 4$$

3. Computing the confidence interval

a. Derivation of the confidence coefficient. The standard normal table (Appendix 3, Table B) is used to obtain the confidence coefficient for estimating a population mean when the population standard deviation σ is known. The **90% confidence coefficient** corresponds to the z value that defines an area encompassing $100(1 - \alpha) = 90\%$ of the population, centered on the population mean μ (Figure 11-2). The area outside the interval is $\alpha = .10$ (i.e., 10%). Hence, the 90% confidence coefficient is $z_{(1 - \alpha/2)} = z_{.95} = 1.645$ (from Table B, Appendix 3). This z value corresponds to an area below (i.e., to the left of) $1 - \alpha/2$, or .95 [i.e., $P(z \leq 1.645) = .95$]. The area between $z = -1.645$ and $z = 1.645 = .90$.

b. Calculation. The 90% confidence interval (CI_μ) for the average weight gain in the population is

$$\begin{aligned} CI_\mu &= \bar{Y} \pm 1.645(\sigma/\sqrt{n}) \\ &= 10.5 \pm 1.645(4) \\ &= 3.92 \text{ to } 17.08 \end{aligned}$$

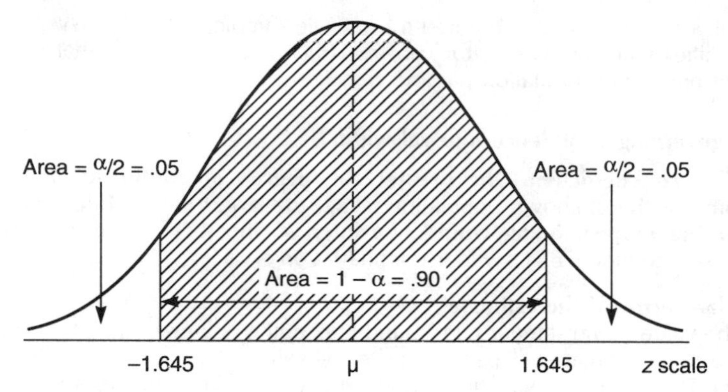

Figure 11-2. Graphic derivation of a 90% confidence coefficient from the standard normal (z) distribution. The desired confidence coefficient, $z_{(1-\sigma/2)} = z_{.95} = 1.645$, corresponds to a z value that defines an area encompassing 90% of the population of point estimates ($\overline{Y}$s), centered on the population mean μ (*shaded area*). [The values tabulated in Table B (see Appendix 3) represent areas to the left of (i.e., below) the corresponding z values on the horizontal axis of the frequency distribution. Thus, the area below $z = 1.645$ is $1 - \alpha/2 = .95$.]

c. **Interpretation.** The researchers conducting the study may be 90% confident that the interval between 3.92 and 17.08 lbs contains the unknown population mean weight gain for infants with the malabsorption syndrome who are fed the new formula for six weeks.

B. Estimating a single population mean μ when the standard deviation σ is unknown. In practice, the population standard deviation σ is rarely known and the sample standard deviation (s) is used to estimate σ. When the sample standard deviation (s) is substituted for σ, s, as well as $\overline{Y}$, varies from sample to sample. To account for this additional source of variability, the required confidence coefficient is obtained using the **student's *t* distribution** (often referred to simply as the t distribution), rather than the standard normal distribution.

1. Features of the *t* distribution

a. **Shape.** As shown in Figure 11-3, the t distribution, like the standard normal distribution, is represented by a bell-shaped curve that is symmetric about a central mean (in this case, the mean = 0). Note that the t distribution curve is flatter than the normal distribution curve, with thicker tails. The thickness of the tails is determined by the number of degrees of freedom.

b. The **degrees of freedom,** abbreviated ***df,*** refer to the number of independent observations used to calculate a quantity. In computing a variance, for example, the divisor is the sample size $n - 1$ [see Ch 1 III B 2 b (1) (a)]; that is, there are ***n* − 1 degrees of freedom** restricting the set of permissible values for the divisor.

(1) The fewer the degrees of freedom, the thicker the tails of the t distribution.

(2) When the number of degrees of freedom is very large, the t distribution is equivalent to the normal distribution.

c. Table C in Appendix 3 lists **values of the *t* distribution** corresponding to five areas under the curve (probabilities), according to degrees of freedom. The subscripts of t (e.g., $t_{.95}$,

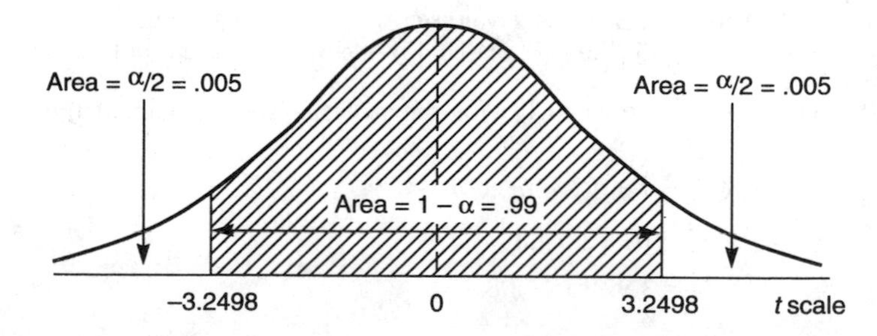

Figure 11-3. Derivation of a 99% confidence coefficient from the t distribution. Note that the required confidence coefficient, 3.2498, corresponds to a t value delineating an area equal to $1 - \alpha = .99$, centered on the population mean (*shaded area*). α, the area outside the confidence interval, is .01, and the area to the left of 3.2498 is $1 - \alpha/2 = .995$.

$t_{.995}$) represent areas to the left of these t values on the horizontal axis of the frequency distribution.

2. **Example.** A psychiatrist carries out a study to evaluate the level of stress experienced by first-year medical students at a particular medical school. Stress levels were monitored in 10 students selected at random from the class by attaching an electrode to the frontalis muscle. The average electromyographic (EMG) activity in the sample of 10 students was found to be 35.8 μvolts, with a standard deviation of 2.5 μvolts. What is the 99% confidence interval estimate of the true mean EMG activity for all students in the class?

3. **Parameters and sample values**
   a. **The parameter to be estimated is μ,** the average EMG activity in the total population of first-year medical students at this medical school.
   b. The **point estimate** of μ is the **sample mean, $\bar{Y}$,** which is 35.8 μvolts.
   c. The **population standard deviation σ** is estimated by the sample standard deviation s, which equals 2.5 μvolts. The **standard error $\sigma_{\bar{y}}$** is estimated as follows [see Ch 10 III A 3 b (3) (b)]:

$$s_{\bar{y}} = s/\sqrt{n} = 2.5/\sqrt{10} = .79$$

4. **Computing the confidence interval**
   a. **Derivation of the confidence coefficient**
      (1) When estimating a single population mean μ, the **degrees of freedom** for the t distribution is defined as $n - 1$, where $n =$ the sample size.
      (2) The appropriate confidence coefficient for a $100(1 - \alpha)\%$ confidence interval is $t_{(1-\alpha/2)}$, that is, the value in Table C, Appendix 3 located at the intersection between the desired coefficient (i.e., $t_{(1-\alpha/2)}$; see Appendix 3, Table C, *columns*) and the degrees of freedom (see Appendix 3, Table C, *rows*). In this example, the 99% confidence coefficient defines a confidence interval encompassing an area of $1 - \alpha = .99$; the corresponding area outside the interval $= \alpha = .01$. In other words, the 99% confidence coefficient is $t_{(1-.005)} = t_{.995} = 3.2498$, found in Table C at the intersection of the column labeled $t_{.995}$* and the row corresponding to $df = 10 - 1 = 9$. The area to the left of (i.e., below) $t = 3.2498$ is $1 - .005 = .995$ (see Figure 11-3).
   b. **Calculation.** The 99% confidence interval (CI_μ) is calculated as

$$CI_\mu = \bar{Y} \pm t(s/\sqrt{n})$$

$$= 35.8 \pm (3.2498)(2.5/\sqrt{10})$$

$$= 33.2 \text{ to } 38.4$$

   c. **Interpretation.** The psychiatrist can be 99% confident that the population mean EMG activity for the first-year medical class lies between 33.2 and 38.4 μvolts.

C. **Estimating the difference between two population means when the population standard deviations are unknown**

1. **Example.** A study was carried out to determine the effect, if any, of pesticide exposure on blood pressure. A random sample of 100 men was selected from a group of agricultural workers known to have been exposed to pesticides. One hundred randomly selected workers with no such exposure comprised the control group. The mean systolic blood pressure for the pesticide cohort was 145 mm Hg, with a sample standard deviation, s_E, of 20 mm Hg (where E represents exposed workers and NE, nonexposed workers). The mean systolic blood pressure in the control group was 120 mm Hg with a sample standard deviation, s_{NE}, of 15 mm Hg. Calculate the 90% confidence interval for the true difference in mean systolic blood pressure between the exposed and nonexposed populations.

2. **Parameters and sample values**
   a. **The parameter to be estimated is $\mu_E - \mu_{NE}$,** the difference in mean systolic blood pressure between the exposed and nonexposed populations.

---

*The confidence coefficient for a $100(1 - \alpha)\%$ confidence interval, $t_{(1-\alpha/2)}$, is equal to that value of t which defines an area equal to .99, *centered* on the population mean (see Figure 11-3, *shaded area*).

b. The **point estimate** of this difference is the difference between the sample means, $\bar{Y}_E - \bar{Y}_{NE} = 145 - 120 = 25$ mm Hg.

c. Calculating the standard error of the estimate

(1) The variation among all possible values of $\bar{Y}_E - \bar{Y}_{NE}$, obtained from repeated samples drawn from the population of interest (i.e., the standard error of the estimate) is given by the formula

$$\sigma_{(\bar{Y}_E - \bar{Y}_{NE})} = \sqrt{\frac{\sigma_E^2}{n_E} + \frac{\sigma_{NE}^2}{n_{NE}}} \quad (11.1)$$

(2) The population standard deviations, σ_E and σ_{NE}, are unknown but they may be estimated by the sample standard deviations, s_E and s_{NE}. If the population standard deviations are **assumed to be equal** (i.e., $\sigma_E = \sigma_{NE} = \sigma$), both s_E and s_{NE} estimate this common unknown population standard deviation. By pooling the information gathered from both samples, a more accurate estimate of the common population standard deviation σ can be obtained. The corresponding **pooled estimate of σ^2, denoted s_p^2,** is computed using the formula

$$s_p^2 = \frac{(n_E - 1)s_E^2 + (n_{NE} - 1)s_{NE}^2}{n_E + n_{NE} - 2} \quad (11.2)$$

$$= \frac{(99)(20^2) + (99)(15^2)}{198}$$

$$= 312.5$$

(3) The pooled variance s_p^2 is a weighted average of the sample variances s_E^2 and s_{NE}^2 (where the "weights" are the sample sizes, n_E and n_{NE}).*

(4) Substituting s_p^2 for s_E^2 and s_{NE}^2 in formula 11.1, the standard error is estimated as

$$s_{(\bar{Y}_E - \bar{Y}_{NE})} = \sqrt{\frac{s_p^2}{n_E} + \frac{s_p^2}{n_{NE}}}$$

$$= \sqrt{\frac{312.5}{100} + \frac{312.5}{100}}$$

$$= 2.5$$

(5) The standard error, $s_{(\bar{Y}_E - \bar{Y}_{NE})}$, estimates the sample-to-sample variability among all possible values of the difference between sample means ($\bar{Y}_E - \bar{Y}_{NE}$), if all possible groups of 100 subjects were drawn from both the exposed and nonexposed populations and the differences between their sample means computed.

3. Computing the confidence interval

a. Derivation of the confidence coefficient. Because the population standard deviations are unknown, the *t* distribution is used to derive the confidence coefficient.

(1) The number of degrees of freedom associated with s_p is calculated as

$$df = n_E + n_{NE} - 2$$

$$= 100 + 100 - 2$$

$$= 198$$

Because Table C (see Appendix 3) contains no values corresponding to $df = 198$, the closest value, $df = 200$, may be used as an approximation.

(2) The 90% confidence coefficient corresponds to $t_{(1-\alpha/2)} = t_{(1-.05)} = t_{.95}$ (Figure 11-4); this value, 1.6525, is found in Table C (see Appendix 3) at the intersection of the column labeled $t_{.95}$ and the row corresponding to $df = 200$.

---

*Obviously, if one sample is much larger than the other, it will contribute substantially more to the final average, s_p^2. When the sample sizes are equal, s_p^2 is the arithmetic average of the sample variances.

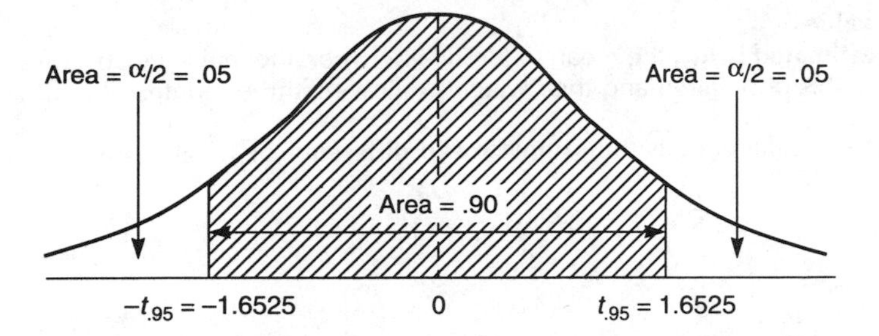

Figure 11-4. Derivation of a 90% confidence coefficient from the t distribution. Note that the required confidence coefficient, 1.6525, corresponds to a t value that defines an area equal to .90, centered on the population mean (*shaded area*). The area below $t = 1.6525$ is equal to $(1 - \alpha/2) = .95$.

b. Calculation. The 90% confidence interval $[\mathrm{CI}_{(\mu_E - \mu_{NE})}]$ is

$$\mathrm{CI}_{(\mu_E - \mu_{NE})} = (\bar{Y}_E - \bar{Y}_{NE}) \pm t[s_{(\bar{Y}_E - \bar{Y}_{NE})}]$$

$$= (145 - 120) \pm 1.6525\sqrt{\frac{312.5}{100} + \frac{312.5}{100}}$$

$$= 20.9 \text{ to } 29.1$$

c. Interpretation. The investigators conducting this study can be 90% confident that the unknown difference in average systolic blood pressure between the exposed and nonexposed populations falls within the interval 20.9 to 29.1 mm Hg.

D. Estimating the mean of the differences in a paired study. Study subjects may be **paired** or **matched** to control for potentially confounding variables as well as to increase experimental precision (see Ch 8 II D 2). Investigators conducting a paired study often wish to estimate the mean, μ_D, of the differences, $d_i = Y_{1i} - Y_{2i}$, between the two members of each matched pair (i.e., the mean of the population of differences for all pairs of subjects, where that population of differences is represented by D).

1. Example. Table 11-1 lists the weights of nine healthy women participating in a study of the effect of a certain oral contraceptive on weight gain, before and three months after beginning the medication.* What is the 95% confidence interval for the mean difference in weight before and after three months of oral contraceptive use for the population of women from which this sample was drawn?

Table 11-1. Body Weights for Nine Women Before and After Three Months of Oral Contraceptive Use

	Weight (in pounds)		
Subject	**Initial**	**3-month**	**Differences (d_i)**
1	120	123	+3
2	141	143	+2
3	130	140	+10
4	150	145	−5
5	135	140	+5
6	140	143	+3
7	120	118	−2
8	140	141	+1
9	130	132	+2

Mean: $\bar{Y}_d = \sum_{i=1}^{9} d_i/n = 19/9 = 2.111$

*In this example, the women serve as their own controls and the pre- and post-medication weights constitute the "matched" pairs.

2. Parameters and sample values

a. The parameter to be estimated is μ_D, the mean difference between the initial weights of the women comprising this population and their weights following three months of oral contraceptive use.

b. The **point estimate** of this difference is the **sample mean difference, $\bar{Y}_d$,** calculated as

$$\bar{Y}_d = \sum_{i=1}^{n} d_i / n = 19/9 = 2.111$$

c. Standard error

(1) The individual variation among all possible values of the differences (σ_D), is estimated by s_d, where

$$s_d^2 = \frac{\sum d_i^2 - \frac{(\sum d_i)^2}{n}}{n-1}$$

$$= \frac{181 - \frac{(19)^2}{9}}{8}$$

$$= 17.611$$

and hence, $s_d = \sqrt{17.611} = 4.197$

(2) The standard error $\sigma_{\bar{Y}_d}$ represents the sample-to-sample variation among all possible values of the point estimate $\bar{Y}_d$ that could be obtained from repeated samples of nine pairs drawn from the population. It is estimated by

$$s_{\bar{Y}_d} = s_d / \sqrt{n}$$

$$= 4.197 / \sqrt{9}$$

$$= 1.399$$

3. Computing the confidence interval

a. Derivation of the confidence coefficient. Since σ_D, the standard deviation of the population of differences, is unknown, the t distribution (Table C, Appendix 3) is used to obtain the confidence coefficient.

(1) The number of degrees of freedom is calculated as

$$df = n - 1 = 9 - 1 = 8$$

(2) The 95% confidence coefficient corresponds to $t_{(1-.05/2)} = t_{.975}$ with 8 degrees of freedom, given in Table C (see Appendix 3) as 2.306.

b. Calculation. The 95% confidence interval (CI_{μ_D}) is calculated using the formula

$$CI_{\mu_D} = \bar{Y}_d \pm (t)(s_{\bar{Y}_d})$$

$$= 2.111 \pm (2.306)(4.197/\sqrt{9})$$

$$= -1.11 \text{ to } 5.34$$

c. Interpretation. The researchers conducting this study can be 95% confident that, on the average, women in this population who receive the oral contraceptive in question for three months will experience a change in their weight ranging from a loss of 1.11 lbs. to a gain of 5.34 lbs.

E. Estimating a single population proportion

1. Use of proportions. Clinical research data are often expressed as the **proportion** of subjects possessing a given characteristic or incurring a particular outcome.

a. When the dependent variable is measured on a **nominal** scale, one of the few appropriate descriptive measures is the proportion of individuals comprising each category (see Ch 1 IV A). For example, for the variable "marital status," the proportions of married, single, widowed, and divorced individuals may be reported. In studies comparing the efficacy of two chemotherapeutic regimens, the response variable is often "mortality status," that is, the proportion of patients still alive five years after initiating therapy.

b. Proportions are also used to describe **ordinal** or **interval/ratio** data (see Ch 1 IV B, C). For example, a study employing systolic blood pressure as the dependent variable may report the proportion of subjects with systolic pressures greater than 160 mm Hg.

c. The confidence interval method may be used to estimate a single population proportion P or the difference between two population proportions $P_1 - P_2$.

2. **Example.** A team of mental health professionals carried out a study to evaluate illegal drug use among medical students at a particular medical school. Fifteen of the fifty students randomly selected to fill out an anonymous questionnaire admitted to taking an illegal drug on at least one occasion during their time at the school. What is the 95% confidence interval estimate of the proportion of students at the school who have used illegal drugs?

3. **Parameters and sample values**

a. **The parameter to be estimated is *P*,** the proportion of students attending this particular medical school who have taken an illegal drug on at least one occasion.

b. The point estimate of P is $\hat{p}$, the proportion of subjects in the sample of 50 students who have used illegal drugs. Thus,

$$\hat{p} = 15/50 = .3$$

c. The **standard error** of $\hat{p}$ is estimated by

$$s_{\hat{p}} = \sqrt{\frac{\hat{p}(1-\hat{p})}{n}} = \sqrt{\frac{(.3)(1-.3)}{50}} = .065$$

4. **Computing the confidence interval**

a. **Derivation of confidence coefficient**

(1) The sample proportion $\hat{p}$ has a frequency distribution that is approximately normally distributed with mean $= P$ (i.e., the population proportion) and standard error $= \sqrt{P(1-P)/n}$ when both $n(P)$ and $n(1-P)$ are greater than to or equal to 5. That is, if all possible samples of n subjects are drawn from the population of interest, the population of sample proportions will follow a normal frequency distribution centered on the true population proportion P.

(2) Consequently, the 95% confidence coefficient, 1.96, is obtained from the standard normal table (Table B, Appendix 3: $z_{(1-\alpha/2)} = z_{(1-.05/2)} = z_{.975} = 1.96$).

b. **Calculation.** The 95% confidence interval (CI_P) is calculated using the formula

$$\text{CI}_P = \hat{p} \pm 1.96\sqrt{\frac{\hat{p}(1-\hat{p})}{n}} = .3 \pm 1.96\sqrt{\frac{.3(1-.3)}{50}} = .17 \text{ to } .43$$

c. **Interpretation.** The research team can be 95% confident that the unknown proportion of students at this medical school who have used illegal drugs at some point during the course of their studies is between 17% and 43%.

F. **Estimating the difference between two population proportions**

1. **Example.** In a study conducted to compare drug and surgical treatments for chronic lower back pain, 80 out of 100 patients treated surgically experienced relief from their pain, while 50 out of 100 patients receiving a nonsteroidal anti-inflammatory agent experienced relief. What is the 99% confidence interval estimate of the difference in the proportion of patients afforded relief from their back pain by surgery and by medication?

2. **Parameters and sample values**
   a. **The parameter to be estimated is $P_S - P_D$,** the difference between the proportion of patients experiencing relief after surgery (P_S) and the proportion experiencing relief after drug treatment (P_D).
   b. The **point estimate** of this difference is $\hat{p}_S - \hat{p}_D$, where $\hat{p}_S$ is the proportion of patients in the sample afforded pain relief after surgery (i.e., 80/100, or .8) and $\hat{p}_D$ is the corresponding proportion of patients in the sample afforded relief by the medication (i.e., 50/100 or .5). Thus, $\hat{p}_S - \hat{p}_D = .8 - .5 = .3$.
   c. The **standard error** associated with $\hat{p}_S - \hat{p}_D$ is estimated by $s_{(\hat{p}_S - \hat{p}_D)}$; it describes the spread of all possible values of $\hat{p}_S - \hat{p}_D$ about the true population difference, $P_S - P_D$:

$$s_{(\hat{p}_S - \hat{p}_D)} = \sqrt{\frac{\hat{p}_S(1 - \hat{p}_S)}{n_S} + \frac{\hat{p}_D(1 - \hat{p}_D)}{n_D}}$$

$$= \sqrt{\frac{(.8)(.2)}{100} + \frac{(.5)(.5)}{100}}$$

$$= .064$$

3. **Computing the confidence interval**
   a. **Derivation of the confidence coefficient.** The 99% confidence coefficient, 2.58, is the closest value in Table B (see Appendix 3) corresponding to $z_{(1-\alpha/2)} = z_{.995}$, where $1 - \alpha = .99$ is the area encompassed by the confidence interval and $\alpha = .01$ is the area outside the interval.
   b. **Calculation.** The 99% confidence interval [$\text{CI}_{(P_S - P_D)}$] is

$$\text{CI}_{(P_S - P_D)} = (\hat{p}_S - \hat{p}_D) \pm 2.58\sqrt{\frac{\hat{p}_S(1 - \hat{p}_S)}{n_S} + \frac{\hat{p}_D(1 - \hat{p}_D)}{n_D}}$$

$$= (.8 - .5) \pm 2.58\sqrt{\frac{(.8)(.2)}{100} + \frac{(.5)(.5)}{100}}$$

$$= .13 \text{ to } .47$$

   c. **Interpretation.** The physician who conducted this study can be 99% confident that the true difference in the proportion of patients afforded relief from their lower back pain by surgery and the proportion experiencing relief after drug treatment lies between 13% and 47% for the populations of patients receiving the two therapies.

III. MATHEMATICAL DERIVATION OF CONFIDENCE INTERVALS. The confidence interval estimate of a population mean μ may be used to illustrate the mathematical derivation of confidence interval estimates for any unknown population parameter.

A. As shown in Figure 11-1, 95% of the population of sample means lies within $1.96\sigma_{\bar{Y}}$ of the population mean μ. That is,

$$P(\mu - 1.96\sigma_{\bar{Y}} \leq \bar{Y} \leq \mu + 1.96\sigma_{\bar{Y}}) = .95$$

Or, in terms of the corresponding z scale,

$$P(-1.96 \leq z \leq 1.96) = .95$$

B. This expression can be mathematically manipulated to obtain a probability statement about μ (i.e., a statement in which μ alone appears as the center term of the inequality).

1. $\bar{Y}$ can be converted to the corresponding standardized (z) value using the formula [see Ch 10 III B 2 a (1)]

$$z = \frac{\bar{Y} - \mu}{\sigma_{\bar{Y}}}$$

Thus,

$$P\left(-1.96 \leq \frac{\bar{Y} - \mu}{\sigma_{\bar{Y}}} \leq 1.96\right) = .95$$

2. To eliminate $\sigma_{\bar{y}}$ from the center term, each term in parentheses can be multiplied by $\sigma_{\bar{y}}$ (an operation carried out on all terms within the parentheses will not change the value of the expression), leaving

$$P(-1.96\sigma_{\bar{y}} \leq \bar{Y} - \mu \leq 1.96\sigma_{\bar{y}}) = .95$$

3. $\bar{Y}$ is then subtracted from each term in parentheses to eliminate it from the center term

$$P(-\bar{Y} - 1.96\sigma_{\bar{y}} \leq -\mu \leq -\bar{Y} + 1.96\sigma_{\bar{y}}) = .95$$

4. Each term in parentheses is then multiplied by -1. Note that multiplying an inequality by a negative number reverses the direction of the inequality

$$P(\bar{Y} + 1.96\sigma_{\bar{y}} \geq \mu \geq \bar{Y} - 1.96\sigma_{\bar{y}}) = .95$$

5. Rewriting this expression using the conventional order for an inequality gives

$$P(\bar{Y} - 1.96\sigma_{\bar{y}} \leq \mu \leq \bar{Y} + 1.96\sigma_{\bar{y}}) = .95$$

That is, the probability that the interval $\bar{Y} \pm 1.96\sigma_{\bar{y}}$ will contain the population mean μ is .95.

PROBLEMS*

11-1. As part of a study to determine the effect of bumetanide on urinary calcium excretion, nine randomly selected men each received an oral dose of .5 mg of the drug. Urine was collected hourly for the next six hours. The mean excretion rate for this sample of nine men was found to be 7.5 mg/hr, with a standard deviation of 6.0 mg/hr.

a. What is the 95% confidence interval estimate of the population mean excretion rate for all men receiving bumetanide? Interpret this interval.

b. Urine was also collected from a random sample of 16 men who did not receive bumetanide. The mean excretion rate for this sample was 6.5 mg/hr, with a standard deviation of 2.0 mg/hr. What is the 95% confidence interval estimate of the difference in excretion rates between the population of men receiving the drug and the population of men who did not receive it?

11-2. The mean serum cholesterol level in a certain population of normal healthy men is 240 mg/dl and the standard deviation is 40 mg/dl. A clinical researcher interested in comparing cholesterol levels in this healthy population with those in men with coronary artery disease measured serum cholesterol levels in a random sample of 100 men who had undergone coronary bypass surgery during the preceding two-year period. The mean serum cholesterol level for this sample was 260 mg/dl.

a. What is the 95% confidence interval estimate of the mean serum cholesterol level for the population of all men undergoing coronary bypass surgery? Assume that the standard deviation of serum cholesterol measurements for this population is the same as that of the healthy population (i.e., 40 mg/dl).

b. Based on this estimate, can the researcher conclude that the mean serum cholesterol level of men undergoing coronary bypass surgery differs from that of healthy men?

11-3. The table below lists serum digoxin levels for nine healthy men between the ages of 20 and 45, determined four hours and eight hours after rapid intravenous injection of the drug.

a. What is the 95% confidence interval estimate of the true mean difference in serum digoxin concentration four hours and eight hours after the injection?

b. A second sample of 11 healthy men from the same age group received an intravenous infusion of digoxin. At the end of eight hours, the mean serum digoxin concentration in this sample was .95 ng/ml, with a standard deviation of .20 ng/ml. What is the 99% confidence interval estimate of the true difference in mean serum digoxin concentration eight hours after injection between the population receiving the drug by rapid intravenous injection and the population receiving the drug by infusion?

Serum Digoxin Concentrations for Nine Men Following Rapid Intravenous Injection

	Serum Digoxin Concentration (ng/ml)		
Subject	**4-hr**	**8-hr**	**Differences (d_i)**
1	1.0	1.0	0.0
2	1.3	1.3	0.0
3	0.9	0.7	−0.2
4	1.0	1.0	0.0
5	1.0	0.9	−0.1
6	0.9	0.8	−0.1
7	1.3	1.2	−0.1
8	1.1	1.0	−0.1
9	1.0	1.0	0.0

Mean: $\overline{Y}_d = \sum_{i=1}^{9} d_i/n = -.6/9.0 = -.067$

Adapted from Duncan RC, Knapp RG, Miller MC III: *Introductory Biostatistics for the Health Sciences*, 2nd ed. Albany, Delmar, 1983, p 112.

*The problems in this section are adapted from Duncan RC, Knapp RG, Miller MC III: *Introductory Statistics for the Health Sciences*, 2nd ed. Albany, Delmar, 1983, pp 76–77, 112–113, 176.

11-4. The reported national success rate for a new surgical procedure to treat chronic angle-closure glaucoma is 90%.

a. If the success rate at a particular ophthalmologic hospital is the same as the national average, how many unsuccessful operations employing the new procedure should a clinical investigator expect to identify in a review of 100 patient records?

b. The investigator ultimately identifies 15 unsuccessful chronic angle-closure glaucoma operations among the 100 cases surveyed. What is the 95% confidence interval estimate of the true proportion of successful operations performed at the ophthalmologic hospital using the new surgical procedure? Does the experience at this hospital parallel the national experience?

11-5. In a study of the relationship between oral contraceptive use and hypertension, eight hypertensive women were identified in a random sample of 40 women using oral contraceptives. Fifteen subjects in a random sample of 60 women using other methods of contraception were found to be hypertensive. What is the 99% confidence interval estimate of the difference in the proportion of hypertensive women among the population using oral contraceptives and the population using other contraceptive methods?

Solutions on p 381.

STUDY QUESTIONS

Directions: Each of the numbered items or incomplete statements in this section is followed by answers or by completions of the statement. Select the **one** lettered answer or completion that is **best** in each case.

1. All of the following statements about a 99% confidence interval estimate of the population mean for a normally distributed variable are true EXCEPT

(A) when the interval is computed for a single sample of n measurements, an investigator can be 99% confident that it includes the unknown population mean μ
(B) an investigator may be 99% confident that the mean of other samples drawn from the population will fall within the given interval
(C) in the long run, if an investigator calculates many confidence interval estimates of the population mean μ, using different samples drawn from the same population, the investigator can expect 99% of these intervals to contain the unknown population mean
(D) a 99% confidence interval is wider than a 95% confidence interval

2. Twenty-five patients with congestive heart failure were weighed before and after receiving a novel diuretic agent, and the average weight loss (i.e., the difference between the two weights) for this sample was found to be 2.00 pounds, with a standard error of 1.38 pounds. The 95% confidence interval estimate of the mean weight loss was computed using the formula

$$2 \pm t(1.38/\sqrt{25}).$$

Which of the following statements accurately describes this 95% confidence interval?

(A) It provides information on the normal range of values for the mean weight loss among patients with congestive heart failure treated with this drug
(B) It is associated with $25+25-2=48$ *df*
(C) It predicts that 95% of patients in the population will register weight losses in the range defined by this interval
(D) It was calculated incorrectly, because standard error was erroneously substituted for standard deviation in the formula for calculating the interval; the correct formula is $2 \pm t(1.38)$

3. The width of a confidence interval is influenced by all of the following factors EXCEPT

(A) the sample size n
(B) the desired level of confidence
(C) the degree of subject-to-subject variation in the response variable
(D) the degree of sample-to-sample variation among sample estimates of the population parameter
(E) the distance between the sample mean and the population mean

4. The width of a confidence interval will decrease when which of the following changes occur?

(A) The desired level of confidence increases
(B) The subject-to-subject variation in values of the response variable increases
(C) The sample size n increases
(D) The precision of the estimate of the population mean decreases

5. The mean plasma potassium level for 25 adult men with primary hyperaldosteronism was found to be 3.35 mEq/L, with a standard deviation of .50 mEq/L. Based on this sample, the 95% confidence interval estimate of the mean plasma potassium level for the population of men with primary hyperaldosteronism is

(A) 3.250–3.450 mEq/L
(B) 3.154–3.546 mEq/L
(C) 2.318–4.382 mEq/L
(D) 3.144–3.556 mEq/L
(E) 3.179–3.521 mEq/L

1-B 4-C
2-D 5-D
3-E

6. In 1990, a random sample of 1000 black men between the ages of 40 and 60 was selected from a particular community. One hundred of these men were diagnosed with coronary heart disease (CHD). The 95% confidence interval estimate of the prevalence of CHD among all black men residing in this community is

(A) .081–.119
(B) .091–.109
(C) .200–.400
(D) .076–.124

7. As part of a study of the association between hypertension and myocardial infarction, 60 out of 200 hypertensive men were found to have experienced a silent myocardial infarction. Among 700 normotensive men, 50 had experienced a silent myocardial infarction. The 95% confidence interval estimate of the difference in the prevalence of silent myocardial infarction among the population of hypertensive men and the population of normotensive men is

(A) .143–.317
(B) .196–.264
(C) .164–.296
(D) .230–.370

8. The mean urinary chloride excretion rate for 25 premature infants treated at a certain neonatal intensive care unit was 170 mEq/day. The standard deviation for the population of all premature infants is known to be 20 mEq/day. The 99% confidence interval estimate of the average population urinary chloride excretion rate is therefore

(A) 162.2–177.8 mEq/day
(B) 118.4–221.6 mEq/day
(C) 166.0–174.0 mEq/day
(D) 159.7–180.3 mEq/day
(E) 158.8–181.2 mEq/day

9. An endocrinologist conducted a study to determine weight loss from excessive perspiration during insulin-induced hypoglycemia in patients receiving β-adrenergic blockers compared to drug-free controls. Twelve patients receiving propranolol experienced a mean cumulative weight loss of 120 g (standard deviation = 10 g), while 11 control patients experienced a mean cumulative weight loss of 70 g (standard deviation = 8 g). The 95% confidence interval estimate of the difference in mean cumulative weight loss during insulin-induced hypoglycemia between the population of patients receiving propranolol and the control population is

(A) 43.5–56.5 g
(B) 42.1–57.9 g
(C) 42.6–57.4 g
(D) 46.2–53.8 g

6-A
7-C
8-D
9-B

Directions: Each group of items in this section consists of lettered options followed by a set of numbered items. For each item, select the **one** lettered option that is most closely associated with it. Each lettered option may be selected once, more than once, or not at all.

Questions 10–13

Match each clinical research study described below with the appropriate formula for computing the required confidence interval.

(A) $\bar{Y} \pm z(\sigma/\sqrt{n})$

(B) $(\bar{Y}_1 - \bar{Y}_2) \pm t\sqrt{\frac{s_p^2}{n_1} + \frac{s_p^2}{n_2}}$

(C) $\hat{p} \pm z\sqrt{\frac{\hat{p}(1-\hat{p})}{n}}$

(D) $(\hat{p}_1 - \hat{p}_2) \pm z\sqrt{\frac{\hat{p}_1(1-\hat{p}_1)}{n_1} + \frac{\hat{p}_2(1-\hat{p}_2)}{n_2}}$

(E) $\bar{Y} \pm t(s/\sqrt{n})$

10. A random sample of 25 students is selected from the freshman class at a certain medical school. Study subjects are asked to report whether they have ever used marijuana. Based on the information obtained from this sample, the researcher conducting the study wishes to estimate, with 95% confidence, the proportion of students in the freshman class who have used marijuana.

11. A second sample of 25 students is chosen from the sophomore class at the medical school observed in question 10. In this instance, the researcher wishes to estimate, with 95% confidence, the difference in the proportion of freshman medical students who have used marijuana and the proportion of sophomore medical students who have done so.

12. As part of a study examining the relationship between obesity and serum cholesterol levels, a pair of cardiologists measure serum cholesterol levels in 1500 obese women and compute the mean and standard deviation for these subjects. The investigators also determine the mean and standard deviation of serum cholesterol levels for 2000 women with normal body weights. They wish to estimate, with 95% confidence, the average serum cholesterol level for the population of all obese women.

13. In addition to the confidence interval estimate described in question 12, the cardiologists conducting this study also wish to estimate, with 95% confidence, the difference in average serum cholesterol levels between the population of obese women and the population of normal women.

Questions 14–18

Match each description with the corresponding confidence interval formula.

(A) $\bar{Y} \pm z(\sigma/\sqrt{n})$

(B) $(\bar{Y}_1 - \bar{Y}_2) \pm t\sqrt{\frac{s_p^2}{n_1} + \frac{s_p^2}{n_2}}$

(C) $\hat{p} \pm z\sqrt{\frac{\hat{p}(1-\hat{p})}{n}}$

(D) $(\hat{p}_1 - \hat{p}_2) \pm 1.96\sqrt{\frac{\hat{p}_1(1-\hat{p}_1)}{n_1} + \frac{\hat{p}_2(1-\hat{p}_2)}{n_2}}$

(E) $\bar{Y} \pm t(s/\sqrt{n})$

14. Estimates the difference between two population means (i.e., $\mu_1 - \mu_2$) when the population standard deviations are unknown

15. Estimates the difference between two population proportions (i.e., $P_1 - P_2$)

16. Estimates a population mean μ when the population standard deviation σ is known

17. Estimates a population proportion P

18. Estimates a population mean μ when the population standard deviation σ is unknown

---

10-C
11-D
12-E
13-B
14-B
15-D
16-A
17-C
18-E

ANSWERS AND EXPLANATIONS

1. The answer is B *[I A 1 b (3), B 2 a–b, C 1]*
A confidence interval provides an estimate of a given population parameter, such as the population mean μ. When a clinical investigator computes a 99% confidence interval estimate of the population mean from a single sample of size n, she may be 99% confident that this interval actually includes the unknown population mean. She can also expect that 99% of such confidence intervals computed from many samples (drawn from the same population) would encompass the true population mean. A given confidence interval, however, provides no information about the expected values of the point estimates obtained from other samples. A 99% confidence interval is wider than a 95% confidence interval because it is based on a larger confidence coefficient (e.g., for a 95% confidence interval, the confidence coefficient $z_{.975} = 1.96$, while for a 99% confidence interval, the confidence coefficient $z_{.995} = 2.58$).

2. The answer is D *[II D 3 b]*
The 95% confidence interval has been computed incorrectly. A confidence interval provides an estimate of a given population parameter; it provides no information on the normal range (i.e., the range encompassing 95% of healthy individuals) of values for a response variable. Because this investigation uses a paired study design, the number of degrees of freedom is $n - 1 = 24$.

3. The answer is E *[I C 1–2]*
It is obvious from the general formula for a confidence interval that the width of the interval will be affected by both the magnitude of the confidence coefficient (i.e., the desired level of confidence) and the standard error of the estimate (i.e., the sample-to-sample variation). Since the standard error is a function of the subject-to-subject variability in the response variable, as well as the sample size n, these factors also influence the width of a confidence interval. The distance between a calculated sample mean and the corresponding population mean is unknown and does not affect the confidence interval estimate.

4. The answer is C *[I B 1, C 1–2]*
The general formula for computing a confidence interval is

$$\text{estimate} \pm (\text{confidence coefficient} \times \text{standard error of estimate})$$

As the sample size n increases or the subject-to-subject variation decreases, the standard error decreases, and, according to the formula, the width of the confidence interval also decreases. As the standard error increases (i.e., the precision of the point estimate decreases), the width of the confidence interval increases.

5. The answer is D *[II B 4]*
The population parameter to be estimated is μ, the mean plasma potassium level for the population of men with primary hyperaldosteronism. The point estimate of μ is $\overline{Y} = 3.35$ mEq/L, the sample mean derived from the population of 25 men. Because the population standard deviation is unknown, the 95% confidence coefficient must be obtained from the t distribution, with the number of degrees of freedom $df = n - 1 = 24$. From Table C (see Appendix 3), the value of $t_{.975}$ with 24 df is 2.0639. Using the general formula for a confidence interval,

$$\text{estimate} \pm (\text{confidence coefficient} \times \text{standard error of estimate})$$

$$\begin{aligned} CI_{\mu} &= \overline{Y} \pm t_{.975}(s/\sqrt{n}) \\ &= 3.35 \pm 2.0639(.5/\sqrt{25}) \\ &= 3.14\text{–}3.56 \text{ mEq/L} \end{aligned}$$

Thus, the researcher conducting this study can be 95% confident that the mean plasma potassium level in men with primary hyperaldosteronism lies between 3.14 and 3.56 mEq/L.*

6. The answer is A *[II E 4]*
The parameter to be estimated is P, the prevalence of coronary heart disease among black men

---

*This question was adapted from Duncan RC, Knapp RG, Miller MC III: *Introductory Biostatistics for the Health Sciences*, 2nd ed. Albany, Delmar, 1983, p 113.

residing in the community in question. The point estimate of P is $\hat{p}$, the prevalence computed from the sample = 100/1000 = .10. The 95% confidence coefficient, $z_{(1-\alpha/2)} = z_{.975} = 1.96$, is obtained from the standard normal table (Table B, Appendix 3). Using the general formula for a confidence interval,

estimate ± (confidence coefficient × standard error of estimate)

$$\text{CI}_P = \hat{p} \pm \sqrt{\frac{\hat{p}(1-\hat{p})}{n}}$$

$$= .10 \pm 1.96\sqrt{\frac{(.10)(.90)}{1000}}$$

$$= .081–.119$$

The study coordinators may be 95% confident that the prevalence of coronary heart disease in the population studied is between .08 and .12.

7. The answer is C *[II F 3]*
The parameter to be estimated in this instance is $P_1 - P_2$, the difference in the prevalence of silent myocardial infarction among hypertensive and normotensive men. The point estimate of this difference is $\hat{p}_1 - \hat{p}_2$, where $\hat{p}_1$ is the prevalence of silent myocardial infarction among the hypertensive men sampled ($\hat{p}_1 = 60/200 = .30$) and $\hat{p}_2$ is the corresponding prevalence in the normotensive sample ($\hat{p}_2 = 50/700 = .07$). The 95% confidence coefficient, $z_{(1-\alpha/2)} = z_{.975}$, is listed in Table B (Appendix 3) as 1.96. The 95% confidence interval is

estimate ± (confidence coefficient × standard error of estimate)

$$\text{CI}_{(P_1-P_2)} = (\hat{p}_1 - \hat{p}_2) \pm 1.96\sqrt{\frac{\hat{p}_1(1-\hat{p}_1)}{n_1} + \frac{\hat{p}_2(1-\hat{p}_2)}{n_2}}$$

$$= (.30 - .07) \pm 1.96\sqrt{\frac{(.30)(.70)}{200} + \frac{(.07)(.93)}{700}}$$

$$= .164–.296$$

The research team conducting the study can be 95% confident that this interval contains the unknown difference in the prevalence of silent myocardial infarction between the two populations.

8. The answer is D *[II A 3]*
The desired population parameter is μ, the mean urinary chloride excretion rate. The point estimate of this parameter is $\overline{Y} = 170$ mEq/day, the average urinary chloride excretion rate in the sample of 25 premature infants. Because the population standard deviation σ is known, the 99% confidence coefficient, $z_{.995} = 2.58$, can be obtained from the standard normal table (Table B, Appendix 3). Using the general formula for calculating a confidence interval,

estimate ± (confidence coefficient × standard error of estimate)

$$CI_\mu = \overline{Y} \pm 2.58(\sigma/\sqrt{n})$$

$$= 170 \pm 2.58(20/\sqrt{25})$$

$$= 159.7–180.3 \text{ mEq/day}$$

The physician who conducted this study can be 99% confident that the mean urinary chloride excretion rate for the population of premature infants lies between 159.7 and 180.3 mEq/day.*

9. The answer is B *[II C 2 c, 3]*
The population parameter to be estimated is $\mu_P - \mu_C$, the difference in mean weight loss during insulin-induced hypoglycemia between the population of patients receiving propranolol and the control population. The point estimate of this difference is $\overline{Y}_P - \overline{Y}_C$, the difference between the sample means (i.e., 120 − 70 = 50 g). The common unknown standard deviation σ is estimated by s_p; the

*This question was adapted from Duncan RC, Knapp RG, Miller MC III: *Introductory Biostatistics for the Health Sciences*, 2nd ed. Albany, Delmar, 1983, p 111.

corresponding pooled variance s_p^2 is

$$s_p^2 = \frac{s_P^2(n_P - 1) + s_C^2(n_C - 1)}{n_P + n_C - 2}$$

$$= \frac{100(12 - 1)+64(11 - 1)}{12 + 11 - 2}$$

$$= 82.857$$

Given that the number of degrees of freedom $df = 12 + 11 - 2 = 21$, the 95% confidence coefficient, $t_{(1-\alpha/2)} = t_{.975}$, can be found in Table C (see Appendix 3); this value is listed as 2.0796. The 95% confidence interval $[CI_{(\mu_P - \mu_C)}]$ is computed using the formula

$$CI_{(\mu_P - \mu_C)} = (\bar{Y}_P - \bar{Y}_C) \pm t_{.975}\sqrt{\frac{s_p^2}{n_P} + \frac{s_p^2}{n_C}}$$

$$= 50.0 \pm (2.0796)(3.7990)$$

$$= 42.1\text{–}57.9 \text{ g}$$

Thus, the endocrinologist who carried out this study can be 95% confident that the computed interval encompasses the true difference in mean weight loss between the population of propranolol-treated patients and the control population.

10–13. The answers are: 10-C *[II E 3 c, 4 b]*, **11-D** *[II F 2 c, 3 b]*, **12-E** *[II B 3 c, 4 b]*, **13-B** *[II C 2 c (4), 3 b]*

The general formula for calculating a confidence interval estimate of a given population parameter is

estimate ± (confidence coefficient × standard error of estimate)

where the estimate is the point estimate computed from a sample, the confidence coefficient is the desired level of confidence, and the standard error refers to the variation among all possible values of the point estimate.

In the first study, the initial unknown population parameter to be estimated is P, the proportion of students in the freshman class at this medical school who have used marijuana. The point estimate of P is $\hat{p}$, the proportion of students in the sample who have used marijuana. The standard error of this estimate is

$$s_{\hat{p}} = \sqrt{\frac{\hat{p}(1 - \hat{p})}{n}}$$

Because the sample proportion $\hat{p}$ follows an approximately normal distribution, the confidence coefficient, $z_{(1-\alpha/2)}$, is derived from the standard normal table (Appendix 3, Table C).

The second unknown population parameter to be estimated in this study is $P_1 - P_2$, the difference between the proportion of freshman medical students (P_1) and sophomore medical students (P_2) who have used marijuana. The sample estimate in this case is $\hat{p}_1 - \hat{p}_2$, the difference between the proportion of marijuana users among the sample of 25 freshmen and the proportion among the sample of 25 sophomores. The standard error is estimated by

$$s_{(\hat{p}_1 - \hat{p}_2)} = \sqrt{\frac{\hat{p}_1(1 - \hat{p}_1)}{n_1} + \frac{\hat{p}_2(1 - \hat{p}_2)}{n_2}}$$

and the confidence coefficient is again obtained from the standard normal table.

The next study wishes to first estimate the mean serum cholesterol level μ for the population of obese women. The point estimate of this parameter is $\bar{Y}$, the mean serum cholesterol level computed for the sample of 1500 obese women. The standard error of the point estimate is given by the formula

$$s_{\bar{Y}} = \sigma/\sqrt{n}$$

which is estimated by $s/\sqrt{n}$ when the population standard deviation σ is unknown. Because σ is unknown, the confidence coefficient, $t_{(1-\alpha/2)}$, is obtained from the t distribution (Table C, Appendix 3).

The final population parameter to be estimated is $\mu_1 - \mu_2$, the difference in the mean serum cholesterol level between the population of obese women and the normal population. The point estimate of this difference is $\overline{Y}_1 - \overline{Y}_2$, the difference in mean cholesterol levels for the two samples. The standard error of this estimate is

$$s_{(\overline{y}_1 - \overline{y}_2)} = \sqrt{\frac{s_p^{\,2}}{n_1} + \frac{s_p^{\,2}}{n_2}}$$

As in question 12, the confidence coefficient is obtained from the t distribution.

14–18. The answers are: 14-B *[II C 3 b]*, **15-D** *[II F 3 b]*, **16-A** *[II A 3 b]*, **17-C** *[II E 4 b]*, **18-E** *[II B 4 b]* The appropriate confidence interval formula for estimating the difference between two population means, $\mu_1 - \mu_2$, is

$$\mathrm{CI}_{(\mu_1 - \mu_2)} = \overline{Y}_1 - \overline{Y}_2 \pm t\sqrt{\frac{s_p^{\,2}}{n_1} + \frac{s_p^{\,2}}{n_2}}$$

where $s_p^{\,2}$ is an estimate of the common unknown population variance σ^2. Because the population standard deviations are unknown, the relevant confidence coefficient must be obtained from the t distribution (Appendix 3, Table C) where $df = n_1 + n_2 - 2$.

The confidence interval formula that estimates the difference between two population proportions, $P_1 - P_2$, is

$$\mathrm{CI}_{(P_1 - P_2)} = (\hat{p}_1 - \hat{p}_2) \pm z\sqrt{\frac{\hat{p}_1(1 - \hat{p}_1)}{n_1} + \frac{\hat{p}_2(1 - \hat{p}_2)}{n_2}}$$

In this instance, the population of differences between sample proportions follows the normal distribution and the confidence coefficient is therefore derived from the standard normal table.

When the population standard deviation σ is known, the confidence interval estimate of the population mean μ is calculated using the formula

$$\mathrm{CI}_{\mu} = \overline{Y} \pm z(\sigma/\sqrt{n})$$

The appropriate confidence coefficient, $z_{(1-\alpha/2)}$, is obtained from the standard normal table (Table B, Appendix 3).

The confidence interval formula for estimating an unknown population proportion P is

$$\mathrm{CI}_P = \hat{p} \pm z\sqrt{\frac{\hat{p}(1 - \hat{p})}{n}}$$

The sample proportion $\hat{p}$ is approximately normally distributed. Consequently, the relevant confidence coefficient, $z_{(1-\alpha/2)}$, is derived from the standard normal table.

When the population standard deviation σ is unknown, the confidence coefficient is derived from the t distribution (Appendix 3, Table C) and the appropriate confidence interval formula for estimating the population mean μ is

$$\mathrm{CI}_{\mu} = \overline{Y} \pm t(s/\sqrt{n})$$

12 Hypothesis Testing: The Basis of Statistical Reasoning

I. HYPOTHESIS TESTING USING INFERENTIAL STATISTICS. All medical research, whether it be a trial of a new drug, a comparison of two surgical techniques, or an assessment of the relationship between diet and heart disease, begins with a **research question**. To answer this question, data are obtained from a sample drawn from the population of interest.

A. The hypothesis. The research question is typically posed as a declarative statement, or **research hypothesis**, postulating the existence of a difference between groups or an association among factors. The hypothesis may be derived from a hunch, an educated guess based on published results, or preliminary observations. If the data obtained from the sample can be shown beyond a reasonable doubt to be consistent with the hypothesis, it is **accepted** as an accurate statement about the population. Otherwise, the hypothesis is **rejected.**

B. Statistical tests of the hypothesis. Because of random variation, even an unbiased sample may not accurately represent the population as a whole. As a result, it is possible that any observed differences or associations may have occurred by chance. The probability that a given outcome is due to chance can be mathematically estimated by means of **inferential statistics**. Statistical testing of a research hypothesis allows a physician to quantify the risk of error involved in making inferences about a population based on the information obtained from a sample. This chapter outlines the principles common to all statistical tests of the hypothesis.

II. THE STATISTICAL REASONING PROCESS: A SIMPLE EXAMPLE. A group of medical students meets every Friday at the local tavern, where they flip a coin to determine who will pay for the weekly refreshments. A member of the group is offered an opportunity to take advantage of his classmates by purchasing an allegedly "loaded" coin. How can the student decide if the coin in question is truly "loaded" or merely an ordinary penny?

A. The student's decision can be framed as the following research question: "For the population of all outcomes that can result from a flip of this particular coin, will heads occur more often than tails?" For the purpose of statistical testing, it is customary to restate this question in terms of two opposing hypotheses.

1. The **alternative hypothesis, H_A**, is the research question stated in a specific testable form.
   a. **H_A** in this case states "**The coin is 'loaded'**; over the long run, more flips will result in heads than tails."
   b. In general, H_A proposes "**there is a treatment effect.**"
2. The **null hypothesis, H_O**, is the opposing supposition.
   a. **H_O** in this case states "**The coin is fair**; over the long run, heads and tails will occur equally often."
   b. In general, H_O proposes "**there is no treatment effect.**"
   c. H_O, therefore, is the "straw man" of statistical hypothesis testing. **If the sample data provide sufficient evidence** to discredit **H_O**, it **can be rejected in favor of H_A.**

B. Obtaining information about the population from a sample. Before purchasing the coin, the student insists on trying it out and bullies the buyer into permitting 10 free throws. These 10 tosses represent a sample of all possible tosses of the coin (the population). The student will base his conclusion about the fairness of the coin on the information provided by this sample. He throws 2 tails (T) and 8 heads (H) [2 T–8 H].

C. Evaluating the evidence against H_O

1. The **sample outcome** (2 T–8 H) can be explained in one of two ways.
   a. **H_A is true**, that is, over the long run, the outcome "heads" will occur more frequently than the outcome "tails."
   b. **H_O is true**, that is, the coin is fair and the sample outcome is due to random chance.
2. **If H_O is true**, the most common result of 10 tosses should be an equal number of heads and tails (5 T–5 H). However, other outcomes, such as 4 T–6 H, 3 T–7 H, or 2 T–8 H, will sometimes be seen. **The more extreme an outcome, the less frequently it will occur.**
3. **The relative frequency distribution** of all outcomes of 10 tosses, **assuming that H_O is true** (i.e., the coin is fair), is shown in Table 12-1. This distribution can be obtained by repeating a series of 10 tosses with a known fair coin an extremely large number of times (e.g., 100,000 times) and tallying the frequency of the occurrence of each outcome. In practice, mathematically derived frequency tables are substituted for this operation.
4. **Formulating a decision rule.** The student decides that an outcome that occurs 5% of the time or less is sufficiently unlikely to be the result of random chance and a fair coin. In statistical language, **he adopts a decision rule based on a level of significance** of .05.

D. Drawing a conclusion.

D. Drawing a conclusion. From Table 12-1, the observed outcome, 2 T–8 H, can be expected to occur less than 5% of the time when a fair coin is tossed 10 times. Therefore, the student concludes that the coin in question is unlikely to be fair, and buys it, anticipating weeks of free refreshments.

1. When the sample data provide sufficient evidence to conclude that H_A is true, the results are said to be **statistically significant** at the chosen level of significance. In other words, it is **unlikely** that the sample outcome is due to random chance.
2. When the sample data do not support H_A (i.e., the observed outcome is **likely** to occur in a population for which H_O is true), the results are reported as **not statistically significant**.

E. Estimating the probability of an error

1. **Type I error.** Even though he decides that the coin is probably "loaded" (on the basis of the observed result, 2 T–8 H), the student realizes that he *may* have lost $10 by mistakenly purchasing a fair coin. If he has indeed falsely rejected H_O and concluded that H_A is true when it is not, he has committed a type I error.
   a. The **probability** of committing a type I error is expressed as α, or

$$\alpha = P(\text{type I error})$$

Table 12-1. Frequency Distribution of Possible Outcomes of 10 Tosses of a Fair Coin

Outcome	Relative Frequency	Cumulative Frequency
0 T–10 H	.00098	.00098
1 T–9 H	.0098	.0108
2 T–8 H	.0439	.0547
3 T–7 H	.1172	.1719
4 T–6 H	.2051	.3770
5 T–5 H	.2460	.6230
6 T–4 H	.2051	.8281
7 T–3 H	.1172	.9453
8 T–2 H	.0439	.9892
9 T–1 H	.0098	.9990
10 T–0 H	.00098	1.0

H = heads; T = tails

b. The type I error rate the student is willing to tolerate is called the **level of significance (α level)** of the statistical test. Customarily, the level of significance is set at .05 or .01. The choice of level of significance depends on the consequences of committing a type I error; in this instance, the consequence is the student paying $10 for an ordinary penny.

c. Decision rules and α levels. By concluding that the coin was "loaded" when the sample of 10 throws yielded an outcome of 2 T–8 H or a more extreme outcome, the student ensured that his probability of committing a type I error was 5% or less. To ensure that his chance of a type I error was 1% or less (i.e., $\alpha = .01$), the student would modify the decision rule to conclude that the coin was "loaded" only if the sample of 10 tosses yielded a result of 1 T–9 H or 0 T–10 H (see Table 12-1). Therefore, through his choice of which outcomes are labeled "unlikely to occur by chance, given a fair coin," the student controls the chance of committing a type I error.

2. Type II error. If the student had obtained an outcome of 3 T–7 H for the 10 tosses of the coin, the decision rule he adopted would have forced him to conclude that the coin was fair. If the coin is actually "loaded," he has failed to accept H_A when it is, in fact, true. This mistake is known as a type II error.

a. The **probability** of committing a type II error is expressed as β, or

$$\beta = P(\text{type II error})$$

b. In other words, β, the probability that the student will falsely conclude that the coin is fair, is the probability that 10 tosses of a "loaded" coin will yield a result of 3 T–7 H.

c. The determination of β is complicated by the fact that **a different relative frequency distribution** of the possible outcomes of 10 coin tosses **exists for every possible loading value.**

(1) If, for example, the coin vendor claims that the loading value is .50 mg in favor of heads, the student must know the frequency distribution of the outcomes of 10 tosses of a coin "loaded" with a weight of .50 mg favoring heads to compute β.

(a) Perhaps the outcome 3 T–7 H (or a more extreme outcome, such as 4 T–6 H) occurs 20% of the time when a coin "loaded" with this weight is tossed 10 times.

(b) In this instance, the probability that the student will commit a type II error is .20.

(2) Changes in the loading value necessitate the need to consult a different frequency distribution and thus result in corresponding alterations in the value of β.

3. The *p*-value versus the α level. A conclusion based on a statistical test is typically reported in conjunction with a *p*-value. The *p*-value and the α level, while similar in terms of the information they symbolize, have slightly different definitions.

a. The *p*-value represents the actual probability of obtaining the particular sample outcome (or one more extreme) from a population for which H_O is true. That is, it is **the probability of a type I error**. The *p*-value, therefore, varies from sample to sample.

b. The **α level** represents the **risk of incurring a type I error that the investigator is willing to tolerate.** It is chosen by the investigator and hence, is independent of the data obtained from any given sample.

c. The *p*-value measures the strength of evidence against H_O.

(1) The **smaller** the *p*-value, the stronger the evidence **against** H_O.

(2) The *p*-value at which evidence against H_O is declared decisive is α, the level of significance.

d. Restating the **decision rule** in terms of the *p*-value and the desired α level:

(1) For the student mentioned in II C 4, the desired α level is .05.

(2) Thus, the decision rule can be stated: **"Reject H_O and conclude that H_A is true if $p \leq \alpha = .05$."**

III. THE STATISTICAL REASONING PROCESS: A CLINICAL EXAMPLE.

The following example illustrates the same procedure as that outlined in II. However, in this instance, data are expressed on an interval/ratio scale.

A. Example. A cardiologist wishes to investigate whether men between 40 and 60 years of age who have had an acute myocardial infarction (AMI) have serum cholesterol levels that are higher, on the average, than those of the general population. From previous work, she knows that the average cholesterol level in the general population is 242 mg/dl and the standard deviation (σ) of cholesterol values is 40 mg/dl.

B. The research question "Do men between 40 and 60 years of age with a history of AMI have a higher average serum cholesterol level than the general population?" **may be framed as a pair of opposing hypotheses.**

1. **H_O** states that there is **no difference** in average serum cholesterol levels between the population of men who have suffered an AMI and the general population. It is written symbolically:

$$H_O: \mu_{MI} = 242 \text{ mg/dl}$$

where μ_{MI} represents the average cholesterol level of the population of men between 40 and 60 years of age who have had an AMI (i.e., the population mean).

2. **H_A** states that the average cholesterol level of the population of men who have had an AMI is **greater than** that of the general population, and is written symbolically:

$$H_A: \mu_{MI} > 242 \text{ mg/dl}$$

C. Obtaining information about the population from a sample. The investigator measures cholesterol levels in a random sample of 100 men between the ages of 40 and 60 who have suffered an AMI, and computes a sample mean ($\bar{Y}$) of 260 mg/dl. Conclusions about the value of the population mean μ_{MI} are based on this sample value.

D. Measuring the disparity between the population parameter and the sample parameter

1. The **test statistic** measures the disparity between the value assumed for the population mean [given that H_O is true (i.e., $\mu_{MI} = 242$ mg/dl)] and the observed sample mean (i.e., $\bar{Y} = 260$ mg/dl) **in units of standard deviation.**

2. **Calculating the test statistic**

   a. The general form of the test statistic is:

$$\text{test statistic} = \frac{\text{sample estimate} - \text{hypothesized population value}}{\text{standard deviation*}}$$

   b. Using this formula, the test statistic in this example is calculated as:

$$\text{test statistic} = \frac{\bar{Y} - \mu_{MI}}{\sigma_{MI}/\sqrt{n}} = \frac{260 - 242}{40/\sqrt{100}} = 4.5$$

   c. In other words, the observed sample mean, $\bar{Y}$, is 4.5 standard deviations greater than the population mean ($\mu_{MI} = 242$ mg/dl) postulated by H_O (and, in this case, 4.5 is a z score).

E. Evaluating the evidence against H_O

1. The disparity between the sample estimate ($\bar{Y}$) and the population parameter hypothesized by H_O (μ_{MI}) can be explained in one of two ways.

   a. **H_A is true,** that is, the sample was drawn from a population whose mean cholesterol level is greater than 242 mg/dl.

   b. **H_O is true,** that is, the disparity is attributable to random chance (known as **sampling error** or **sampling variation**).

2. The **larger the disparity** between the observed sample value and the hypothesized population value for the parameter in question, the **greater the evidence against H_O.**

3. **The probability** of obtaining a sample mean ($\bar{Y}$) as large as 260 mg/dl, when the true population mean is 242 mg/dl, **is determined by comparing the observed value of $\bar{Y}$ with the frequency distribution of all possible values** of $\bar{Y}$ that could be obtained from repeated

*The standard deviation in the denominator is the standard error of the estimate. When hypotheses concern population means, the standard error of the estimate is the standard error of the mean, or $\sigma_{\bar{y}} = \sigma/\sqrt{n}$. In this example, $\sigma_{MI} = \sigma_{GENERAL} = 40$ mg/dl.

samples drawn from a population **for which H_O is true.** This probability is the same as that of computing a test statistic greater than or equal to 4.5 when H_O is true.

a. This frequency distribution could be generated by repeatedly drawing samples of 100 men from a population **whose true mean is known to be 242 mg/dl**, calculating the average cholesterol value for each sample, and tallying the relative frequency of occurrence of each value of the sample mean. However, in actual practice, mathematically derived frequency tables are substituted for this operation.

b. For example, the frequency distribution of sample means follows a normal distribution (see Ch 10 III A 3 c). Since, in this instance, the test statistic is the standardized sample mean (i.e., *z* score), the standard normal table (Appendix 3, Table B) provides the frequency distribution of potential values of the test statistic.

c. Less than .01% of the values of the test statistic should exceed 4.5 by random chance when H_O is true (Figure 12-1). Or, expressed on the $\bar{Y}$ scale, **the probability of obtaining a sample mean equal to or larger than 260 mg/dl by random chance from a population whose true mean cholesterol level is 242 mg/dl is less than .01% (i.e., $p \leq .0001$).**

d. The investigator sets $\alpha = .01$. In other words, she decides to accept H_A as true (and declare $\mu_{MI} > 242$ mg/dl) only if there is a less than 1% chance of obtaining a sample mean as large or larger than 260 mg/dl by random chance from a population whose mean is 242 mg/dl. The investigator's choice can be expressed as the following **decision rule.**

(1) If $p \leq .01$, reject H_O and conclude that the sample data support H_A; report **"results are statistically significant"** (i.e., unlikely to be due to chance).

(2) If $p > .01$, do not reject H_O; report **"results are not statistically significant"** (i.e., likely to be due to chance).

F. Drawing a conclusion

1. Since $p \leq .01$, the chosen level of significance, the investigator **rejects H_O** in favor of H_A. This result is **statistically significant at the 1% level of significance.**

2. In clinical terms, the investigator can conclude that the average serum cholesterol level in men between 40 and 60 years of age with a history of AMI is **significantly higher than that of the general population** ($p \leq .0001$).

3. The chance that this conclusion is incorrect (i.e., that the test declared H_A true when H_O is true and the sample was drawn from a population whose mean cholesterol level is equal to 242 mg/dl) is less than or equal to .01%. In other words, **the probability that the investigator has committed a type I error is less than .0001.**

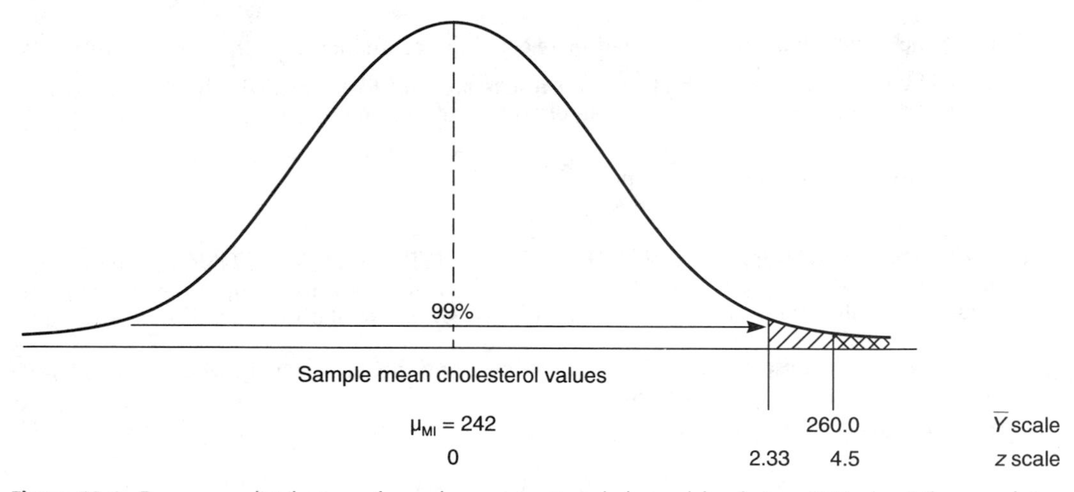

Figure 12-1. Frequency distribution of sample mean serum cholesterol levels ($n = 100$) about the population mean $\mu_{MI} = 242$ mg/dl. The area of the *hatched and cross-hatched regions* under the curve = .01, and the area of the *cross-hatched region* alone $\leq .0001$. Values of $\bar{Y}$ and *z* (the test statistic) that define these areas are indicated on the horizontal axis.

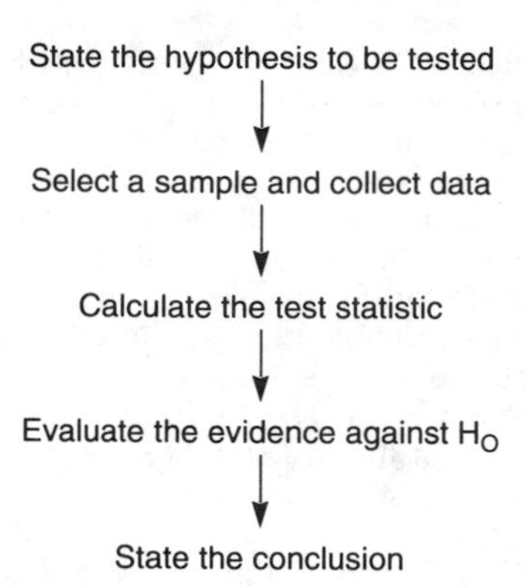

Figure 12-2. Flowchart depicting the five steps that constitute all statistical tests of hypothesis.

IV. GENERAL STEPS IN STATISTICAL HYPOTHESIS TESTING. The statistical reasoning process described in II and III can be broken down into five discrete sequential steps. These steps, summarized below and depicted in flowchart form in Figure 12-2, are the basis of all statistical tests of hypothesis.

A. State the hypothesis to be tested. As discussed in II A 1–2 and III B 1-2, the research question is typically rephrased in the form of two opposing hypotheses, the null hypothesis, H_O, and the alternative hypothesis, H_A.

B. Select a sample and collect data. Values observed in a sample drawn from the population of interest are used to estimate the population parameter (e.g., the population mean).

C. Calculate the test statistic. The test statistic is a measure of the difference between the population parameter proposed by H_O and its actual sample estimate, expressed in terms of standard deviation.

D. Evaluate the evidence against H_O. The test statistic is compared with its frequency distribution to determine the probability of obtaining this value by random chance when H_O is true (i.e., the *p*-value).

1. The level of significance, α, is the point at which the evidence is sufficient to declare H_O false.

2. The significance level is used to derive a **decision rule,** a procedure defining the conditions under which H_O is accepted or rejected.

E. State the conclusion. H_O is accepted or rejected in accordance with the decision rule.

1. If **H_O is rejected in favor of H_A**, the results are said to be **statistically significant** (i.e., unlikely to be due to chance) at the chosen level of significance.

2. If **H_O is not rejected**, the results are reported as **not statistically significant** (i.e., likely to be due to chance) at the chosen level of significance.

V. THE QUANTIFICATION OF UNCERTAINTY IN HYPOTHESIS TESTING. A key feature of the statistical testing procedure is that it permits the investigator to **quantify the risk of error** associated with inferences about a population based on information obtained from a sample.

A. The type I error (false positive). The predetermined risk of accepting H_A when, in fact, H_O is true, is α, the risk of a **type I error.**

1. **Notation.** In probability notation,

$$\begin{aligned}\alpha &= P(\text{reject } H_O \mid H_O \text{ true}) \\ &= P(\text{conclude } H_A \text{ true} \mid H_O \text{ true}) \\ &= P(\text{type I error})\end{aligned}$$

2. Analogy of statistical tests to diagnostic tests

a. Diagnostic tests must answer the question "Are the test results for a particular patient more likely to be part of the population of test values associated with healthy individuals or the population of values associated with diseased patients (see Ch 3 I)?"

(1) The answer to this question depends on the location of the **positivity criterion** (see Ch 3 I B and Ch 4) for the diagnostic test.

(2) Because most diagnostic tests are less than perfect, a certain percentage of healthy patients are erroneously labeled as diseased **(false positives)** and a certain percentage of diseased patients are falsely declared disease-free **(false negatives).** The selection of a positivity criterion represents **a trade-off between the false positive rate** (FPR; Figure 12-3, *solid area*) **and the false negative rate** (FNR; see Figure 12-3, *hatched area*).

b. Analogously, the statistical test outlined in III addresses the question "Did the sample mean ($\overline{Y}$) come from a population of sample means clustering about μ_0 (H_O true; Figure 12-4, *left curve*) or did it come from a population of sample means clustered about μ_1 (H_A true; see Figure 12-4, *right curve*)?"

(1) The closer $\overline{Y}$ (or the equivalent test statistic) lies on the horizontal axis to μ_0, the more likely H_O is true (i.e., $\mu_{MI} = 242$ mg/dl). Conversely, the farther $\overline{Y}$ is to the right on the horizontal axis, the more likely H_A is true (i.e., $\mu_{MI} > 242$ mg/dl).

(2) Thus, as with a diagnostic test, the answer to the question posed in V A 2 b depends upon the selection of a **positivity criterion**, the value on the horizontal axis that defines **how far** to the right $\overline{Y}$ (or the equivalent test statistic) must lie to conclude that H_A is true.

(3) The choice of a value for the positivity criterion represents **a trade-off between the possibility of rejecting H_O when it is actually true** (false positive rate; see Figure 12-4, *solid area*) **and accepting H_O when it is actually false** (false negative rate; see Figure 12-4, *hatched area*).

(4) For both diagnostic tests and statistical tests, the location of the positivity criterion is determined by the false positive rate (in the case of a statistical test, α, the level of significance) that the investigator is willing to tolerate.

3. Choosing a level of significance

a. Prior to data collection, the investigator must decide on the level of significance that represents a tolerable risk of committing a type I error.

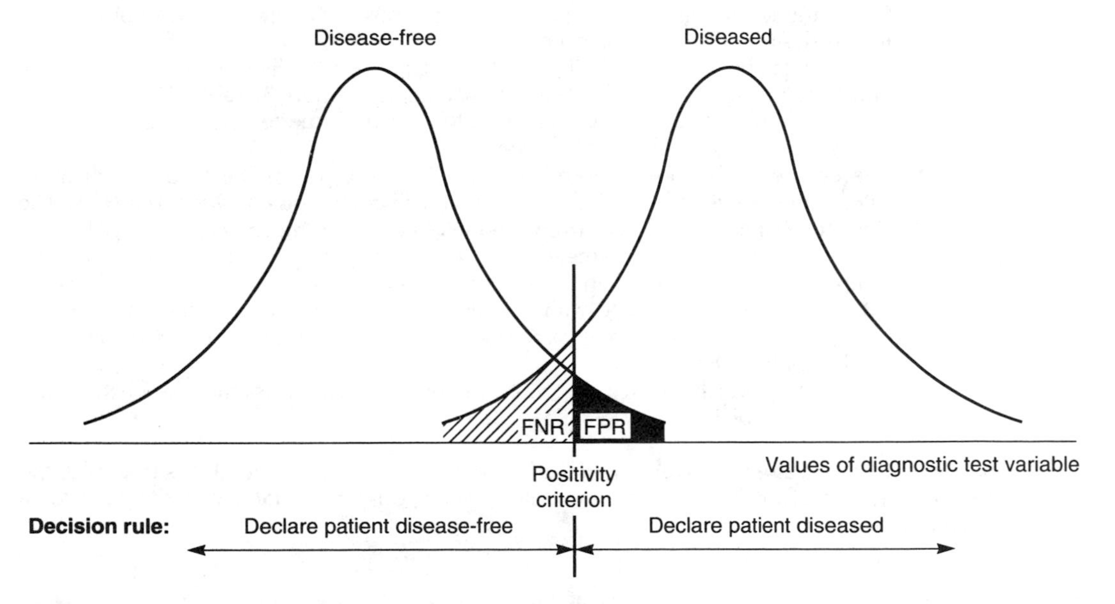

Figure 12-3. Graphic representation of the errors associated with a diagnostic test. The *curve on the left* represents the frequency distribution of diagnostic test values for a disease-free population; the *curve on the right* represents the frequency distribution of diagnostic test values for a diseased population. *FNR* = false negative rate; *FPR* = false positive rate.

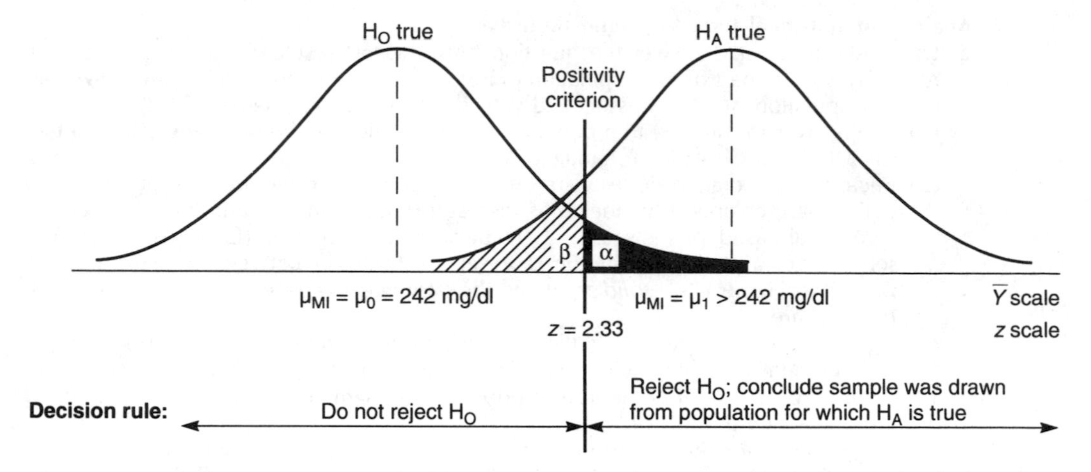

Figure 12-4. Graphic representation of the errors associated with a statistical test when the level of significance $\alpha = .01$. The *curve on the left* represents the frequency distribution of the population of sample mean cholesterol levels when the null hypothesis (H_O) is true, while the *curve on the right* represents the frequency distribution of the population of sample mean cholesterol levels when the alternative hypothesis (H_A) is true. The *solid area* corresponds to the risk of a type I error (α), that is, the probability of accepting H_A when H_O is true. The *hatched area* corresponds to the risk of a type II error (β), that is, the probability of accepting H_O when H_A is true.

(1) In the example discussed in III, the investigator set $\alpha = .01$. By choosing this value for the positivity criterion, she can conclude that men with a history of AMI have a higher average serum cholesterol level than the general population (i.e., H_A is true) only if the chance of error associated with this conclusion is 1% or less.

(2) When comparing a new procedure or medication to an established treatment, it is often desirable to select a very low value for α (typically, $\alpha = .01$), especially if the new treatment is known to be associated with a substantially greater risk of serious side effects.

b. Formulating a decision rule based on the location of the positivity criterion

(1) The positivity criterion is located so that the area under the frequency distribution curve and to the right of (i.e., greater than) the positivity criterion value corresponds to the desired level of significance.

(2) For the example described in III, $\alpha = .01$. The area under the standard normal curve and to the right of $z = 2.33$ is equal to .01 (see Appendix 3, Table B). Thus, in this case, the positivity criterion corresponds to a value of the test statistic $z = 2.33$ (see Figure 12-4).

(3) The choice of this positivity criterion can be expressed as the following **decision rule: If the calculated test statistic $z \geq 2.33$, reject H_O and conclude that H_A is true (i.e., the sample was drawn from a population whose mean $\mu > 242$ mg/dl).**

(4) In the population of sample means ($\bar{Y}$s) centered about μ_0 (H_O true), only 1% of the sample means fall more than 2.33 standard deviations from the population mean.

(a) If the z value associated with a particular sample mean is greater than or equal to 2.33, the chance that it came from a population for which $\mu_{MI} = \mu_0$ (H_O true) is less than 1%.

(b) In other words, if H_A ($\mu_{MI} > 242$ mg/dl) is accepted as true, the chance of a type I error is 1% or less.

B. The type II error (false negative). The risk of accepting H_O when, in fact, H_A is true, is β, the risk of a **type II error.** The type II error rate is analogous to the **false negative rate** for a diagnostic test.

1. Notation. In probability notation,

$$\begin{aligned} \beta &= P(\text{conclude } H_O \text{ true} \mid H_O \text{ false}) \\ &= P(\text{conclude } H_O \text{ true} \mid H_A \text{ true}) \\ &= P(\text{type II error}) \end{aligned}$$

2. For the example given in III, β is the probability that the investigator will fail to conclude that men with a history of AMI have a higher average serum cholesterol level than men in the general population, when, in fact, they do have a higher average level. In other words, if the population of men who have had an AMI do have higher average serum cholesterol levels than men in the general population (i.e., H_A is indeed true), the statistical test should fail to conclude that H_A is true ($\beta \times 100$)% of the time (see Figure 12-4, *hatched area*).

3. **Calculation of β: limitations**
   a. In practice, the calculation of the probability of a type II error is complicated by the fact that H_A does not typically pose an explicit value for the population parameter μ_1. For example, in III B, H_O states that $\mu_{MI} = \mu_0 = 242$ mg/dl, while H_A merely states that $\mu_{MI} > \mu_0$.
   b. **Unless the exact location of μ_1 on the horizontal axis is known, the exact magnitude of β cannot be determined.**
   c. The requirement that an explicit value be specified for μ_{MI} under H_A is analogous to the requirement for a specified loading value in the example outlined in II E 2. For each possible value of μ_{MI} specified by H_A, a different value of β can be computed.

C. Power

1. **Definition.** Power is the ability of a statistical test to detect a difference of a specified magnitude (known as a **clinically important difference**), given that this difference exists in the populations being compared. That is, it is the **probability that a statistical test will reject H_O given that H_O is false.**

2. **Notation.** In probability notation,

$$\text{power} = P(\text{reject } H_O | H_O \text{ false})$$
$$= P(\text{conclude } H_A \text{ true} | H_A \text{ true})$$

3. **Properties**
   a. Power is the **complement** of β, the type II error rate:

$$\text{power} = 1 - \beta$$

   b. Unlike α and β, power is **not the risk of a particular error.** Instead, it is the **probability that a statistical test will reach a particular correct conclusion** (e.g., the conclusion that H_A is true when, in fact, it actually is).
   c. The power of a statistical test is analogous to the **sensitivity (i.e., true positive rate)** of a diagnostic test.

4. For the example described in III, power is the probability that the statistical test will lead the investigator to conclude that the mean serum cholesterol level for the population of men with a history of AMI is higher than that found in the general population when this is indeed the case.

5. In Figure 12-5, power is the area under the right-hand curve (H_A true) to the **right** of the positivity criterion.

D. Clinically important difference (effect size)

1. **Definition.** The **clinically important difference,** or **effect size,** specifies the difference between two population parameters that is meaningful from a clinical perspective. Two parameters that differ by less than the clinically important difference are presumed equivalent from a biological standpoint.

2. The stipulation of a clinically important difference allows an investigator to assign a value to μ_1, the population mean proposed by H_A; to rewrite H_O and H_A in terms of this value; and to calculate power and β.

3. **Example.** For the cholesterol study reported in III, the consensus among colleagues consulted by the investigator is that a difference in average serum cholesterol levels between two comparison groups must be at least 10 mg/dl to be biologically meaningful.
   a. The two opposing hypotheses can be rewritten in terms of the specified effect size.
      (1) H_O states: "There is no difference between the average serum cholesterol level in the population of men who have had an AMI and the level in the general population," or **H_O: $\mu_{MI} = \mu_0 = 242$ mg/dl.** In terms of the effect size, H_O states

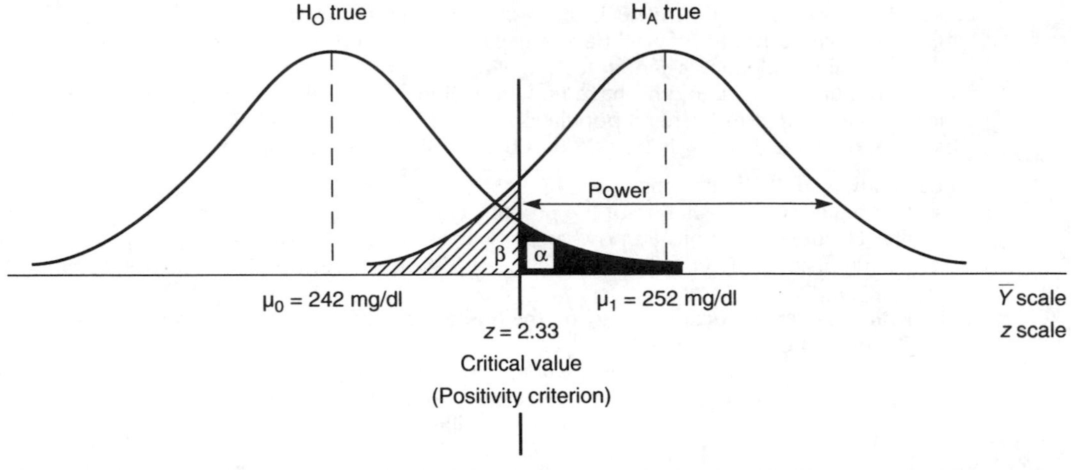

Figure 12-5. Graphic representation of the relationship between the level of significance, α (*solid area*), the probability of a type II error, β (*hatched area*), and power (*open area of righthand curve*), based on an effect size of 10 mg/dl (see V D 3). The *curve on the left* represents the frequency distribution of the population of sample means when the null hypothesis (H_O) is true, while the *curve on the right* represents the frequency distribution of the population of sample means ($\overline{Y}$s) when the alternative hypothesis (H_A) is true.

that a difference in the average cholesterol level of at least 10 mg/dl between the population of men who have suffered an AMI and the general population does not exist.

(2) H_A states: "The mean serum cholesterol level of men with a history of AMI is 10 mg/dl higher than that found in the general population," or $\mathbf{H_A: \mu_{MI} = \mu_1 = 252}$ **mg/dl.**

b. The values for α, β, and power can also be defined in terms of the specified effect size.

(1) The level of significance α is the probability that the statistical test will conclude that serum cholesterol levels in the population of men with a history of AMI and the general population differ by 10 mg/dl, when, in fact, such a difference **does not exist**, or

$$\alpha = P(\text{conclude that a 10 mg/dl difference exists} \mid \text{no such difference exists})$$
$$= P(\text{reject } H_O \mid H_O \text{ true and } H_A \text{ false})$$

(2) The type II error β is the probability that the statistical test will fail to identify a difference of at least 10 mg/dl in serum cholesterol levels for the two populations when, in fact, such a difference exists, or

$$\beta = P(\text{fail to find a 10 mg/dl difference} \mid \text{10 mg/dl difference exists})$$
$$= P(\text{conclude } H_O \text{ true} \mid H_O \text{ false and } H_A \text{ true})$$

(3) Power is the probability that the statistical test will detect a difference of 10 mg/dl between the two populations when such a difference actually exists, or

$$\text{power} = P(\text{conclude that a 10 mg/dl difference exists} \mid \text{10 mg/dl difference exists})$$
$$= P(\text{reject } H_O \mid H_O \text{ false and } H_A \text{ true})$$

c. Computing power

(1) The sample mean ($\overline{Y}$) that corresponds to the positivity criterion $z = 2.33$ [see V A 3 b (2)] is

$$z = \frac{\overline{Y} - \mu}{\sigma/\sqrt{n}}$$
$$2.33 = \frac{\overline{Y} - 242}{40/\sqrt{100}}$$
$$\overline{Y} = 242 + 2.33(4)$$
$$= 251.32$$

In other words, the positivity criterion $z_0 = 2.33$ on the z scale is equivalent to a positivity criterion of $\overline{Y}_0 = 251.32$ on the $\overline{Y}$ scale.

(2) The **decision rule** [see V A 3 b (3)] adopted by the investigator can be reformulated in terms of this value for $\overline{Y}$ as follows: **"If the observed sample mean $\overline{Y} \geq 251.32$ mg/dl, reject H_O and conclude that H_A is true (i.e., conclude that the sample was drawn from a population whose mean $\mu > 242$ mg/dl)."**

(3) In Figure 12-5, β and power are equal to the hatched and open areas, respectively, under the frequency distribution for "H_A true" (see Figure 12-5, *right curve*). Or, in probability notation,

$$\beta = P(\text{conclude } H_O \text{ true} \mid H_A \text{ true})$$
$$= P(\overline{Y} \leq 251.32 \mid \mu_1 = 252 \text{ mg/dl})$$
$$\text{power} = P(\text{reject } H_O \mid H_A \text{ true})$$
$$= P(\overline{Y} \geq 251.32 \mid \mu_1 = 252 \text{ mg/dl})$$

(4) The value of the positivity criterion, $z_0 = 2.33$, is a standard normal value relative to a population mean of 242 mg/dl (H_O true; see Figure 12-5, *left curve*). The areas corresponding to β and power are those under the right-hand curve in Figure 12-5 (H_A true; $\mu_1 = 252$ mg/dl). The z value corresponding to the area to the left of the positivity criterion $\overline{Y}_0 = 251.32$ mg/dl is the standard normal value relative to a population mean of 252 mg/dl (H_A true; see Figure 12-5, *right curve*).*

(5) Therefore, the positivity criterion $\overline{Y}_0 = 251.32$ is converted to a standardized value (z score) for the frequency distribution for which H_A is true (i.e., the distribution with a population mean $\mu_1 = 252$ mg/dl), using the formula†

$$z_1 = \frac{\overline{Y} - \mu}{\sigma / \sqrt{n}}$$
$$= \frac{251.32 - 252}{40 / \sqrt{100}} = -.17$$

(6) Thus,

$$\beta = P(\overline{Y} \leq 251.32)$$
$$= P(z_1 \leq -.17)$$
$$= .4325 \text{ (see Appendix 3, Table B), or } .43$$

and power is computed as

$$\text{power} = 1 - \beta$$
$$= 1 - .43 = .57$$

(7) Interpretation. There is a 43% chance that the statistical test will fail to detect a 10 mg/dl difference in the average cholesterol levels of the population of men with a history of AMI and the general population, given that a difference of this magnitude actually exists. Alternatively, the chance that the statistical test will detect a 10 mg/dl difference in the average cholesterol levels of the two populations, given that this difference exists, is only 57%.

E. Summary of test outcomes. Table 12-2 summarizes the possible outcomes of a statistical test, using a 2 x 2 table analogous to those used to summarize the performance characteristics of a diagnostic test (see Table 2-2 and Table 3-2).

---

*The value of the positivity criterion on the $\overline{Y}$ scale (i.e., $\overline{Y}_0 = 251.32$ mg/dl) is constant for both curves depicted in Figure 12-5. The corresponding z score, however, differs depending on whether it refers to the standardized distance from μ_0 (H_O true; see Figure 12-5, *left curve*) or the standardized distance from μ_1 (H_A true; see Figure 12-5, *right curve*).

†The population standard deviation σ is assumed to be equal for both populations.

Table 12-2. A 2 × 2 Table Summarizing Possible Outcomes of a Statistical Test

Statistical Decision	Data from a Population for which: H_O True	H_O False and H_A True
Do Not Reject H_O	Correct decision	Incorrect decision, or type II error *P*(type II error) = β
Reject H_O	Incorrect decision, or type I error *P*(type I error) = α	Correct decision *P*(correct decision) = power

VI. DESIGN AND INTERPRETATION OF CLINICAL RESEARCH STUDIES

A. Manipulating uncertainty. The investigator's choice of α, effect size, and sample size prior to data collection dictate the power of the statistical test used to analyze the data as well as the risk of committing a type I or a type II error. As discussed in V A 2 b (3), the selection of these values represents a compromise between rejecting H_O when it is actually true and accepting H_O when it is false.

1. Effect of altering the level of significance (α) given a fixed sample size *n*

a. As the **value of α is decreased** (e.g., from .05 to .01), the **positivity criterion is shifted to the right** along the horizontal axis, defining a smaller area under the frequency distribution curve (Figure 12-6A, *hatched area*). As a result:

(1) The value of β (i.e., the risk of failing to reject H_O when it is false) **increases** (see Figure 12-6B, *hatched area*)

(2) Power = (1 − β), the open area in Figure 12-6B, **decreases**

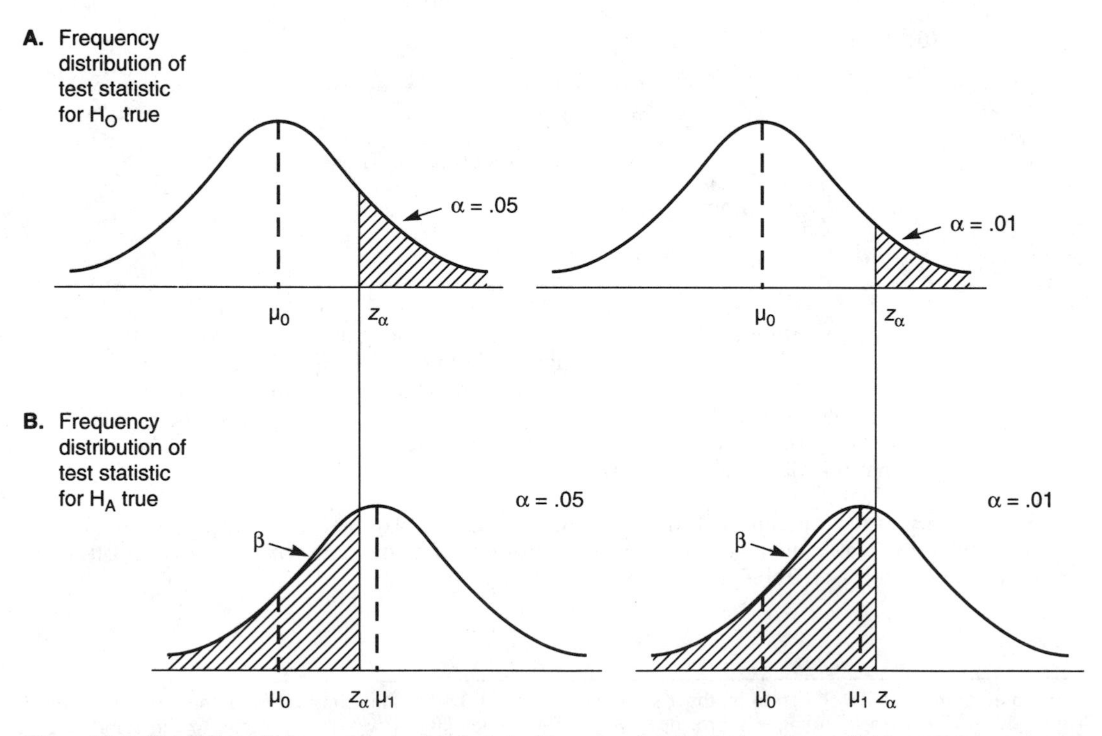

Figure 12-6. The effect of decreasing the level of significance α from .05 (*A, curve on left*) to .01 (*A, curve on right*). As α decreases (*A, hatched area*), β increases (*B, hatched area*). (Reprinted from Duncan RC, Knapp RG, Miller MC III: *Introductory Biostatistics for the Health Sciences*, 2nd ed. Albany, Delmar, 1983, p 105.)

b. Conversely, as the **value of α is increased:**

(1) β decreases

(2) Power increases

2. Effect of altering effect size given fixed sample size n and level of significance α

a. As the clinically important **effect size is increased,** that is, as the distance between μ_0 (the population mean hypothesized by H_O) and μ_1 (the population mean hypothesized by H_A) is **increased, β decreases** and **power increases**. In other words, as the magnitude of the clinically important difference ($\mu_0 - \mu_1$) is increased:

(1) The statistical test is increasingly likely to detect this difference (i.e., power increases)

(2) The test is less likely to fail to detect it (β decreases)

b. As the **effect size ($\mu_0 - \mu_1$) is decreased:**

(1) β increases (i.e., the statistical test is more likely to fail to detect the specified difference)

(2) Power decreases (i.e., the smaller the chance the test will detect the difference)

3. Effect of altering sample size. For a fixed level of significance (α), as the **sample size** n is **increased, β decreases** and **power increases.**

a. Example. The investigator conducting the serum cholesterol study discussed in III repeats the study in a sample of 400 men with a history of AMI.

(1) Increasing the value of n from 100 to 400 people **decreases** the value of $\overline{Y}_0$ corresponding to the positivity criterion $z_0 = 2.33$ (assuming that $\alpha = .01$ and the clinically important difference = 10 mg/dl). Thus,

$$z_0 = \frac{\overline{Y}_0 - \mu_0}{\sigma/\sqrt{n}}$$

$$2.33 = \frac{\overline{Y}_0 - 242}{40/\sqrt{400}}$$

$$\overline{Y}_0 = 242 + 2.33(2)$$

$$= 246.66$$

(2) β and power are recalculated to reflect this new value for $\overline{Y}_0$:

$$\beta = P(\text{conclude } H_O \text{ true} \mid H_A \text{ true})$$

$$= P(\overline{Y} \leq 246.66 \mid \mu_1 = 252)$$

$$= P\left(z \leq \frac{246.66 - 252}{40/\sqrt{400}}\right)$$

$$= P(z \leq -2.67)$$

$$= .0038$$

$$\text{power} = P(\text{reject } H_O \mid H_A \text{ true})$$

$$= 1 - .0038$$

$$= .9962$$

(3) Thus, increasing the sample size from $n = 100$ to $n = 400$ **decreases** the probability of a type II error (β) from .4325 to .0038 and **increases** the chance of detecting a clinically important difference of 10 mg/dl, given that such a difference actually exists (i.e., **power**), from .5675 to .9962.

(4) The effect of increasing the sample size from $n = 100$ to $n = 400$ is illustrated graphically in Figure 12-7.

(a) Because **the spread** of the population of sample means ($\overline{Y}$s) about the population mean μ **is smaller** for the larger sample size,* the shape of the frequency

---

*Recall that this spread is $\sigma/\sqrt{n}$, the standard error.

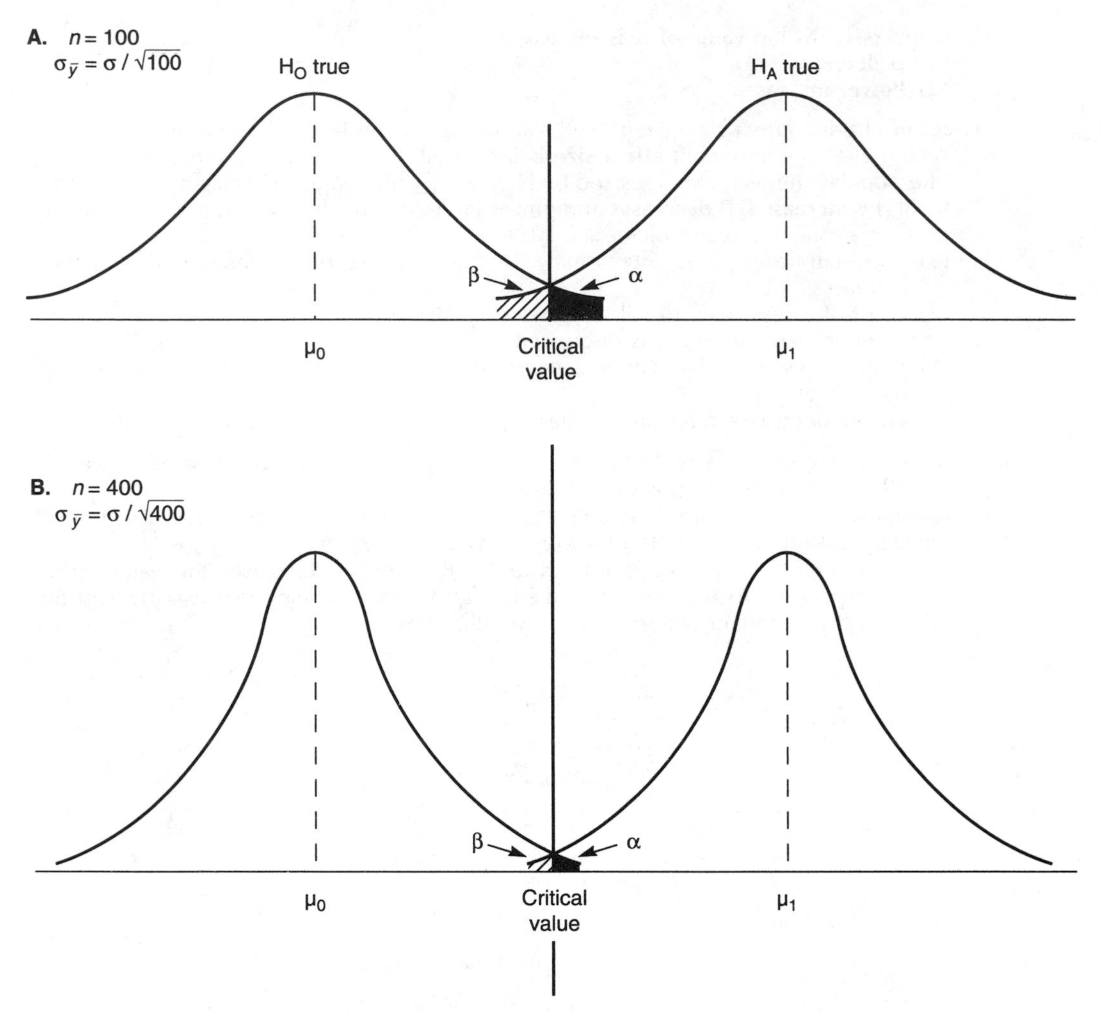

Figure 12-7. The effect of increasing sample size from $n = 100$ (*A*) to $n = 400$ (*B*) on the level of significance α (*solid area*), the probability of a type II error, β (*hatched area*), and power (*open area*). The *curves on the left* represent the frequency distribution of the test statistic when the null hypothesis (H_O) is true, while the *curves on the right* represent the frequency distribution when the alternative hypothesis (H_A) is true.

distribution curve is more compressed (i.e., less of the area under the curve is contained in the tails). Consequently, both α and β decrease and power increases.

(b) To maintain the same level of significance (α) for the larger sample size, the positivity criterion must be shifted to the left, further decreasing β and increasing power.

b. By selecting an appropriate sample size, an investigator can **ensure a specified power** for the statistical test that will be used to analyze the data.

(1) To determine the proper sample size, the investigator must specify the following:

(a) The desired type I error (i.e., false positive rate)

(b) The desired type II error rate (i.e., false negative rate) or, alternatively, the desired power of the test

(c) The effect size (i.e., the clinically important difference in population values)

(d) The standard deviation σ of the response variable in the population under investigation (obtained from previous work, the medical literature, a pilot study, or derived from an educated guess)

(2) Selecting a sample size for a specific statistical test, based on these factors, is discussed in detail in Chapter 16

B. The proper interpretation of α, β, and power

1. The level of significance α is sometimes mistakenly interpreted to mean that if H_O is rejected, there is a $(1 - \alpha) \times 100\%$ probability that H_A is true.
   - **a.** For example, in the serum cholesterol study (using $\alpha = .01$), rejection of H_O **does not** support the conclusion that there is a 99% chance that the mean serum cholesterol level in the population of men who have suffered an AMI is greater than $\mu_0 = 242$ mg/dl (i.e., that H_A is true).
   - **b.** Rather, the **correct interpretation of the meaning of** α is "If the mean serum cholesterol level in the population of men with a history of AMI is 242 mg/dl (i.e., H_O is true), the statistical test should reject this hypothesis, at most, $.01 \times 100\%$ of the time."

2. In other words, α, β, and power represent the **performance characteristics** of a particular statistical test under different circumstances (H_O true or false), **not probability statements** about the true value of a population parameter. Thus, α, β, and power are analogous to the false positive rate, false negative rate, and sensitivity of a diagnostic test, respectively, and are **not** equivalent to the predictive value of the test [i.e., $P(\text{reject } H_O \text{ on basis of test results} \mid H_O \text{ true}) \neq P(H_O \text{ true} \mid H_O \text{ rejected on basis of test results})$].

C. Meaning of "statistically significant" and "not statistically significant" results. Statistical significance is not synonymous with **biologic** or **clinical** relevance. Conversely, the failure to demonstrate statistical significance does not rule out the existence of a clinically important difference between two populations.

1. **Determining statistical significance**
   - **a.** Any difference, however small, may be found statistically significant (i.e., unlikely to have occurred by random chance) **if the sample size *n* is sufficiently large.** Since as *n* increases, power also increases, even very small effect sizes can be detected.
     - **(1)** For example, in the serum cholesterol study, if the sample size had been $n = 5000$, a difference between the population of men with a history of AMI and the general population as small as 2 mg/dl could have been detected with a power of 90%.
     - **(2)** However, a difference of this small magnitude, while statistically significant, may not be **clinically important.**
   - **b. The *p*-value and the magnitude of an experimental effect**
     - **(1)** A **small *p*-value** indicates that the observed sample result is unlikely to have occurred by random chance in a population for which H_O is true. While a small *p*-value therefore provides evidence that **some difference** between two populations exists, it does not necessarily imply that this difference is large.
     - **(2) The *p*-value is a function of the following two variables.**
       - **(a) The sample size *n*.** As ***n* increases, *p* decreases.**
         - **(i)** For example, in a study of the relationship between serum cholesterol values and AMI in men, 1000 men with a history of AMI are sampled and found to have a mean serum cholesterol level $\overline{Y}_{MI} = 247$ mg/dl (compared to a level of 242 mg/dl in the general population). Assuming that the standard deviation $\sigma = 40$ mg/dl, the test statistic is calculated as

$$z_1 = \frac{\overline{Y} - \mu_0}{\sigma/\sqrt{n}} = \frac{247 - 242}{40/\sqrt{1000}} = 3.95, \ \boldsymbol{p < .0001}$$

         - **(ii)** A concurrent study, conducted in women, finds that the average serum cholesterol level in 20 women with a history of AMI is $\overline{Y}_{MI} = 260$ mg/dl. Again, assuming the standard deviation $\sigma = 40$ mg/dl, the test statistic is

$$z_2 = \frac{\overline{Y} - \mu_0}{\sigma/\sqrt{n}} = \frac{260 - 242}{40/\sqrt{20}} = 2.01, \ \boldsymbol{p < .02}$$

(iii) Although the p-value reported for men is much smaller than that reported for the study conducted in women, the actual estimate of the disparity between individuals suffering an AMI and the general population is much smaller in men (i.e., $247 - 242 = 5$ mg/dl) than in women (i.e., $260 - 242 = 18$ mg/dl). The smaller p-value reported for the study conducted in men **does not,** therefore, indicate that AMI has a greater effect on serum cholesterol levels in men than in women.

(b) The standard deviation σ. As **σ increases, p** also **increases.** In other words, when the standard deviation is large, even substantial differences between the sample estimate and the hypothesized population value may be attributable to random chance.

2. The interpretation of negative results

a. When the sample size n is small, a statistical test is more likely to attribute an observed experimental difference to random chance.

(1) **Example.** The serum cholesterol study is conducted in a sample of 10 men who have suffered an AMI, found to have an average serum cholesterol value ($\bar{Y}$) of 260 mg/dl. The test statistic in this instance is

$$z = \frac{\bar{Y} - \mu_0}{\sigma/\sqrt{n}} = \frac{260 - 242}{40/\sqrt{10}} = 1.42, \ \boldsymbol{p < .08}$$

(2) According to the decision rule adopted in III E 3 d, p is not less than α (i.e., z is not greater than 2.33) and, hence, H_O cannot be rejected.

(3) Using the formulas for calculating β and power given in V D 3 c (assuming the effect size = 10 mg/dl),

$$\beta = P(\text{conclude } H_O \text{ true} \mid H_O \text{ false}) = .9357, \text{ or } .94$$

$$\text{power} = 1 - \beta = .06$$

(4) That is, when $n = 10$, the statistical test will fail to detect a clinically important difference of 10 mg/dl in population means 94% of the time when such a difference actually exists. Or, in terms of power, the test will correctly reject H_O only 6% of the time.

(5) Few physicians would believe a reported conclusion when the error associated with this conclusion is 94%. Thus, not only is the observed result not statistically significant, it is **inconclusive,** that is, unlikely to accurately reflect the existence or absence of a difference in population means.

b. A statistical test may fail to reject H_O because:

(1) H_O is actually true

(2) The power of the test is too low to detect a false H_O

c. Therefore, **accepting H_O** is warranted only when **β is low and power is high.**

(1) When **β is high** and **power low,** an inconclusive result (i.e., one that fails to reach statistical significance) should be reported as **"Results are not statistically significant; the data do not provide sufficient evidence to support H_A."**

(2) Similarly, when β and power are unknown (as is the case for some statistical tests lacking an extensive tabular summary of values for the appropriate frequency distribution), the failure to detect a statistically significant difference does not indicate that such a difference does not exist.

PROBLEMS

Problems 12-1 through 12-4

Use the following description of a clinical research study to solve problems 12-1 through 12-4.

A cardiologist wishes to compare an experimental antihypertensive agent with a commonly used standard drug treatment. She knows from prior research that the average systolic blood pressure in patients receiving the standard drug is 130 mm Hg. Because she has only 16 patients available for participation in the study, she wishes to investigate the effect of this sample size on the type II error rate and the power of the statistical test to be used for data analysis. The statistical test is to be carried out with a level of significance $\alpha = .01$, and it is assumed that $\sigma_{new} = 20$ mm Hg.

This investigator frames the research question in terms of two opposing hypotheses.

H_O: $\mu_{new} = 130$ mm Hg; that is, there is no difference in mean systolic blood pressure between patients receiving the new drug and those receiving the standard treatment.

H_A: $\mu_{new} < 130$ mm Hg (e.g., $\mu_{new} = 120$ mm Hg); that is, the mean systolic blood pressure in patients receiving the new drug is lower than that of patients taking the standard drug.

12-1. If the new drug actually does lower systolic blood pressure by 10 mm Hg over the standard treatment, what is the power of the statistical test to detect a difference of this magnitude?

12-2. The investigator determines that the average systolic blood pressure in the sample of 16 patients is 119.5 mm Hg.

a. What conclusions can she draw from the sample result, using a level of significance $\alpha = .01$?

b. Determine the p-value for this test.

c. Interpret α, β, power, and the p-value for this test.

12-3. If the level of significance α is increased from .01 to .10, how does this affect β, the chance of a type II error? Interpret the new values of α and β.

12-4. What is the value of β, the probability of a type II error, if the investigator wishes to detect a difference in systolic blood pressure of 5 mm Hg, given that such a difference actually exists in the population?

12-5. Clinical researchers working for pharmaceutical company A claim that their new antihypertensive drug "significantly lowers systolic blood pressure (compared to a standard drug) with $p < .001$." Rival pharmaceutical manufacturer B claims that its new antihypertensive drug "significantly lowers systolic blood pressure with $p < .05$." Is company A's drug more effective than company B's? Explain.

Solutions on p 384.

STUDY QUESTIONS

Directions: Each of the numbered items or incomplete statements in this section is followed by answers or by completions of the statement. Select the **one** lettered answer or completion that is **best** in each case.

Questions 1–3

In a report summarizing the results of a study comparing serum sodium levels in normotensive patients and newly diagnosed hypertensive patients prior to dietary sodium restriction, a team of clinical researchers states that the difference in mean serum sodium levels was not statistically significant at the 5% level of significance. In addition, the researchers report that, for the sample sizes employed in this study, the power of the chosen statistical test to detect a difference in mean serum sodium levels of at least 5 mEq/L between the two patient populations is .20.

1. Assume that serum sodium levels in the two populations do, in fact, differ by at least 5 mEq/L. The probability that the statistical test will fail to detect this difference could then be described by all of the following statements EXCEPT

(A) this probability is equal to .80
(B) this probability increases as the level of significance is decreased
(C) this probability represents the type II error rate
(D) this probability is smaller than the probability that the test will fail to detect a difference in serum sodium levels of at least 10 mEq/L

2. All of the following statements about the power of the chosen statistical test are true EXCEPT

(A) the reported value for power suggests that the research team should not conclude that "no difference exists between the mean serum sodium levels of the two populations"
(B) the power of the test will decrease if the probability of a type II error increases
(C) this value represents a 20% chance that the statistical test will detect a difference in serum sodium levels between the two populations of at least 5 mEq/L, given that such a difference actually exists
(D) this value represents the probability of detecting a difference of at least 5 mEq/L in mean serum sodium levels when H_O is true
(E) given the reported value of power, a statistically significant difference in the mean serum sodium levels of the two patient populations may have been observed if the investigators had used a larger sample size

3. If the investigators repeat the study with a larger sample size, they would expect which of the following changes to occur?

(A) The probability that they will detect a difference in mean serum sodium levels between normotensive and hypertensive patients, given that no such difference actually exists, increases
(B) The probability that they will fail to detect a difference in mean serum sodium levels between the two populations, given that such a difference does exist, decreases
(C) The power of the statistical test will decrease
(D) The level of significance, α, will increase
(E) The probability that the statistical test will have a false negative outcome will increase

---

1-D
2-D
3-B

4. Which of the following is a result of carrying out a given statistical test of hypothesis at a level of significance $\alpha = .01$?

(A) The probability of accepting H_O when it is false is .01
(B) H_O is more likely to be rejected than if the investigator had chosen a level of significance $\alpha = .05$
(C) The probability of observing a statistically significant result is .01
(D) The probability of rejecting H_O when it is true is .01
(E) H_O is true at least 1% of the time

Questions 5–6

A child psychiatrist conducts a study to compare anxiety levels in hospitalized children whose parents participate in their routine care (e.g., bathing and feeding) and those whose parents are prohibited by hospital regulations from participating in such care. Anxiety levels were determined for five children in each of the two comparison groups, using a standardized psychiatric rating scale. At a research conference, the psychiatrist reports that "the difference in average anxiety scores for the two groups was not statistically significant at the 5% level of significance."

5. Which of the following statements can be made about the investigator's conclusion?

(A) The anxiety level experienced by children whose parents participate in their routine care is the same as that experienced by children whose parents are prohibited from doing so
(B) The p-value for the chosen statistical test is $p < .05$
(C) The observed difference in anxiety levels between the two comparison groups is unlikely to have occurred as a result of random chance
(D) Because the failure of this test to detect a true difference between the comparison groups may be attributable to the low power of the chosen statistical test, the results of the study are inconclusive

6. Which statement correctly describes the chance of error associated with the investigator's conclusion?

(A) Because the sample size was small, the chance of error associated with this conclusion is likely to be low
(B) The chance that no difference in anxiety levels exists between the two groups of children is 5%
(C) The chance of error is equal to the p-value, which is greater than .05
(D) The chance of error is less than 5%
(E) The chance of error is β (the probability of a type II error), which is likely to be large

7. All of the following factors may explain the failure of a statistical test to detect a significant difference of a specified magnitude (effect size) between two comparison groups EXCEPT

(A) low power of the statistical test
(B) small sample size
(C) such a difference actually does not exist
(D) high false negative rate for the statistical test
(E) little subject-to-subject variation in the response variable

8. All of the following factors influence the probability that a statistical test will detect a difference of a specified magnitude (effect size), given that the difference actually exists, EXCEPT

(A) the false positive rate selected by the investigator
(B) the level of significance α
(C) the sample size
(D) the degree of subject-to-subject variation in the response variable
(E) the magnitude of the population parameters being compared

---

4-D 7-E
5-D 8-E
6-E

Directions: Each item below contains four suggested answers of which **one or more** is correct. Choose the answer

A if **1, 2, and 3** are correct
B if **1 and 3** are correct
C if **2 and 4** are correct
D if **4** is correct
E if **1, 2, 3, and 4** are correct

Questions 9–12

A pharmaceutical company wishes to evaluate the effectiveness of a novel antihypertensive agent. The project leader responsible for the drug decides that it will be submitted for regulatory approval only if it can be shown to lower average systolic blood pressure by at least 10 mm Hg compared to a standard drug. She determines that by administering the new drug and the standard drug to two groups of 100 subjects and choosing a level of significance $\alpha = .01$, she can be 90% certain of detecting the relevant 10 mm Hg difference, given that such a difference actually exists.

9. Which of the following statements accurately describe the power of the statistical test used in this study?

(1) It is greater than the power of the test would be if the test were carried out at a level of significance $\alpha = .05$
(2) It is defined as the probability of detecting the 10 mm Hg difference in systolic blood pressure between the comparison groups, given that this difference exists
(3) It is unaffected by the standard deviation in systolic blood pressure readings
(4) It is equal to .90

10. If the two comparison groups comprise 300 subjects each, rather than 100, which of the following changes occur?

(1) The value of power increases
(2) The value of α increases
(3) The value of β decreases
(4) The probability of concluding that a 10 mm Hg difference in systolic blood pressure exists, when, in fact, it does not, increases

11. Which of the following statements accurately describe the probability that the statistical test will fail to detect a 10 mm Hg difference in mean systolic blood pressure between individuals receiving the new drug and those receiving the standard drug, given that such a difference exists?

(1) It is defined as β, the probability of a type II error
(2) It decreases as the level of significance α increases
(3) It decreases as the standard deviation in values of the response variable decreases
(4) It is equal to .10

12. At a management conference to decide the fate of the new drug, the project leader reports that "the new drug significantly lowers systolic blood pressure compared to the standard drug ($p < .01$)." This p-value indicates which of the following?

(1) The mean systolic blood pressure of individuals receiving the new drug differs from that of individuals receiving the standard drug by at least 10 mm Hg
(2) The observed difference in sample means is likely to be due to random chance (sampling error)
(3) H_O should not be rejected
(4) The probability of obtaining the observed difference in the sample means (or a more extreme difference) by random chance from two populations whose means are equal is less than 1%

9-C
10-B
11-E
12-D

ANSWERS AND EXPLANATIONS

1–3. The answers are: 1-D *[V B, D; VI A 1]*, **2-D** *[V A, C; VI C 2]*, **3-B** *[V A–C; VI A 3]*
The probability that the statistical test will fail to detect a difference of a specified clinically important magnitude, given that such a difference actually exists in the study population, is β, the probability of a type II error. Because β decreases as the effect size increases (given a fixed sample size and level of significance), increasing the effect size from 5 mEq/L to 10 mEq/L will result in a new value of β that is *smaller* than the original probability. In other words, the value of β (i.e., the probability) described by the researchers is larger than the probability associated with a 10 mEq/L effect size.

The value of β also decreases if the sample size increases (given a fixed sample size and level of significance), but increases if the level of significance α decreases. Since $\beta = 1 - \text{power}$, $\beta = .80$.

The probability of rejecting the null hypothesis (H_O) when it is in fact true (type I error) is α, the level of significance. Power is the probability of rejecting H_O when it is false. In this instance, the research team has a 20% chance of concluding that mean serum sodium levels in the normotensive and hypertensive populations differ by at least 5 mEq/L when such a difference truly exists. This low power may account for their failure to observe a statistically significant result. If, however, they had employed a larger sample size, they may have found such a significant difference, since power increases as the sample size increases. Because power is the complement of β, the probability of a type II error, power decreases as β increases.

The probability that the statistical test will detect a difference in the mean serum sodium levels between the population of normotensive patients and the population of hypertensive patients when such a difference does not actually exist is α, the level of significance of the test. Conversely, the probability that the test will fail to find such a difference when it does exist is β, the false negative rate. As the sample size increases, α and β decrease and power increases.

4. The answer is D *[V A, B; VI A 1 b, B]*
The level of significance α is the probability of rejecting H_O when it is true (in this case, $\alpha = .01$). The probability of accepting H_O when it is false is β; this probability cannot be determined unless the value of the population parameter posed by the alternative hypothesis (H_A) is known. A statistical test carried out at the $\alpha = .05$ level of significance is more likely to reject H_O than a test carried out at the $\alpha = .01$ level of significance because an increase in α results in a corresponding increase in power and decrease in β, the false negative rate. The values α and β represent statistical test performance characteristics analogous to the false positive rate and false negative rate of a diagnostic test. They do not provide an estimate of how often a given H_O is true or false; that is, they cannot be equated with the predictive value of a diagnostic test.

5–6. The answers are: 5-D *[VI C 1–2]*, **6-E** *[V B; VI C 2]*
A result that is not statistically significant does not necessarily warrant acceptance of the null hypothesis, H_O (i.e., the comparison groups experience the same level of anxiety). Low power of the statistical test may account for a failure to detect a clinically important difference. When a result is not statistically significant, random chance is a likely explanation for any observed difference in sample means. In this example, the *p*-value is $>.05$. The small sample size also suggests that the result is likely to be caused by random chance.

An experimental result is reported as "not statistically significant" when H_O is not rejected. The chance of error associated with failing to reject H_O is β, the probability of a type II error. Although β has not been specified, it is likely to be large because of the small sample size used in this study (β increases as the sample size decreases). The level of significance α and the *p*-value (in this instance, $p > .05$) specify the probability of a type I error (rejecting H_O when it is true). The level of significance α is a measure of statistical test performance given that H_O is true, not the probability that H_O is actually true.

7. The answer is E *[VI C 1 b (2) (b), 2 a–c]*
Obviously, a statistical test may fail to detect a difference in a given population parameter among comparison groups because no such difference actually exists. Low power or small sample size may also account for a nonstatistically significant result. The smaller the subject-to-subject variation in values of the response variable, the higher the power of the test and hence, the more likely it will detect a difference among comparison groups.

8. The answer is E *[V C; VI A]*
The probability that a statistical test will detect a true difference between two population parameters is

the power of that test. Power is unaffected by the size of the population means. It is increased as the level of significance α increases, as the sample size increases, as the effect size increases, or as the subject-to-subject variation in values of the response variable decreases.

9–12. The answers are: 9-C (2, 4) *[V C; VI A 1, 3 a (4) (a)]*, **10-B (1, 3)** *[VI A 3]*, **11-E (all)** *[V B, C 3 a; VI A 1 b (1), 3 a (4) (a)]*, **12-D (4)** *[II E 3; VI C 1 b (1)]*
Power in this case is defined as the probability of detecting a difference in mean systolic blood pressure between the two populations of at least 10 mm Hg when such a difference actually exists; this probability has been determined by the project leader to be .90. Power decreases as the standard deviation increases and the level of significance decreases. Therefore, power is less when $\alpha = .01$ than when $\alpha = .05$.

As the sample size increases, α and β decrease and power increases. The probability that a difference between the population means exists when, in fact, it does not, is α, the level of significance.

The probability that the project leader will fail to detect a difference in systolic blood pressure of at least 10 mm Hg between the population receiving the new drug and that receiving the standard drug, given that such a difference exists, is β, or a type II error. The value of β is calculated here as $1 - \text{power} = .10$. This value decreases as the level of significance α increases or as the standard deviation in values of the response variable decreases.

The reported p-value is the probability of obtaining the observed difference in the sample means (or a more extreme difference) by random chance from two populations whose means do not differ. A small p-value, such as that obtained in this study, indicates that random chance is an unlikely explanation for the observed difference in sample means, but does not necessarily mean that the magnitude of the difference is large. Because the results of this study are statistically significant, the null hypothesis (H_O) is rejected in favor of the alternative hypothesis (H_A).

13
Tests of Statistical Significance: Chi-square Procedures

I. SELECTING AN APPROPRIATE STATISTICAL TEST: A DECISION TREE APPROACH

A. The remaining chapters focus on **the statistical tests of hypothesis most frequently encountered in the medical literature**. Each test is discussed in the context of the five general steps of statistical hypothesis testing outlined in the preceding chapter (see Ch 12 IV); the discussions include a **summary of the study designs for which the test is appropriate**.

B. Selection of an appropriate statistical test

1. **The decision tree approach.** As demonstrated in Chapter 5, the decision tree, a "map" of the available courses of action, can provide a systematic framework for selecting among several therapeutic or diagnostic options. In a similar fashion, the decision tree can be used to structure the selection of a statistical test appropriate to a given study design (Figure 13-1). The choice of a statistical test should **always precede** data collection.

2. **Branch points.** The branch points of the tree represent characteristics of the study that govern the selection process.
   a. The **initial branch point** (see Figure 13-1) deals with the **type of research question** posed by the study. A research question may evaluate either of the following:
      (1) **Associations** [typically between a putative (i.e., suspected) risk factor and a disease] (see Figure 13-1A)
      (2) **Differences** between comparison groups [e.g., comparing therapeutic or diagnostic options] (see Figure 13-1B)
   b. The **second branch point** groups studies according to the **type of data** collected in the study. Study data may be categorical (i.e., in the form of **counts**) or **continuous** (i.e., in the form of quantitative measurements).
      (1) **Categorical** (also known as **count**, or **frequency**) **data** define the numbers of subjects comprising each division of a **categorical** variable. Data measured on all three types of rating scales (see Ch 1 I A) may be converted to counts.
         (a) **Nominal scale** (see Ch 1 I A 1). Table 13-1 lists the numbers of patients in a particular inner city medical clinic with and without sickle cell anemia, according to blood type. For example, ten patients with sickle cell anemia have blood type O, while 30 healthy patients have this blood type.
         (b) **Ordinal scale** (see Ch 1 I A 2). In Table 13-2, lung cancer patients are cross-classified according to disease severity (stages I–III) and smoking status.
         (c) **Interval/ratio scale** (see Ch 1 I A 3). A cardiologist defined five systolic blood pressure intervals, then allocated 200 patients with a history of acute myocardial infarction, previously classified as type A or type B personalities, to the appropriate interval. Table 13-3 shows the number of patients in each blood pressure category according to personality type (type A or type B).
      (2) **Continuous data,** or **quantitative measurements** made on a continuous variable, reflect the "amount" of that variable. Continuous data may be measured on an **interval/ratio** scale or on an **ordinal** scale.
   c. The **third branch point** corresponds to the **type of study design**.
      (1) **Observational designs** include **cohort, case–control,** and **cross-sectional** studies (see Ch 8 II B 2).
      (2) **Experimental studies,** unlike observational designs, randomly allocate subjects to comparison groups (see Ch 9 I A 1).

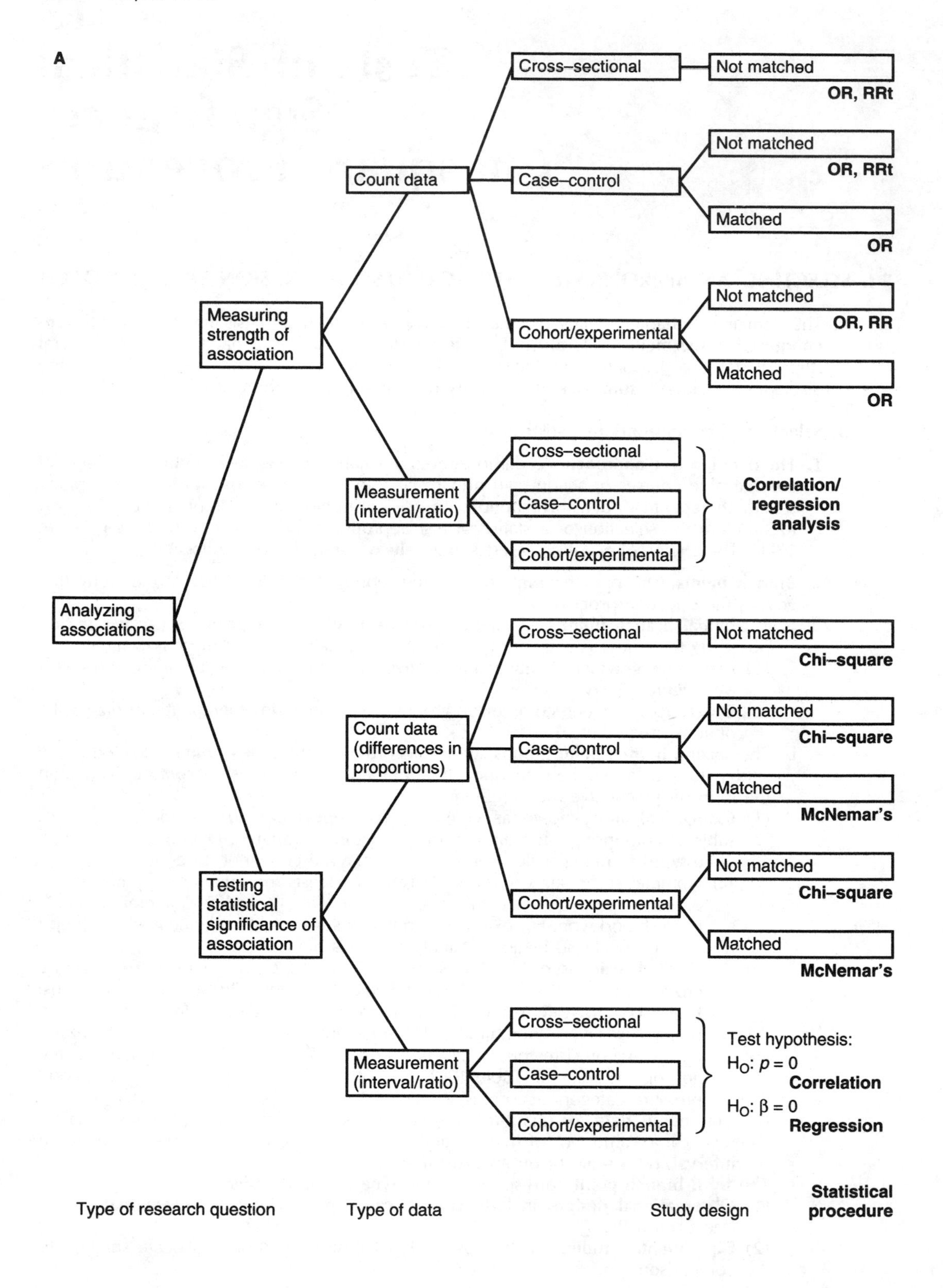
A
Analyzing associations
Measuring strength of association
Testing statistical significance of association
Count data
Measurement (interval/ratio)
Count data (differences in proportions)
Measurement (interval/ratio)
Cross–sectional
Case–control
Cohort/experimental
Not matched
Matched
OR, RRt
OR, RRt
OR
OR, RR
OR
Correlation/ regression analysis
Chi–square
Chi–square
McNemar's
Chi–square
McNemar's
Test hypothesis:
H_O: $p = 0$
Correlation
H_O: $\beta = 0$
Regression
Type of research question
Type of data
Study design
Statistical procedure

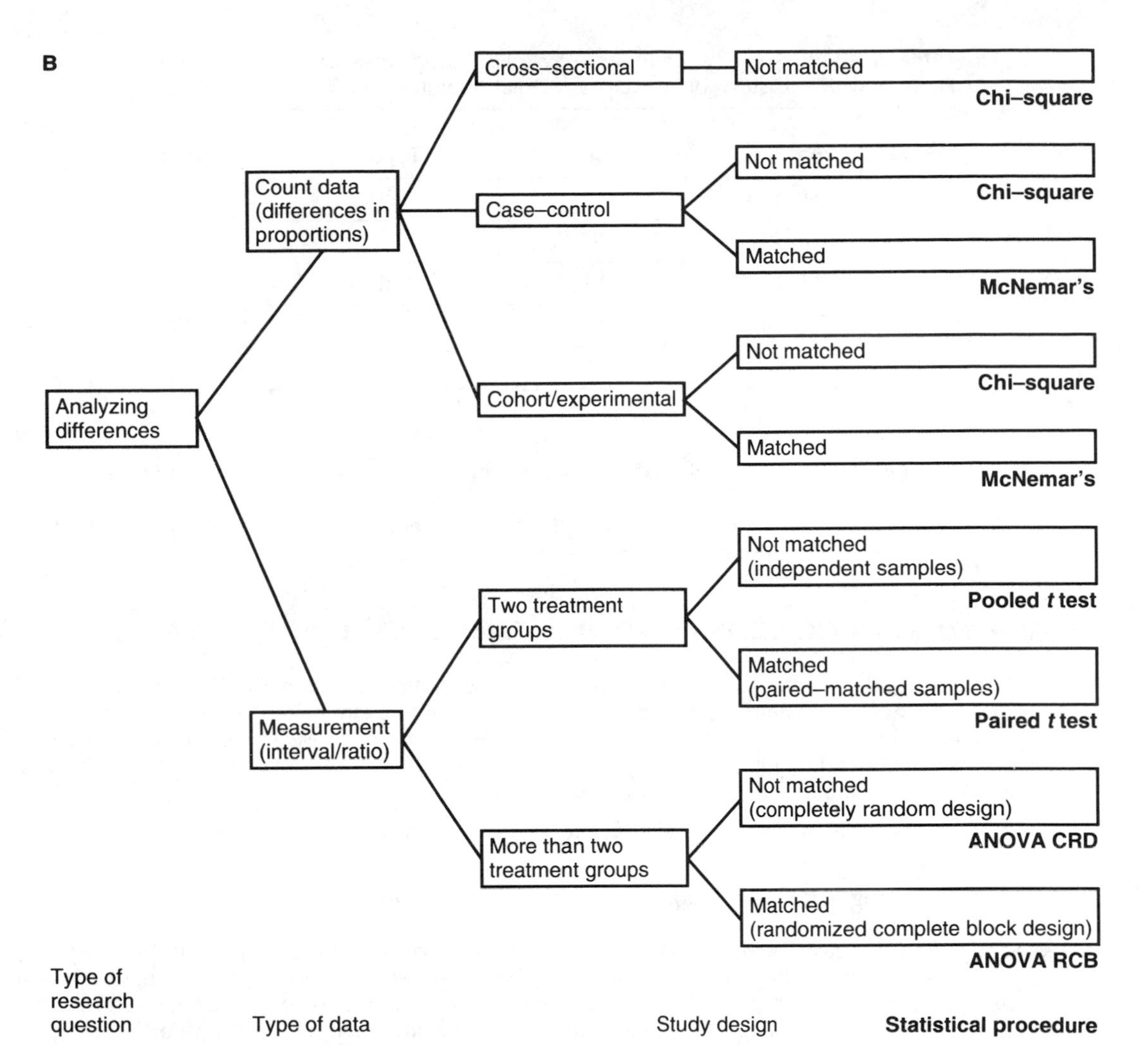

Figure 13-1. Decision trees used to choose an appropriate statistical test when the research question is concerned with (*A*) evaluating an association between two variables or (*B*) analyzing differences among comparison groups. Of the statistical procedures shown, only the chi-square test and the McNemar's test are discussed in Chapter 13; the other procedures are covered in Chapters 14–17. *OR* = odds ratio; *RR* = relative risk; *RRt* = relative risk with restrictive conditions; *ANOVA* = analysis of variance; *RCB* = randomized complete block design; *CRD* = completely random design.

Table 13-1. Occurrence of Sickle Cell Anemia (SCA) in Patients with Each of the Four Major Blood Types

Blood Type	SCA Present	SCA Absent	Totals
O	10	30	40
A	12	18	30
B	15	15	30
AB	13	12	25

Table 13-2. Disease Severity and Smoking Status in Lung Cancer Patients

Disease Severity	Smokers	Nonsmokers
Stage 1	60	40
Stage 2	75	25
Stage 3	80	20

Table 13-3. Personality Type and Systolic Blood Pressure (BP) in 200 Patients with a History of Acute Myocardial Infarction

	Personality Type		
Systolic BP (mm Hg)	**Type A**	**Type B**	**Totals**
81–100	10	20	30
101–120	20	60	80
121–160	40	20	60
>160	20	10	30
Totals	90	110	200

d. Another branch point related to study design corresponds to the **number of comparison groups**.

e. The final branch point asks whether study subjects comprise **matched** (**paired**) or **independent** (**not matched**) samples.

II. ANALYZING ASSOCIATIONS INVOLVING FREQUENCY (COUNT) DATA

A. Epidemiological studies often seek to identify an **association between** a projected **risk factor and a disease**. For example, an epidemiologist may want to examine the relationship between cigarette smoking and premature death from cardiovascular disease or to determine if exposure to a certain industrial chemical can be linked to the occurrence of congenital heart defects. Such studies address two questions:

1. Is the association **statistically significant,** that is, is it unlikely to have resulted from random chance?
2. How **strong is the association;** that is, is it of clinical or practical importance?

B. The chi-square test. When the data are in the form of **counts,** the chi-square test is used to check for a statistically significant association between two variables. The chi-square test is also appropriate for studies in which a research question regarding a putative association between two variables can be phrased in terms of a **difference between proportions** (e.g., Does the incidence of breast cancer vary according to the amount of fat in the diet?).

C. Application of the chi-square test to specific study designs. The following examples illustrate the use of the chi-square test to evaluate the statistical significance of an association involving frequency data, following the five sequential steps comprising all tests of hypothesis (see Ch 12 IV). In each example, the decision tree approach (see I B 1) is used to justify the selection of the chi-square procedure.

1. **Cross-sectional study: independent samples.** In a cross-sectional study, the total number of study subjects is **fixed** by the investigator prior to data collection. Subjects are then simultaneously classified as diseased or disease-free and exposed or nonexposed (see Ch 8 II B 2 d).
   a. A clinical investigator hypothesizes a relationship between the stage of malignant ovarian tumors at the time of diagnosis and the perceived stress level of patients. He wishes to test this hypothesis at the $\alpha = .05$ level of significance.
   b. Testing the hypothesis
   (1) State the hypothesis to be tested.
   (a) The **null hypothesis** (H_O) states: "Ovarian cancer severity and stress level are unrelated" (i.e., there is **no association** between the two).
   (b) The **alternative hypothesis** (H_A) states: "The severity of ovarian cancer at the time of diagnosis is related to the patient's perceived level of stress" (i.e., there is an **association** between the two).
   (2) Select a sample and collect data. The investigator identifies 1000 patients with ovarian cancer from the records maintained by a statewide tumor registry program.

He then categorizes these patients according to disease severity (mild = stage I; moderate = stages II and III; and severe = stage IV) and perceived stress level (1 = none, 2 = mild, 3 = moderate, and 4 = severe). The results are shown in Table 13-4. The numbers of patients in the 12 cells comprising this table are known as **observed,** or **sample, frequencies,** denoted symbolically as O_{ij}.

(3) Calculate the test statistic.

(a) This study seeks to identify an **association** between two variables (ovarian cancer severity and perceived stress level), using a **cross-sectional study design** to collect **frequency** data from **independent samples**. Hence, the chi-square test is the appropriate statistical procedure (Figure 13-2).

(b) The chi-square statistic (denoted χ_c^2) measures the **disparity between** the **observed frequencies** (obtained from the sample) and the **expected frequencies when H_O is true**.

(c) Calculation of expected frequencies (E_{ij})

(i) When **no association** exists between ovarian cancer severity and perceived stress level, the probability that an individual will have mild ovarian cancer and also have a perceived stress level of 1 can be calculated using the multiplication rule for independent (i.e., unassociated) events (see Ch 2 III B 2 b) as:

$$P(\text{mild disease and stress level} = 1) = P(\text{mild disease})P(\text{stress level} = 1)$$

(ii) The probability that a patient will have mild ovarian cancer is given in Table 13-4 as 880/1000 = .88. The probability that a patient will report a stress level = 1 is given as 410/1000 = .41. Thus,

$$P(\text{mild disease and stress level} = 1) = (880/1000)(410/1000)$$
$$= .3608$$

(iii) The number of patients expected to have mild ovarian cancer and a perceived stress level = 1 when H_O is true is obtained by multiplying the probability computed above by the total number of individuals in the sample. Thus, the number of patients **expected** to comprise cell (1,1) **if H_O is true,** denoted E_{11}, is

$$E_{11} = P(\text{mild disease and stress level} = 1) \times 1000$$
$$= (880/1000)(410/1000)(1000)$$
$$= 360.8$$

(iv) The expected frequencies for the remaining cells of Table 13-4 are calculated in a similar fashion. All expected and observed frequencies for this example are given in Table 13-5; note that both frequencies should add up to the same total.

Table 13-4. Disease Severity and Perceived Stress Level in Ovarian Cancer Patients

Disease Severity	Stress Level 1	2	3	4	Totals
Mild	362	60	141	317	880
Moderate	29	5	15	21	70
Severe	19	5	6	20	50
Totals	410	70	162	358	1000*

*Underlined values are fixed prior to data collection.

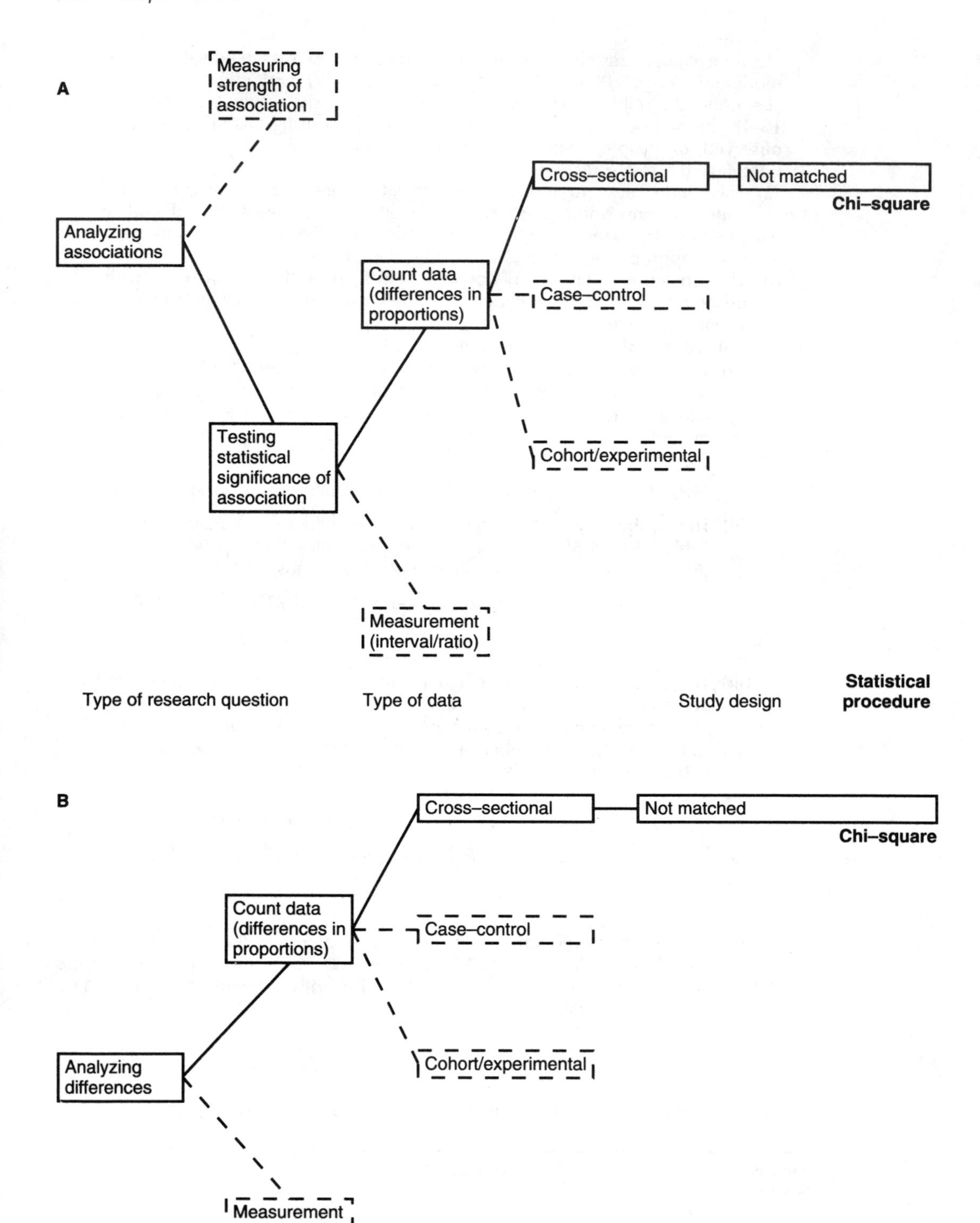

Figure 13-2. Decision tree pathways (derived from Figure 13-1) used to select a statistical test for frequency data collected from independent samples in a cross-sectional study. These pathways illustrate how to choose a test (*A*) when associations are being analyzed and (*B*) when differences are being analyzed. *Dotted lines* indicate decision tree paths not taken.

Table 13-5. Disease Severity and Perceived Stress Levels in Ovarian Cancer Patients: Observed versus Expected Frequencies

Disease Severity	Stress Level 1	2	3	4	Totals
Mild	362 (360.8)*	60 (61.6)	141 (142.56)	317 (315.04)	880
Moderate	29 (28.7)	5 (4.9)	15 (11.34)	21 (25.06)	70
Severe	19 (20.5)	5 (3.5)	6 (8.1)	20 (17.9)	50
Totals	410	70	162	358	1000†

*Numbers in parentheses represent the calculated expected frequencies.
†Underlined values are fixed prior to data collection.

(v) A simplified formula for computing the expected cell frequencies is

$$E_{ij} = \frac{(\text{total for row } i)(\text{total for column } j)}{\text{grand total}}$$

(d) The general formula for computing the chi-square statistic is

$$\chi_c^2 = \sum_{\text{all } ij} \frac{(O_{ij} - E_{ij})^2}{E_{ij}}$$

That is, for each cell, the difference between the expected frequency and the observed frequency is squared; this result is then divided by the expected frequency. The results of this operation are summed for all the cells to derive the test statistic. Hence, the test statistic for this example is

$$\chi_c^2 = \frac{(362 - 360.8)^2}{360.8} + \frac{(60 - 61.6)^2}{61.6} + \cdots + \frac{(20 - 17.9)^2}{17.9}$$
$$= 3.46$$

(4) Evaluate the evidence against H_O.

(a) An examination of the formula for calculating the chi-square statistic reveals that the value of the statistic increases as the disparity between the observed and expected frequencies increases. Therefore, the **larger the value of the test statistic, the greater the evidence against H_O.**

(b) Frequency distribution of χ_c^2

(i) Table A (see Appendix 3) tabulates the frequency distribution of possible values of the chi-square statistic when H_O is true.

(ii) Degrees of freedom (*df*). The number of degrees of freedom, which defines the relevant row in Table A (see Appendix 3), is computed using the formula

$$df = (\text{rows} - 1)(\text{columns} - 1)$$

For this example, $df = (3 - 1)(4 - 1) = 6$.

(c) Calculating the *p*-value. The *p*-value is the actual probability of obtaining a test statistic equal to or greater than $\chi_c^2 = 3.46$ by random chance when H_O is true.

(i) From Table A, Appendix 3 (row $df = 6$), this probability is greater than 10%, that is, $p > .10$.

(ii) That is, the calculated value of χ_c^2 (3.46) is less than $\chi_{.90}^2 = 10.645$. Since the area to the right of 10.645 is 10%, the area to the right of $\chi_c^2 = 3.46$ is greater than 10%.

(d) Decision rule. Based on the chosen level of significance $\alpha = .05$, the investigator formulates the following decision rule: "**Reject H_O if $p \leq .05$.**"

(5) **State the conclusion.**

(a) **Statistical conclusion.** Because $p > .05$, the decision rule adopted by the investigator does not permit him to reject H_O in favor of H_A.

(b) **Medical interpretation.** In other words, the experimental evidence is insufficient to demonstrate the existence of an association between ovarian cancer severity and perceived stress level during the five years preceding diagnosis at the .05 level of significance (i.e., the association is **not statistically significant**).

(c) **Chance of error.** The chance of error associated with the investigator's failure to disprove H_O is β, which is unknown.

$$\beta = P(\text{conclude } H_O \text{ true} \mid H_O \text{ false})$$
$$= P(\text{fail to conclude association exists} \mid \text{association exists})$$

(d) **Power of the test.** Power, the probability that a statistical test will detect an association of specified magnitude, given that association actually exists in the population, is also unknown.

$$\text{power} = P(\text{conclude association exists} \mid \text{association exists})$$
$$= 1 - \beta$$

(e) Because power and β are unknown, the observed result is inconclusive (i.e., if an association between ovarian cancer severity and stress exists, the test failed to demonstrate it). Note however, that the conclusion "no such association actually exists" is not justified; low power of the statistical test may account for its failure to reject H_O [see Ch 12 VI C 2 b (2)].

2. **Cohort or experimental study design: independent samples.** The details of the chi-square test are the same for both the cohort and experimental study designs. The principal distinction between these two designs is that in an experimental study, subjects are **randomly assigned** to the comparison groups, while in a cohort study, subjects **self-select** their group status (see Ch 8 II B 1).

a. **Features.** In cohort and experimental studies, the investigator can control or manipulate the number of subjects who are exposed or not exposed to the putative risk factor (i.e., **exposure is fixed** by the investigator); **disease status**, on the other hand, **is random**. For example, in an experimental study, the investigator controls the exposure of study subjects to the risk factor through their random assignment to comparison groups. In a cohort study, an investigator may select fixed numbers of subjects from the populations of exposed or nonexposed individuals. Alternatively, a single sample of subjects may be selected from a heterogeneous population and classified at the beginning of the study according to their level of exposure.

b. **Example.** The director of community health for a certain state observes that women living in rural parts of the state have a higher rate of miscarriage than women living in urban areas. Farming is the primary occupation in rural areas, relying heavily on the use of a particular organophosphorus pesticide. The director hypothesizes that the rate of miscarriage is related to exposure to this pesticide, and wishes to test this hypothesis at the .01 level of significance.

c. **Testing the hypothesis**

(1) **State the hypothesis to be tested.** The research question may be formulated in either of two ways.

(a) The opposing hypotheses are stated in terms of an **association** between pesticide exposure and the frequency of miscarriage. Using this conceptual framework, $\mathbf{H_O}$ is: "There is **no association** between exposure to the pesticide and the frequency of miscarriage" (i.e., the two are independent); and $\mathbf{H_A}$ states: "There is an **association** between pesticide exposure and the frequency of miscarriage."

(b) Since exposure is fixed in a cohort study, the response variable may be expressed as the **proportion** of study subjects in each comparison group who contract the disease in question. Thus, the research question may also be formulated in terms of a **difference in proportions,** or the incidence of miscarriage. [Incidence can be used because measures of disease frequency, such as prevalence and incidence, represent **proportions** (see Ch 7 I A 1, B 1)].

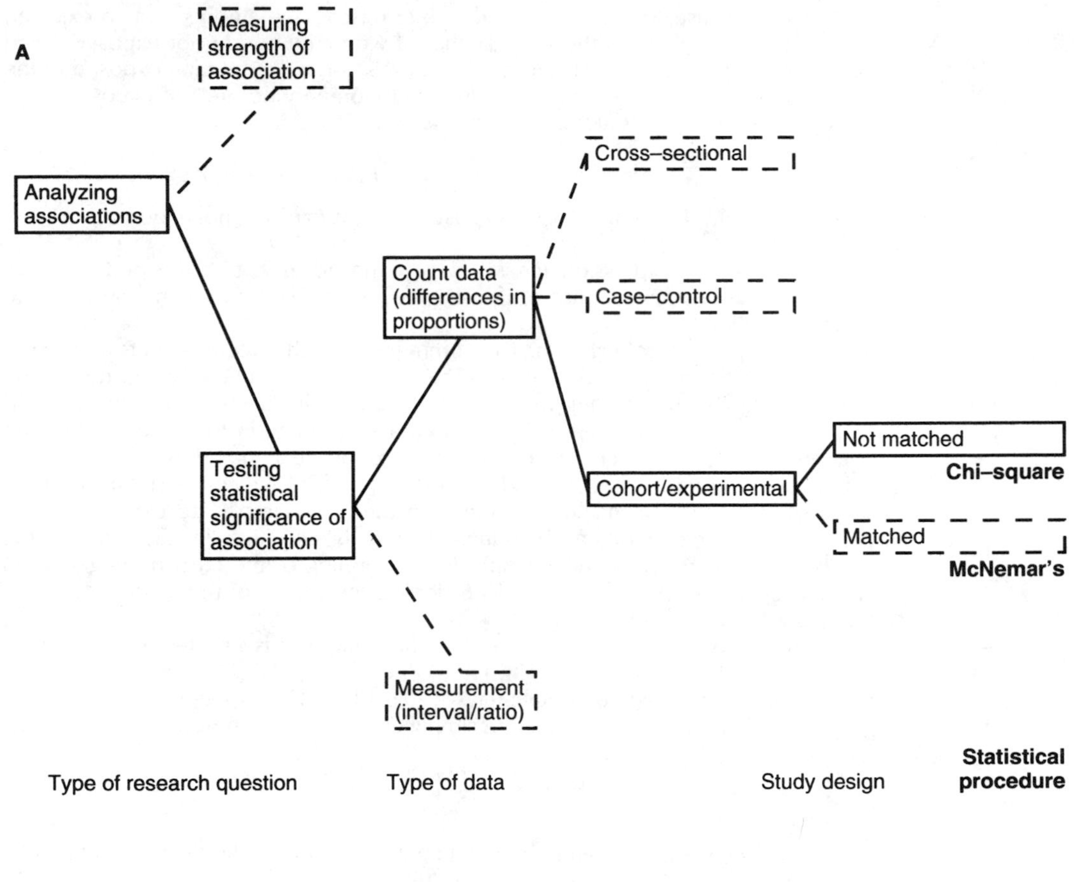

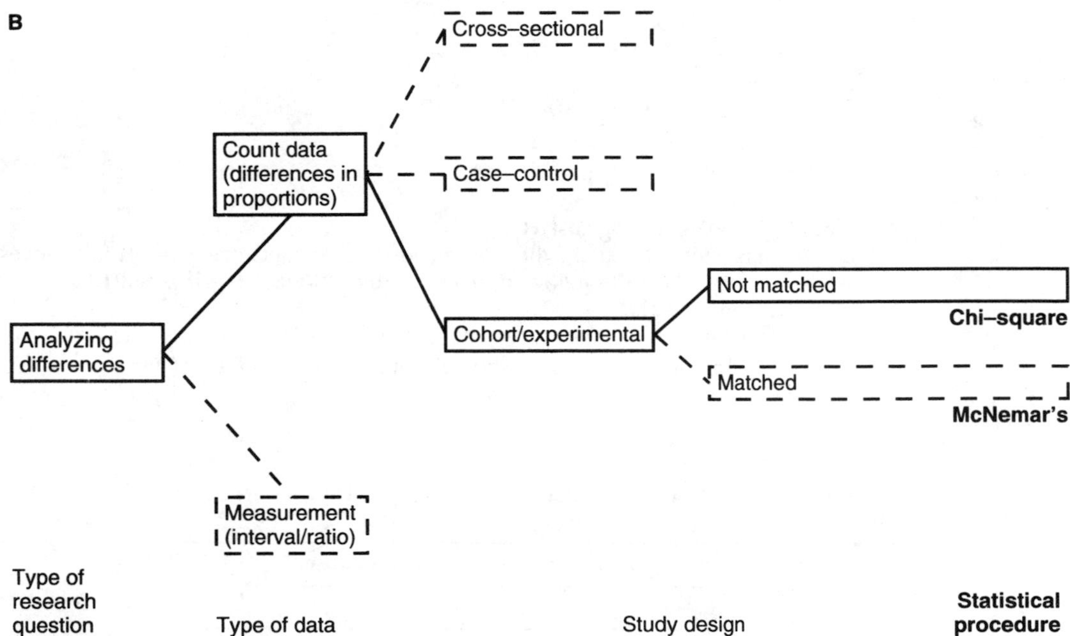

Figure 13-3. Decision tree pathways (derived from Figure 13-1) used to select a statistical test for frequency data collected from independent samples in a cohort or experimental study. These pathways illustrate how to choose a test (*A*) when associations are being analyzed and (*B*) when differences are being analyzed. *Dotted lines* indicate decision tree paths not taken.

(i) In this case, $\mathbf{H_O}$ is: "The incidence of miscarriage among women exposed to the pesticide is the same as that of women who are not exposed," and $\mathbf{H_A}$ can be stated: "The incidence of miscarriage in women exposed to the pesticide differs from the incidence in women who are not exposed."

(ii) In terms of conditional probabilities,

$$\mathbf{H_O}: P(\text{miscarriage} \mid \text{exposed}) = P(\text{miscarriage} \mid \text{not exposed})$$

$$\mathbf{H_A}: P(\text{miscarriage} \mid \text{exposed}) \neq P(\text{miscarriage} \mid \text{not exposed})$$

where the expression $P(\text{miscarriage} \mid \text{exposed})$ is read, "the probability of miscarriage given that a woman has been exposed to the pesticide in question."

(2) Select a sample and collect data. The community health director selects a group of 100 women residing in a rural area of the state who become pregnant during the month of March (crop dusting with the pesticide typically occurs between April and August). She also selects a group of 100 women residing in the state capital who become pregnant in March. Because the two groups are assumed to be comparable with respect to potentially confounding variables such as age and obstetrical history, study subjects are not matched. Subjects in the rural sample are exposed to the pesticide as a matter of course during their daily activities. In each group, the number of miscarriages between April (the beginning of the crop dusting season) and December is recorded. Table 13-6 depicts the results of the study.

(3) Calculate the test statistic.

(a) Using the decision tree approach, the chi-square test is selected as appropriate to this study design (Figure 13-3).

(b) Calculation of expected frequencies (Table 13-7). The expected frequencies for each cell are calculated using the simplified formula given in II C 1 b (3) (c) (v):

$$E_{ij} = \frac{(\text{total row } i)(\text{total column } j)}{\text{grand total}}$$

(c) Using the general formula for computing the chi-square statistic and Table 13-7,

$$\chi_c^2 = \sum_{\text{all } ij} \frac{(O_{ij} - E_{ij})^2}{E_{ij}}$$

$$= \frac{(30-20)^2}{20} + \frac{(10-20)^2}{20} + \frac{(70-80)^2}{80} + \frac{(90-80)^2}{80}$$

$$= 12.5$$

(4) Evaluate the evidence against $\mathbf{H_O}$.

(a) Decision rule. Based on the chosen level of significance $\alpha = .01$, the investigator formulates the following decision rule: "**Reject $\mathbf{H_O}$ if $p \leq .01$.**"

(b) Calculating p-value

(i) The p-value is the area under the curve representing the frequency distribution of χ_c^2 lying to the right of the calculated test statistic, $\chi_c^2 = 12.5$ (Figure 13-4).

Table 13-6. Incidence of Miscarriage among Women Exposed to an Agricultural Pesticide

	Exposed	Not Exposed	Totals
Miscarriage	30	10	40
No Miscarriage	70	90	160
Totals	100*	100	200

*Underlined values are fixed prior to data collection.

Table 13-7. Incidence of Miscarriage among Women Exposed to an Agricultural Pesticide: Observed versus Expected Frequencies

	Exposed	Not Exposed	Totals
Miscarriage	30 (20)*	10 (20)	40
No Miscarriage	70 (80)	90 (80)	160
Totals	100†	100	200

*Numbers in parentheses represent the calculated expected frequencies.
†Underlined values are fixed prior to data collection.

(ii) The number of degrees of freedom is

$$df = (r-1)(c-1)$$
$$= (2-1)(2-1)$$
$$= 1$$

(iii) The row corresponding to $df = 1$ in Table A (see Appendix 3) gives the area to the right of $\chi_c^2 = 7.879$ as $1 - .995 = .005$. Because the test statistic, $\chi_c^2 = 12.5$, falls to the right of 7.879 on the horizontal axis of the frequency distribution of χ_c^2 (see Figure 13-4), the area to the right of χ_c^2 is less than .005 (i.e., $p < .005$).

(iv) Therefore, the probability of obtaining a test statistic equal to or greater than 12.5 by random chance from a population for which H_O is true is less than .005.

(5) State the conclusion.

(a) Statistical conclusion. Because $p < .01$ (i.e., $p < \alpha$), H_O is **rejected** at the $\alpha = .01$ level of significance.

(b) Medical interpretation

(i) The incidence of miscarriage in women exposed to the pesticide differs significantly from the incidence in women who are not exposed. That is, **the difference** in the incidence of miscarriage between the two groups **is statistically significant at the 1% level of significance.**

(ii) Alternatively, it can be concluded that **an association exists** between pesticide exposure and subsequent miscarriage.

(c) Chance of error. The probability that H_O has been erroneously rejected is α, the risk of a type I error; the value selected as the level of significance guarantees that this risk is less than 1%. The actual probability that the director has concluded that the incidence of miscarriage differs between women exposed to the pesticide and those who have not, when, in fact, such a difference does not exist, is less than .5% (i.e., $p < .005$).

3. Case–control study: independent samples. In a case–control study, disease status is **fixed** (i.e., determined by the investigator prior to data collection), while prior exposure to the suspected risk factor (the response variable) is **random**.

a. Example. An epidemiologist employed by the state discussed in the previous example

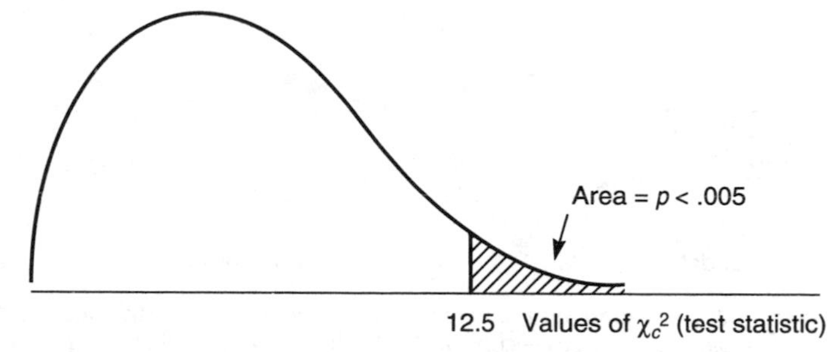

Figure 13-4. Frequency distribution of all possible values of χ_c^2 obtained by repeated sampling of a population for which H_O is true. The *shaded area* corresponds to $p < .005$; $\chi_c^2 = 12.5$ is the calculated test statistic.

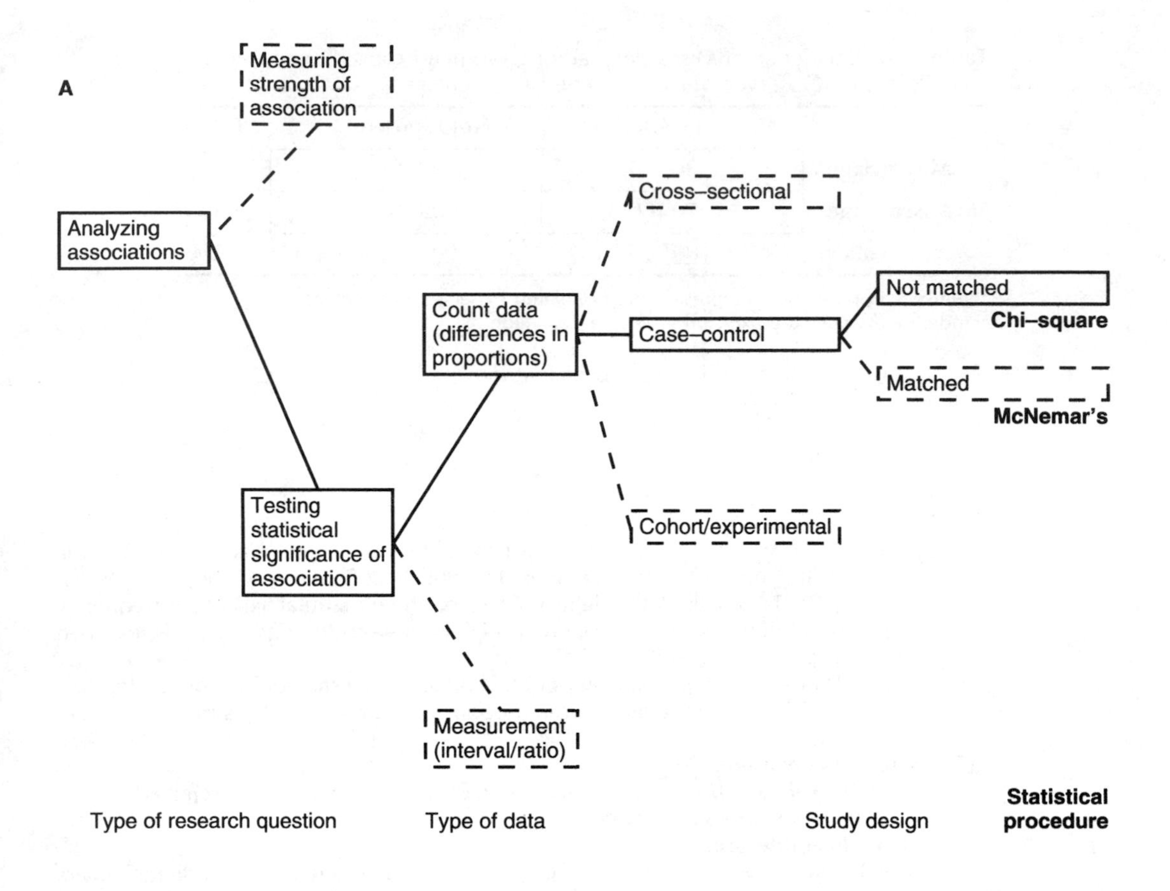

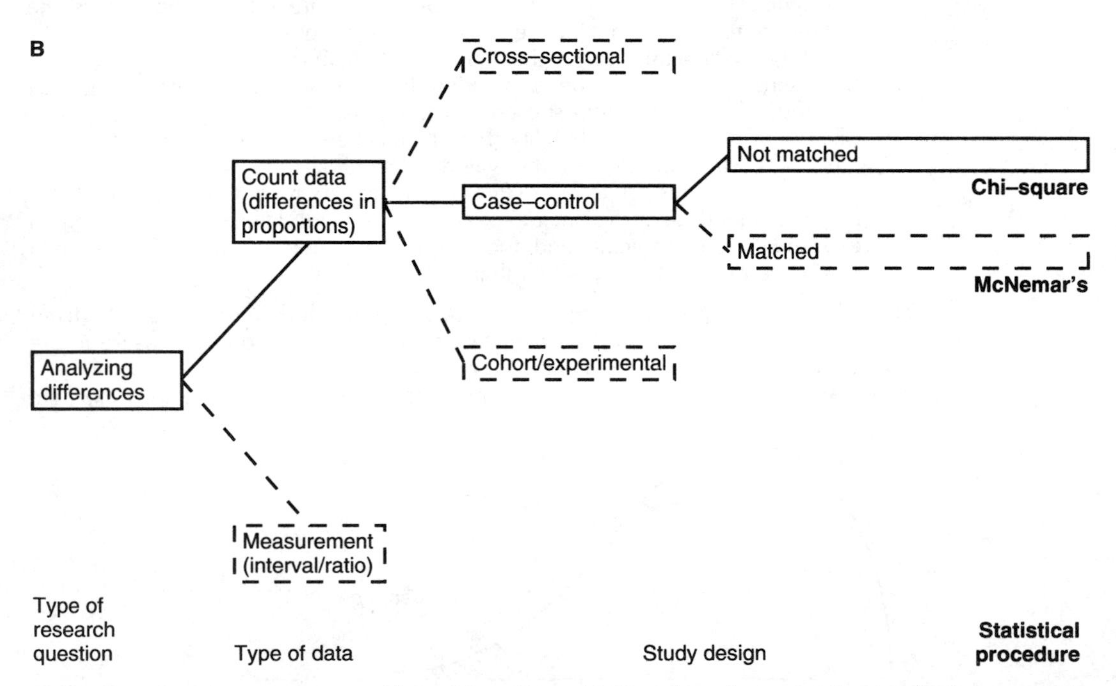

Figure 13-5. Decision tree pathways (derived from Figure 13-1) used to select a statistical test for frequency data collected from independent samples in a case–control study. These pathways illustrate how to choose a test (*A*) when associations are being analyzed and (*B*) when differences are being analyzed. *Dotted lines* indicate decision tree paths not taken.

conducts a second study, using a case–control design, to examine the association between pesticide exposure and miscarriage. In this instance, she wishes to test the hypothesis that such an association exists at the $\alpha = .01$ level of significance.

b. Testing the hypothesis

(1) State the hypothesis.

(a) In terms of an association, $\mathbf{H_O}$ would be stated: "There is **no association** between exposure to the pesticide and miscarriage," and $\mathbf{H_A}$ would be: "An **association** between these two factors exists."

(b) In terms of a **difference between proportions** (exposure rate), $\mathbf{H_O}$ is stated: "The rate of exposure to the pesticide does not differ between women who have experienced a miscarriage and those who have not," and $\mathbf{H_A}$ is: "The population of women who have had a miscarriage and the population of women who have not differ in their rate of exposure to the pesticide.

(c) In probability notation,

$$\mathbf{H_O}: P(\text{exposed} \mid \text{miscarriage}) = P(\text{exposed} \mid \text{no miscarriage})$$

$$\mathbf{H_A}: P(\text{exposed} \mid \text{miscarriage}) \neq P(\text{exposed} \mid \text{no miscarriage}).$$

(2) Select a sample and collect data. The epidemiologist selects 40 women who experienced a miscarriage between September and March of the following year (i.e., following the annual crop dusting season) from the records of a large hospital serving both rural and urban residents. She also selects 160 women who had full-term pregnancies, culminating in a normal delivery at this hospital, during the same time period. The 200 subjects are interviewed to determine their prior exposure to the pesticide. Subjects in the two groups are not matched on the basis of any confounding variable. The results of the study are presented in Table 13-8.

(3) Calculate the test statistic.

(a) Using the decision tree approach, the chi-square test is selected as appropriate to this study design (Figure 13-5).

(b) The expected frequencies for each cell are computed as in the preceding example and depicted in Table 13-9.

(c) Using the general formula for computing the chi-square statistic,

$$\chi_c^2 = \sum_{\text{all } ij} \frac{(O_{ij} - E_{ij})^2}{E_{ij}}$$

$$= \frac{(30-20)^2}{20} + \frac{(10-20)^2}{20} + \frac{(70-80)^2}{80} + \frac{(90-80)^2}{80}$$

$$= 12.5$$

(4) Evaluate the evidence against $\mathbf{H_O}$. As in the preceding example, the p-value associated with $\chi_c^2 = 12.5$, with 1 degree of freedom, is less than .005.

(5) State the conclusion.

(a) Statistical conclusion. Since the p-value ($p < .005$) is less than $\alpha = .01$, H_O is rejected at the $\alpha = .01$ level of significance.

(b) Medical interpretation. The rate of exposure to the pesticide in the population of women who experienced a miscarriage differs significantly from the rate of exposure in the population of women who had a full-term pregnancy.

Table 13-8. Pesticide Exposure among Women Experiencing a Miscarriage and Women Experiencing a Normal Pregnancy

	Exposed	Not Exposed	Totals
Miscarriage	30	10	40*
No Miscarriage	70	90	160
Totals	100	100	200

*Underlined values are fixed prior to data collection.

Table 13-9. Pesticide Exposure among Women Experiencing a Miscarriage and Women Experiencing a Normal Pregnancy: Observed versus Expected Frequencies

	Exposed	Not Exposed	Totals
Miscarriage	30 (20)*	10 (20)	40†
No Miscarriage	70 (80)	90 (80)	160
Totals	100	100	200

*Numbers in parentheses represent the calculated expected frequencies.
†Underlined values are fixed prior to data collection.

(c) **Chance of error.** Because H_O has been rejected, only the type I error rate is relevant. The probability that the epidemiologist conducting this study has concluded that the rates of exposure to the pesticide differ for women who experienced a miscarriage and those who did not, when the rates are not different, is less than .01 (specifically, $p < .005$).

4. **Cohort, experimental, and case–control study designs: matched (paired) samples.** The chi-square procedure used with matched samples (see Ch 8 II D 2) is similar to that described above for independent samples, except that the data are tabulated somewhat differently and a different formula (known as **McNemar's test**) is used to calculate χ_c^2. The general procedure used for matched samples is illustrated below with a prospective cohort study.

a. **Example.** The director of community health described in II C 2 b conducts yet another study examining the relationship between pesticide exposure and miscarriage. In this study, she wishes to use matched samples to test the hypothesis that an association exists between these two factors at the $\alpha = .01$ level.

b. **Testing the hypothesis**

(1) **State the hypothesis to be tested.** H_O and H_A are the same as those formulated in the cohort study conducted with independent samples [see II C 2 c (1)].

(2) **Select a sample and collect data.**

(a) Again, the incidence of miscarriage in 100 women residing in a rural part of the state who become pregnant in March (prior to the annual crop dusting) is compared to the incidence in 100 women residing in the state capital who also become pregnant in this month. However, in this study, the women in the two cohorts are **matched** on the basis of age and history of prior miscarriage. That is, for each 30-year-old woman in the rural group who has experienced one miscarriage prior to the study, the control (urban) group includes a matching 30-year-old woman who has also had one prior miscarriage.

(b) For the purpose of data analysis, these two subjects are "mates." That is, statistical testing in the matched (paired) study is carried out using the **matched pair** rather than the individual subject.

(c) The results of the study are shown in Table 13-10.

(i) In cell (1,1), the value labeled *a* (here, $a = 5$) represents the number of pairs in which both members of the pair experienced a miscarriage. Similarly, cell (1, 2), labeled *b*, represents the number of pairs in which the exposed

Table 13-10. Incidence of Miscarriage among Women Exposed to an Agricultural Pesticide: A Cohort Study Using Matched Samples

	Not Exposed		
Exposed	Miscarriage	No Miscarriage	Totals
Miscarriage	5 *a*	*b* 32	37 ($a + b$)
No Miscarriage	3 *c*	*d* 60	63 ($c + d$)
Totals	8 ($a + c$)	92 ($b + d$)	$N = 100$

partner experienced a miscarriage and the control member of the pair did not.

(ii) The grand total (N) represents the total number of pairs, rather than the total number of subjects (see Table 13-6).

(3) Calculate the test statistic.

(a) The decision tree procedure identifies **McNemar's test** as the appropriate statistical test for this study (Figure 13-6). This variant of the chi-square test is used only when both the **independent and dependent variable are dichotomous** (e.g., in this study, there are two levels of exposure and two possible outcomes). Other statistical procedures are employed with polychotomous variables.

(b) Formula. In McNemar's test, the test statistic, χ_c^2, is calculated using the formula

$$\chi_c^2 = \frac{(|b - c| - 1)^2}{b + c}$$

(c) Thus, for this example,

$$\chi_c^2 = \frac{(|b - c| - 1)^2}{b + c}$$

$$= \frac{(|32 - 3| - 1)^2}{3 + 32}$$

$$= 22.4$$

(4) Evaluate the evidence against H_O.

(a) The test statistic follows the χ_c^2 frequency distribution (see Table A, Appendix 3).

(b) Because McNemar's test applies only to dichotomous variables (i.e., the data are displayed in a 2×2 table), the number of degrees of freedom $[df = (r - 1)(c - 1)]$ is always equal to 1.

(c) Decision rule. Based on the chosen level of significance $\alpha = .01$, the investigator formulates the following decision rule: "**Reject H_O if $p \leq .01$.**"

(d) Calculating p-value. According to Table A (see Appendix 3), the area to the right of the calculated value for the test statistic ($\chi_c^2 = 22.4$, with 1 degree of freedom) is less than .005. Thus, $p < .005$.

(5) State the conclusion.

(a) Statistical conclusion. Because the p-value ($p < .005$) is less than $\alpha = .01$, H_O is rejected at the .01 level of significance.

(b) Medical interpretation. There is a statistically significant difference in the probability of miscarriage between the exposed and nonexposed populations [i.e., P(miscarriage | exposed) $\neq$ P(miscarriage | not exposed)]. In other words, there is a statistically significant association between miscarriage and exposure to the pesticide.

(c) The chance that the statistical test has concluded that a difference exists in the probability of miscarriage for women exposed to the pesticide and those who were not, when such a difference does not exist (i.e., the risk of a type I error) is less than .5% ($p < .005$)

D. Technical considerations

1. One-tailed versus two-tailed tests of hypothesis

a. A statistical test that seeks to define the probability that a significant difference exists between two outcomes, **regardless of the direction of that difference,** is known as a **two-tailed test**. For example, in the cohort study discussed in II C 2 b, H_A proposes that the miscarriage rates between the exposed and nonexposed populations are **different** (i.e., the investigator is interested in detecting either [P(miscarriage | exposed) > P(miscarriage | not exposed)] *or* [P(miscarriage | exposed) < P(miscarriage | not exposed)].

b. A statistical test that seeks to define the probability that a significant difference **exists in a given direction** is known as a **one-tailed test**.

(1) In II C 2 b, a one-tailed test might have been used to test H_A: "The incidence of

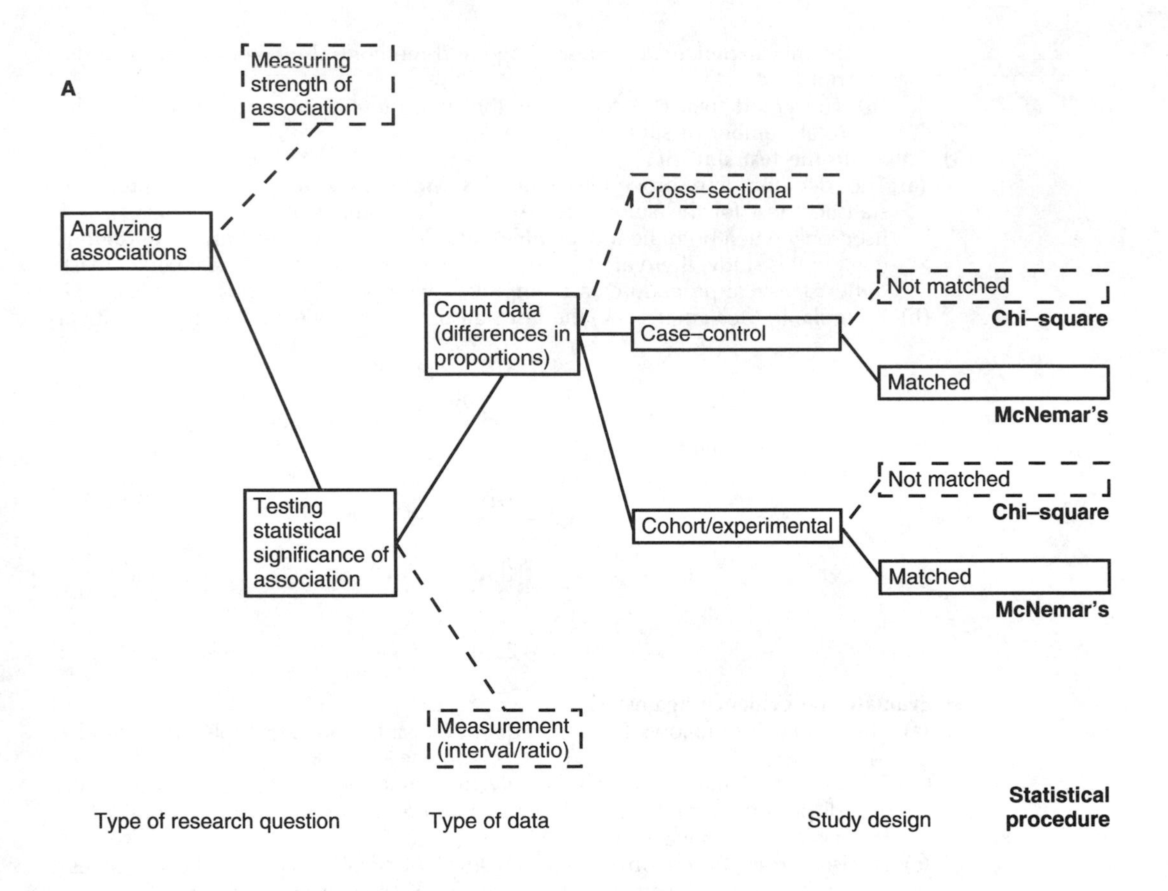

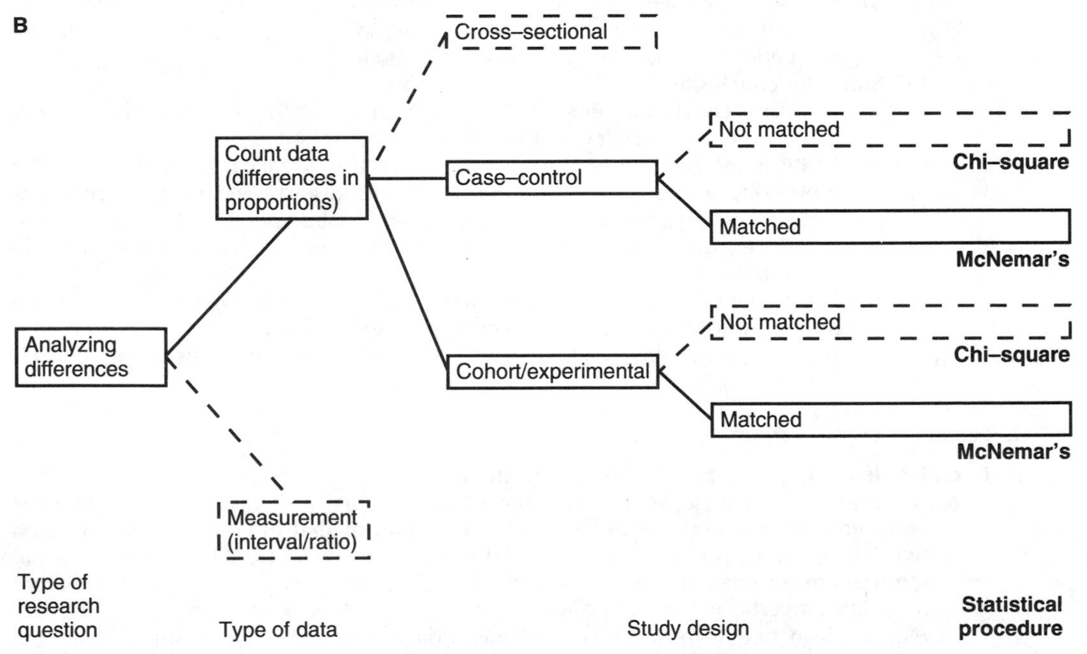

Figure 13-6. Decision tree pathways (derived from Figure 13-1) used to select a statistical test for frequency data collected from matched (paired) samples in a cohort, experimental, or case–control study. These pathways illustrate how to choose a test (*A*) when associations are being analyzed and (*B*) when differences are being analyzed. *Dotted lines* indicate decision tree paths not taken.

miscarriage in the exposed population is **higher** than that in the nonexposed population."

(2) The p-value for a one-tailed test is one-half that obtained for a two-tailed test. [Table A (see Appendix 3) lists two-tailed values for the χ_c^2 statistic.]

c. A one-tailed test is more powerful than a two-tailed test.

d. A one-tailed test should **always** be selected prior to, not after, data collection.

2. **Validity.** The chi-square test is invalid for small expected frequency values.
   a. Considerable disagreement exists, however, with regard to what constitutes "small." One popular rule states that all expected frequencies should be greater than one.
   b. Categories may be collapsed to increase the expected frequencies to acceptable values. For example, five classes of disease severity—grave, severe, moderate, mild, and asymptomatic—may be collapsed into two: severe and not severe.

3. **Interpretation of negative results and power of the test.** A failure to achieve statistical significance may represent the fact that a hypothesized association truly does not exist in the populations from which the samples were drawn (i.e., H_O is true) or it may be attributable to insufficient power of the statistical test to detect an association which exists.
   a. **Without information on power (or β), results that are not statistically significant are inconclusive** (see Ch 12 VI C 2).
   b. Mathematical difficulties associated with computing the area under the curve representing the frequency distribution of H_A complicate the acquisition of information on the power of chi-square tests. Cohen provides a number of tables relating power, β, and sample size for the chi-square test.* These tables are particularly useful for selecting an appropriate sample size (see Ch 16 IV).

4. A small p-value for the chi-square test indicates that **the evidence in favor of an association is strong,** but **does not correspond to the magnitude of the association**. Statistical measures that define the strength of an association are described in the next chapter.

BIBLIOGRAPHY

Cohen J: *Statistical Power Analysis in the Behavioral Sciences.* New York, Academic Press, 1977.

Duncan RC, Knapp RG, Miller MC III: *Introductory Biostatistics for the Health Sciences,* 2nd ed. Albany, Delmar, 1983.

Fleiss JL: *Statistical Methods for Rates and Proportions,* 2nd ed. New York, John Wiley, 1981.

Moore DS: *Statistics: Concepts and Controversies* 2nd ed. New York, WH Freeman, 1979, p 323.

*Cohen J: *Statistical Power Analysis in the Behavioral Sciences.* New York, Academic Press, 1977.

PROBLEMS*

13-1. In a study of the effect of dental hygiene instruction on the subsequent occurrence of dental caries in children, 100 children were randomly selected from the patient population of a pediatric dental clinic. Fifty children were randomly designated to receive instruction in dental hygiene and fifty were randomly assigned to the control (no instruction) group. Six months later, the number of new cavities was tabulated for each child. Use the results of this study, summarized in the table below, to answer the following questions.

	Number of New Cavities			
	0–1	**2–3**	**4–5**	**Totals**
Instruction	30 (25)*	15 (15)	5 (10)	50
No Instruction	20 (25)	15 (15)	15 (10)	50
Totals	50	30	20	100

*Numbers in parentheses represent the calculated expected frequencies.

a. What type of design was used in this study? What is the response (outcome) variable?
b. Test the hypothesis that dental hygiene instruction is related to the number of new cavities occurring during the succeeding six months at the $\alpha = .05$ level of significance (follow the five steps of statistical hypothesis testing outlined in Ch 12 IV). Report and interpret the *p*-value.

13-2. A team of cardiologists conducted a study to investigate the association between oral contraceptive use and hypertension. Forty women using oral contraceptives and sixty women using other methods of contraception were identified from the records of participating physicians. The proportion of hypertensive patients in each group was recorded and is shown in the table below.

	Number of Women		
	Hypertensive	**Normotensive**	**Totals**
Oral Contraceptives	8 (9.2)*	32 (30.8)	40
Other	15 (13.8)	45 (46.2)	60
Totals	23	77	100

*Numbers in parentheses represent the calculated expected frequencies.

a. What type of design was employed in this study?
b. Test the hypothesis that the proportion of hypertensive women among those who use oral contraceptives differs from the proportion of hypertensive women among those who rely on other contraceptive methods at the 1% level of significance. Report and interpret the *p*-value.
c. What is the chance of error associated with the conclusion drawn in problem 13-2b?

13-3. A psychiatrist wanted to study the occurrence of undesirable side effects associated with the pharmacological management of panic disorder. He identified sixty patients receiving desipramine for the treatment of panic disorder from the records of a community mental health clinic and randomly selected thirty of these patients to receive alprazolam in addition to their antidepressant medication. The reaction time of patients in both groups was assessed in a

*The problems in this section are adapted from Duncan RC, Knapp RG, Miller MC III: *Introductory Biostatistics for the Health Sciences*, 2nd ed. Albany, Delmar, 1983, pp 176–177.

standard test of psychomotor impairment and rated as satisfactory or not satisfactory. Twelve patients receiving desipramine alone performed satisfactorily, compared to eight of those receiving both desipramine and alprazolam.

a. What type of study design was used in this investigation?

b. Test the hypothesis that the motor performance of patients who receive desipramine and alprazolam differs significantly from patients who receive desipramine alone at the 5% level of significance. Report and interpret the *p*-value.

c. What is the chance of error associated with the conclusion drawn in problem 13-3b?

Solutions on p 387.

STUDY QUESTIONS

Directions: Each of the numbered items or incomplete statements in this section is followed by answers or by completions of the statement. Select the **one** lettered answer or completion that is **best** in each case.

Questions 1-5

A physician wants to know if the number of male esophageal cancer patients diagnosed with multiple primary tumors differs from the proportion of female esophageal cancer patients with the same diagnosis. She selects random samples of 60 male and 40 female esophageal cancer patients and records the number in each sample diagnosed with multiple primary tumors. Forty men and 10 women with the tumors are identified.

1. What is H_O for this study?

(A) The proportion of male esophageal cancer patients with multiple primary tumors is greater than that of female esophageal cancer patients with such tumors
(B) An association exists between gender and the presence of multiple primary tumors
(C) The proportion of male esophageal cancer patients diagnosed with multiple primary tumors does not differ from that of female esophageal cancer patients diagnosed with multiple primary tumors
(D) The proportion of men with esophageal cancer does not differ from that of women with esophageal cancer
(E) The proportion of men with esophageal cancer differs from that of women with esophageal cancer

2. The calculated value of the test statistic is

(A) .65
(B) 16.67
(C) 14.04
(D) .72
(E) none of the above

3. How many degrees of freedom are associated with the test statistic?

(A) 1
(B) 2
(C) 3
(D) 4
(E) Cannot be determined from the data

4. If the tabulated critical value of the chi-square statistic for the 5% level of significance is 3.84, what is the most appropriate conclusion that can be drawn from this study?

(A) The proportion of male esophageal cancer patients diagnosed with multiple primary tumors differs from the proportion of female esophageal cancer patients diagnosed with such tumors ($p < .05$)
(B) The proportion of male esophageal cancer patients diagnosed with multiple primary tumors does not differ from the proportion of female esophageal cancer patients diagnosed with such tumors ($p > .05$)
(C) It is 95% certain that the proportion of male esophageal cancer patients diagnosed with multiple primary tumors equals the proportion of female patients with the same diagnosis
(D) The investigator can be 95% certain that more men than women have esophageal cancer
(E) The observed sample difference is likely to be due to random chance

5. Which of the following statements is an accurate interpretation of the *p*-value associated with the study conclusion?

(A) The probability of obtaining the given sample results by random chance, from populations in which no difference in the proportion of esophageal cancer patients diagnosed with multiple primary tumors exists, is less than 5%
(B) The observed difference in sample frequencies is likely to be due to random chance
(C) The sample sizes are too small to detect a significant difference in frequency
(D) The investigator can be certain that the proportion of men diagnosed with multiple primary tumors differs from the proportion of women with the same diagnosis
(E) The chance of error associated with the failure of the statistical test to detect a significant difference in the number of men and women diagnosed with multiple primary tumors is less than 5%

1-C
2-B
3-A
4-A
5-A

Questions 6–11

A psychologist postulates that the occurrence of speech defects in elementary school children is related to their socioeconomic background. He selects samples of children from each of three socioeconomic groups (upper-income, middle-income, and low-income) and records the number of children between the ages of 6 and 12 in each group with a speech defect. In analyzing the data collected in the study, he computes a chi-square test statistic equal to 1.1; the tabulated critical value for the chosen level of significance ($\alpha = .05$) is 5.99.

6. How many degrees of freedom are associated with the test statistic?

(A) 1
(B) 2
(C) 4
(D) 6
(E) Cannot be determined from the data

7. What is the most appropriate conclusion that can be drawn from the results of this study?

(A) Socioeconomic status and the occurrence of speech defects are unrelated
(B) There is a statistically significant association between the occurrence of speech defects and socioeconomic status in the study population ($p < .05$)
(C) The association between the occurrence of speech defects and socioeconomic status in the study population is not statistically significant at the 5% level of significance
(D) The observed association between socio economic status and the occurrence of speech defects is unlikely to be due to random chance

8. The p-value associated with this conclusion is described by which of the following statements?

(A) It is less than .05
(B) It is greater than the maximum probability of a type I error that the investigator is willing to tolerate
(C) It is the probability that the statistical test will fail to detect an association between socioeconomic status and the occurrence of speech defects, given that such an association actually exists in the study population
(D) It is the power of the statistical test
(E) It cannot be determined from the data

9. What is the chance of error associated with this conclusion?

(A) Less than 5%
(B) β, which is not specified
(C) The probability of committing a type I error
(D) .95
(E) $1 - \beta$

10. All of the following statements about the study conclusion are true EXCEPT

(A) the observed association is likely to be due to random chance
(B) the failure to detect a statistically significant association may be due to low power of the statistical test
(C) a strong association between socioeconomic status and the occurrence of speech defects is likely
(D) the failure to detect a significant association between speech defects and socioeconomic status may be due to the fact that such a difference truly does not exist in the study population

11. Valid reasons for the failure of the statistical test to detect a significant association between the occurrence of speech defects and socioeconomic status include all of the following EXCEPT

(A) low power of the statistical test
(B) insufficient sample size
(C) no such association exists in the study population
(D) the small p-value associated with the study conclusion

6-B 9-B
7-C 10-C
8-B 11-D

ANSWERS AND EXPLANATIONS

1–5. The answers are: 1-C *[II C 1 b (1); Ch 12 II A]*, **2-B** *[II C 1 b (3)]*, **3-A** *[II C 1 b (4)]*, **4-A** *[II C 1 b (5)]*, **5-A** *[II C 1 b (5); Ch 12 II E 3; VI C 1 b]*
By definition, the null hypothesis (H_O) proposes that no association exists between the comparison groups. In this study, the two comparison groups are men with esophageal cancer and women with this form of cancer; the two groups are to be compared with respect to the proportion of patients diagnosed with multiple primary tumors. Hence H_O states: "There is no difference between the proportion of male esophageal cancer patients diagnosed with multiple primary tumors and the proportion of female esophageal cancer patients with the same diagnosis."

The following table presents the results of this study, as well as the expected frequencies computed for each cell.

	Number of Patients		
	Multiple Tumors Present	**Multiple Tumors Absent**	**Totals**
Men	40 (30)*	20 (30)	60
Women	10 (20)	30 (20)	40
Totals	50	50	100

*Numbers in parentheses represent the calculated expected frequencies.

Using the formula for calculating the chi-square statistic,

$$\chi_c^2 = \sum_{\text{all } ij} \frac{(O_{ij} - E_{ij})^2}{E_{ij}}$$
$$= \frac{(40-30)^2}{30} + \frac{(20-30)^2}{30} + \frac{(10-20)^2}{20} + \frac{(30-20)^2}{20}$$
$$= 16.67$$

The number of degrees of freedom associated with the chi-square statistic is

$$df = (r-1)(c-1)$$
$$= (2-1)(2-1)$$
$$= 1$$

The area to the right of the critical value of the chi-square statistic (3.84) is $\alpha = .05$. Since the calculated test statistic, $\chi_c^2 = 16.67$, exceeds the tabulated critical value (i.e., falls to the right of this value on the horizontal axis of the chi-square frequency distribution), the area to the right of the test statistic is less than .05. This area corresponds to the *p*-value, the probability of obtaining a value for χ_c^2 equal to or greater than 16.67 by random chance when H_O is true. That is, $p < .05$ ($p < \alpha$) and H_O is therefore rejected at the 5% level of significance. The physician conducting this study can conclude that the proportion of male esophageal cancer patients diagnosed with multiple primary tumors differs in a statistically significant fashion from the proportion of female esophageal cancer patients diagnosed with multiple primary tumors. The probability that this difference is due to random chance is less than 5%.

The *p*-value is the probability of obtaining a difference in the number of male and female patients with multiple primary tumors, equal to or greater than that observed, by random chance from populations for which no such difference exists. A small *p*-value indicates that the observed sample difference is unlikely to be due to random chance. All statistical conclusions represent statements about a population based on information obtained from a sample and thus involve some degree of uncertainty. The chance of error associated with a failure of the statistical test to detect a significant difference in the frequencies of interest is β, the probability of a type II error. The value of β is not specified in this study.

6–11. The answers are: 6-B *[II C 1 b (4)]*, **7-C** *[II C 1 b (4), (5)]*, **8-B** *[II C 1 b (5); Ch 12 II E 3]*, **9-B** *[II C 1 b (5); Ch 12 II E 2; V B]*, **10-C** *[II C 1 b (5)]*, **11-D** *[II C 1 b (5); Ch 12 VI C 2 b]*
The contingency table summarizing the data obtained in this study would have three columns (one each for upper-income, middle-income, and low-income socioeconomic group) and two rows (labeled defect present and defect absent, respectively). Thus, the number of degrees of freedom associated with the test statistic is

$$
\begin{aligned}
df &= (r - 1)(c - 1) \\
&= (2 - 1)(3 - 1) \\
&= 2
\end{aligned}
$$

Since the calculated test statistic, $\chi_c^2 = 1.1$, does not exceed the tabulated critical value, $\chi_c^2 = 5.99$, the p-value for the test is $p > .05$. This p-value is greater than the chosen level of significance $\alpha = .05$. Therefore, the null hypothesis (H_O) cannot be rejected at the 5% level of significance. This failure to achieve statistical significance does not guarantee that H_O is true, and may have occurred because of low power of the statistical test. The probability that a test statistic greater than or equal to that computed from this sample could have been obtained by random chance for a population for which H_O is true is greater than 5%.

The p-value is greater than 5%. Hence, it exceeds the level of significance α, the predetermined risk of a type I error the investigator is willing to tolerate. The probability that the statistical test will fail to detect an association that actually exists is β, the probability of a type II error (not α or p). Power, the complement of β, is the probability that a statistical test will detect an association between study variables that exists in the populations from which the samples were drawn.

The chance of error associated with the failure to reject H_O is β, which is not specified for this study. While α, the probability of a type I error, is chosen by the investigator (in this study, $\alpha = .05$, or 5%), β, as well as its complement, power, are often difficult to compute in the absence of information about the frequency distribution of H_A; these values are therefore often unknown.

The statistical test failed to reject H_O. Therefore, the data do not support the hypothesis that an association exists between socioeconomic status and the occurrence of speech defects. Results that fail to achieve statistical significance indicate that an observed association is likely to be due to random chance.

The failure of the statistical test to detect a significant association between socioeconomic status and the occurrence of speech defects may indicate that such an association does not exist in the population from which the sample was drawn. Alternatively, it may be attributable to low power of the statistical test. Because power decreases as the sample size decreases, a small sample size may have contributed to the failure of the test to detect a statistically significant association. A small p-value would lead to rejection of H_O and the conclusion that a statistically significant association does exist between study variables.

14
Quantifying Risk

I. DEFINING THE MAGNITUDE OF ASSOCIATIONS

A. Overview. After a statistically significant association between a risk factor and a disease has been identified (see Ch 13 II), a clinical investigator typically wishes to determine the **magnitude** of that association—that is, to **quantify the risk** resulting from exposure.

B. Measures of risk

1. **Measures of frequency** describe **absolute risk** (also known as incidence) in terms of **basic risk statements**. A **basic risk statement** has the following characteristics.
   - **a.** It expresses the probability that a specified outcome (e.g., a particular disease) will occur within a population.
   - **b.** It can be expressed as a **conditional probability**. For example, the risk of toxic shock syndrome (TSS) in women who use tampons can be written $P(\text{TSS}|\text{tampon use}) = P(\text{D}+|\text{E}+) = .00006$ (where D+ = diseased and E+ = exposed).
   - **c.** It **requires a denominator**. Thus, the statement, "Last year, 503 deaths occurred as a result of commercial airline accidents," does not define the risk of dying in a commercial airline accident. To calculate this risk, the number of deaths (503) must be divided by the number of people at risk (i.e., the number of airline passengers in the year in question).
2. **Measures of association** utilize **comparative risk statements** to study why certain events occur.
   - **a.** A comparative **risk** statement defines the **risk** of developing a disease **among individuals exposed** to a suspected risk factor **compared to those with no such exposure.**
   - **b. Comparative frequency** measures include **relative risk,** or **risk ratio** (see II A) and **odds ratio,** or **relative odds** (see II B).
     - **(1)** For frequency data (i.e., data in the form of counts), the decision tree analysis outlined in Chapter 13 I B, identifies relative risk and the odds ratio as appropriate for measuring the magnitude of a statistically significant association between a risk factor and a disease (Figure 14-1).
     - **(2)** These measures provide a **quantitative** assessment of the magnitude of the association between a putative risk factor and a disease.
     - **(3) Comparative risk measures,** such as relative risk, **do not provide information on absolute risk**. For example, an epidemiological report stated that a child receiving aspirin during a viral illness is 11 times more likely to develop Reye's syndrome than a child who does not receive aspirin; that is, the risk of developing Reye's syndrome is 11 times greater in children given aspirin.* While this statement of comparative risk does indicate the strength of the association between Reye's syndrome and aspirin use, it provides no information on the absolute risk of developing the condition in children receiving aspirin [i.e., $P(\text{Reye's syndrome}|\text{aspirin use})$]. The absolute risk may actually be extremely small, compared to the substantial degree of comparative risk.

---

* Centers for Disease Control: Reye's syndrome—Ottio, Michigan. *Morbidity Mortality Weekly Report* 29:532–539, 1980.

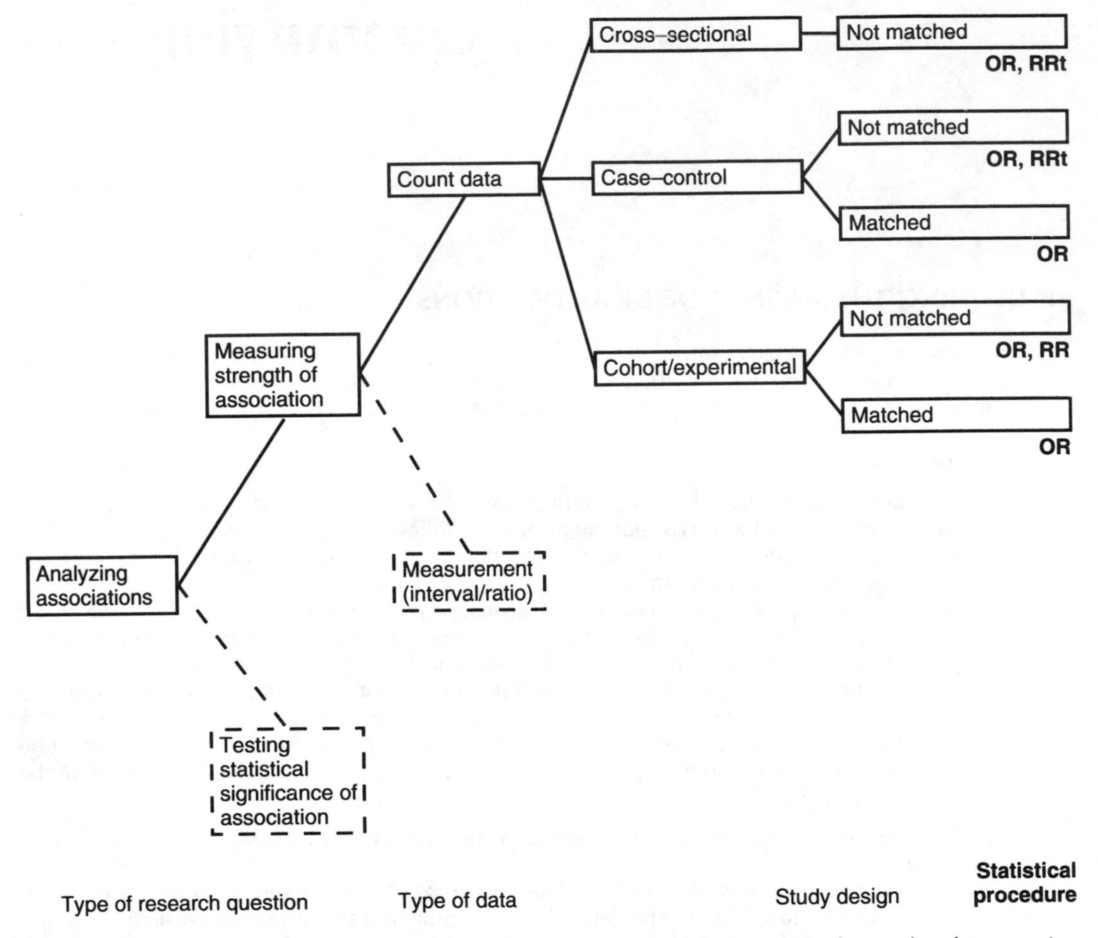

Figure 14-1. Decision tree pathways (derived from Figure 13-1) used to select a statistical procedure for measuring the strength of an association when the data are in the form of counts. *Dotted lines* indicate decision tree paths not taken. *OR* = odds ratio; *RR* = relative risk; *RRt* = relative risk with restrictive conditions.

3. **Measures of potential impact** quantify the possible **consequences of exposure** to a risk factor for a population. These measures are discussed in III and include **attributable risk, population attributable risk,** and **population attributable fraction**. They have important uses in health care planning and health education because of their preventive potential. That is, by estimating the extent to which a disease is related to a specific risk factor, health care planners can predict the effectiveness of control or prevention programs.

II. MEASURES OF ASSOCIATION (COMPARATIVE RISK)

A. Relative risk (RR)

1. **Definition.** RR compares the probability of an outcome among individuals who have a specified characteristic or who have been exposed to a given risk factor to the probability of that outcome among individuals who lack the characteristic or who have not been exposed. In other words, it is the **ratio of the incidence of the outcome among exposed individuals to the incidence among nonexposed individuals**.

2. **Calculation**
   a. The data to be used in calculating RR are first summarized in a 2 × 2 table (Table 14-1).

Table 14-1. A 2 × 2 Table for Calculating the Relative Risk (RR) of Contracting a Disease Given Exposure to a Risk Factor

	D+ (Cases)	D− (Controls)	Totals
E+	*a*	*b*	$a + b$
E−	*c*	*d*	$c + d$
Totals	$a + c$	$b + d$	$N = a + b + c + d$

D+ = diseased; D− = disease-free; E+ = exposed; E− = nonexposed.

b. Formula. RR is then computed using the formula

$$\text{RR} = \frac{\textbf{incidence} \text{ of disease in } \textbf{exposed} \text{ group}}{\textbf{incidence} \text{ of disease in } \textbf{nonexposed} \text{ group}}$$

$$= \frac{\text{risk of disease if exposed}}{\text{risk of disease if not exposed}}$$

$$= \frac{P(\text{D}+|\text{E}+)}{P(\text{D}+|\text{E}-)} = \frac{a/(a+b)}{c/(c+d)} \qquad (14.1)$$

3. Interpretation

a. The disease (or other health-related outcome) is **RR times more likely to occur** among those exposed to the suspected risk factor than among those with no such exposure.

b. The larger the value of RR, the stronger the association between the disease in question and exposure to the risk factor.

c. Values of RR close to 1 indicate that the disease and exposure to the risk factor are **unrelated** (i.e., the risk of occurrence is the same for both exposed and nonexposed individuals).

d. Values of RR less than 1 indicate a **negative association** between the risk factor and the disease (i.e., a protective rather than detrimental effect).

4. RR and study design

a. RR can be directly calculated only in a cohort or experimental study.

b. Because incidence cannot be estimated from a case–control study,* RR cannot be calculated directly. Under some circumstances, the RR in a case–control study can be estimated by the odds ratio (see II B 3 b).

5. Example. In the cohort study described in Chapter 13 II C 2 b, the incidence of miscarriage among 100 women exposed to a particular organophosphorus pesticide was shown to differ significantly from the rate among 100 women who were not exposed (chi-square test; $p < .005$).

a. Calculating RR

(1) Table 14-2 is a 2 × 2 table summarizing the results of this study.

(2) Since this is a cohort study, the row (exposure) totals $(a + b)$ and $(c + d)$ were **fixed** by the investigator, while the column (disease) totals $(a + c)$ and $(b + d)$ are

Table 14-2. Incidence of Miscarriage among Women Exposed to an Agricultural Pesticide

	Miscarriage (D+)	No Miscarriage (D−)	Totals
Exposed (E+)	30 *a*	*b* 70	100 $(a + b)$
Not Exposed (E−)	10 *c*	*d* 90	100 $(c + d)$
Totals	40 $(a + c)$	160 $(b + d)$	200

* In a case–control study, the numbers of diseased and disease-free subjects are fixed by the investigator prior to data collection. Therefore, overall or conditional rates of disease [i.e., $P(\text{D}+)$, $P(\text{D}-)$, $P(\text{D}+|\text{E}+)$, $P(\text{D}+|\text{E}-)$, $P(\text{D}-|\text{E}-)$] cannot be estimated from this type of study.

random. Therefore, the following population probabilities can be estimated:

$P(D+|E+)$ = the incidence of miscarriage in the exposed population = $a/(a+b)$

$P(D+|E-)$ = the incidence of miscarriage in the nonexposed population = $c/(c+d)$

$P(D-|E+)$ = the incidence of normal, full-term pregnancies in the exposed population = $b/(a+b)$

$P(D-|E-)$ = the incidence of normal, full-term pregnancies in the nonexposed population = $d/(c+d)$

(3) Substituting in formula 14.1,

$$\text{RR} = \frac{P(D+|E+)}{P(D+|E-)} = \frac{a/(a+b)}{c/(c+d)} = \frac{30/100}{10/100} = 3$$

b. Interpretation. A pregnant woman exposed to the pesticide is three times more likely to miscarry than a pregnant woman who is not exposed. That is, the **risk** of miscarriage **is three times greater** among those exposed to the pesticide than those who are not exposed.

B. Odds ratio (OR)

1. Definition

a. OR, another measure of comparative risk, compares the **odds** that a disease (or other health-related outcome) will occur among individuals who have a particular characteristic or who have been exposed to a risk factor to the odds that the disease will occur in individuals who lack the characteristic or who have not been exposed.

b. Probability versus odds

(1) The **odds** (O) that a given outcome will occur is defined as

$$O = \frac{P(\text{outcome will occur})}{1 - P(\text{outcome will occur})} = \frac{P(\text{outcome will occur})}{P(\text{outcome will } \textbf{not} \text{ occur})}$$

(2) An **odds** of A to B (written A:B or A/B) **in favor of** an outcome indicates that the **probability** that this outcome will occur is $A/(A+B)$.

2. Calculation

a. As in the calculation of RR, the data to be used in the calculation of OR may be summarized in a 2×2 table (see Table 14-1).

b. Formulas

(1) Independent samples (i.e., subjects not matched)

(a) The odds that a person who is exposed to the risk factor (E+) will have the disease (D+), that is, $O_{D+|E+}$, is

$$O_{D+|E+} = \frac{\text{probability of } \textbf{disease} \text{ if exposed}}{\text{probability of } \textbf{no disease} \text{ if exposed}} = \frac{P(D+|E+)}{P(D-|E+)} = \frac{a/(a+b)}{b/(a+b)} = a/b$$

(b) The odds that a person who is not exposed (E−) to the risk factor will have the disease (D+), that is $O_{D+|E-}$, is

$$O_{D+|E-} = \frac{\text{probability of } \textbf{disease} \text{ if not exposed}}{\text{probability of } \textbf{no disease} \text{ if not exposed}}$$

$$= \frac{P(D+|E-)}{P(D-|E-)}$$

$$= \frac{c/(c+d)}{d/(c+d)}$$

$$= c/d$$

(c) The formula for computing the OR for independent samples is

$$OR = \frac{\text{odds that } \textbf{exposed} \text{ individual will have disease}}{\text{odds that } \textbf{nonexposed} \text{ individual will have disease}}$$

$$= \frac{P(D+|E+)/P(D-|E+)}{P(D+|E-)/P(D-|E-)} \quad (14.2)$$

$$= \frac{O_{D+|E+}}{O_{D+|E-}}$$

$$= \frac{a/b}{c/d}$$

$$= (ad)/(bc) \quad (14.3)$$

(2) Matched (paired) samples. The formula for computing OR in a study in which subjects have been matched on the basis of one or more potentially confounding variables is

$$OR = \frac{\text{odds that } \textbf{exposed} \text{ individual will have disease}}{\text{odds that } \textbf{nonexposed} \text{ individual will have disease}}$$

$$= b/c \quad (14.4)$$

3. Interpretation

a. The **odds of having the disease** in question are **OR times greater** among those exposed to the suspected risk factor than among those with no such exposure.

b. For **rare** diseases (e.g., most chronic diseases, which have a prevalence of less than 10%), **OR approximates RR.**

(1) That is, the risk of the disease is approximately **OR times greater** among those exposed to the suspected risk factor than among those with no such exposure.

(2) This is true because $RR = [P(D+|E+)]/[P(D+|E-)] = [a/(a+b)]/[c/(c+d)]$, and when a disease occurs infrequently in the population, a and c are small relative to b and d. Thus a/b approximates $a/(a+b)$ and c/d approximates $c/(c+d)$.

c. The larger the value of OR, the stronger the association between the disease in question and exposure to the risk factor.

d. When the value of OR is close to 1, the disease and exposure to the risk factor are **unrelated,** that is, the odds that an exposed individual will have the disease are the same as those for a nonexposed individual.

e. Values of OR less than 1 indicate a negative association (i.e., a protective effect) between the risk factor and the disease.

4. OR and study design

a. Since $P(D+|E+)$, $P(D+|E-)$, $P(D-|E+)$, and $P(D-|E-)$ can be estimated from the results of a **cohort or an experimental** study, **OR can be calculated directly** when these study designs are employed.

b. OR in a case–control study

(1) Because the conditional probabilities of disease $P(D+|E+)$, $P(D+|E-)$, $P(D-|E+)$, and $P(D-|E-)$ cannot be calculated directly from the results of a case–control study, formula 14.2 cannot be used with this study design.

(2) Estimates of the conditional probability of exposure [e.g., $P(E+|D+)$] can be obtained in a case–control study. Using Bayes' Rule (see Ch 3 III C), a formula for computing OR can be derived in terms of the conditional probabilities of exposure. This derivation is discussed in detail in IV.

c. Calculation

(1) Thus, for a **cohort or experimental study,**

$$\text{OR} = \frac{\text{odds that } \textbf{exposed} \text{ individual will have disease}}{\text{odds that } \textbf{nonexposed} \text{ individual will have disease}}$$

$$= \frac{O_{D+|E+}}{O_{D+|E-}}$$

$$= \frac{P(D+|E+)/P(D-|E+)}{P(D+|E-)/P(D-|E-)}$$

(2) While for a **case–control study,**

$$\text{OR} = \frac{\text{odds that a } \textbf{diseased} \text{ individual has been exposed}}{\text{odds that a } \textbf{healthy} \text{ individual has been exposed}}$$

$$= \frac{O_{E+|D+}}{O_{E+|D-}}$$

$$= \frac{P(E+|D+)/P(E-|D+)}{P(E+|D-)/P(E-|D-)} \tag{14.5}$$

(a) Using the notation adopted for -Table 14-1, the numerator of this expression can be rewritten:

$$\frac{P(E+|D+)}{P(E-|D+)} = \frac{a/(a+c)}{c/(a+c)}$$

$$= a/c$$

and the denominator can be rewritten:

$$\frac{P(E+|D-)}{P(E-|D-)} = \frac{b/(b+d)}{d/(b+d)}$$

$$= b/d$$

(b) Thus, the equation for a case–control study becomes:

$$\text{OR} = \frac{P(E+|D+)/P(E-|D+)}{P(E+|D-)/P(E-|D-)}$$

$$= \frac{a/c}{b/d}$$

$$= ad/bc$$

(3) Hence, although the underlying conditional probability relationship defining OR differs for the case–control and cohort study designs, the same simplified calculational formula can be used in both cases.

d. Interpretation

(1) The OR calculated in a case–control study is interpreted in the same way as the OR calculated in a cohort or experimental study. That is, the odds of having the disease are **OR times greater** among individuals exposed to the risk factor than among those not exposed. Stated another way, it is **OR times more likely** that a diseased individual has been exposed to the risk factor than a healthy individual.

(2) **For rare diseases,** OR may be interpreted as an estimate of RR (i.e., the risk of developing the disease in question is OR times greater among individuals exposed to the risk factor than among those with no such exposure). Only under these circumstances does a **case–control study** provide a measure of the strength of an association **comparable** to that obtained from a cohort or cross-sectional study.

5. Examples

a. Case–control study, independent (unmatched) samples. Chapter 13 II C 3 a described a case–control study investigating the relationship between pesticide exposure in pregnant women and the subsequent incidence of miscarriage. In this study, the exposure rate among 40 women experiencing a miscarriage was shown to differ at the $\alpha = .01$ level of significance ($p < .005$) from the rate among 160 women experiencing a normal full-term pregnancy.

(1) Calculating OR

(a) Table 14-3 is a 2×2 table summarizing the results of this study.

(b) Using formula 14.5,

$$\text{OR} = \frac{P(\text{E}+|\text{D}+)/P(\text{E}-|\text{D}+)}{P(\text{E}+|\text{D}-)/P(\text{E}-|\text{D}-)}$$

$$= \frac{(30/40)/(10/40)}{(70/160)/(90/160)}$$

$$= \frac{(30)(90)}{(10)(70)}$$

$$= 3.86$$

(2) Interpretation

(a) The **odds of miscarrying are 3.86 times greater** for women exposed to the pesticide than for those with no such exposure.

(b) In other words, it is **3.86 times more likely to find prior exposure** to the pesticide among women experiencing a miscarriage than among women experiencing a normal full-term pregnancy.

(c) If the prevalence of miscarriage in the study population is low, the risk of miscarriage after exposure to the pesticide is approximately 3.86 times greater than the risk when no such exposure has occurred (i.e., $\text{RR} \approx 3.86$).

b. Case–control study, matched samples. In a second case–control study of the association between pesticide exposure and miscarriage, 100 women who experienced a miscarriage after the annual crop dusting season (cases) were identified from the records of a hospital serving rural and urban residents of a particular county. One hundred women who had a full-term pregnancy, culminating in a normal delivery (controls), were identified from the same records. The cases and controls were matched on the basis of age and prior history of miscarriage. All subjects were interviewed to determine their exposure to the pesticide during the previous crop dusting season.

(1) Calculating OR

(a) The data obtained in this study are summarized in Table 14-4. Using McNemar's test (see Ch 13 II C 4), the association between pesticide exposure

Table 14-3. Pesticide Exposure among Women Experiencing a Miscarriage and Women Experiencing a Normal Pregnancy

	Miscarriage (D+)	No Miscarriage (D−)	Totals
Exposed (E+)	30	70	100
Not Exposed (E−)	10	90	100
Totals	40	160	200

Table 14-4. Pesticide Exposure and Miscarriage: A Case–Control Study with Matched Subjects

	Controls (No Miscarriage)		
Cases (Miscarriage)	**Exposed (E+)**	**Not Exposed (E−)**	**Totals**
Exposed (E+)	5	32	37
Not Exposed (E−)	3	60	63
Totals	8	92	100

and miscarriage is statistically significant at the $\alpha = .01$ level of significance ($p < .005$).

(b) Using Formula 14.4,

$$\begin{aligned} OR &= b/c \\ &= 32/3 \\ &= 10.7 \end{aligned}$$

(2) Interpretation

(a) The odds in favor of pesticide exposure are 10.7 times greater for women experiencing a miscarriage (cases) than for women experiencing a normal full-term pregnancy (controls).

(b) Stated another way, the odds that a pregnant woman who is exposed to the pesticide will miscarry are 10.7 times greater than the odds that a pregnant woman with no such exposure will do so.

(c) The prevalence of miscarriage in the population is not low. Hence, the odds ratio cannot be used in this instance to estimate RR.

6. Confidence intervals for OR

a. Definition. A 95% confidence interval for the OR indicates that an investigator can be 95% confident that the calculated interval encompasses the true OR in the populations from which the samples were drawn.

b. Calculation. Woolf's method* first calculates the 95% confidence interval for the natural logarithm (*ln*) of OR. The upper and lower limits of the confidence interval estimate of the OR are then derived as the antilogs of the upper and lower limits of the 95% confidence interval for *ln* OR.

(1) Using the general formula for computing a confidence interval (see Ch 11 I B 1), the 95% confidence interval estimate of *ln* OR can be calculated as:

$$\begin{aligned} \text{endpoints of interval} &= \text{estimate} \pm (\text{confidence coefficient} \times \text{standard error of estimate}) \\ &= \ln OR \pm 1.96\sqrt{\frac{1}{a}+\frac{1}{b}+\frac{1}{c}+\frac{1}{d}} \end{aligned}$$

(2) The upper limit (UL) and lower limit (LL) of the 95% confidence interval estimate of the OR are the antilogs of the upper and lower limits of the confidence interval estimate of *ln* OR:

$$UL_{OR} = e^{UL_{\ln OR}}$$

$$LL_{OR} = e^{LL_{\ln OR}}$$

c. Example. In II B 5, the odds of miscarrying were calculated to be 3.86 times greater among women exposed to a certain pesticide than among women with no such

* Woolf B: On estimating the relation between blood group and disease. *Ann Hum Genet* 19:251–253, 1955.

exposure (i.e., OR = 3.86). The 95% confidence interval estimate of *ln* OR, ($CI_{ln\,OR}$), is

$$CI_{ln\,OR} = ln(3.86) \pm 1.96\sqrt{\frac{1}{30} + \frac{1}{70} + \frac{1}{10} + \frac{1}{90}}$$
$$= 1.35 \pm .78$$
$$= .57 \text{ to } 2.13$$

and the 95% confidence interval estimate of OR (CI_{OR}) is

$$CI_{OR} = e^{.57} \text{ to } e^{2.13}$$
$$= 1.8 \text{ to } 8.4$$

In other words, the epidemiologist conducting this study may be 95% confident that the OR for the population of women from which the study sample was drawn falls between 1.8 and 8.4.

7. Testing hypotheses involving OR

a. A confidence interval estimate of OR can be used to test the **significance** of an association between a disease and a suspected risk factor.

(1) When **OR = 1,** there is **no association** between the risk factor and the disease.

(2) Therefore, the two opposing hypotheses can be written as follows.

(a) The null hypothesis, H_O, is, "No association exists between the disease and the risk factor, or $\mathbf{H_O}$**: OR = 1.**

(b) The alternative hypothesis, H_A, is "The disease is associated with exposure to the risk factor, or $\mathbf{H_A}$**:** $\mathbf{OR \neq 1}$**.**

(3) **If 1 is in the calculated confidence interval,** $\mathbf{H_O}$ **is not rejected** and the results are reported: "An association between the risk factor and the disease cannot be demonstrated from the data" (rather than "There is no association between the risk factor and the disease").

(4) **If the calculated confidence interval does not include 1,** $\mathbf{H_O}$ **is rejected** and the results are reported: "A statistically significant association (for a 95% confidence interval, $p < .05$) exists between the risk factor and the disease."

b. The epidemiologist conducting the case–control study of the association between pesticide exposure and miscarriage described in II B 5 wishes to determine the statistical significance of this association. In II B 6 c, she calculated the 95% confidence interval estimate of the OR as 1.8 to 8.4. Since this interval does not include 1, H_O (i.e., "No association between pesticide exposure and miscarriage exists") is rejected. The investigator concludes that there is a statistically significant association ($p < .05$) between these two variables.

III. MEASURES OF POTENTIAL IMPACT.

The computation of measures of potential impact is based on the assumption that a cause-and-effect relationship between the disease (or other health-related outcome) and exposure to the risk factor has been established.

A. Attributable risk (AR; also known as risk difference, excess risk, and rate difference)

1. Definition. AR defines the excess risk of disease that can be **ascribed to exposure to the risk factor,** over and above that experienced by people who are not exposed. It thus provides an estimate of the number of cases of the disease that might be prevented if exposure to the risk factor were eliminated and is useful for determining the magnitude of the public health problem posed by such exposure.

2. Calculation. AR is calculated using the formula

$$\text{AR} = \text{incidence of disease among those exposed} - \text{incidence of disease among those not exposed}$$
$$= P(D+|E+) - P(D+|E-) \quad (14.6)$$

3. Interpretation. The risk of developing the disease is **increased by AR** for those individuals exposed to the risk factor.

4. Since the computation of AR requires estimates of disease incidence in the exposed and nonexposed populations [i.e., $P(D+|E+)$ and $P(D+|E-)$], AR can be directly calculated only in a cohort or an experimental study.

5. **Example**
   a. For the cohort study described in II A 5 (see Table 14-2),

$$\begin{aligned} AR &= P(D+|E+) - P(D+|E-) \\ &= (30/100) - (10/100) \\ &= .20 \end{aligned}$$

   b. Thus, assuming that pesticide exposure is causally related to the occurrence of miscarriage, the increased risk of miscarriage that can be attributed to such exposure is .20. If a pregnant woman is exposed to the pesticide, her risk of miscarriage is increased by .20.

B. Population attributable risk (PAR)

1. **Definition.** PAR is a measure of the excess risk of disease in a **population** that can be solely attributed to the risk factor.

2. **Calculation.** PAR is obtained by multiplying the AR by the prevalence of exposure to the risk factor in the population [$P(E+)$]:

$$PAR = AR \times P(E+) \tag{14.7}$$

3. PAR can be obtained in a cohort or experimental study **if an independent estimate of the proportion of people in the population who have been exposed to the risk factor** is available.

4. **Example**
   a. In the cohort study discussed in II A 5 (see Table 14-2), AR was shown to be .20.
   b. The frequency of pesticide exposure in the population cannot be estimated from the data summarized in Table 14-2, since, in a cohort study, the numbers of exposed and nonexposed subjects are chosen by the investigator.* Assuming that $P(E+)$ is estimated as .50 (e.g., based on a report in the medical literature),

$$\begin{aligned} PAR &= (.20)(.50) \\ &= .10 \end{aligned}$$

   c. The **excess incidence (i.e., risk)** of miscarriage in the population that is associated with pesticide exposure is .10.

C. Attributable fraction in exposed (AF_E, also known as attributable risk percent in the exposed, etiologic fraction in exposed, or simple etiologic fraction)

1. **Definition.** The attributable fraction in the exposed is the difference in the incidence of disease between people who are exposed to the risk factor and those who are not exposed, divided by the rate among those exposed.

2. **Calculation.** AF_E is calculated using the formula

$$\begin{aligned} AF_E &= \frac{P(D+|E+) - P(D+|E-)}{P(D+|E+)} \qquad (14.8) \\ &= \frac{AR}{P(D+|E+)} \end{aligned}$$

Provided the $RR \geq 1$, formula 14.8 may be reduced by dividing both numerator and denominator by $P(D+|E-)$:

$$AF_E = \frac{RR - 1}{RR}$$

* If $P(E+)$ is known prior to the study, the numbers of subjects in the exposure groups can be selected to represent the underlying population exposure rates.

3. Example

a. For the cohort study examining the relationship between pesticide exposure and the incidence of subsequent miscarriage discussed in the preceding two examples (see II A 5 and Table 14-2),

$$\begin{aligned} AF_E &= \frac{P(D+|E+) - P(D+|E-)}{P(D+|E+)} \\ &= \frac{AR}{P(D+|E+)} \\ &= \frac{(30/100) - (20/100)}{(30/100)} \\ &= .33 \end{aligned}$$

b. In other words, 33% of the excess miscarriages that occurred among those exposed to the pesticide were attributable to this exposure.

D. Attributable fraction in total population (AF_T, also known as population attributable risk percent, total etiologic fraction, or population attributable fraction)

1. Definition. The attributable fraction in the population is the **proportion of the total risk of disease in a population that can be attributed to exposure** to a suspected risk factor.

2. Calculation. AF_T is calculated using the formula

$$\begin{aligned} AF_T &= \frac{\text{population attributable risk}}{P(D+)} \\ &= \frac{P(E+)(AR)}{P(D+)} \\ &= \frac{P(E+)(RR - 1)}{1 + P(E+)(RR - 1)} \end{aligned} \qquad (14.9)$$

where $RR = [P(D+|E+)]/[P(D+|E-)]$ and $P(E+)$ is the proportion of people in the population who have been exposed to the risk factor.

3. AF_T and exposure

a. When the exposure rate in the general population [i.e., $P(E+)$] and the RR are low, only a small proportion of excess cases of the disease can be attributed to exposure to the risk factor. Conversely, when a large proportion of the general population has been exposed (e.g., cigarette smokers) and the value of RR is high, a large proportion of the excess cases of the disease can be ascribed to exposure to the risk factor.

b. If the proportion of individuals in the general population who are exposed to the risk factor declines, AF_T decreases as well, even if RR is unchanged.

c. Because AF_T depends upon $P(E+)$, the rate of exposure to the risk factor, its use for comparative purposes is limited to populations whose rates of exposure are approximately equal.

4. AF_T and study design

a. The calculation of AF_T in a cohort or experimental study requires an independent estimate of $P(E+)$ if the number of exposed subjects is fixed.

b. AF_T cannot be calculated in a case–control study unless the following is true.

(1) The frequency of disease occurrence, $P(D+)$, is low. In this case, OR provides an estimate of RR.

(2) The subjects comprising the control (healthy) group are a **random sample** of the healthy population. In this instance, $P(E+|D-) \approx P(E+)$; that is, the exposure rate among the subjects in the control sample approximates the overall exposure rate for the general population.

c. When these two assumptions have been met, $P(E+|D-)$ can be substituted for $P(E+)$ and OR can be substituted for RR in formula 14.9.

5. Example

a. To calculate AF_T for the cohort study presented in II A 5 (see Table 14-2), it is assumed that the sample is representative of the target population in terms of the proportion of pregnant women exposed to the risk factor (pesticide), that is, it is assumed that the exposure rate in the target population $P(E+) = .5$. RR was computed in II A 5 a as 3. Substituting these values in formula 14.9,

$$AF_T = \frac{[P(E+)](RR-1)}{1+[P(E+)(RR-1)]}$$

$$= \frac{(.5)(3-1)}{1+[(.5)(3-1)]}$$

$$= .5$$

b. Approximately 50% of the excess risk of miscarriage in the population is attributable to pesticide exposure. That is, approximately 50% of the miscarriages in the total population could have been prevented by eliminating exposure to the pesticide.

IV. DERIVATION OF THE ODDS RATIO (OR) IN CASE–CONTROL STUDIES

A. Calculation of OR for a cohort or experimental study

1. The **odds that an individual exposed to the risk factor will have the disease,** $O_{D+|E+}$, is defined as

$$O_{D+|E+} = \frac{P(D+|E+)}{P(D-|E+)}$$

2. The **odds that an individual who has not been exposed to the risk factor will have the disease,** $O_{D+|E-}$, is defined as

$$O_{D+|E-} = \frac{P(D+|E-)}{P(D-|E-)}$$

3. The OR is

$$OR = \frac{O_{D+|E+}}{O_{D+|E-}}$$

B. Using Bayes' Rule to rewrite $O_{D+|E+}$ and $O_{D+|E-}$

1. Using Bayes' Rule (see Ch 3 III C),

$$P(D+|E+) = \frac{P(E+|D+)P(D+)}{P(E+)}$$

$$P(D-|E+) = \frac{P(E+|D-)P(D-)}{P(E+)}$$

2. Thus, $O_{D+|E+}$ can be rewritten

$$O_{D+|E+} = \frac{P(D+|E+)}{P(D-|E+)}$$

$$= \frac{P(E+|D+)P(D+)}{P(E+|D-)P(D-)}$$

3. Similarly,

$$P(D+|E-) = \frac{P(E-|D+)P(D+)}{P(E-)}$$

$$P(D-|E-) = \frac{P(E-|D-)P(D-)}{P(E-)}$$

4. Thus, $O_{D+|E-}$ can be rewritten

$$O_{D+|E-} = \frac{P(D+|E-)}{P(D-|E-)}$$

$$= \frac{P(E-|D+)P(D+)}{P(E-|D-)P(D-)}$$

5. Therefore, OR can be rewritten as

$$OR = \frac{O_{D+|E+}}{O_{D+|E-}}$$

$$= \frac{[P(E+|D+)P(D+)]/[P(E+|D-)P(D-)]}{[P(E-|D+)P(D+)]/[P(E-|D-)P(D-)]}$$

6. $P(D+)$ and $P(D-)$ cancel, leaving an expression in terms of the probabilities of exposure, which can be estimated from the results of a case–control study:

$$OR = \frac{P(E+|D+)/P(E-|D+)}{P(E+|D-)/P(E-|D-)}$$

$$= \frac{O_{E+|D+}}{O_{E+|D-}}$$

BIBLIOGRAPHY

Centers for Disease Control: Reye's syndrome—Ottio, Michigan. *Morbidity Mortality Weekly Report* 29:532–539, 1980.

Cornfield J, Haenszel W: Some aspects of retrospective studies. *J Chron Dis* 11:523–534, 1960.

Fletcher RH, Fletcher SW, Wagner EH: *Clinical Epidemiology: the Essentials*. Baltimore, Williams and Wilkins, 1982, p 178.

Gehlbach SH: *Interpreting the Medical Literature: A Clinician's Guide*. Lexington, MA, DC Health, 1982.

Kleinbaum DG, Kupper LL, Morgenstein H: *Epidemiologic Research*. Belmont, CA, Lifetime Learning Publications, 1982.

Levin ML, Bertell R: Simple estimation of population attributable risk from case–control studies. *Am J Epidemiol* 108:78–79, 1978.

Mantel N, Haenszel W: Statistical aspects of the analysis of data from retrospective studies of disease. *J Natl Cancer Inst* 22:719–748, 1959.

Rothman KJ: *Modern Epidemiology*. Boston, Little, Brown, 1986.

Woolf B: On estimating the relation between blood group and disease. *Ann Hum Genet* 19:251–253, 1955.

PROBLEMS

14-1. A team of clinical researchers hypothesized that major electrocardiographic (ECG) abnormalities are a risk factor for death from coronary heart disease (CHD). In a study designed to test this hypothesis, 47 black men between 40 and 64 years of age at initial examination who had major ECG abnormalities and 144 black men in the same age group with no ECG abnormalities were recruited as subjects. Both groups were followed for 20 years and deaths from CHD recorded. The table below summarizes the results of this study.

	Death from CHD		
ECG Abnormality	**Yes**	**No**	**Totals**
Present	8	39	47
Absent	10	134	144
Totals	18	173	191

Adapted from Strogatz et al: Electrocardiographic abnormalities and mortality among middle-aged black men and white men of Evans County, Georgia. *J Chron Dis* 40:149–155, 1987.

a. What type of study design was used?

b. Calculate the absolute risk of dying from CHD for black men between the ages of 40 and 64 who have major ECG abnormalities.

c. Calculate and interpret the RR of death from CHD associated with presence of major ECG abnormalities.

d. Calculate and interpret the OR.

e. What is the 95% confidence interval estimate of the OR? Interpret this interval.

f. Test the hypothesis that there is a statistically significant association between death from CHD and the presence of major ECG abnormalities. Report the p-value.

g. Calculate and interpret the attributable risk associated with the presence of major ECG abnormalities.

14-2. A study is carried out to evaluate the association between alcohol consumption and death from all causes. The investigator conducting the study classifies 4550 subjects according to their monthly alcohol consumption: abstainers, light drinkers (1–30 drinks per month), moderate drinkers (31–90 drinks per month), and heavy drinkers (91 or more drinks per month). Study participants are followed for a 10-year period, and deaths from all causes are recorded in the table below.

	Condition of Subject at End of Study:		
Alcohol Consumption	**Alive**	**Dead**	**Totals**
Abstain	300	900	1200
Light	500	2000	2500
Moderate	120	380	500
Heavy	175	175	350
Totals	1095	3455	4550

a. What type of design was used for this study?

b. Estimate the absolute risk of death from any cause for a heavy drinker in this population.

c. Calculate the RR of death from any cause for a heavy drinker compared to a light drinker. Does RR provide information about the absolute risk of death from any cause for a heavy drinker?

d. Calculate the RR of death from any cause for a heavy drinker compared to an abstainer.

e. Calculate the RR of death from any cause for a moderate drinker compared to a light drinker.

f. Calculate and interpret the OR for heavy drinkers compared to light drinkers.

g. Calculate and interpret the 95% confidence interval estimate of the OR.

h. Is there a statistically significant association between alcohol consumption and death from all causes? Report the *p*-value. What is the chance of error associated with the conclusion?

i. Calculate and interpret the attributable risk for heavy drinking compared to abstinence.

14-3. To study the possible association between oral contraceptive use and the occurrence of rheumatoid arthritis (RA), an investigator selected 100 women with a confirmed diagnosis of RA and 200 women undergoing treatment at the same medical facility for other musculoskeletal conditions as subjects. The medical records of all subjects (prior to the date of first diagnosis) were reviewed for evidence of oral contraceptive use. The results are summarized in the table below.

	Rheumatoid Arthritis		
Oral Contraceptive Use	**Present**	**Absent**	**Totals**
User	40	120	160
Nonuser	60	80	140
Totals	100	200	300

a. What type of study design was employed?

b. Calculate and interpret the OR. Can RR be directly calculated from the results of this study? Why?

c. Calculate the 95% confidence interval for the OR.

d. Is there a statistically significant association between oral contraceptive use and occurrence of rheumatoid arthritis?

14-4. In a study similar to that described in problem 14-3, cases and controls are matched with respect to age and year of registry at the particular medical facility. The table below shows the results of this study.

	Controls (RA Absent)		
Cases (RA Present)	**OC Use**	**No OC Use**	**Totals**
OC Use	20	20	40
No OC Use	50	10	60
Totals	70	30	100

RA = rheumatoid arthritis; OC = oral contraceptive

a. Calculate and interpret the OR.

b. What is the purpose of matching in this study?

Solutions on p 389.

STUDY QUESTIONS

Directions: Each of the numbered items or incomplete statements in this section is followed by answers or by completions of the statement. Select the **one** lettered answer or completion that is **best** in each case.

1. All of the following statements about the absolute risk of a disease are true EXCEPT

(A) absolute risk is synonymous with the incidence of the disease
(B) absolute risk is the probability that a healthy individual will develop the disease during a specified time period
(C) absolute risk is the underlying rate from which RR and AR are derived
(D) absolute risk is the ratio of the incidence of the disease among those exposed to the relevant risk factor to the incidence of the disease among those with no such exposure

2. RR measures which of the following?

(A) The probability that a person who is exposed to a certain risk factor will develop the disease in question
(B) How much more likely it is that a patient who has the disease has been exposed to a particular risk factor compared to a healthy individual
(C) The incidence of a disease
(D) The magnitude of the association between a disease (or other health-related outcome) and a suspected risk factor

3. An occupational safety officer knows that the RR of non-Hodgkin's lymphomas following exposure to a particular industrial chemical is 12.5. What can he conclude from this information?

(A) A worker who must routinely handle large quantities of the chemical in question has a very high probability of developing a malignant lymphoma
(B) It is unlikely that the observed association between exposure to the chemical and non-Hodgkin's lymphomas is due to random chance
(C) The incidence of non-Hodgkin's lymphomas is very high among workers exposed to the chemical
(D) A worker who must routinely handle large quantities of the chemical is 12.5 times more likely to develop a malignant lymphoma than a worker who is not exposed to the chemical

4. All of the following statements about attributable risk (AR) are true EXCEPT

(A) AR is a measure of the absolute risk (incidence) that can be attributed to a particular risk factor
(B) AR is calculated as $[P(D+|E+)] - [P(D+|E-)]$, where D and E represent "disease" and "exposure," respectively
(C) AR is the excess risk of the disease experienced by those exposed to the risk factor
(D) AR can be directly computed only in a case–control study

5. OR is an estimate of RR only if which of the following conditions exists?

(A) The disease is relatively rare in the general population (in both the exposed and nonexposed populations)
(B) The incidence rates are computed from the results of a prospective cohort study
(C) The data were collected using a case–control study
(D) The rate of exposure to the risk factor is relatively low among both cases and controls

---

1-D
2-D
3-D
4-D
5-A

Questions 6–7

The table below depicts data on a food poisoning outbreak following a back-to-school party attended by 200 medical students.

	Ate Food			Did Not Eat Food		
	Ill	**Not Ill**	**Totals**	**Ill**	**Not Ill**	**Totals**
Barbecue	90	30	120	20	60	80
Coleslaw	67	33	100	43	57	100
Totals	157	63	220	63	117	180

6. What is the RR of developing food poisoning after barbecue consumption?

(A) 90/120
(B) 20/80
(C) .75/.25
(D) 90/30

7. What is the RR of developing food poisoning after eating coleslaw?

(A) 1.56
(B) 2.03
(C) .75
(D) 3.0

Questions 8–11

In a case–control study examining the relationship between developmental disorders and prenatal exposure to cocaine, the hospital records of 1000 infants diagnosed with a developmental disorder and 1000 control infants were inspected for proven maternal cocaine abuse. As the following table shows, of the 1000 children with a developmental disorder, 800 were born to mothers known to have abused cocaine during their pregnancy, compared to 300 of the control infants.

	Developmental Disorder		
Maternal Cocaine Use	**Present**	**Absent**	**Totals**
Present	800	300	1100
Absent	200	700	900
Totals	1000	1000	2000

8. What is the absolute risk of developmental disorders among infants exposed to cocaine in utero?

(A) 9.33
(B) 3.27
(C) .73
(D) .80
(E) It cannot be computed from the given data

9. What is the OR of a developmental disorder given exposure to cocaine?

(A) 9.33
(B) 3.27
(C) .73
(D) .80

10. What is the RR of a developmental disorder given exposure to cocaine?

(A) 9.33
(B) 3.27
(C) .73
(D) .80
(E) RR cannot be computed from the given data

11. What is the prevalence of developmental disorders?

(A) .50
(B) .55
(C) .73
(D) .80
(E) Cannot be computed from the data

6-C 9-A
7-A 10-E
8-E 11-E

Questions 12–15

In a cross-sectional study of the relationship between smoking and anxiety, 1000 people were simultaneously classified according to smoking status (smoker or nonsmoker) and current level of anxiety (high or low). As is summarized in the table below, 300 of these individuals were found to have a high level of anxiety, 500 were identified as smokers, and 200 were smokers who also reported a high level of anxiety.

	Anxiety Level		
	High	**Low**	**Totals**
Smoker	200	300	500
Nonsmoker	100	400	500
Totals	300	700	1000

12. What is the incidence of high anxiety levels among the study participants?

(A) .30
(B) .50
(C) .40
(D) .20
(E) This cannot be computed from the given data

13. What is the prevalence of high anxiety levels?

(A) .30
(B) .50
(C) .40
(D) .20
(E) This cannot be computed from the given data

14. What is the RR of high anxiety for smokers compared to nonsmokers?

(A) 2.67
(B) 2.0
(C) .30
(D) .40
(E) RR is impossible to estimate unless the prevalence and incidence of high anxiety levels are approximately equal

15. What is the OR?

(A) 3.3
(B) 2.67
(C) 2.0
(D) .20
(E) OR cannot be computed from the given data

12-E
13-A
14-E
15-B

Questions 16–19

One hundred children known to have been exposed to high levels of lead during the first 12 months of life were followed for 15 years; 40 developed an affective disorder. A similar group of 100 children who were not exposed to high lead levels during the first 12 months of life were also followed over the same time period. Five of these children developed an affective disorder. The data regarding the relationship between lead exposure and the disorder are summarized in the following table.

	Affective Disorder		
	Present	**Absent**	**Totals**
Exposed to Lead	40	60	100
Not exposed to Lead	5	95	100
Totals	45	155	200

16. What is the incidence of affective disorders among those exposed to high levels of lead during the first 12 months of life?

(A) .20
(B) .50
(C) .40
(D) .225
(E) This cannot be computed from the given data

17. What is the RR of developing an affective disorder for those exposed to high levels of lead during the first 12 months of life, compared to those with no such exposure?

(A) 12.67
(B) 8.0
(C) .23
(D) .40
(E) This cannot be computed from the given data

18. What is the OR for affective disorders among those exposed to high levels of lead during the first 12 months of life, compared to those with no such exposure?

(A) 12.67
(B) 8.0
(C) .23
(D) .40
(E) OR cannot be computed from the given data

19. What is the attributable risk (AR) for affective disorders given lead exposure?

(A) .40
(B) .05
(C) .35
(D) 8.0
(E) AR cannot be computed from the given data

16-C
17-B
18-A
19-C

ANSWERS AND EXPLANATIONS

1. The answer is D *[I B 1; II A 1; III A 1]*
The ratio of the incidence of disease among those exposed to a given risk factor and the incidence among those with no such exposure is the relative, not the absolute, risk (i.e., the RR). Absolute risk, which is synonymous with the cumulative incidence of a disease, is the probability that a healthy person will develop the disease in question during a specified time period. Attributable risk (AR) is the difference in incidence (i.e., in absolute risk) between those exposed to the suspected risk factor and those not exposed. Both AR and RR are derived from incidence rates, that is, from measures of absolute risk.

2. The answer is D *[II A 1, 3; III A 1]*
Relative risk (RR) measures how much more likely an individual who is exposed to a risk factor is to develop a given disease than an individual who is not exposed. The higher the value of RR, the greater the association between the disease and the risk factor. RR alone provides no estimate of the probability of developing the disease (i.e., absolute risk, or incidence). Attributable risk (AR) measures how much more likely it is that a patient who contracts a disease has been exposed to a risk factor.

3. The answer is D *[II A 3]*
The reported value of relative risk (RR) indicates that a chemical worker who is exposed to this carcinogenic chemical is 12.5 times more likely to develop a lymphoma than a co-worker who is not routinely exposed. RR tells the occupational safety worker nothing about the probability (i.e., absolute risk, or incidence) of non-Hodgkin's lymphomas. To determine if the observed association is likely to be due to random chance, it is necessary to conduct an appropriate statistical analysis.

4. The answer is D *[III A 1–4]*
Attributable risk (AR), the difference in the incidence of a disease among those subjects exposed to a risk factor and the incidence among nonexposed subjects, can be directly calculated only from the results of a cohort study (incidence rates cannot be calculated directly in a case–control study). It provides a measure of the excess risk of disease that can be attributed to exposure to the risk factor, and is calculated using the formula $[P(D+|E+)] - [P(D+|E-)]$.

5. The answer is A *[II A 4, B 3 b]*
The odds ratio (OR) approximates relative risk (RR) when the outcome or disease is relatively rare among the general population. The approximation does not depend upon a low rate of exposure among controls. However, the exposure rate for the control group must be representative of that experienced by the general population. In contrast to RR, OR can be calculated for all study designs.

6–7. The answers are: 6-C *[II A 2 b]*, **7-A** *[II A 2 b]*
The relative risk (RR) of food poisoning following barbecue consumption is the ratio of the incidence of food poisoning among those students who ate the barbecue (i.e., the exposed group) and the incidence among those students who did not (i.e., the nonexposed group):

$$\begin{aligned} RR &= \frac{P(D+|E+)}{P(D+|E-)} \\ &= \frac{P(\text{ill}|\text{ate barbecue})}{P(\text{ill}|\text{did not eat barbecue})} \\ &= \frac{90/120}{20/80} \\ &= \frac{.75}{.25} \\ &= 3.00 \end{aligned}$$

Similarly, the RR of food poisoning after eating coleslaw is the ratio of the incidence of food poisoning among those students who ate coleslaw and the incidence among those students who did not:

$$\text{RR} = \frac{P(\text{ill}|\text{ate coleslaw})}{P(\text{ill}|\text{did not eat coleslaw})}$$
$$= \frac{67/100}{43/100}$$
$$= 1.56$$

8–11. The answers are: 8-E *[I B 1 a–c; II A 4 b]*, **9-A** *[II B 4 c (2), b]*, **10-E** *[II A 4, B 3 b]*, **11-E** *[Ch 8 II B 2 c (5), d]*
Measures of incidence cannot be directly calculated from the results of a case–control study. Hence, the absolute risk, which is the probability that an infant born to a mother who abused cocaine will have a developmental disorder (i.e., the incidence of such disorders), cannot be determined from the given data.

The odds ratio (OR) is calculated using the data in the table accompanying questions 8–11 and the following formula:

$$\text{OR} = ad/bc$$
$$= \frac{(800)(700)}{(300)(200)}$$
$$= 9.33$$

An infant with a developmental disorder is 9.33 times more likely to have a mother who abused cocaine during her pregnancy than a normal healthy infant.

The incidence of developmental disorders, $P(\text{D}+|\text{E}+)$, can only be directly computed in a cohort study. Thus, the data obtained in this study cannot be used to calculate relative risk (RR), which is the ratio of the incidence of developmental disorders among infants exposed to cocaine in utero and those without such exposure. If it was known that developmental disorders occur infrequently among this population, then OR would be an estimate of RR (i.e., an infant exposed to cocaine in utero would be 9.33 times more likely to have a developmental disorder than one who was not), but no information about the frequency of the disorders was given.

Disease prevalence cannot be computed from the results of a case–control study because the investigator controls the number of study subjects who have the disease (cases) and the number who do not (controls). Prevalence is estimated from the results of a cross-sectional study.

12–15. The answers are: 12-E *[I B 1 a–c; Ch 8 II B 2 d (3) (a)]*, **13-A** *[Ch 7 I A 2]*, **14-E** *[II A 4 a]*, **15-B** *[II B 2]*
Incidence cannot be calculated in a cross-sectional study. Estimates of incidence require a healthy cohort followed forward in time to determine the occurrence of new cases of the disease in question. Cases identified in a cross-sectional study represent prevalent (existing) cases, rather than incident cases.

Prevalence is defined as the proportion of cases of a disease (or other health-related outcome) among a given population. Hence, in this instance, the prevalence of high anxiety is the proportion of study subjects who were classified as highly anxious:

$$\text{prevalence} = 300/1000$$
$$= .30$$

Because relative risk (RR) is a ratio of incidence rates (in exposed compared to nonexposed individuals), it cannot be directly estimated from the results of a cross-sectional study. If, however, prevalence and incidence are approximately equal (a requirement that is difficult both to satisfy and to confirm), prevalence may be substituted for incidence in the calculation of RR.

Using the data in the table accompanying questions 12–15, the odds ratio (OR) is

$$\text{OR} = ad/bc$$
$$= \frac{(200)(400)}{(300)(100)}$$
$$= 2.67$$

The odds that an individual is a smoker are 2.67 times greater if the person is highly anxious than if he or she is not. Stated another way, the odds in favor of a high level of anxiety are 2.67 times greater among smokers than nonsmokers.

16–19. The answers are: 16-C *[I B 1; Ch 7 I B 1 c]*, **17-B** *[II A 1–2]*, **18-A** *[II B 2 b (1)]*, **19-C** *[III A 2]*
The incidence of affective disorders is the probability that a child will develop such a disorder following exposure to high lead levels during the first 12 months of life, that is, $P(D+|E+)$, where D+ represents the presence of an affective disorder and E + represents exposure to lead. In this instance, 40 of the 100 children known to have been exposed to high lead levels during the first 12 months of life subsequently developed an affective disorder. Thus,

$$\text{incidence} = 40/100 = .40$$

Relative risk (RR) is the ratio of the incidence of affective disorders among those children exposed to high lead levels and the incidence of affective disorders among nonexposed children:

$$RR = \frac{P(D+|E+)}{P(D+|E-)} = \frac{40/100}{5/100} = 8.0$$

The odds ratio (OR) is calculated using the formula

$$OR = \frac{O_{D+|E+}}{O_{D+|E-}}$$

where

$$O_{D+|E+} = \text{odds of affective disorder if exposed to lead} = \frac{P(D+|E+)}{P(D-|E+)} = \frac{40/100}{60/100} = 40/60$$

$$O_{D+|E-} = \text{odds of affective disorder if not exposed} = \frac{P(D+|E-)}{P(D-|E-)} = \frac{5/100}{95/100} = 5/95$$

Thus, using the formula given above,

$$OR = \frac{40/60}{5/95} = \frac{(40)(95)}{(60)(5)} = 12.67$$

The odds that a child will develop an affective disorder after exposure to lead as an infant are 12.67 times greater than the odds of developing an affective disorder in the absence of such exposure. If affective disorders are rare, OR can also be interpreted as RR. That is, the risk of developing an affective disorder is 12.67 times greater for a child exposed to lead than one who is not.

Attributable risk (AR) defines the excess risk of developing an affective disorder that can be attributed to lead exposure, over and above that which occurs in the absence of such exposure. It is calculated as the difference in the incidence of affective disorders among children exposed to lead during the first 12 months of life and children who are not:

$$AR = [P(D+|E+) - P(D+|E-)] = (40/100) - (5/100) = .35$$

15
Tests of Statistical Significance: Regression and Correlation

I. ANALYZING ASSOCIATIONS THAT INVOLVE INTERVAL/RATIO (I.E., CONTINUOUS) DATA

A. Overview of useful statistical procedures. Chapters 13 and 14 discussed statistical procedures that allow clinical investigators to analyze associations involving **frequency data**. This chapter outlines comparable procedures used for **continuous data** (i.e., data measured on an interval/ratio scale; Figure 15-1). Both types of procedures enable clinical investigators to determine:

1. The **statistical significance** of an observed association
2. The **magnitude** of the association
3. The amount of **variation** in the response variable that is **attributable** to a putative risk factor

B. Correlation analysis and **regression analysis** are two procedures **used to analyze associations involving continuous, or interval/ratio, data**.

1. **Correlation analysis** measures the **strength** of the association between two study variables. The term "correlation analysis," as used in this chapter, refers to **Pearson's product moment correlation coefficient** (also known as **Pearson's *r***).
2. **Regression analysis** derives a **prediction equation** for estimating the value of one variable given the value of the second.
3. Correlation and regression analyses **function similarly for all study designs**.

C. Data representation and organization

1. **The data** used in a correlation or regression analysis consist of **pairs of measurements** made on the same unit of observation (most often, the same study subject). Each member of the pair corresponds to one of the **two study variables.** For example, in a study of the relationship between hypertension and blood cholesterol levels, systolic blood pressure and serum cholesterol value are the pair of measurements to be assessed for each study subject and to be represented by the two study variables.
2. **Notation.** The pairs are denoted symbolically (X_i, Y_i), where the subscript i represents the ith study subject. By convention, X typically represents the **independent variable**, while Y represents the **dependent,** or response, **variable**.
3. **Variation among study types**
   a. **In epidemiological studies,** the independent variable X is often a suspected risk factor (e.g., a low-fiber diet) and the dependent variable Y is the occurrence of disease or other health-related outcome (e.g., the occurrence of colorectal cancer).
   b. **In experimental studies,** values of the independent variable are fixed by the investigator rather than determined by nature. For example, in a study of the efficacy of a new antiviral drug in preventing the recurrence of herpes simplex infections, the investigator selects the dosages (the independent variable) to be administered. Thus, dosage is a fixed, not a random, variable.
   c. **Correlation vs. regression**
      (1) Correlation analysis is restricted to studies in which both variables are random, or determined by nature (e.g., epidemiological studies).

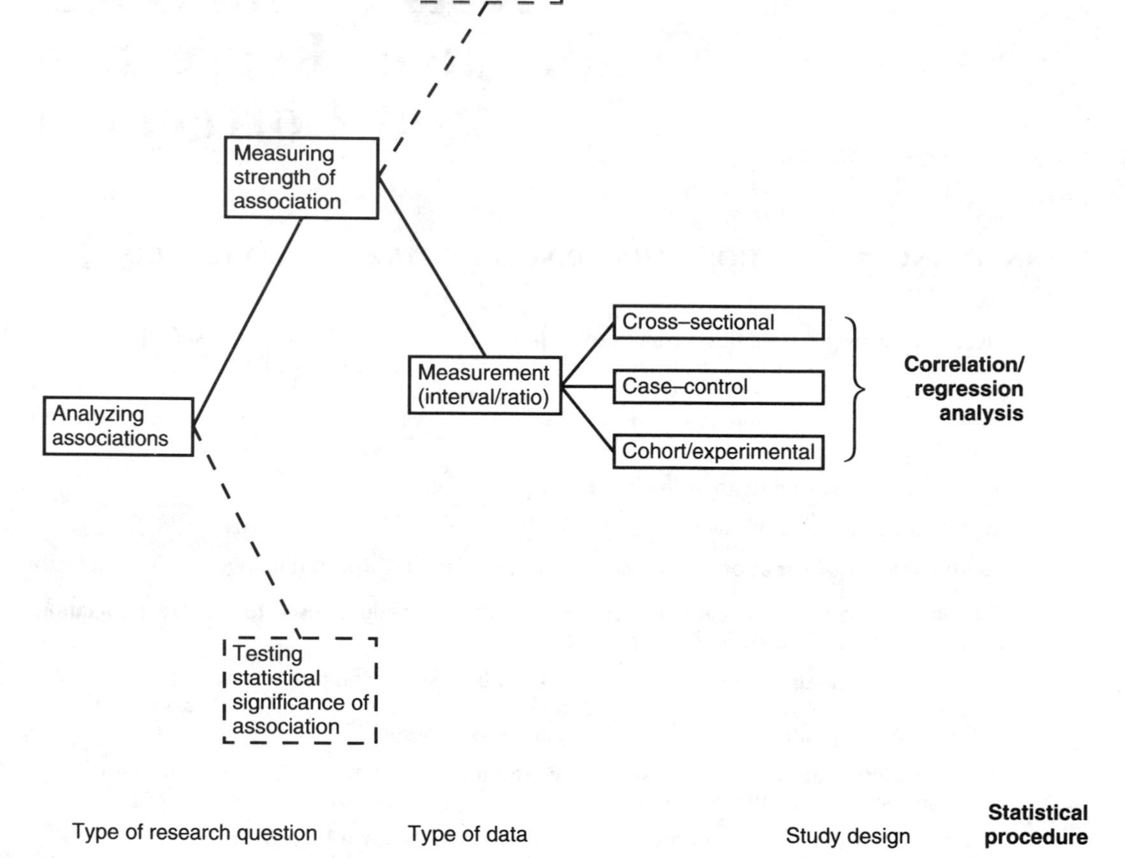

Figure 15-1. Decision tree pathways (derived from Figure 13-1) used to select a statistical test for analyzing associations involving interval/ratio data. *Dotted lines* indicate decision tree paths not taken.

(2) Regression analysis may be used in either of the following situations:
 (a) When one variable is fixed and one is random (**the classic regression model**)
 (b) When both variables are random (**the correlation model**)

(3) When both variables are random, the terms "dependent variable" and "independent variable" are irrelevant.

4. **Tabular representation** of data pairs prior to correlation or regression analysis is shown in Table 15-1.

5. **Graphic representation: the scatter diagram.** Quantitative data obtained in a study of the association between two variables can be graphically displayed in a **scatter diagram**. Each pair (X_i,Y_i) of values (see Table 15-1) is represented in the scatter diagram by a dot located at the point (X,Y).

 a. **Choosing the axes**
  (1) When both variables are random variables, the choice of which axis is labeled X and which Y is arbitrary.
  (2) When regression analysis is to be used to predict the value of one variable from the value of the other, the variable to be "predicted" (i.e., the dependent variable) is plotted on the Y axis.

 b. **Example.*** In a study of the relationship between plasma amphetamine levels and amphetamine-induced psychosis, 10 chronic amphetamine abusers underwent psychiatric evaluation and were assigned a psychosis intensity score. At the same time, plasma

*This example is adapted from Duncan RC, Knapp RG, Miller MC III: *Introductory Biostatistics for the Health Sciences*, 2nd ed. Albany, Delmar, 1983, p 133.

Table 15-1. Tabular Representation of Data for Correlation or Regression Analysis

Subject	Measurement on Variable 1 (*X*)	Measurement on Variable 2 (*Y*)
1	X_1	Y_1
2	X_2	Y_2
3	X_3	Y_3
4	X_4	Y_4
:	:	:
n	X_n	Y_n

amphetamine levels in these patients were determined. The results are shown in Table 15-2; Figure 15-2 is the scatter diagram of this data.

c. **The relationship** between *X* and *Y* may be described by a straight line or by a more complex curvilinear relationship. Alternatively, the scatter diagram may show that the two variables are unrelated.

   (1) **Linear relationships** may be either positive (i.e., as the values of one variable increase, the values of the other variable increase as well) or negative (i.e., as the values of one variable increase, the values of the other decrease). For example, the relationship between plasma amphetamine level and psychosis intensity is a positive linear relationship (see Figure 15-2).

Table 15-2. Psychosis Intensity Scores and Plasma Amphetamine Levels for 10 Chronic Amphetamine Abusers

Patient	Psychosis Intensity Score (*Y*)	Plasma Amphetamine mg/ml (*X*)
1	10	150
2	30	300
3	20	250
4	15	150
5	45	450
6	35	400
7	50	425
8	15	200
9	40	350
10	55	475

Adapted from Duncan RC, Knapp RG, Miller MC III: *Introductory Biostatistics for the Health Sciences*, 2nd ed. Albany, Delmar, 1983, p 133.

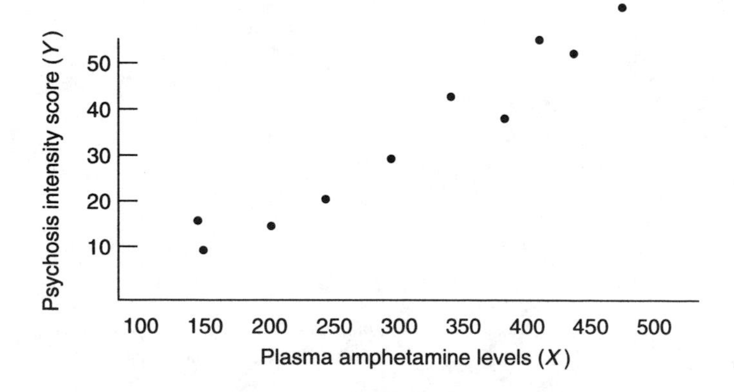

Figure 15-2. Scatter diagram relating psychosis intensity to plasma amphetamine levels in 10 chronic amphetamine abusers.

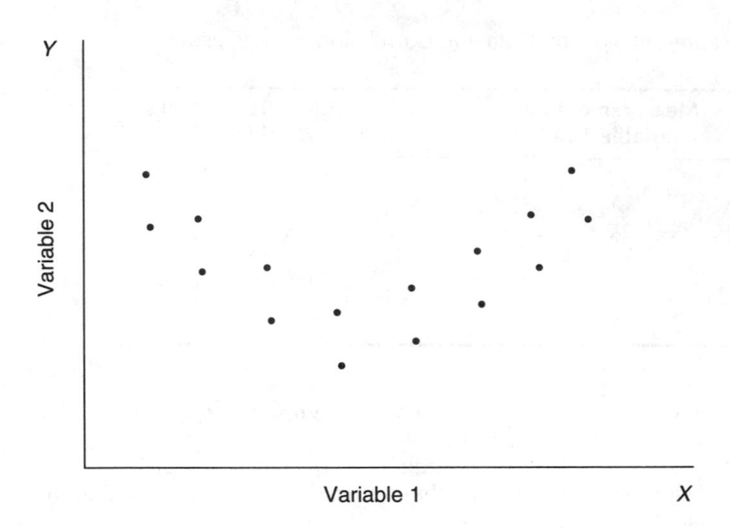

Figure 15-3. Scatter diagram illustrating a nonlinear relationship between two study variables.

(2) **The relationship between X and Y may be nonlinear, or curvilinear** (Figure 15-3). For example, the relationship between age and death rate is nonlinear. During the neonatal period, the death rate is high; it decreases, becomes relatively stable through middle age, and then rises again during old age.

(3) **When X and Y are unrelated,** the data pairs are randomly distributed (Figure 15-4). For example, no relationship, either linear or nonlinear, exists between foot size and IQ.

6. **Adapting nonlinear data.** Both correlation and regression analysis assume the existence of a linear relationship between the two study variables. When the data do not conform to this requirement, the investigator has two options.
   a. **Linear transformation of the data.** In some cases, a linear relationship can be created by an appropriate transformation of the data. For example, in pharmacological studies, the dose–response relationship is typically curvilinear, while the corresponding log dose–response relationship is linear. The common logarithm, natural logarithm, and square root are commonly used transformations.
   b. **Nonlinear regression analysis.** Statistical methods that describe a curvilinear relationship between study variables (nonlinear regression methods) exist. However, because these methods are considerably more complicated, every effort should be made to identify a suitable linear transformation of the data.

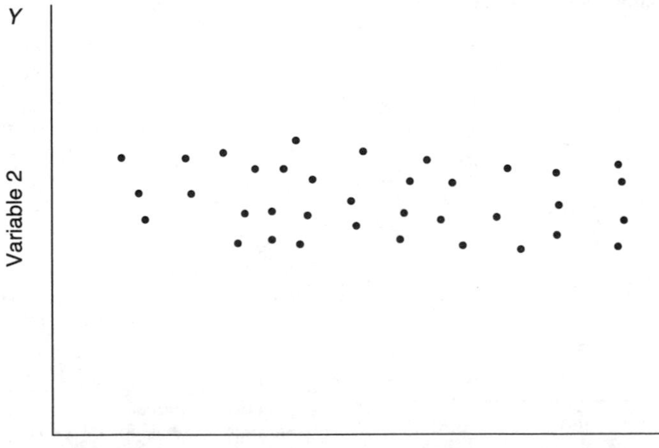

Figure 15-4. Scatter diagram from a study of two unrelated variables.

II. CORRELATION ANALYSIS

A. Assessing the strength of the association between two variables. The **correlation coefficient,** denoted symbolically as ***r***, defines both the **strength** and the **direction** of the linear relationship between two variables.

1. Characteristics of the correlation coefficient

a. The correlation coefficient is an index number between −1 and +1.

(1) When *r* = −1, the two variables have a **perfect negative linear relationship.** In this case, all points in the scatter diagram fall exactly on a straight line that slopes downward from left to right.

(2) When *r* = +1, the study variables have a **perfect positive linear relationship.** In this case, all points in the scatter diagram fall exactly on a straight line that slopes upward from left to right.

(3) When *r* = 0, there is not a linear relationship between the study variables. The relationship may be **nonlinear** (see Figure 15-3); alternatively, the study variables may be **unrelated** (see Figure 15-4). For example, in Figure 15-4, the relationship between the two variables is described by a line whose slope is equal to 0, that is, a change in one variable has no effect on the other.

b. The better the points on the scatter diagram approximate a straight line, **the greater the magnitude of *r*.**

c. The correlation coefficient calculated for a **sample** drawn from a population of interest (***r***) is an **estimate of the population correlation coefficient,** denoted symbolically as ***ρ***. The population correlation coefficient is a measure of the linear association between the study variables for all members of the population. In other words, ***r* is the statistic** that estimates the **population parameter *ρ*.**

2. Calculation

a. Formulas

(1) The correlation coefficient, r, is defined by the formula

$$r = \frac{\frac{1}{n}\sum_{i=1}^{n}(X_i - \bar{X})(Y_i - \bar{Y})}{s_x s_y} \qquad (15.1)$$

where $\bar{X}$ is the mean of the observed values for the X variable, $\bar{Y}$ is the mean of the observed values of the Y variable, s_x is the standard deviation of the values of the X variable, s_y is the standard deviation of the values of the Y variable, and n is the number of pairs of measurements. Thus,

$$\sum_{i=1}^{n}(X_i - \bar{X})(Y_i - \bar{Y}) = (X_1 - \bar{X})(Y_1 - \bar{Y}) + (X_2 - \bar{X})(Y_2 - \bar{Y}) + \cdots + (X_n - \bar{X})(Y_n - \bar{Y})$$

(a) The correlation coefficient is a **dimensionless value** (the units of measurement in the numerator and denominator cancel). Thus, r is unaffected by changes in the units, provided the measurements are made on the same subjects. For example, if the units of weight and height are changed from pounds and inches to grams and centimeters, r remains the same.

(b) When the relationship between X and Y is **positive,** values of X above their mean $\bar{X}$ tend to be paired with above average values of Y, while values of X below $\bar{X}$ tend to be paired with below average values of Y. Therefore, the product $(X_i - \bar{X})(Y_i - \bar{Y})$ is more often positive than negative, leading to a positive value for r (note that s_x and s_y are **always** positive).

(c) When the relationship between the X and Y variables is **negative**, values of X above the mean $\bar{X}$ tend to be paired with values of Y below their mean $\bar{Y}$. Therefore, the product $(X_i - \bar{X})(Y_i - \bar{Y})$ is more often negative than positive, leading to a negative value of r.

(2) For calculational purposes, formula 15.1 may be rewritten:

$$r = \frac{S_{xy}}{\sqrt{(S_{xx})(S_{yy})}} \qquad (15.2)$$

where

$$S_{xy} = \sum_{i=1}^{n} X_i Y_i - \frac{\sum_{i=1}^{n} X_i \sum_{i=1}^{n} Y_i}{n} \qquad (15.3)$$

$$S_{xx} = \sum_{i=1}^{n} X_i{}^2 - \frac{\left(\sum X_i\right)^2}{n} \qquad (15.4)$$

$$S_{yy} = \sum_{i=1}^{n} Y_i{}^2 - \frac{\left(\sum Y_i\right)^2}{n} \qquad (15.5)$$

b. Example

(1) Using the data in Table 15-2 and assuming Y = psychosis intensity score and X = plasma amphetamine level:

$$\sum Y_i = 315 \qquad \sum X_i = 3150$$

$$\sum Y_i{}^2 = 12{,}225 \qquad \sum X_i{}^2 = 1{,}128{,}750$$

$$\overline{Y} = 31.5 \qquad \overline{X} = 315$$

$$s_y = 15.995 \qquad s_x = 123.153$$

$$\sum X_i Y_i = 116{,}375$$

(2) Substituting in formulas 15.3, 15.4, and 15.5,

$$S_{xy} = \sum XY - \frac{\left(\sum X\right)\left(\sum Y\right)}{n}$$

$$= 116{,}375 - \frac{(315)(3150)}{10}$$

$$= 17{,}150$$

$$S_{yy} = \sum Y^2 - \frac{\left(\sum Y\right)^2}{n}$$

$$= 12{,}225 - \frac{(315)^2}{10}$$

$$= 2302.5$$

$$S_{xx} = \sum X^2 - \frac{\left(\sum X\right)^2}{n}$$

$$= 1{,}128{,}750 - \frac{(3150)^2}{10}$$

$$= 136{,}500$$

(3) Substituting these values in formula 15.2,

$$r = \frac{S_{xy}}{\sqrt{(S_{xx})(S_{yy})}}$$

$$= \frac{17{,}150}{\sqrt{(136{,}500)(2302.5)}}$$

$$= .97$$

(4) Thus, the correlation between psychosis intensity and plasma amphetamine level is $r = .97$, a nearly perfect positive correlation.

B. Assessing the statistical significance of an association

1. Using an appropriate statistical test, it is possible to address the question "Is there a statistically significant linear relationship between two study variables in the **population** from which the samples were selected?" As with other statistical tests of hypothesis, the answer to this question revolves around the most likely explanation for the disparity between the sample estimate (i.e., the sample correlation coefficient r) and its corresponding population parameter (i.e., the population correlation coefficient ρ).

2. **Example.** The psychiatrist conducting the study of the relationship between amphetamine-induced psychosis and plasma amphetamine levels described in I C 5 b wishes to determine whether a statistically significant association between these variables exists in the population from which the sample was selected. That is, he wishes to know whether it is likely that a correlation coefficient of the magnitude $r = .97$ or greater would be obtained from a sample of 10 subjects by random chance when the population correlation coefficient (ρ) is equal to 0. The investigator follows the five steps of statistical hypothesis testing outlined in Chapter 12 IV.

a. **State the hypothesis.** The null hypothesis (H_O) is: "Plasma amphetamine levels are not linearly related to psychosis intensity in the population of patients from which the 10 study participants were selected (i.e., $\mathbf{H_O: \rho = 0}$)," and the alternative hypothesis (H_A) is: "There is a linear relationship between plasma amphetamine levels and psychosis intensity in the population from which the study subjects were selected (i.e., $\mathbf{H_A: \rho \neq 0}$)."

b. **Select a sample and collect data** (see Table 15-2 and Figure 15-2).

c. **Calculate the test statistic.** The test statistic measures the disparity between the observed **sample** correlation coefficient ($r = .97$) and the value assumed for the **population** correlation coefficient by H_O ($\rho = 0$). The test statistic for this example is the sample correlation coefficient, $r = .97$. [Alternatively, both the test statistic $t = r\sqrt{(n-2)(1-r^2)}$ may be calculated and statistical significance assessed using a t test (see Ch 16).]

d. **Evaluate the evidence against H_O**

(1) **Frequency distribution of the test statistic.** The frequency distribution of all possible values of the test statistic when H_O is true is provided by Table H in Appendix 3. The number of degrees of freedom (df) associated with the frequency distribution of the test statistic is $\mathbf{df = n - 2}$, where n equals the number of pairs of observations. Here, $df = 10 - 2 = 8$.

(2) **Calculating the *p*-value.** The p-value is the probability of obtaining the calculated value of the test statistic by random chance when H_O is true. In Table H (see Appendix 3), the value for $\alpha = .01$, $df = 8$ is .7646. Since $r = .97$ falls to the right of .7646, $p < .01$. Hence, in this example, the probability of obtaining a sample correlation coefficient equal to or greater than $r = .97$ from a population with a correlation coefficient $\rho = 0$ is less than 1%.

(3) **Decision rule.** Prior to data collection, the psychiatrist chooses a 5% level of significance ($\alpha = .05$). Based on this level of significance, he derives the following decision rule: "**If $p < .05$, reject H_O.**"

e. **State the conclusion**

(1) Because, in this example, the calculated value of r was .97, the following can be concluded.

(a) **Statistical conclusion.** Since $p \leq .05$, H_O is rejected at the $\alpha = .05$ level of significance.

(b) **Clinical interpretation.** There is a statistically significant association between plasma amphetamine level and psychosis intensity in the population of patients from which the 10 study subjects were selected ($p < .01$).

(c) **Chance of error.** When H_O is rejected, the chance of error (the type I error rate) is given by the p-value or by the level of significance, α. In this case, the chance of error associated with concluding that the population correlation coefficient $\rho \neq 0$ (i.e., the chance that a **sample** correlation coefficient equal to or greater than .97 would be obtained by random chance from a population whose correlation coefficient $\rho = 0$) is less than 1% ($p < .01$).

(2) If, however, the calculated value of r had been .48 (i.e., $p > .05$), the following could be concluded.

(a) **Statistical conclusion.** In this instance, since $p > \alpha$, H_O is not rejected at the $\alpha = .05$ level of significance.

(b) **Clinical interpretation.** The results are not statistically significant ($p > .05$); the data do not support the existence of a significant linear association between plasma amphetamine levels and psychosis intensity at the 5% level of significance in the study population.

(c) **Chance of error.** When H_O is not rejected, the chance of error is β, the type II error rate:

$$\beta = P(H_O \text{ not rejected} | H_O \text{ false})$$
$$= P(\text{fail to detect statistically significant association} | \text{association exists})$$

Information on β is unavailable for this example.

3. **Power of the test.** The power of the statistical test is defined as the probability that the test will detect a statistically significant association between the two study variables, given that such a difference actually exists.

a. The value of power depends on the effect size, the sample size, and the chosen level of significance (see Ch 12 VI A).

b. A failure to reject H_O may be due to low power of the statistical test, as well as to the absence of an association between the study variables. When power is low, the statistical test may simply be too weak to detect an association of the specified magnitude in the population from which the study sample was drawn.

c. For large sample sizes, power is typically high and small deviations of ρ from 0 (i.e., small effect sizes) can be detected by the statistical test. Therefore, an estimate of the **magnitude of the population correlation coefficient ρ** is an important adjunct to the demonstration of statistical significance. This estimate is provided by r, the sample correlation coefficient, and r^2, the coefficient of determination.

C. **Coefficient of determination (r^2)**

1. **Definition.** The coefficient of determination, r^2, measures the **proportion of the variation** in one variable **that can be attributed** to, or explained by, variation in the second variable. That is, r^2 is that proportion of the variance in one variable that can be explained by its linear relationship to the other.

a. The coefficient of determination, therefore, is the counterpart of attributable risk for interval/ratio data (see Ch 14 III A).

b. While r^2 defines the **magnitude** of the association, it does not define the **direction.** The sign of r indicates whether the association is positive or negative.

2. **Characteristics of r^2**

a. **When $r^2 = 0$** (r, of course, also equals 0), none of the variation in Y can be attributed to changes in X.

b. **When $r^2 = 1$,** all of the variation in Y is attributable to its linear relationship with X.

3. **Graphic representation (Figure 15-5)**

a. Figure 15-5A illustrates a perfect positive correlation (i.e., $r = +1$ and $r^2 = 1$, or 100%). The linear relationship between X and Y accounts for all of the observed variation in Y. There is no variation in the Y values for a fixed value of X because all of the points representing the data pairs fall on a straight line.

b. In Figure 15-5B, there is variation among the Y values at fixed values of X. This variation in Y may be attributed in part to its linear relationship with X. Other factors—including **systematic variation** resulting from the relationship between Y and other unknown variables, and **random subject-to-subject variation**—account for the remainder of the variation. By specifying the proportion of the total variation in Y that is attributable to its linear relationship with X, r^2 provides a mathematical tool for separating these sources of variation.

4. **When r is statistically significant, r^2 is as well.** It may be concluded that the proportion of variation in Y that is attributable to X differs significantly from 0 in the **population** from which the sample was drawn.

5. **Example.** In the study of the association between amphetamine-induced psychosis and plasma amphetamine levels, $r = .97$ and $r^2 = (.97)^2 = .94$. That is, 94% of the variation in psychosis intensity can be attributed to variations in plasma amphetamine level.

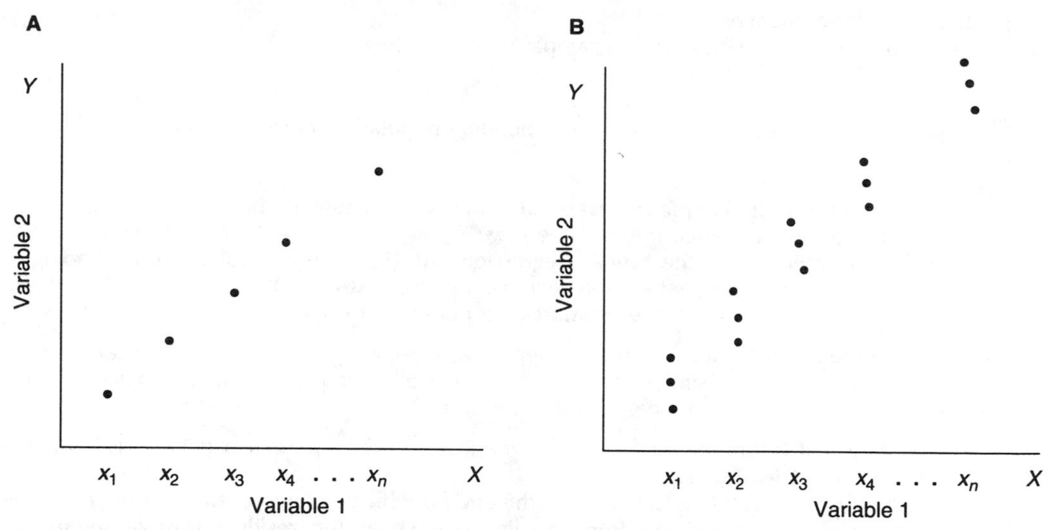

Figure 15-5. Graphic representation of r^2. In (*A*), the linear relationship between *X* and *Y* accounts for *all* of the observed variation in *Y*, and $r^2 = 1$. In (*B*), the variation in *Y* is only partially attributable to the linear relationship between *X* and *Y*; this proportion of the total variation is described by r^2, which lies between 0 and 1.

D. The proper interpretation of *r*

1. **Correlation does not imply causality.** The existence of a statistically significant correlation between study variables **does not** prove that a cause-and-effect relationship exists between them.

2. **A statistically significant correlation** between two study variables **does not imply that the association is clinically important.**
   a. Statistical significance merely indicates that the calculated value of *r* is unlikely to have resulted from random chance when the population correlation coefficient $\rho = 0$.
   b. When the sample size is large, the correlation may achieve statistical significance even though the actual deviation of ρ from 0 is small.
   c. The size of the *p*-value indicates the likelihood that an association exists; it does not specify the magnitude of that association. **The value of *r* (or, preferably, of r^2) is the best indicator of the magnitude of the association in the population.**

3. **Failure to demonstrate the statistical significance** of a given value of *r* may be due to:
   a. The absence of a linear relationship between *X* and *Y* (i.e., H_O is true and the population correlation coefficient $\rho = 0$)
   b. Low power of the statistical test

4. **Nonsensical,** or **spurious, correlations** may be obtained when **averages** or **aggregate data for groups of subjects** are substituted for pairs of measurements on individual subjects (known as the **ecological fallacy**).
   a. **Example.** A study examining the relationship between aggregate mortality from coronary artery disease in 18 countries and the average wine consumption in each country reported an unexpectedly strong negative correlation between the two variables.
   b. Correlations derived from such aggregate studies may disappear when data for individual subjects are analyzed.

III. REGRESSION ANALYSIS

A. Derivation of the regression line

1. **The goal of regression analysis** is to derive a linear equation that best fits a set of data pairs (X_i, Y_i), represented as points on a scatter diagram. This equation can be used to **predict** values of the response variable (*Y*) for given values of the independent variable (*X*).

2. Form of the equation

a. The general equation for the **sample regression line** is

$$\hat{Y} = b_0 + b_1 X$$

b. The general equation for the corresponding **population regression line** is

$$Y = \beta_0 + \beta_1 X$$

c. The **slope of the sample regression line (b_1)** is the change in the average value of Y for every one unit change in the value of X.

d. The ***Y* intercept of the sample regression line (b_0)** is the Y value corresponding to $X = 0$; it is the point where the regression line crosses the Y axis.

e. The sample regression line **estimates** the population regression line.

3. Graphic representation. Figure 15-6 depicts a regression line for a hypothetical group of four data points. No single line passes through all four points simultaneously, but the regression line represents the "best fit" to the data.

4. The statistical technique for finding the line that best fits a particular data set is known as the **method of least squares.**

a. The "best-fitting" line is defined as **the one for which the sum of the squared distances** of all of the data points from the line (known as the **residual sum of squares) is minimized.**

(1) The vertical deviation of any point (X_i, Y_i) from the regression line (the **estimated error** or **residual**), is defined as

$$\begin{aligned} d_i &= Y_i - \hat{Y}_i \\ &= Y_i - (b_0 + b_1 X_i) \end{aligned} \qquad (15.6)$$

where Y_i denotes the observed value of Y at the given value of X and $\hat{Y}_i$ represents the corresponding value of Y derived from the regression equation.

(2) Therefore, the sum of the squared distances of each of the points from the line is

$$\begin{aligned} \sum d_i^2 &= \sum (Y_i - \hat{Y}_i)^2 \\ &= \sum (Y_i - b_0 - b_1 X_i)^2 \end{aligned}$$

(3) The values of b_0 (Y intercept) and b_1 (slope) that describe the line with the smallest residual sum of squares (i.e., the line for which $\Sigma\, d_i^2$ is minimized) are obtained using the principles of calculus.

b. Calculating the slope and intercept of the regression line

(1) The formula for the slope of the regression line is

$$b_1 = \frac{S_{xy}}{S_{xx}} \qquad (15.7)$$

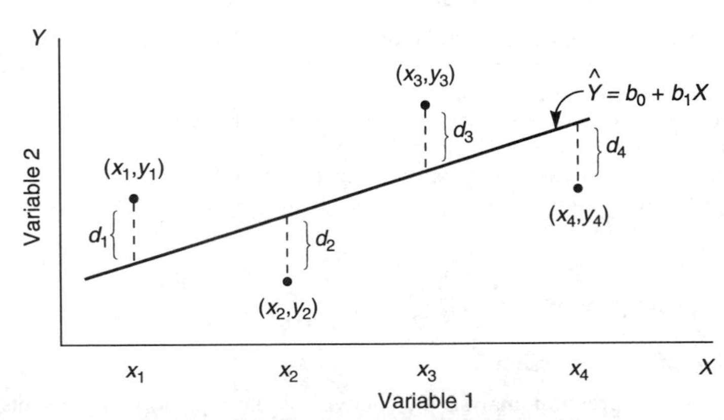

Figure 15-6. Graphic representation of the line that best fits four hypothetical data points (X_i, Y_i). d_i = the vertical deviation of the point (X_i, Y_i) from the line (the error or residual).

where S_{xy} and S_{xx} are given by formula 15.3 and formula 15.4, respectively [see II A 2 a (2)].

(2) The formula for the Y intercept of the regression line is

$$b_0 = \overline{Y} - b_1\overline{X} \quad (15.8)$$

where $\overline{X}$ represents the mean of the X values and $\overline{Y}$ the mean of the Y values.

c. Example. The psychiatrist conducting the investigation of the relationship between plasma amphetamine levels and amphetamine-induced psychosis described in I C 5 b (see Table 15-2) also examines this data using regression analysis. Using the values for S_{xy}, S_{xx}, $\overline{X}$, and $\overline{Y}$ computed during correlation analysis (see II A 2 b), the slope and Y intercept of the regression line are calculated as

$$b_1 = \frac{S_{xy}}{S_{xx}} = \frac{17{,}150}{136{,}500}$$

$$= .126$$

$$b_0 = \overline{Y} - b_1\overline{X}$$

$$= 31.5 - (.126)(315)$$

$$= -8.08$$

The regression equation relating plasma amphetamine level (X) to psychosis intensity (Y) is therefore

$$\hat{Y} = -8.08 + .126X$$

d. Describing the variation around the regression line

(1) The standard deviation of the data points about the regression line is called the **standard error of the estimate**, denoted symbolically $s_{y.x}$.

(2) The formula for computing the standard error of the estimate is

$$s_{y.x} = \sqrt{\frac{\text{SSE}}{n-2}} \quad (15.9)$$

where n = the number of (X_i, Y_i) pairs and SSE is the **error** or **residual sum of squares,** defined as

$$\text{SSE} = \sum_{i=1}^{n} d_i^2 = \sum_{i=1}^{n} (Y_i - \hat{Y}_i)^2$$

For computational purposes, this expression can be rewritten

$$\text{SSE} = S_{yy} - b_1 S_{xy} \quad (15.10)$$

where b_1 is the slope of the regression line and S_{yy} and S_{xy} are derived from formula 15.5 and formula 15.3, respectively.

(3) Example. For the regression line derived in III A 4 c (see Table 15-2),

$$\text{SSE} = S_{yy} - b_1 S_{xy}$$

$$= 2302.5 - (.126)(17{,}150)$$

$$= 147.756$$

Thus, using formula 15.9,

$$s_{y.x} = \sqrt{\frac{\text{SSE}}{n-2}}$$

$$= \sqrt{\frac{147.756}{8}}$$

$$= 4.3$$

The standard error of the estimate, $s_{y.x} = 4.3$, is the standard deviation of the observed psychosis intensity scores (Ys) around the regression line $\hat{Y} = -8.08 + .126X$.

B. Assessing the statistical significance of the linear relationship between X and Y

1. As in correlation analysis, it is possible to test the statistical significance of the linear relationship between two study variables in the **population** from which the samples were drawn. The research question in this instance is "Does the slope of the population regression line (β_1) differ from 0?"

2. **Example.** The statistical significance of the relationship between plasma amphetamine levels and psychosis intensity can be determined using the five steps of statistical hypothesis testing.

   a. **State the hypothesis.**

      (1) H_O states "There is no linear relationship between plasma amphetamine level (X) and psychosis intensity (Y) in the population of interest; the slope of the population regression line is 0 (i.e., $\mathbf{H_O}$**:** $\boldsymbol{\beta_1 = 0}$)."

      (2) H_A states "Plasma amphetamine level and psychosis intensity are linearly related in the population of interest; the slope of the population regression line differs significantly from 0 (i.e., $\mathbf{H_A}$**:** $\boldsymbol{\beta_1 \neq 0}$)."

   b. **Select a sample and collect data** (see Table 15-2). In III A 4 c, the **sample regression line** was shown to be

$$\hat{Y} = -8.08 + .126X$$

   The sample values $b_0 = -8.08$ and $b_1 = .126$ are estimates of the corresponding values of the population Y intercept β_0 and slope β_1.

   c. **Calculate the test statistic.**

      (1) The formula for computing the test statistic is derived from the general formula for a test statistic (see Ch 12 III D 2),

$$\text{test statistic} = \frac{\text{sample estimate} - \text{hypothesized population value}}{\text{standard deviation of sample estimate}}$$

$$t = \frac{b_1 - \beta_1}{\text{standard error of } b_1}$$

$$= \frac{b_1 - \beta_1}{s_{y.x}/\sqrt{S_{xx}}} \qquad (15.11)$$

      where the denominator (i.e., the **standard error of** $\boldsymbol{b_1}$), measures the spread of all possible values of the sample slope b_1 about the true population slope β_1.

      (2) Substituting the previously calculated values for b_1 (see III A 4 c), S_{xx} [see II A 2 b (2)], and $s_{y.x}$ [see III A 4 d (3)] in formula 15.11,

$$t = \frac{.126 - 0}{4.3/\sqrt{136{,}500}}$$

$$= 10.8$$

   d. **Evaluate the evidence against $\mathbf{H_O}$.**

      (1) **Frequency distribution of the test statistic.** The frequency distribution of the test statistic when H_O ($\beta_1 = 0$) is true is the ***t* distribution** (Table C, Appendix 3). The number of degrees of freedom is $df = n - 2$, where n is the number of (X_i, Y_i) pairs; for this example, $df = 10 - 2 = 8$.

      (2) **Calculating the *p*-value**

         (a) A test of hypothesis is either **one-tailed** or **two-tailed**, depending on whether H_A specifies a direction [i.e., whether H_A specifies $\beta_1 > 0$ (one-tailed), $\beta_1 < 0$ (one-tailed), or $\beta_1 \neq 0$ (two-tailed)].

         (b) Because H_A ($\beta_1 \neq 0$) does not specify a particular direction, this example uses a two-tailed test of significance.

         (c) Using Table C (see Appendix 3), with $df = 8$, the area to the right of the calculated test statistic $t = 10.8$ is found to be less than .005. This value is doubled to obtain the appropriate p-value for the two-tailed test. Thus, $p < .01$.

      (3) **Decision rule.** Based on a 5% level of significance ($\alpha = .05$), chosen prior to data collection, the investigator formulates the following decision rule: **"Reject $\mathbf{H_O}$ if $\boldsymbol{p \leq .05}$."**

e. State the conclusion.

(1) **Statistical conclusion.** Because $p \leq .05$, H_O is rejected at the $\alpha = .05$ level of significance.

(2) **Clinical interpretation.** A statistically significant linear relationship between plasma amphetamine levels and psychosis intensity exists in the population from which the sample of 10 patients was drawn. In other words, the population slope β_1 differs significantly from 0.

(3) **Chance of error.** When H_O is rejected, the chance of error is given by the p-value. In this instance, the chance that the test has detected a linear association between the two study variables when, in fact, such a relationship does not exist in the population is less than 1% ($p < .01$).

(4) If, rather than 10.8, the calculated value of the test statistic had been 1.3 ($p > .20$), H_O would not have been rejected (i.e., the existence of a linear relationship between the study variables cannot be demonstrated). The chance of error associated with this conclusion is β, the type II error rate.* The value of β is unknown in this example; hence, the failure to achieve statistical significance may be due to the absence of a linear relationship between study variables in the population or to low power of the statistical test.

C. Coefficient of determination (r^2)

1. Definition. The **coefficient of determination (r^2)** defines the proportion of the total variation in values of the dependent variable (Y) that can be attributed to its linear relationship with the independent variable (X).

2. Calculation

a. Formula. The coefficient of determination is calculated using the formula

$$r^2 = \frac{\text{sum of squares due to regression}}{\text{total variation in } Y}$$

$$= \frac{\text{SSR}}{\text{SST}}$$

where $\text{SSR} = b_1 S_{xy}$ and $\text{SST} = S_{yy}$. Note that SST = SSR + SSE and $\text{SSE} = \text{SST} - \text{SSR} = S_{yy} - b_1 S_{xy}$. Thus,

$$r^2 = \frac{(b_1)(S_{xy})}{S_{yy}} \quad (15.12)$$

b. Example. Substituting the previously computed values of b_1 (see III A 4 c), S_{xy}, and S_{yy} [see II A 2 b (2)] for the study of the association between plasma amphetamine levels and psychosis intensity in formula 15.12,

$$r^2 = \frac{(.126)(17{,}150)}{2302.5}$$

$$= .940$$

That is, 94% of the variation in psychosis intensity (Y) is attributable to the variation in plasma amphetamine level (X).

c. Note that the value of r^2 computed in this fashion is the same as that computed by squaring the value of the correlation coefficient r (see II C 5).

3. When β_1 differs significantly from 0, r^2 is also statistically significant (i.e., $r^2 \neq 0$).

D. Predicting *Y* from the regression line. Provided it can be demonstrated that the sample regression line $Y = b_0 + b_1 X$ **adequately describes the relationship between *X* and *Y*,** this equation can be used to predict the value of the response variable Y for a specified value of the independent variable X. For example, a regression equation describing the linear relationship between MCAT scores (X) and freshman GPA (Y) at a certain medical school can be used to predict the academic performance of an incoming student.

---

*The Greek letter β (with no subscript), used to denote the probability of a type II error, should not be confused with β_1, the slope of the population regression line, or β_0, the Y intercept of this line.

1. **Evaluating the adequacy of the regression line.** Two criteria determine the suitability of the regression equation for predicting values of the dependent variable.
   a. A statistically significant linear relationship between X and Y can be demonstrated in the population from which the sample was drawn.
   b. A sufficiently large proportion of the variation in Y can be accounted for by its linear relationship to X. The value of r^2, the coefficient of determination, provides this information.

2. **Example**
   a. In III B 2, the association between plasma amphetamine level and psychosis intensity in the population from which 10 study subjects were drawn was shown to be statistically significant ($p < .01$).
   b. The coefficient of determination for this study ($r^2 = .94$; see III C) indicates that 94% of the variation in psychosis intensity (Y) can be attributed to variation in plasma amphetamine level (X).
   c. Hence, according to the two criteria for evaluating the adequacy of the regression line, the sample regression equation $\hat{Y} = -8.08 + .126X$ appears suitable for predicting psychosis intensity from a given plasma amphetamine level.
   d. A new patient has a plasma amphetamine level of 410 mg/ml. Her **predicted** psychosis intensity score is

$$\begin{aligned}\hat{Y} &= -8.08 + .126X \\ &= -8.08 + (.126)(410) \\ &= 43.4\end{aligned}$$

E. **Using regression analysis correctly**

1. The guidelines for the proper use of correlation techniques are also applicable to regression analysis (see II D).

2. The sample regression equation **should not be used to predict values of *Y* outside the range of values for *X* observed in the sample.** Beyond this range, the nature of the relationship between the two variables is unknown. For example, a regression equation relating height (Y) and age (X) for a sample of boys between the ages of 6 and 18 could not be used to accurately predict the height of a 50-year-old man or an 18-year-old woman.

PROBLEMS*

15-1. A pharmaceutical company conducted a pilot study to evaluate the relationship between three doses of a new hypnotic agent and sleeping time. The results of this study are shown in the table below.

Subject	Sleeping Time (hrs; Y)	Dose (mM/kg; X)
1	4	3
2	6	3
3	5	3
4	9	10
5	8	10
6	7	10
7	13	15
8	11	15
9	9	15
	$\bar{Y} = 8.0$	$\bar{X} = 9.3$
	$\Sigma Y = 72.0$	$\Sigma X = 84.0$
	$\Sigma Y^2 = 642.0$	$\Sigma X^2 = 1002.0$
	$\Sigma XY = 780.0$	

Adapted from Duncan RC, Knapp RG, Miller MC III: *Introductory Biostatistics for the Health Sciences,* 2nd ed. Albany, Delmar, 1983, p 132.

a. What is the regression equation relating dose (X) to sleeping time (Y)?

b. Test the hypothesis that there is a statistically significant linear dose–response relationship at the $\alpha = .05$ level of significance.

c. What is the estimated sleeping time for a patient receiving 12 mM/kg of the new drug?

15-2. A clinical researcher studying the effects of stress on diastolic blood pressure subjected nine monkeys to increasing levels of electric shock for two minutes as they attempted to obtain food by pressing a bar. Diastolic blood pressure readings were recorded for all the monkeys at the end of this two-minute stress period and are summarized in the following table.

Subject	Shock intensity (μvolts; X)	Systolic Blood Pressure (mm Hg; Y)
1	30	125
2	30	130
3	30	120
4	50	150
5	50	145
6	50	160
7	70	175
8	70	180
9	70	180
	$\Sigma X = 450$	$\Sigma Y = 1365$
	$\bar{X} = 50$	$\bar{Y} = 151.6667$
	$\Sigma X^2 = 24{,}900$	$\Sigma Y^2 = 211{,}475$
	$\Sigma XY = 71{,}450$	
	$(\Sigma X)(\Sigma Y) = 614{,}250$	

Adapted from Duncan RC, Knapp RG, Miller MC III: *Introductory Biostatistics for the Health Sciences,* 2nd ed. Albany, Delmar, 1983, p 133.

*The problems in this section are adapted from Duncan RC, Knapp RG, Miller MC III: *Introductory Biostatistics for the Health Sciences,* 2nd ed. Albany, Delmar, 1983, pp 132–134.

a. What is the equation for the regression line relating blood pressure to shock intensity?

b. Does a significant linear relationship exist between diastolic blood pressure and shock intensity (use $\alpha = .05$)?

c. Calculate and interpret r^2.

d. What is the predicted diastolic blood pressure reading for a 60 μvolt shock?

15-3. The table below lists grade point averages (GPA) at the end of the first two years of basic science instruction and scores on Part I of the National Board Examination for 12 medical students.

Student	GPA (X)	Score On National Board Exam (Y)
1	3.35	620
2	2.37	445
3	3.13	445
4	3.10	560
5	1.94	295
6	3.00	570
7	2.85	415
8	1.96	430
9	2.98	560
10	2.55	515
11	2.23	430
12	1.95	435
	$\overline{X} = 2.6175$	$\overline{Y} = 476.6667$
	$\Sigma X = 31.41$	$\Sigma Y = 5720$
	$\Sigma X^2 = 85.1323$	$\Sigma Y^2 = 2{,}816{,}050$
	$\Sigma XY = 15{,}357.55$	

Adapted from Duncan RC, Knapp RG, Miller MC III: *Introductory Biostatistics for the Health Sciences*, 2nd ed. Albany, Delmar, 1983, p 134.

a. What is the correlation between GPA and examination score?

b. Calculate and interpret r^2.

c. Is the correlation coefficient computed in problem 15-3a statistically significant at the $\alpha = .05$ level of significance? Report and interpret the p-value.

d. What is the equation for the regression of GPA on examination score?

e. What is the predicted examination score for a student with a GPA = 3.0?

Solutions on p 394.

STUDY QUESTIONS

Directions: Each of the numbered items or incomplete statements in this section is followed by answers or by completions of the statement. Select the **one** lettered answer or completion that is **best** in each case.

Questions 1–3

Researchers at a certain medical school conduct a study of 100 students at their institution to examine the association between grade point average (GPA) and a number of other potential indicators of academic performance. The table below lists the correlation coefficient and the associated *p*-value for the relationship between medical school GPA and each of these factors.

	r	*p*-value
SAT Score	−.08	$p > .10$
MCAT Score	.40	$p < .001$
Undergraduate Nonscience GPA	−.11	$p > .10$
Undergraduate Science GPA	.20	$p < .05$
Interview Score	.25	$p < .01$

1. Which of the following statements best describes the relationship between medical school GPA and MCAT scores?

(A) The observed association between these two variables is likely to be the result of random chance
(B) An improvement in an applicant's MCAT score will translate into improved academic performance in medical school
(C) Students with high medical school GPAs also tend to have high MCAT scores
(D) Students with high MCAT scores will excel in medical school
(E) MCAT scores are unrelated to academic performance in medical school

2. Which of the following statements best describes the relationship between medical school GPA and SAT scores?

(A) The population correlation coefficient (ρ) differs significantly from zero
(B) An improvement in an applicant's SAT scores will lead to diminished academic performance in medical school
(C) There is a statistically significant negative association between medical school GPA and SAT scores
(D) The observed correlation between medical school GPA and SAT scores is likely to be attributable to random chance
(E) High SAT scores indicate that a student will perform well in medical school

3. Which variable appears to be causally related to medical school GPA?

(A) MCAT score
(B) Interview score
(C) Undergraduate science GPA
(D) None of the above

1-C
2-D
3-D

Questions 4–8

For each of 10 patients with emphysema, a clinical researcher tabulated the number of years the patient smoked and the attending physician's subjective evaluation of the extent of lung damage (measured on a scale of 0 to 100). She determined that the regression equation describing the relationship between the number of years the patient smoked (X) and the extent of lung damage (Y) was

$$\hat{Y} = 11.24 + 1.31X$$

4. All of the following statements about H_O in a statistical test of the association between smoking duration and lung damage are true EXCEPT

(A) H_O may be written, "The slope of the population regression line is zero."
(B) H_O may be written, "There is a statistically significant association between smoking duration and the extent of lung damage for the sample of 10 patients."
(C) H_O may be written, "No linear relationship exists between smoking duration and lung damage in the population of patients from which the sample was drawn."
(D) H_O may be written, "$\beta_1 = 0$."

5. Given that the calculated test statistic is 4.9 and the tabulated critical value for $\alpha = .05$ is 3.18, the most appropriate conclusion about the relationship between smoking duration and the extent of lung damage is

(A) there is no linear relationship between the extent of lung damage and the number of years a patient has smoked ($p > .05$)
(B) there is a statistically significant relationship between the extent of lung damage and smoking duration ($p > .05$)
(C) there is a statistically significant relationship between the extent of lung damage and smoking duration ($p < .05$)
(D) the observed association between lung damage and the number of years a patient has smoked is likely to be due to random chance

6. What is the predicted extent of lung damage for a patient who has smoked for 30 years?

(A) 39.3
(B) 54.0
(C) 30.0
(D) 50.5
(E) This cannot be determined from the data

7. A statistically significant positive correlation ($p < .05$) between the number of years an emphysema patient has smoked and the extent of lung damage indicates which of the following?

(A) The longer a given patient has smoked, the more likely he or she is to develop emphysema
(B) Patients who have smoked longer tend to incur a greater degree of lung damage
(C) Smoking for a longer period of time causes increased lung damage
(D) The observed association between the extent of lung damage and smoking duration is likely to be the result of random chance
(E) There is a strong association between the extent of lung damage and smoking duration among emphysema patients

8. A postdoctoral associate repeats this study and observes a coefficient of determination, r^2, equal to .25. Based on this observation, all of the following statements about the relationship between smoking duration and the extent of lung damage are true EXCEPT

(A) the association between these two study variables is statistically significant
(B) the correlation coefficient for this relationship is equal to .50
(C) the sample of emphysema patients may have been drawn from a population having a correlation coefficient $\rho = 0$
(D) twenty-five percent of the variation in lung damage can be attributed to variation in smoking duration

---

4-B
5-C
6-D
7-B
8-A

9. A study is conducted to examine the relationship between daily caffeine intake and systolic blood pressure. The calculated correlation coefficient r measures which of the following?

(A) The extent to which caffeine intake and blood pressure are causally related
(B) The degree of association between the two study variables
(C) The likelihood that caffeine intake and systolic blood pressure are mutually exclusive
(D) The statistical significance of the association between daily caffeine intake and blood pressure

10. In an investigation of the relationship between study habits and test performance, the correlation coefficient relating the number of hours spent in the anatomy laboratory to scores on the final anatomy examination was computed as .914 ($p < .01$). What is the most appropriate conclusion that can be drawn from this result?

(A) Spending more time in the lab improves performance on the final exam
(B) Spending more time in the lab detracts from performance on the final exam
(C) Students who spend more time in the lab tend to do worse on the final exam
(D) Students who spend more time in the lab tend to do better on the final exam
(E) The observed association between time spent in the lab and test performance is likely to be due to random chance

---

9-B
10-D

ANSWERS AND EXPLANATIONS

1–3. The answers are: 1-C *[II A 1 a (2), B 2 d (2)]*, **2-D** *[II B 2 d (2)]*, **3-D** *[II D 1]*
Since the correlation coefficient, $r = .40$, is positive, the association between GPA and MCAT scores is also positive—that is, high medical school GPAs tend to be associated with high MCAT scores. Assuming a level of significance $\alpha = .05$, the p-value associated with this correlation (i.e., $p < .001$) indicates that the observed association between MCAT scores and medical school GPA is statistically significant. Thus, it is unlikely (a less than .1% chance) to be the result of random chance.

The p-value associated with the correlation between medical school GPA and SAT scores (i.e., $p > .10$) indicates that the correlation coefficient is not statistically significant. As a result, it is likely that a sample yielding a calculated correlation coefficient of this magnitude could be drawn by random chance from a population in which the correlation coefficient ρ is equal to zero, and it is impossible to conclude that ρ differs significantly from zero.

The existence of a statistically significant correlation between two variables cannot be equated with causality. An observed association, even if it is statistically significant, may be due to a biased comparison. Other unidentified variables may cause a given performance indicator and medical school GPA to "travel together"; that is, they may confound the observed association.

4–8. The answers are: 4-B *[III B 2 a (1)]*, **5-C** *[III B 2 d (2), e (1)]*, **6-D** *[III D]*, **7-B** *[III B 2 e (2); II D 1]*, **8-A** *[III C]*
The null hypothesis (H_O) is the hypothesis of "no association." In other words, it specifies that no linear relationship exists between the two study variables (i.e., the slope of the population regression line, β_1, is equal to zero). In contrast, the alternative hypothesis (H_A) specifies that a significant linear relationship does exist between the study variables.

Since the calculated test statistic, 4.9, exceeds the tabulated critical value of 3.18, the area to the right of the test statistic under the frequency distribution when H_O is true (i.e., the p-value) is less than .05. The p-value is less than the specified level of significance, $\alpha = .05$. Thus, H_O is rejected, indicating that a significant linear relationship exists in the study population between smoking duration and the extent of lung damage.

The value of Y (lung damage) can be predicted from the regression equation, setting $X = 30$ years,

$$\begin{aligned}\hat{Y} &= 11.24 + 1.31X \\ &= 11.24 + (1.31)(30) \\ &= 50.54, \text{ which can be rounded to } 50.5\end{aligned}$$

A statistically significant correlation indicates the existence of a linear relationship between two study variables that is unlikely to have occurred as a result of random chance. Statistical significance cannot be equated with causality. Therefore, it is impossible to infer that smoking for a longer period of time causes increased lung damage. The strength of the association cannot be deduced merely from a statistically significant correlation.

The coefficient of determination r^2 defines the amount of variation in Y that can be attributed to variation in X. It provides no information on statistical significance. This information is provided by the p-value, which measures the likelihood that the sample of emphysema patients was drawn from a population with a correlation coefficient $\rho = 0$. If $r^2 = .25$, r, the correlation coefficient, is equal to .50.

9. The answer is B *[II A]*
The value of the correlation coefficient indicates the degree of association between two study variables. A p-value (from an appropriate statistical test of hypothesis) is needed to determine if the association between the variables is statistically significant (i.e., unlikely to be due to random chance). Association does not imply causation.

10. The answer is D *[II A 1 a (2), B 2 e]*
The correlation coefficient is positive, indicating a positive association between the two study variables. Thus, students who spend more time in the anatomy lab tend to perform better on the final examination. A statistically significant correlation means that the observed association is unlikely to be the result of random chance, but it does not guarantee that the study variables are causally related.

16 Tests of Statistical Significance: Paired and Pooled *t* Tests

I. ANALYZING DIFFERENCES AMONG POPULATION MEANS. Procedures for evaluating the significance of differences among comparison groups are among the most commonly used statistical techniques in medical research. Such procedures vary according to the type of data being studied.

A. Categorical data. When the data are in the form of **counts** (i.e., frequency data), **chi-square** procedures are used to test for differences in population proportions (see Ch 13).

B. Continuous data. When a study involves **interval/ratio** (i.e., continuous) data and a comparison of only two groups, the ***t* test** is the appropriate statistical procedure for evaluating the significance of the difference between the two population means. Although the *t* test is most often associated with experimental studies, it may also be used with observational study designs.

II. TESTING THE EQUALITY OF TWO MEANS: INDEPENDENT SAMPLES

A. The pooled *t* test (sometimes referred to as the **student *t* test** or the **independent sample *t* test**) is used to assess the statistical significance of the difference between two population means in a study based on data obtained from **independent samples** (Figure 16-1).

1. The test assumes that values of the dependent variable follow a normal distribution and that the two populations have the same variance (i.e., $\sigma_1 = \sigma_2$).

2. In practice, the *t* test yields a valid result in most situations, unless the two variances differ considerably.

B. Example. A clinical researcher postulates that weight-bearing exercise prevents the development of osteoporosis by increasing the secretion of calcitonin, a hormone that inhibits bone resorption. He wishes to test this hypothesis by comparing blood levels of calcitonin in subjects who exercise to those in subjects who do not.

1. **Testing the hypothesis.** This research question can be examined using the standard five-step procedure for statistical hypothesis testing outlined in Chapter 12 IV.
   a. **State the hypothesis.** Two opposing hypotheses are devised.
      (1) The null hypothesis, H_O, is: "Mean blood calcitonin levels in the population of individuals who engage in weight-bearing exercise (μ_E) are equal to those in the population of individuals who do not (μ_C) [$\mathbf{H_O: \mu_E = \mu_C}$ (or $\mu_E - \mu_C = 0$)]."
      (2) The alternative hypothesis, H_A, is: "The mean blood calcitonin level in the population of individuals who engage in weight-bearing exercise is **greater than** that of individuals who do not [$\mathbf{H_A: \mu_E > \mu_C}$ (or $\mu_E - \mu_C > 0$)]." Because H_A specifies a direction, a one-tailed statistical test must be used (see Ch 13 II D 1).
   b. **Select a sample and collect data.** To control for any confounding age or sex differences in calcitonin secretion, the investigator limits the study to women between the ages of 25 and 40. He recruits 200 normal healthy women in this age range and randomly assigns 100 to the exercise group and 100 to the control group. Those assigned to the exercise group walk on a treadmill at a moderate pace for one hour, while control subjects rest quietly during this period. At the end of the hour, blood calcitonin levels are determined in both groups. Table 16-1 summarizes the results.

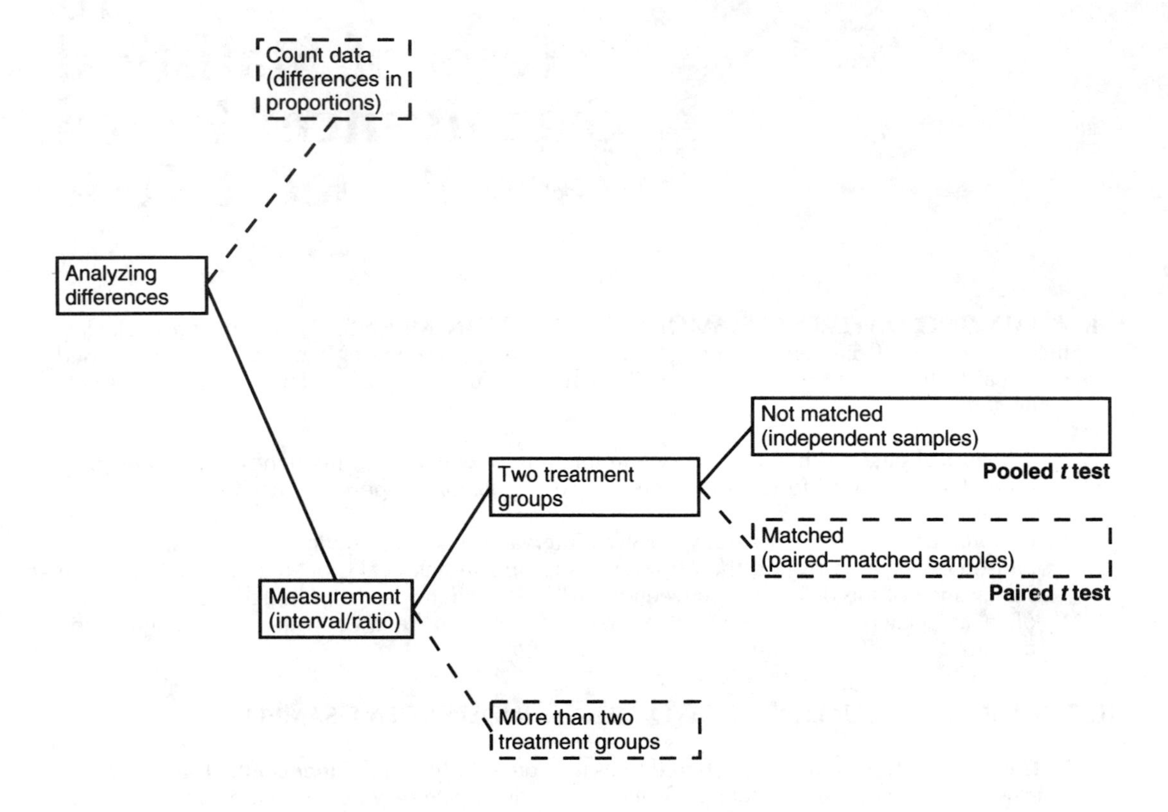

Figure 16-1. Decision tree pathways (derived from Figure 13-1) used to select a statistical procedure for analyzing the difference between two population means in a study using independent samples. *Dotted lines* indicate decision tree paths not taken.

c. **Calculate the test statistic.** The formula for t, the test statistic, is derived from the general formula for computing a test statistic (see Ch 12 III D 2):

$$\text{test statistic} = \frac{\text{sample estimate} - \text{hypothesized population value}}{\text{standard deviation of sample estimate}}$$

$$t = \frac{(\overline{Y}_1 - \overline{Y}_2) - (\mu_1 - \mu_2)}{s_{(\overline{Y}_1 - \overline{Y}_2)}} \tag{16.1}$$

$$= \frac{(\overline{Y}_1 - \overline{Y}_2) - 0}{\sqrt{\dfrac{s_p^{\,2}}{n_1} + \dfrac{s_p^{\,2}}{n_2}}} \tag{16.2}$$

Table 16-1. The Effect of Weight-bearing Exercise on Calcitonin Secretion (μg/dl) in Normal Women between the Ages of 25 and 40

	Exercise Group	**Control Group**
Sample Size	$n_1 = 100$	$n_2 = 100$
Sample Mean	$\overline{Y}_1 = 0.58$	$\overline{Y}_2 = 0.53$
Sample Standard Deviation	$s_1 = 0.21$	$s_2 = 0.19$

where $\overline{Y}_1 - \overline{Y}_2$ represents the difference of the two sample means, n_1 and n_2 represent the two sample sizes, and $s_{(\overline{Y}_1 - \overline{Y}_2)}$ is the standard error of the estimate, which measures the variation of all possible values of $(\overline{Y}_1 - \overline{Y}_2)$ about their center $(\mu_1 - \mu_2)$.

(1) The term s_p in formula 16.2 represents the **pooled standard deviation** (i.e., s_p is an estimate of the common population standard deviation σ, given that $\sigma_1 = \sigma_2 = \sigma$).

(a) The **pooled variance,** s_p^2, is computed using the formula:

$$s_p^2 = \frac{(n_1 - 1)s_1^2 + (n_2 - 1)s_2^2}{n_1 + n_2 - 2} \tag{16.3}$$

Thus, for this example, s_p^2 is

$$s_p^2 = \frac{(100 - 1)(.21)^2 + (100 - 1)(.19)^2}{100 + 100 - 2}$$

$$= .0401$$

and the standard error of the estimate is

$$s_{(\overline{Y}_1 - \overline{Y}_2)} = \sqrt{\frac{s_p^2}{n_1} + \frac{s_p^2}{n_2}}$$

$$= .028$$

(b) Note that when $n_1 = n_2$, s_p^2 is a simple arithmetic average of s_1^2 and s_2^2. If $n_1 = n_2$, s_p^2 is a **weighted** average, where n_1 and n_2 constitute the weights. The use of a weighted average allows the investigator to compensate for the greater contribution of the larger sample to the pooled variance.

(2) The value of the test statistic for the current example is obtained by substituting in formula 16.2:

$$t = \frac{(\overline{Y}_1 - \overline{Y}_2) - 0}{\sqrt{\frac{s_p^2}{n_1} + \frac{s_p^2}{n_2}}}$$

$$= \frac{.05 - 0}{.028}$$

$$= 1.78$$

d. Evaluate the evidence against H_O.

(1) Frequency distribution of the test statistic. When σ_1 and σ_2 are unknown, the test statistic follows the ***t* distribution** (Table C, Appendix 3), with $n_1 + n_2 - 2$ degrees of freedom (*df*). In this example, $df = 100 + 100 - 2 = 198$, which can be approximated by $df = 200$.

(2) The *p*-value, the probability of obtaining a test statistic equal to or greater than $t = 1.78$ by random chance when H_O is true, is the area to the right of 1.78 under the frequency distribution curve (Figure 16-2).

(a) For $df = 200$, Table C (see Appendix 3) gives the area to the right of $t = 1.6525$ as .05. Because the test statistic, $t = 1.78$, lies to the right of $t = 1.6525$ on the horizontal axis of the frequency distribution (see Figure 16-2), the area to the right of the test statistic is less than .05; that is, $p < .05$.

(b) If a two-tailed, rather than a one-tailed, hypothesis (i.e., H_A: $\mu_1 \neq \mu_2$) were being tested, the *p*-value would have to be doubled; in this instance, $p < .10$.

(3) Decision rule. Based on a 5% level of significance ($\alpha = .05$), chosen prior to data collection, the investigator formulates the following decision rule: "**If $p \leq .05$, reject H_O.**"

e. State the conclusion.

(1) Statistical conclusion. Since $p \leq .05$, H_O is rejected at the 5% level of significance; the difference between the two population means is statistically significant.

(2) Clinical interpretation. Mean blood levels of calcitonin are significantly higher in the population of individuals who engage in weight-bearing exercise for one hour than in the population of individuals who do not exercise ($p < .05$).

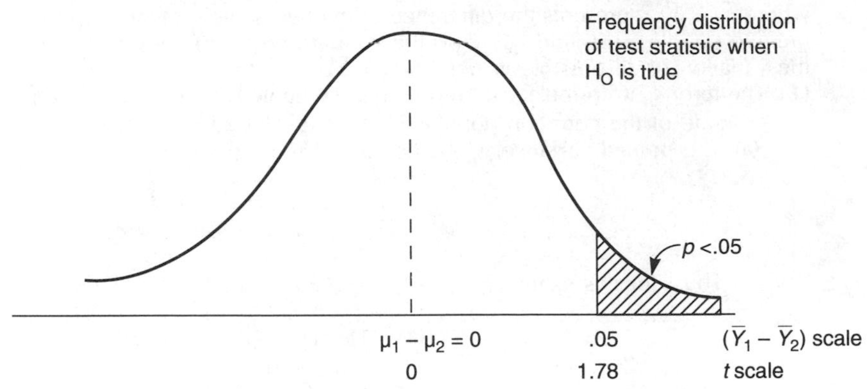

Figure 16-2. Graphic representation of the *p*-value in a study of the effect of weight-bearing exercise on blood calcitonin levels.

(3) **Chance of error.** The chance that the statistical test has led to the conclusion that $\mu_E > \mu_C$ (i.e., that H_A is true), when, in fact, the population means are equal (i.e., H_O is true) is 5% or less.

2. **Estimating the magnitude of the difference between the two means**
   a. **Statistical significance does not necessarily imply clinical relevance.** Even small differences between the two population means may prove significant when the statistical test is sufficiently powerful to detect this difference.
   b. **A 95% confidence interval** (see Ch 11) provides an estimate of the magnitude of the difference between the population means.
      (1) For this calcitonin study, the 95% confidence coefficient corresponds to $t_{(1-.05/2)} = t_{.975}$ with 200 (actually 198) degrees of freedom, given in Table C (see Appendix 3) as 1.9719.
      (2) Therefore, using the general formula for a confidence interval,

$$\text{endpoints of interval} = \text{estimate} \pm (\text{confidence coefficient} \times \text{standard error of estimate})$$
$$= (\bar{Y}_1 - \bar{Y}_2) \pm (1.9719) s_{(\bar{Y}_1 - \bar{Y}_2)}$$
$$= (.58 - .53) \pm (1.9719)(.028)$$
$$= 0.0 \text{ to } 0.1\ \mu g/dl$$

      (3) Thus, the investigator conducting this study can be 95% confident that the difference in mean calcitonin blood levels between the population of individuals who engage in weight-bearing exercise and the population of individuals who do not is included in the interval 0.0 to 0.1 μg/dl.

III. TESTING THE EQUALITY OF TWO MEANS: MATCHED SAMPLES

A. **The paired *t* test** is used to assess the statistical significance of the difference between two population means in a study involving **matched (paired) samples** (Figure 16-3).

1. **Study subjects** may be matched in three ways.
   a. Subjects may be **self-paired,** that is, they serve as their own controls, receiving the experimental treatment on one occasion and the control treatment on another.
   b. The subjects may be **twins or litter mates,** with one twin assigned to each comparison group.
   c. The subjects may be **"artificially" matched** with respect to one or more factors suspected of having an effect on the study variable. One member of the matched pair is assigned to each comparison group.
2. **Rationale for matching**
   a. In observational studies, matching is used **to control potentially confounding variables.**

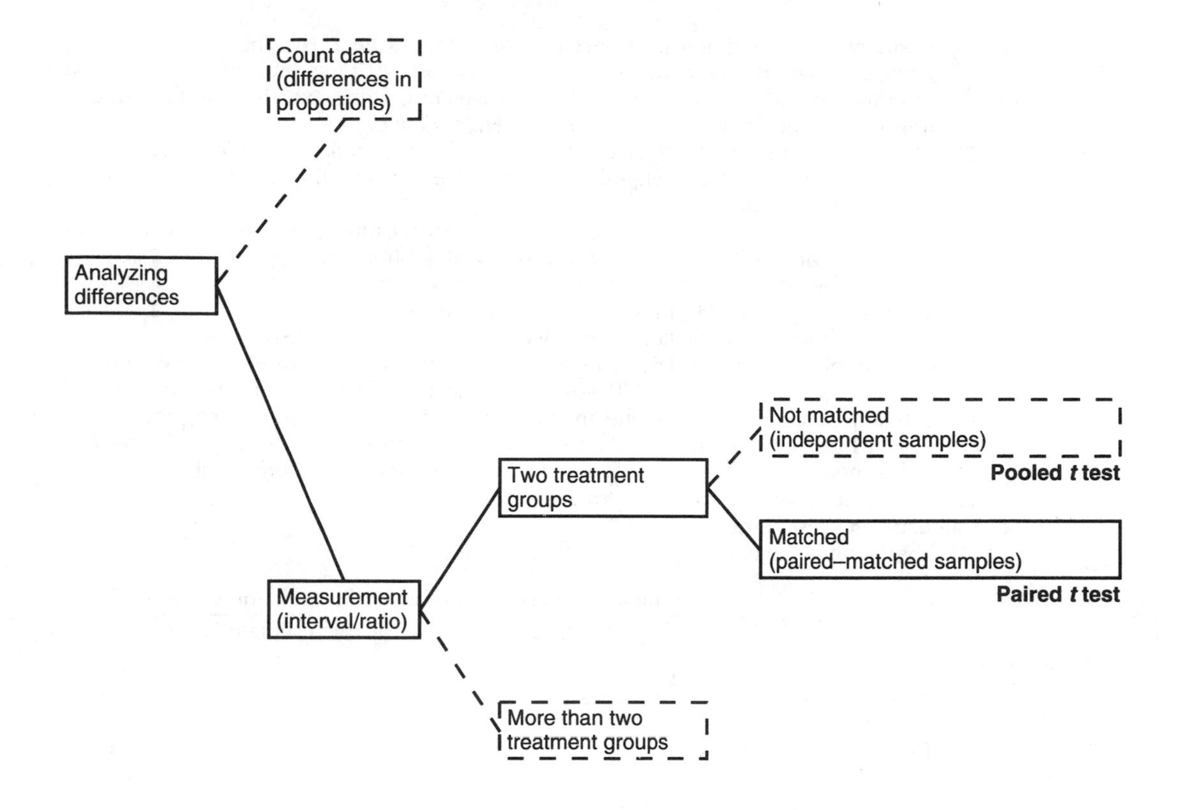

Figure 16-3. Selection of the paired *t* test as the appropriate statistical procedure for analyzing the difference between two population means in a study using matched samples. *Dotted lines* indicate decision tree paths not taken.

b. In **experimental studies,** matching is often used **to increase the precision of estimates and the power of statistical tests** (confounding variables are controlled through randomization).

(1) Matching **increases precision** by **decreasing the width of the confidence interval** estimate of a population parameter.

(2) Matching **increases both precision and power** by **reducing the standard error of the estimate**.

(a) The extent of this reduction is related to the **degree of association** between the paired measurements.

(b) Power is increased only when the reduction in the standard error is sufficient to offset the loss of degrees of freedom ($df = n - 1$ for the paired t test versus $df = n_1 + n_2 - 2$ for the independent sample t test).

B. Example. The researcher discussed in II B replicates his study of the effect of weight-bearing exercise on blood levels of calcitonin, using matched pairs of subjects to control for the potential confounding effect of age.

1. Testing the hypothesis. Again, the investigator tests the hypothesis by carrying out the five steps of statistical hypothesis testing.

a. State the hypothesis. In a paired study, the focus of attention is the population of all possible differences, $d_i = Y_{1i} - Y_{2i}$.

(1) When the mean blood calcitonin level of the population of individuals who engage in weight-bearing exercise equals that of the control (no exercise) population, the

mean of the population of differences, μ_D, should be zero. Thus, H_O would be written, "There is no difference in the mean blood calcitonin level of the population of individuals who engage in one hour of weight-bearing exercise and the population of individuals who do not," that is, $\mathbf{H_O: \mu_D = 0}$.

(2) $\mathbf{H_A}$ is: "The mean blood calcitonin level of the population of individuals who engage in one hour of weight-bearing exercise exceeds that of the population of individuals who do not."

(a) If, as in this example, H_A states that the mean of the first population is greater than that of the second, then the order of subtraction is $d_i = Y_{1i} - Y_{2i}$ and H_A may be written $\mathbf{H_A: \mu_D > 0}$.

(b) If $d_i = Y_{2i} - Y_{1i}$, H_A may be written $\mathbf{H_A: \mu_D < 0}$.

(c) The order of subtraction is irrelevant, but it must be consistent.

b. Select a sample and collect data. The investigator recruits 200 more healthy women to serve as subjects. From these 200 volunteers, he identifies 50 pairs of age-matched subjects and randomly assigns one member of each pair to the exercise group (who walk on a treadmill for one hour) and the other to the control group. As in the first study, he measures blood levels of calcitonin at the end of the one-hour test period. Table 16-2 summarizes the results of this study.

c. Calculate the test statistic.

(1) The formula for computing the test statistic is

$$\text{test statistic} = \frac{\text{sample estimate} - \text{hypothesized population value}}{\text{standard deviation of sample estimate}}$$

$$t = \frac{\bar{Y}_d - \mu_0}{s_d/\sqrt{n}} \tag{16.4}$$

(2) The standard deviation of the n differences (s_d) is obtained using the formula

$$s_d = \sqrt{\frac{\sum d_i^2 - \left(\sum d_i\right)^2 / n}{n-1}} \tag{16.5}$$

Here, $s_d = .156$

(3) Substituting in formula 16.4, the test statistic for this example is

$$t = \frac{\bar{Y}_d - \mu_0}{s_d/\sqrt{n}} = \frac{.05 - 0}{.156/\sqrt{50}} = 2.3$$

Table 16-2. The Effect of Weight-bearing Exercise on Calcitonin Secretion (μg/dl) in Age-matched Subjects

	Sample 1 (from population 1)	**Sample 2 (from population 2)**	**Sample of Differences (from population of differences)**
	Y_{11}	Y_{21}	$d_1 = Y_{11} - Y_{21}$
	Y_{12}	Y_{22}	$d_2 = Y_{12} - Y_{22}$
	.	.	.
	.	.	.
	Y_{1n}	Y_{2n}	$d_n = Y_{1n} - Y_{2n}$
Mean:	$\bar{Y}_1 = \sum Y_{1i}/n$	$\bar{Y}_2 = \sum Y_{2i}/n$	$\bar{Y}_d = \sum d_i/n$
	Exercise Group	**Control Group**	**Differences**
	$\bar{Y}_1 = 0.58$	$\bar{Y}_2 = 0.53$	$\bar{Y}_d = .05$
	$s_1 = .21$	$s_2 = .19$	$s_d = .156$
	$n = 50$	$n = 50$	$n = 50$

d. Evaluate the evidence against H_O.

(1) Frequency distribution of the test statistic. The *t* distribution (Table C, Appendix 3) describes the frequency distribution of the test statistic when H_O is true, with $n - 1$ degrees of freedom. In this study, $df = 50 - 1 = 49$.

(2) The *p*-value is the probability of obtaining a test statistic equal to or greater than $t = 2.3$ by random chance when H_O is true. Although values for $df = 49$ are not given in Table C (see Appendix 3), $.01 < p < .025$ for both $df = 45$ and $df = 50$.

(3) Decision rule. The investigator formulates the following decision rule, based on a level of significance $\alpha = .05$: "**If $p \leq .05$, reject H_O.**"

e. State the conclusion.

(1) Statistical conclusion. Since $p \leq .05$, H_O is rejected at the $\alpha = .05$ level of significance.

(2) Clinical interpretation. There is a statistically significant increase in the mean blood level of calcitonin in the population of individuals who engage in one hour of weight-bearing exercise compared to the population of individuals who do not exercise ($p < .025$, using $df = 50$ as the best approximation).

(3) Chance of error. The probability that the statistical test has detected a significant difference in blood calcitonin level between the exercise group and the control group when no such difference actually exists is less than 2.5%.

2. Estimating the magnitude of the difference between the two population means. A 95% confidence interval may be used to estimate the mean of the population of differences in blood calcitonin levels between individuals who exercise and those who do not (μ_D).

a. Confidence coefficient. The 95% confidence coefficient corresponds to $t_{(1-.05/2)} = t_{.975}$ with 50 degrees of freedom (as an approximation of $df = 49$), given in Table C (see Appendix 3) as 2.0086.

b. Using the general formula for a confidence interval,

$$\text{endpoints of interval} = \text{estimate} \pm (\text{confidence coefficient} \times \text{standard error of estimate})$$

$$= \bar{d} \pm (t)(s_{\bar{d}}) \qquad (\text{where } s_{\bar{d}} = s_d/\sqrt{n})$$

$$= .05 \pm (2.0086)(.156/\sqrt{50})$$

$$= .01 \text{ to } .09\ \mu\text{g/dl}$$

c. Therefore, the investigator can be 95% confident that the true difference between the mean blood level of calcitonin in the population of individuals who engage in one hour of weight-bearing exercise and the mean blood level in the population of individuals who do not is included in the interval .01 to .09 μg/dl.

IV. SELECTING AN APPROPRIATE SAMPLE SIZE. Selection of the proper sample size enables an investigator to maximize available resources while guaranteeing a specified value for power. When a *t* test will be used to analyze the data, Tables D and E (see Appendix 3) can be used to determine sample size.

A. To determine the sample size corresponding to a given value for power, the investigator must first specify:

1. The level of significance α

2. The desired value of power, or its complement β (the type II error rate)

3. The standard deviation of the response variable in the study population. The population standard deviation σ may be **estimated** in any of the following ways:

a. From previous studies

b. From values published in the medical literature

c. From the sample standard deviation obtained in a pilot study conducted with a small number of subjects

d. From the following approximation (given that the population of sample values follows a **normal distribution**), that is,

$$\sigma \approx \frac{\text{range of values for response variable}}{6}$$

4. **The clinically important difference (i.e., effect size),** a threshold value defining the magnitude of the difference between the two means that is considered clinically relevant
   a. The specification of an effect size permits the investigator to designate a value for μ_1, the value of the population mean specified by H_A (see Ch 12 V D 2).
   b. To calculate sample size, the absolute effect size to be detected by the statistical test must first be converted to units of standard deviation (denoted symbolically as Δ):

$$\Delta = \frac{|\text{absolute effect size}|}{\sigma}$$

   where σ is the standard deviation in values of the response variable.
   (1) Thus, for the **pooled (independent sample) *t* test,**

$$\Delta = \frac{|\mu_1 - \mu_2|}{\sigma}$$

   where $\mu_1 - \mu_2$ represents the absolute magnitude of the clinically important difference in population means and σ is the common population standard deviation (the pooled t test assumes $\sigma_1 = \sigma_2$).
   (2) Similarly, for the **paired *t* test,**

$$\Delta = \frac{|\text{absolute effect size}|}{\sigma_D}$$

   where σ_D denotes the standard deviation of the population of differences (d_i's).

B. Examples

1. **Pooled *t* test.** The investigator conducting the study of calcitonin secretion and weight-bearing exercise described in II B decides that the difference in blood calcitonin level between the population of individuals who engage in one hour of weight-bearing exercise and the population of individuals who do not must be at least .10 μg/dl to be clinically relevant. He wishes to determine how many subjects must be assigned to the exercise and control groups to be 90% certain that the clinically important difference of .10 μg/dl will be detected, given that this difference or one greater actually exists between the two population means.
   a. The **researcher first specifies** the following values:
      (1) **The level of significance** $\alpha = .05$. As stated in II B 1 a (2), a one-tailed test will be used to compare mean blood levels of calcitonin for the two populations.
      (2) **Power** = .90. β, the complement of power, is therefore = 1 − .90 or .10.
      (3) The **standard deviation** σ. Based on prior research, σ is estimated as .20 μg/dl.
      (4) The effect size in units of standard deviation (Δ)
         (a) Based on the given absolute effect size of .10 μg/ml and the specified value of σ,

$$\Delta = \frac{|\text{absolute effect size}|}{\sigma} = .10/.20 = .50$$

         (b) That is, the investigator wishes to detect a difference of .50 standard deviations between the two population means, if such a difference exists.
   b. Next, Table E (see Appendix 3) is used to determine the correct sample size. Values in the body of the table represent $n_1 = n_2 = n$, the number of subjects required in each comparison group (the groups are assumed to be of equal size).
      (1) The relevant **panel** of this table corresponds to $\alpha = .05$ (one-tailed test).
      (2) Within this panel, the **column** labeled "$\beta = .10$" is located.
      (3) The appropriate sample size, $n = 70$, is located at the intersection of this column and the **row** labeled "$\Delta = 0.50$."
   c. Hence, $n_1 = n_2 = 70$ subjects are needed to guarantee 90% power in detecting a difference between the two population means of at least .10 μg/dl. The researcher's use of $n_1 = n_2 = 100$ subjects (see Table 16-1) may therefore be viewed as an unnecessary expenditure of both patient time and investigator resources.

2. Paired *t* test.* In a study to determine the effect of a certain oral contraceptive on weight gain, a gynecologist plans to weigh a sample of healthy women before and after three months of oral contraceptive use (i.e., subjects will be **self-paired;** the pre- and postmedication weights constitute the matched pairs). She determines that a weight gain of less than 5 pounds is not clinically relevant (i.e., the absolute effect size = 5 pounds). Prior to conducting the study, she wishes to know how many subjects should be included to guarantee that the statistical test will have at least a 90% chance (power) of detecting a 5-pound difference in average weight before and after three months of oral contraceptive use.

a. First, **the researcher specifies** the following values.

(1) The level of significance $\alpha = .05$. A one-tailed test will be used to analyze the data.

(2) Power = .90. β, the complement of power, is therefore $1 - .90 = .10$.

(3) The standard deviation σ. Based on preliminary research, the researcher assumes that the differences in the before and after weights will range from a minimum of 0 (no weight gain) to a maximum of 24 pounds and that the weight gain measurements are approximately normally distributed. She therefore estimates σ from the approximation

$$\sigma \approx \frac{\text{range of values for response variable}}{6}$$

$$\approx \frac{24 - 0}{6}$$

$$\approx 4$$

(4) The effect size in units of standard deviation (Δ)

(a) Based on the absolute effect size of 5 pounds and the estimated value of σ,

$$\Delta = \frac{|\text{absolute effect size}|}{\sigma_D}$$

$$= 5/4$$

$$= 1.25$$

(b) In other words, the gynecologist conducting this study must choose a sample size n that will ensure that the statistical test has a 90% chance of detecting a difference of at least 1.25 standard deviations between the before and after weight measurements, given that such a difference truly exists.

b. Table D (see Appendix 3) is used to determine the correct sample size n, where n = the number of pairs.

(1) The relevant panel of the table corresponds to $\alpha = .05$ (one-tailed test).

(2) Within this panel, the column labeled "$\beta = .10$" is located.

(3) The correct number of pairs, $n = 7$, is located at the intersection of this column and the row labeled "$\Delta = 1.3$."

c. Thus, if $n = 7$ pairs of before and after measurements are taken in this study, the paired t test has a 90% chance (power) of detecting a difference of at least 5 pounds, given that such a difference actually exists.

d. Sample size and the interpretation of negative results

(1) The gynecologist conducts the study as outlined above with seven women; the results are not statistically significant. She calculates the following:

$$\beta = P(\text{do not reject } H_O | H_O \text{ false})$$

$$= P(\text{test fails to detect 5 pound difference} | \text{5 pound difference exists in population})$$

$$= .10$$

*This example is adapted from Duncan RC, Knapp RG, Miller MC III: *Introductory Biostatistics for the Health Sciences*, 2nd ed. Albany, Delmar, 1983, p 114.

Power = P(reject $H_O | H_O$ false)

= P(test will detect 5 pound difference|5 pound difference exists in population)

= .90

(2) Due to the high value of power (and low probability of a type II error), the investigator may be willing to conclude that a difference of at least 5 pounds between the before and after weight measurements truly does not exist. That is, she may conclude that H_O is true, with a chance of error = 10%.

PROBLEMS*

16-1. In a study of the effect of bumetanide on urinary calcium excretion, nine randomly selected men each received an oral dose of .5 mg of the drug. Urine was collected hourly for the next six hours from these men, as well as for a sample of 16 randomly selected men who did not receive bumetanide. The mean urinary calcium excretion rate for the drug-treated group was 7.5 mg/hr, with a standard deviation of 6.0 mg/hr, while that of the control group was 6.5 mg/hr, with a standard deviation of 2.0 mg/hr.

a. Test the hypothesis that the mean calcium excretion rate differs for the bumetanide-treated and control groups (use a level of significance $\alpha = .05$).

b. What is the 95% confidence interval estimate of the difference in mean calcium excretion rate between the population of patients receiving bumetanide and the control population?

16-2. The table below presents data on cumulative weight loss from excessive sweating during insulin-induced hypoglycemia for 12 patients treated with propranolol and 11 control patients.

	Propranolol	Control
Sample Size	$n_1 = 12$	$n_2 = 11$
Sample Mean	$\bar{Y}_1 = 120$ g	$\bar{Y}_2 = 70$ g
Sample Standard Deviation	$s_1 = 10$	$s_2 = 8$

Adapted from Duncan RC, Knapp RG, Miller MC III: *Introductory Biostatistics for the Health Sciences*, 2nd ed. Albany, Delmar, 1983, p 112.

a. Using a 5% level of significance, test the hypothesis that the mean cumulative weight loss differs between the two groups.

b. What is the 95% confidence interval estimate of the difference in mean cumulative weight loss following insulin-induced hypoglycemia for the population of patients receiving propranolol and the control population?

Problems 16-3 through 16-6

The table below lists serum digoxin levels for nine healthy men between the ages of 20 and 45 following a rapid intravenous injection of the drug. Drug levels were determined four hours and eight hours after the injection. Use the data in this table to solve problems 16-3 through 16-6.

	Serum Digoxin Concentration (ng/ml)		
Subject	**4-hr**	**8-hr**	**Differences (d_i)**
1	1.0	1.0	0.0
2	1.3	1.3	0.0
3	0.9	0.7	−0.2
4	1.0	1.0	0.0
5	1.0	0.9	−0.1
6	0.9	0.8	−0.1
7	1.3	1.2	−0.1
8	1.1	1.0	−0.1
9	1.0	1.0	0.0
			$\bar{Y}_d = -0.067$
			$s_d = .071$

Adapted from Duncan RC, Knapp RG, Miller MC III: *Introductory Biostatistics for the Health Sciences*, 2nd ed. Albany, Delmar, 1983, p 112.

*The problems in this section are adapted from Duncan RC, Knapp RG, Miller MC III: *Introductory Biostatistics for the Health III Sciences*, 2nd ed. Albany, Delmar, 1983, pp 76, 112–114. Confidence intervals of various population parameters in problems 1, 3, and 5 were calculated in Chapter 11.

16-3. **a.** Test the hypothesis that serum digoxin levels at four hours differ significantly from serum digoxin levels at eight hours (use a 5% level of significance).

b. What is the 95% confidence interval estimate of the difference between mean serum digoxin levels four hours and eight hours after injection?

16-4. The pair of physicians who carried out this study decide that serum digoxin concentrations in the population of four-hour and the population of eight-hour measurements must differ by at least .04 ng/ml to be clinically relevant. If they wish to replicate the study with 80% certainty that the statistical test will detect this clinically important .04 ng/ml difference in population means, given that such a difference exists, how many subjects should be included in the study? (Use the data in the table as pilot data and a 5% level of significance for the statistical test.)

16-5. A second sample of 11 healthy men, drawn from the same population as the sample described in problem 16-3, received an intravenous infusion of digoxin. At the end of eight hours, the mean serum digoxin concentration in this sample was .95 ng/ml, with a standard deviation of .20 ng/ml.

a. Using a 5% level of significance, test the hypothesis that serum digoxin levels are higher eight hours after rapid injection than eight hours after intravenous infusion.

b. What is the 99% confidence interval estimate of the population mean difference in serum digoxin concentration eight hours after rapid intravenous injection and eight hours after intravenous infusion?

16-6. The research team decides that the difference in serum digoxin concentration eight hours after rapid injection and eight hours after intravenous infusion must be at least .10 ng/ml to be clinically relevant.

a. Using the data in problem 16-5 as pilot data, what sample size is necessary to ensure that the statistical test will detect the clinically important difference of .10 ng/ml with 90% certainty, given that this difference actually exists between the two populations?

b. Estimate the power of the test for the pilot study, using a 1% level of significance. How does this value influence the interpretation of the result obtained in problem 16-5a?

Solutions on p 398.

STUDY QUESTIONS

Directions: Each of the numbered items or incomplete statements in this section is followed by answers or by completions of the statement. Select the **one** lettered answer or completion that is **best** in each case.

1. A study of anterior poliomyelitis patients with unilateral involvement of an extremity is conducted to determine if a significant difference in average blood flow exists between the normal and affected limbs. After measuring blood flow (cc/min/100 cc of limb volume) in both the normal and affected limbs for 200 such patients, the investigators report a difference between average blood flow in the affected and normal limbs ($p < .00001$). All of the following statements about this p-value are true EXCEPT

(A) the magnitude of the p-value may be related to the large sample size
(B) because the p-value is extremely small, the difference in average blood flow between the normal and affected limbs is likely to be very large
(C) the small p-value indicates that it is highly unlikely that a sample of patients with the observed difference in average blood flow between the normal and affected limbs could be obtained by random chance from a population in which there is no such difference
(D) the magnitude of the p-value indicates that the difference in average blood flow between the normal and affected limbs is statistically significant

2. A pediatrician wishes to evaluate the efficacy of an experimental formula in infants who fall below the tenth percentile in weight for their age group. Fifty such infants will be randomly assigned to receive either the experimental formula or a standard commercially available formula for three months. Weight gains for both groups will be recorded at the end of this time period in an attempt to answer the research question "Is the average weight gain higher among infants receiving the experimental formula than among those receiving the standard formula?" After conducting the study, the pediatrician reports at a research conference that the results are "not statistically significant ($p > .05$)." All of the following statements about this conclusion are true EXCEPT

(A) the average weight gain in infants fed the experimental formula is the same as the average weight gain among infants fed the standard formula
(B) a clinically important difference in average weight gain among the two groups of low-weight infants may exist, but the power of the test may have been too low to detect it
(C) the chance of a type II error may be high
(D) H_O is not rejected and the chance of error associated with this conclusion is unknown
(E) the observed difference in average weight gain between the experimental and control groups is likely to be due to random chance

---

1-B
2-A

3. In a study to assess community health services, patients who are treated at a particular rural health facility are asked whether they favor replacing the facility with a team of home health care workers or allowing it to remain open. The health care planners conducting the study wish to know if there is a statistically significant difference in the average distance from the facility between those who favor replacing it with a home health care team and those who wish to retain the facility. The researchers compute a test statistic equal to 28.4 ($p < .00001$). All of the following statements about this result are true EXCEPT

(A) the observed difference in average distance from the facility between those who favor replacing it and those who favor keeping it open is unlikely to have occurred by random chance
(B) there is a statistically significant difference in the average distance from the facility between those who favor replacing it and those who favor keeping it
(C) the two-tailed test used in this study was appropriate
(D) the difference in average distance from the facility between those who favor replacing it and those who favor closing it is large

4. To compare the efficacy of two drugs in lowering serum cholesterol levels, a physician measures serum cholesterol levels in a sample of men receiving drug A and in a sample of men receiving drug B and learns that the average serum cholesterol level is lower in those receiving drug B. Plausible explanations for the observed difference in the response to the two drugs include all of the following EXCEPT

(A) drug B is superior to drug A in lowering cholesterol levels among the population of potential recipients
(B) the observed difference occurred by random chance
(C) extraneous, uncontrolled differences (other than drug treatment) between the sample of men receiving drug A and the sample of men receiving drug B account for the observed difference in efficacy
(D) drug A may have been selectively administered to patients whose serum cholesterol levels were more refractory to drug therapy
(E) the observed difference may be attributed to a small p-value for the statistical test

Questions 5–9

Administrators at a Veterans Administration (VA) hospital wished to know whether the average stay at their hospital differed from the average stay at a nearby general hospital. A random sample of 100 records from the VA hospital for the year 1991 revealed a mean hospital stay of 6.6 days, with a standard deviation of 2.5 days. A random sample of 100 records taken from the files of the general hospital revealed a mean hospital stay of 5.7 days, with a standard deviation of 2.0 days.

5. What statistical procedure is most appropriate for analyzing the data obtained in this study?

(A) Paired t test
(B) Chi-square test
(C) Correlation analysis
(D) Pooled (independent sample) t test
(E) It cannot be determined from the data given

6. What is H_O for the statistical test?

(A) "The length of the average stay of patients at the VA hospital in 1991 differs from the length of the average stay of patients at the general hospital"
(B) "Hospital stays for patients at the VA hospital in 1991 are longer, on the average, than those of patients at the general hospital"
(C) "There is no difference between the length of the average stay at the VA hospital in 1991 and the length of the average stay at the general hospital"
(D) "There is a positive association between the length of the average stay at the VA hospital in 1991 and the length of the average stay at the general hospital"
(E) H_O cannot be formulated from the given information

3-D
4-E
5-D
6-C

7. Given that the calculated test statistic for this study is 2.81 and the tabulated critical value for $\alpha = .025$ is 1.97, what is the most appropriate conclusion?

(A) Reject H_O and conclude that there is a statistically significant difference between the average stay at the VA hospital and at the general hospital in 1991 ($p > .05$)
(B) Reject H_O and conclude that there is no difference between the average stay at the VA hospital in 1991 and the average stay at the general hospital ($p < .05$)
(C) Reject H_O and conclude that the average stay at the VA hospital in 1991 differs significantly from that at the general hospital ($p < .05$)
(D) Do not reject H_O; conclude that there is no difference between the average stay at the VA hospital in 1991 and the average stay at the general hospital ($p > .05$)
(E) Do not reject H_O and conclude that the average stay at the VA hospital in 1991 is longer than that at the general hospital ($p > .05$)

8. Which of the following statements accurately describes the level of significance α?

(A) It specifies the probability of concluding that the length of the average stay in 1991 differs between the two hospitals, when, in fact, there is no such difference among the population of all patients treated at these hospitals
(B) It is divided by two when H_A specifies a one-tailed test
(C) It specifies the probability of rejecting a false H_O
(D) It is typically equal to .025
(E) It is a measure of the probability of a type II error

9. The 95% confidence interval estimate of the difference in average stay at the VA hospital in 1991 and at the general hospital is .27 to 1.53 days. What is the correct interpretation of this interval?

(A) Ninety-five percent of the time, the difference between the average stay at the two hospitals (calculated from random samples of 100 records) will fall between .27 and 1.53 days
(B) If the population of all 1991 hospital records were analyzed, the difference in the average stay between the two hospitals would fall between .27 and 1.53 days
(C) The health care planners conducting this study can be 95% confident that the interval .27 to 1.53 days includes the true difference in average stay for the populations of patients treated at the two hospitals in 1991
(D) Ninety-five percent of all samples of 100 records that could be drawn from the populations of all records at both hospitals will yield mean stays that differ by .27 to 1.53 days
(E) This interval provides no information on the size of the difference in average stay between the two hospitals

---

7-C
8-A
9-C

Directions: Each group of items in this section consists of lettered options followed by a set of numbered items. For each item, select the **one** lettered option that is most closely associated with it. Each lettered option may be selected once, more than once, or not at all.

Questions 10–12

For each clinical study described below, select the type of *t* test most suitable for analyzing the data.

(A) Paired *t* test
(B) Pooled (independent sample) *t* test
(C) A *t* test is inappropriate for this study

10. Thirty men between the ages of 30 and 55 participated in an experiment to study the effectiveness of a particular diet and exercise program in reducing serum cholesterol levels. Baseline cholesterol levels were determined for all study participants prior to initiating the program and again three months later. The research question is "Are average serum cholesterol levels lower after the three-month program of diet and exercise?"

11. A study is carried out to compare the effect of hydrochlorothiazide and a placebo on systolic blood pressure in men between the ages of 25 and 65 who have borderline hypertension. Fifty men diagnosed with borderline hypertension received a placebo for one month, then hydrochlorothiazide for one month. Systolic blood pressure was recorded for all subjects at the end of the month of placebo therapy and at the end of the month of drug therapy. The investigator hypothesizes that there is a statistically significant difference between the two measurements.

12. A research team conducts a study of traffic accident victims to determine if blood alcohol levels in samples taken from the left femoral artery and the left coronary artery differ. Blood samples are drawn during post-mortem examination of 25 traffic accident victims. The research question is "Does the average blood alcohol level in the samples drawn from the left femoral artery differ significantly from the average blood alcohol level in the samples drawn from the left coronary artery?

10-A
11-A
12-A

ANSWERS AND EXPLANATIONS

1. The answer is B *[II B 2 a; III B 1 d (2), e (1); Ch 12 VI C 1 b]*
A small *p*-value indicates that a statistically significant difference (i.e., one unlikely to have occurred by random chance) exists between average blood flow in the affected limbs and average blood flow in the normal limbs (assuming $\alpha = .05$), but provides no information on the magnitude of that difference. Even small differences in sample means may result in a small *p*-value when the sample size is sufficiently large.

2. The answer is A *[II B 1 e, 2 a; Ch 12 VI C 2]*
While a failure to achieve statistical significance [and hence reject the null hypothesis (H_O)] may be due to the absence of a clinically important difference between the two population means, it may also signify low power of the statistical test. Both β (the type II error rate) and its complement, power, are unknown for this study. When the power of the test is unknown, a failure to reject H_O cannot be taken as an indicator of its validity. Such a failure does, however, indicate that the observed sample difference is likely to have occurred by random chance.

3. The answer is D *[II B 1, 2 a; Ch 12 VI C 1 b]*
The small *p*-value reported for this study does not necessarily indicate that the difference in average distance from the medical facility between the two comparison groups is large. Assuming a 5% level of significance, the reported *p*-value indicates that the observed difference in the average distance from the facility is statistically significant (i.e., unlikely to be due to random chance). Because the research question does not specify direction, a two-tailed test is appropriate.

4. The answer is E *[II B 1 d (2); Ch 12 VI C 1 b]*
The magnitude of the *p*-value is not an explanation for the observed difference in the response to the two drugs. Mean serum cholesterol levels may have been lower in the men receiving drug B because this drug is actually superior to drug A, as a result of random chance, or due to a biased comparison. Uncontrolled confounding variables may also account for the observed difference between the comparison groups.

5–9. The answers are: 5-D *[II A]*, **6-C** *[II B 1 a]*, **7-C** *[II B 1 d, e]*, **8-A** *[II B 1 e; Ch 12 II E 1; V A; Ch 13 II D 1]*, **9-C** *[II B 2 b]*
Because this study involves the comparison of two independent samples [the Veteran's Administration (VA) hospital and the general hospital], the pooled *t* test is best suited to the analysis of the data.

The null hypothesis (H_O) is the hypothesis of "no difference." Therefore, for this study, H_O is: "There is no difference in the average stay at the VA hospital and the average stay at the general hospital in 1991," and the alternative hypothesis (H_A) is: "The average stay at the VA hospital in 1991 differed from that at the general hospital." Because no direction is specified, this is a two-tailed hypothesis.

The calculated test statistic, $t = 2.81$, lies to the right of the tabulated critical value for $\alpha = .025$, $t = 1.97$ on the horizontal axis of the frequency distribution. Therefore, the area to the right of the test statistic (the *p*-value) is less than .025. Because H_A does not specify direction, the test is a two-tailed test and the *p*-value must be doubled (i.e., $p < .05$). This *p*-value is less than $\alpha = .05$. Therefore, H_O is rejected at the 5% level of significance. There is a statistically significant difference in the average stay at the VA hospital in 1991 and the average stay at the general hospital.

The level of significance α is the probability of rejecting H_O when it is actually true (i.e., the probability of a type I error). It is divided by two for a two-tailed, not a one-tailed, test. Although α is customarily equal to .05 or .01, it can be any value chosen by the investigator.

The stated confidence interval estimates, with 95% confidence, the difference in the average hospital stays among the population of patients treated at the VA hospital and those treated at the general hospital in 1991. It provides no information about future sample results.

10–12. The answers are: 10-A *[III A]*, **11-A** *[III A]*, **12-A** *[III A]*
The study of the effect of a three-month diet and exercise regimen on serum cholesterol values uses a self-paired design to assess cholesterol values in the same subject. Therefore, the paired *t* test is appropriate.

Patients in the next study serve as their own controls, receiving first the placebo and then the drug. The paired t test is the most suitable statistical procedure for this study.

In the study comparing blood alcohol levels in the left femoral and left coronary artery, both samples are drawn from the same subject. Thus, the paired t test is appropriate.

If any of these studies had measured the particular risk factor in two separate groups, or independent samples, then the pooled t test would have been appropriate.

17
Tests of Statistical Significance: Analysis of Variance

I. ANALYZING DIFFERENCES AMONG MULTIPLE COMPARISON GROUPS

A. The multiple comparison problem. Although, as discussed in Chapter 16 I B, the *t* test is appropriate for evaluating the difference between *two* population means, it is not appropriate when more than two means are being compared.

1. If **multiple *t* tests** are used to compare several pairs of means, **the probability of erroneously detecting at least one significant difference** among the set of comparisons **is larger than the level of significance,** α, selected for each individual test.

2. **Example.** A clinical researcher wishes to compare five population means. For this example, the population means are assumed to be equal (i.e., the null hypothesis, $\mathbf{H_O}$: $\mu_1 = \mu_2 = \mu_3 = \mu_4 = \mu_5$, is true).
   a. If **separate *t* tests** are used to compare each of the 10 possible pairs of five means, and if each test uses a level of significance $\alpha = .05$, then, for each test, $\alpha = P(\text{reject } H_O | H_O \text{ true})$ and $P(\text{fail to reject } H_O | H_O \text{ true}) = 1 - .05 = .95$. That is, for each *t* test, the probability of a correct conclusion (e.g., failing to reject a true H_O) is .95.
   b. Assuming that the tests are independent, the **probability that all 10** will lead to a **correct decision** (i.e., a true H_O is not rejected in any of the 10 tests) is

$$P(\text{fail to reject } H_O \text{ for all 10 tests} | H_O \text{ true}) = (.95)^{10}$$
$$= .5987$$

   c. The **probability of rejecting H_O for at least one** of the 10 tests is

$$P(\text{reject } H_O \text{ for at least 1 of the 10 tests} | H_O \text{ true})$$
$$= 1 - P(\text{fail to reject } H_O \text{ for all 10 tests} | H_O \text{ true})$$
$$= 1 - (.95)^{10}$$
$$= .4013$$
$$= P(\text{type I error for the set of 10 tests})$$

   d. Thus, there is a **40% chance that at least one of the 10 *t* tests will detect** a statistically significant **difference** between a pair of population means **that,** in fact, **does not exist.**

B. Analysis of variance (*F* test)

1. **Purpose**
   a. The analysis of variance evaluates the equality of several population means **without inflating the type I error rate.**
   b. The analysis of variance allows the investigator to break down, or **partition, the total variation** in a set of data **into its component parts** (see I B 4).

2. **Study type**
   a. **Experimental studies** (i.e., those in which study subjects are **randomly assigned to comparison or treatment groups**) are the study type most often associated with the analysis of variance.
   b. **Observational studies** (i.e, those in which subjects are not randomly assigned but

"**self-select**" into groups) also can be evaluated using the analysis of variance. However, more **caution must be exercised** in interpreting the results, particularly in drawing conclusions about causality.

3. **Types of variation in clinical measurements** (see also Ch 1 I C)
   a. **Systematic,** or **"controllable," variation** follows a predictable pattern (e.g., the variation in diastolic blood pressure across different age groups).
   b. **Random variation** (also known as **true biologic variation** or **"uncontrollable" variation**) is the cumulative variation of many unknown factors, each of which contributes a small random effect. Most biological models assume that random effects are just as likely to occur in a positive as in a negative direction.
4. **Partition of total variation in a set of clinical measurements.** Table 17-1 is the schematic representation of a series of hypothetical diastolic blood pressure readings for three populations of patients, each treated with a different antihypertensive drug (Y_{ij} denotes the diastolic blood pressure of the jth person in the ith treatment group). The **total variation** among all the Y_{ij} measurements **can be partitioned into two components:**

   total variation = (variation due to drug treatments) + (experimental error)

   a. **Variation due to treatments** is the **systematic** variation in values of the response variable that can be attributed to **experimental manipulation** (e.g., the three drug therapies of Table 17-1).
   b. **Experimental error** is the variation among measurements carried out on **experimental units treated in a similar fashion.**
      (1) An **experimental unit** is the element that receives one application of the treatment. Human or animal subjects constitute one common type of experimental unit; other examples include health education programs, hospitals, or substance abuse treatment facilities.
      (2) **Experimental error** contains true random variation plus any systematic variation attributable to sources other than those incorporated into the study design.
      (3) **A measure of experimental error** can be obtained by **pooling information on the variation among subjects within each of the populations** under comparison. Experimental error is therefore sometimes referred to as the **variation within groups.**
         (a) The **greater the heterogeneity** within groups, the **greater the experimental error.**
         (b) For example, experimental error will be inflated if subjects in the three treatment groups depicted in Table 17-1 are of widely disparate ages, or if blood pressure readings are made at different times of the day.
5. **The test statistic** computed in the analysis of variance is:

$$F = \frac{\text{variation due to treatments}}{\text{experimental error}}$$

A large value for this ratio indicates that the treatments have a statistically significant effect on the response variable (i.e., at least one of the differences among the population means is statistically significant).

Table 17-1. Schematic Representation of Blood Pressure Measurements in Patients Receiving Three Antihypertensive Drugs

	Population 1 (Drug A)	Population 2 (Drug B)	Population 3 (Drug C)
	Y_{11}	Y_{21}	Y_{31}
	Y_{12}	Y_{22}	Y_{32}
	Y_{13}	Y_{23}	Y_{33}
	⋮	⋮	⋮
Population Mean	μ_1	μ_2	μ_3
Population Standard Deviation	σ_1	σ_2	σ_3

II. THE COMPLETELY RANDOM DESIGN (CRD)

A. Characteristics

1. CRD uses **independent samples** to compare multiple population means. **No restrictions are imposed** on the random assignment of subjects to treatment groups (Figure 17-1).
2. **All systematic sources of variation** in the response variable, other than the variation attributable to the treatments, are **included in the experimental error term**.
3. The CRD **should not be used if** the experimenter anticipates **substantial subject-to-subject variation** in the response variable (in the absence of the treatments).

B. Example. A pharmaceutical company wishes to compare the efficacy of three angiotensin-converting enzyme (ACE) inhibitors in reducing diastolic blood pressure. That is, the project manager overseeing the study needs to know if a statistically significant difference in diastolic blood pressure exists among the populations of patients receiving drugs A, B, and C respectively. She answers this question using the five-step procedure for statistical hypothesis testing outlined in Chapter 12 IV.

1. **State the hypothesis.**
   a. H_O is: "Average diastolic blood pressure does not differ for the populations of patients receiving drug A, drug B, and drug C; that is, $\mathbf{H_O: \mu_A = \mu_B = \mu_C}$."

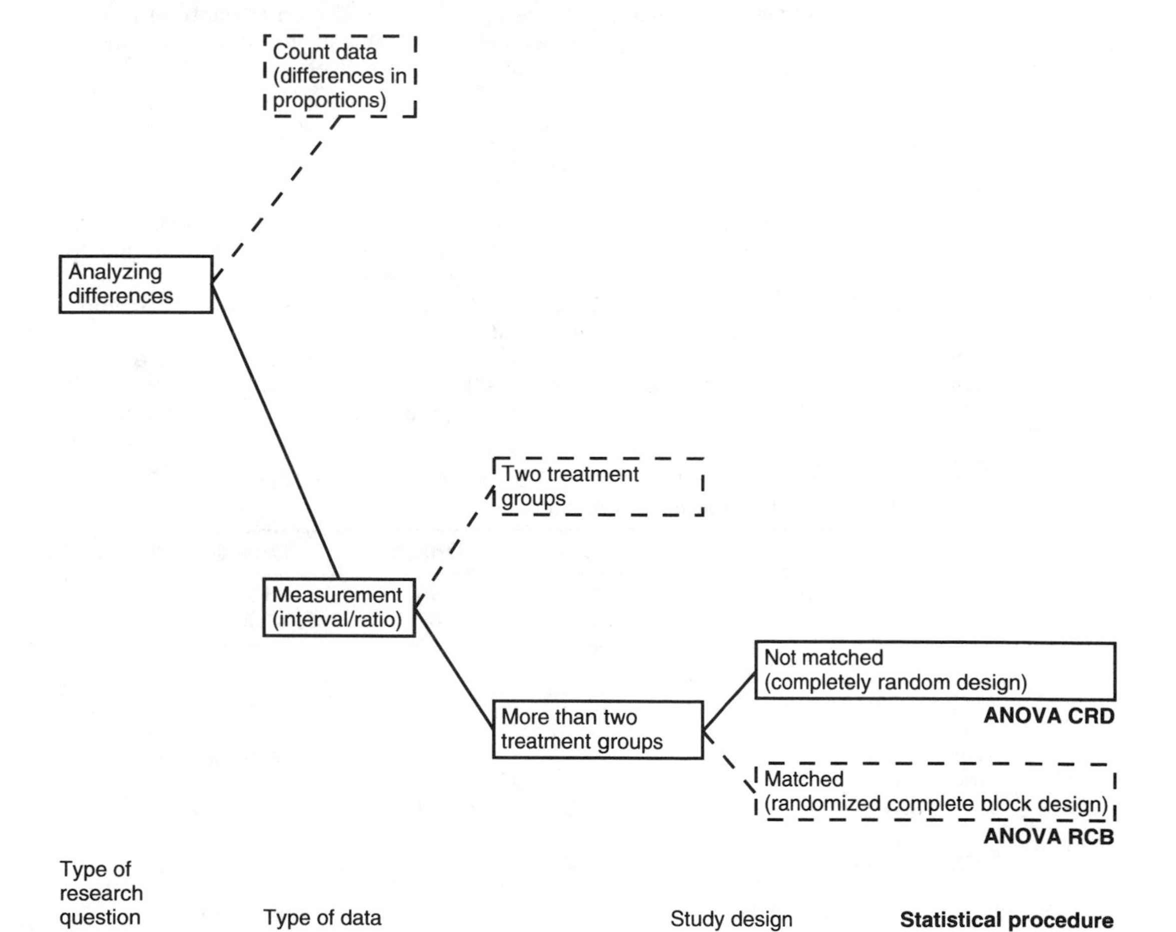

Figure 17-1. Decision tree pathways (derived from Figure 13-1) used to select the analysis of variance (*ANOVA*) for the completely random design (*CRD*) as the appropriate statistical procedure for analyzing differences among several population means in a study based on independent samples. *Dotted lines* indicate decision tree paths not taken.

b. The alternative hypothesis ($\mathbf{H_A}$) is: "Average diastolic blood pressure differs among the three populations; that is, **at least two of the population means are unequal.**" H_A **does not** state that **all** of the population means differ from one another, nor does it specify which means are different.

2. Select a sample and collect data.

a. To reduce subject-to-subject variation in diastolic blood pressure (i.e., experimental error), the project manager decides to limit the study to white men between the ages of 25 and 40. Fifteen men in this age range are recruited to participate in the study.

b. Five of these 15 subjects are randomly designated to receive drug A, five to receive drug B, and five to receive drug C. Table 17-2 summarizes the results of the study. Although the CRD does not demand that **sample sizes** be **equal, the power of the statistical test is greatest** when they are.

3. Calculate the test statistic.

a. The general formula for calculating the test statistic F is

$$F = \frac{\text{variation due to treatments}}{\text{experimental error}}$$

(1) The **denominator** (experimental error) is an **estimate of the subject-to-subject variation** within each of the treatment populations (σ^2), **regardless of whether H_O or H_A is true.***

(2) When H_O is true, the **numerator** of the test statistic is also **an estimate of σ^2.**

(3) In other words, when there is no treatment effect, the variation due to treatments should equal the subject-to-subject variability within a given treatment population. That is,

$$F = \frac{\sigma^2}{\sigma^2} = 1$$

(4) When H_A is true, the numerator of the test statistic no longer approximates σ^2. In other words, when there is a significant treatment effect, the variation due to treatments should be much larger than the experimental error and the test statistic should be much larger than 1. **The greater the value of the test statistic, therefore, the more compelling the evidence against H_O.**

b. Intermediate calculations in the computation of the test statistic are organized in an **analysis of variance (ANOVA) table (Table 17-3).**

(1) Within the table, the total variation in the response variable is partitioned into its

Table 17-2. Diastolic Blood Pressures in Patients Treated with Three Angiotensin-converting Enzyme (ACE) Inhibitors

	Drug A	Drug B	Drug C
	100	90	110
	105	80	90
	95	80	80
	110	75	100
	90	90	95
Sample Size:	$n_1 = 5$	$n_2 = 5$	$n_3 = 5$
Sample Totals:	$T_1 = 500$	$T_2 = 415$	$T_3 = 475$
Sample Means:	$\bar{Y}_1 = 100$	$\bar{Y}_2 = 83$	$\bar{Y}_3 = 95$
Sample Standard Deviations:	$s_1 = 7.91$	$s_2 = 6.71$	$s_3 = 11.18$

*Like the t test, the analysis of variance assumes that $\sigma_1^2 = \sigma_2^2 = \sigma_k^2 = \sigma^2$, that is, that the subject-to-subject variation is the same within all of the treatment populations.

Table 17-3. General Format of the Analysis of Variance (ANOVA) Table in a Study Using the Completely Random Design (CRD)

Source of Variation	*df*	SS	MS	*F*
Among Treatment Groups	df_{tr}	SST	MST	$F = \text{MST}/\text{MSE}$
Error (Within Treatment Groups)	df_e	SSE	MSE	
Total	df_{total}	SS_{total}		

two components: variation due to treatments (numerator of test statistic) and experimental error (denominator of test statistic).

(2) Both numerator and denominator of the test statistic represent **variances**, and may be defined as

$$\text{variance} = \text{SS}/df$$

where SS denotes the **sum of squares** and *df* represents the associated number of **degrees of freedom.** The second and third columns in the body of the ANOVA table depict the values of SS and *df* for the total variation and each of its two components.

(3) The two variances used to derive the test statistic are also known as **mean squares** (see Table 17-3, *column 4*); they are obtained by dividing the sums of squares by their respective degrees of freedom.

(a) Thus, the **mean square for treatments (MST)**, or the variation due to treatments, is

$$\text{MST} = \text{SST}/df_{tr}$$

where SST = treatment sum of squares and df_{tr} = degrees of freedom associated with treatment.

(b) Similarly, the **mean square for error (MSE),** or the variation in the response variable that can be attributed to experimental error, is

$$\text{MSE} = \text{SSE}/df_e$$

where SSE and df_e are the sum of squares and degrees of freedom associated with experimental error.

(i) MSE is an estimate of the common variance in each treatment population ($\sigma^2 = \sigma_1^2 = \sigma_2^2 = \cdots \sigma_k^2$).

(ii) If only two treatments are to be compared (i.e., $k - 2$), MSE is equivalent to s_p^2, the pooled variance used in the denominator of the pooled *t* test.

(4) The final column of the ANOVA table records the value of the test statistic (known as the ***F* statistic** or **variance ratio**), which is the ratio of the two mean squares. Thus,

$$F = \frac{\text{variation due to treatments}}{\text{experimental error}}$$

$$= \text{MST}/\text{MSE}$$

(5) Table 17-4 is the completed ANOVA table for this example. The calculated test statistic, $F = 4.92$, is given in the last column of this table. The computational formulas and associated calculations for this example may be found in section V.

Table 17-4. Analysis of Variance (ANOVA) Table for a Study Comparing the Effect of Three Angiotensin-converting Enzyme (ACE) Inhibitors on Diastolic Blood Pressure (Independent Samples)

Source of Variation	*df*	SS	MS	*F*
Among Treatments	2	SST = 763.33	$\text{MST} = \frac{763.33}{2} = 381.67$	$\frac{\text{MST}}{\text{MSE}} = 4.92$
Error	12	SSE = 930.00	$\text{MSE} = \frac{930}{12} = 77.50$	
Total	14	$SS_{total} = 1693.33$		

4. **Evaluate the evidence against H_O.**
   a. **Frequency distribution of the test statistic.** The ***F* distribution** (Tables F and G, Appendix 3) is the frequency distribution of possible values of the test statistic that could be obtained from samples drawn from populations with equal means (H_O true). To locate a critical value in Table F or G (see Appendix 3), it is necessary to specify the degrees of freedom for the numerator (df_{tr}) and for the denominator (df_e) of the test statistic. These values are provided by the ANOVA table. For this example, Table 17-4 lists $df_{tr} = 2$ and $df_e = 12$.
   b. **Calculating the *p*-value.** Table F corresponds to $\alpha = .05$ and Table G to $\alpha = .01$ (see Appendix 3). In each table, a particular critical value is located at the intersection of df_{tr} and df_e. For $df_{tr} = 2$ and $df_e = 12$, Table F gives the critical value for $\alpha = .05$ as 3.89 and Table G gives the critical value for $\alpha = .01$ as 6.93. The calculated test statistic, $F = 4.92$, lies to the **right** of 3.89 but to the **left** of 6.93; that is, $.01 < p < .05$, or simply $p < .05$.
   c. **Decision rule.** With the customary $\alpha = .05$ level of significance, the decision rule becomes : **"Reject H_O if $p \leq .05$."** The choice of a 5% level of significance indicates that the project leader is willing to tolerate a 5% risk of erroneously rejecting H_O, concluding that the drugs differ in efficacy when, in fact, they do not.
5. **State the conclusion.**
   a. **Statistical conclusion.** Since $p \leq .05$, H_O is rejected at the $\alpha = .05$ level of significance.
   b. **Clinical interpretation.** There is a statistically significant difference in mean diastolic blood pressure among the populations receiving the three ACE inhibitors (i.e., the **difference between at least two** of the population means **is significant;** $p < .05$).
   c. The **chance of error** associated with rejecting H_O (i.e., the risk of a type I error) is 5% or less ($p < .05$).
   d. When the calculated value of F is statistically significant, one of several **multiple comparison tests** (see IV) can be used after the analysis of variance to determine which pairs of treatment means are different.
   e. If the calculated test statistic had been **less than the critical value,** $F = 3.89$, the p-value would be greater than .05.
      (1) Consequently, H_O would not be rejected at the $\alpha = .05$ level of significance (i.e., the results **would not be statistically significant**).
      (2) The chance of error associated with this conclusion is β, the risk of a type II error. Because β is unknown, the failure to achieve statistical significance indicates only that the data are insufficient to demonstrate a difference between the three drugs, **not** that such a difference does not actually exist.

C. **CRD** includes all systematic sources of variation in the experimental error term. Therefore, the design should be avoided if subject-to-subject variation is expected to be great.

1. **Example.** In the current example, age is systematically related to the response variable, diastolic blood pressure. Therefore, if the study participants differ widely in age, the investigator might anticipate considerable subject-to-subject variation in blood pressure, even in the absence of any drug treatment. This systematic variation in blood pressure measurements resulting from age differences among the subjects inflates the value of the experimental error term.

2. **Restriction** (see Ch 8 II D 1) offers the researcher a way of increasing the homogeneity of the experimental units. However, the stricter the inclusion criteria for study participants, the less generalizable the study results. For example, by limiting the ACE inhibitor trial to white men between the ages of 30 and 45, the project manager can reduce the variation in diastolic blood pressure attributable to age, race, and gender—but any observed drug effects may be limited to this population.

III. THE RANDOMIZED COMPLETE BLOCK (RCB) DESIGN

A. Characteristics

1. The RCB design uses **matched samples** to compare multiple population means. That is, in the RCB design, **experimental error is reduced** by grouping the study subjects into **homogeneous blocks** on the basis of a variable (other than the treatment variable) known or suspected of systematically affecting the response variable (Figure 17-2).

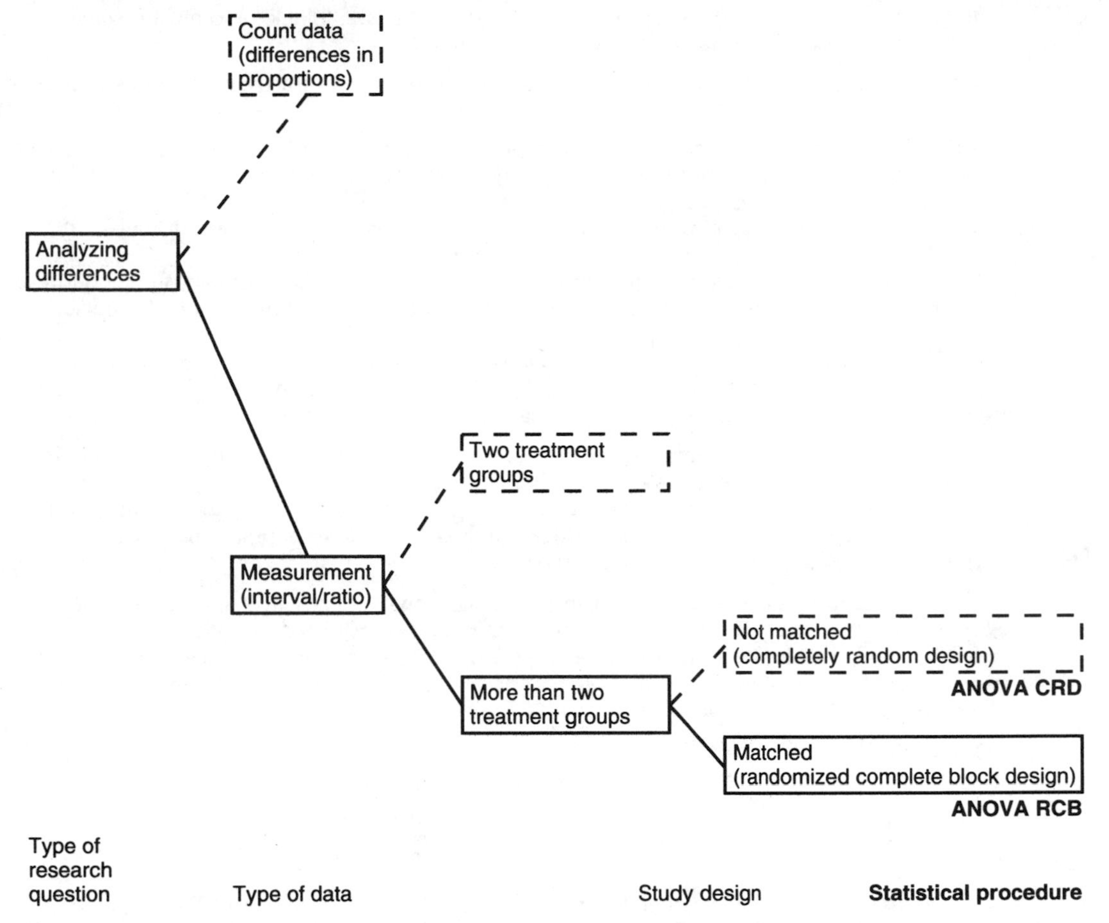

Figure 17-2. Decision tree pathways (derived from Figure 13-1) used to select the analysis of variance (*ANOVA*) for the randomized complete block (*RCB*) design as the appropriate statistical procedure for analyzing differences among several population means in a study in which subjects have been grouped into homogeneous strata or "blocks" prior to randomization. *Dotted lines* indicate decision tree paths not taken.

2. In the RCB design, the total variation in values of the response variable is **partitioned into three components**: **variation attributable to the treatments, variation attributable to the blocking variable** [see III C 3 b (1)], and **experimental error**. Thus,

   total variation = (variation due to treatments)

   + (variation due to blocking variable) + (experimental error)

3. The RCB design results in a **more powerful statistical test** than the CRD **only if the reduction in experimental error is large enough** to compensate for the corresponding loss of degrees of freedom associated with the test statistic.

4. **Blocking and study type**
   a. In **experimental studies,** blocking is used to **increase the power** of the statistical test, by reducing the variation attributable to experimental error.
   b. In **observational studies,** blocking is also a tool for **controlling the effect of a potentially confounding variable**.

5. When subjects clearly can be **grouped** into homogeneous blocks with respect to **a variable other than the treatment variable,** the RCB design should be chosen.

B. Methods of blocking

1. **Self-matching.** As in the paired study (see Ch 16 III A 1), self-matching involves repeated

measurements on the same subject. That is, the **subject is the "block"** and must receive, on different occasions, **all the treatments** under comparison.

2. **Natural matching.** The subjects may **naturally constitute several homogeneous blocks,** such as litters of laboratory rats.
3. **Artificial matching.** Subjects may also be grouped by the investigator on the basis of a **blocking variable** (a variable other than the treatment variable that contributes to the variation in the response variable).
   a. In experimental studies, the blocking variable is usually an **inherent** (rather than assigned) characteristic, such as age, race, or sex.
   b. Within each block, subjects are randomly assigned to treatments. Obviously, therefore, the number of subjects in each block must be equal to the number of treatments.

C. **Example.** Because age is known to affect diastolic blood pressure systematically, the pharmaceutical project manager described in II B decides to replicate the study comparing the three ACE inhibitors, grouping the subjects into blocks on the basis of age.

1. **State the hypothesis.** The opposing hypotheses are the same as those formulated in the first study. Namely, H_O is: "Average diastolic blood pressure does not differ for the populations of patients receiving drug A, drug B, and drug C (**H_O: $\mu_A = \mu_B = \mu_C$**);" and **H_A** is: "Average diastolic blood pressure **differs among the three populations**; that is, the difference between at least two population means is not equal to zero."
2. **Select a sample and collect data.** The project manager defines five age blocks: 20–29 years, 30–39 years, 40–49 years, 50–59 years, and 60 years or older. Within each of these age ranges, there is little variation in diastolic blood pressure. Three subjects are randomly assigned to the treatments in each block, one to each of the three ACE inhibitors. Table 17-5 summarizes the results of the study.
3. **Calculate the test statistic.**
   a. As in the CRD, the test statistic is calculated using the formula

$$F = \frac{\text{variation due to treatments}}{\text{experimental error}}$$

   b. Table 17-6 depicts the general format of the ANOVA table for the RCB design.
      (1) The general layout of the table, as well as the intermediate calculations, are identical to the CRD, with the exception of an additional row for the variation in the response variable due to differences among the blocks (**among blocks** variation).
      (2) The systematic variation in response due to the blocking variable is therefore removed from the experimental error.
      (3) Table 17-7 is the completed ANOVA table for the current example (see V for the computational formulas and associated calculations). The test statistic, listed in the

Table 17-5. Diastolic Blood Pressures for Five Age Groups: The Effect of Three Angiotensin-converting Enzyme (ACE) Inhibitors

Age Range Blocks (Yrs)	Diastolic Blood Pressure Readings (mm Hg)			Block Totals	Block Means
	Drug A	Drug B	Drug C		
20–30	100	90	110	$B_1 = 300$	$\overline{B}_1 = 100$
30–40	105	80	90	$B_2 = 275$	$\overline{B}_2 = 91.7$
40–50	95	80	80	$B_3 = 255$	$\overline{B}_3 = 85.0$
50–60	110	75	100	$B_4 = 285$	$\overline{B}_4 = 95.0$
>60	90	90	95	$B_5 = 275$	$\overline{B}_5 = 91.7$
Sample Size	5	5	5	$N = 15$	
Treatment Totals	$T_1 = 500$	$T_2 = 415$	$T_3 = 475$		
Treatment Means	$\overline{Y}_1 = 100$	$\overline{Y}_2 = 83$	$\overline{Y}_3 = 95$		

Table 17-6. General Format for the Analysis of Variance (ANOVA) Table in a Study Using the Randomized Complete Block (RCB) Design

Source of Variation	*df*	SS	MS	*F*
Among Treatments	df_{tr}	SST	$MST = SST/df_{tr}$	$F = MST/MSE$
Among Blocks	df_{b1}	SSB		
Error	df_e	SSE	$MSE = SSE/df_e$	
Total	df_{total}	SS_{total}		

last column of the table, is

$$F = MST/MSE$$
$$= 381.67/71.25$$
$$= 5.36$$

4. **Evaluate the evidence against H_O.**
   a. **Frequency distribution of the test statistic.** The frequency distribution for the test statistic is the ***F* distribution** (Tables F and G, Appendix 3), with degrees of freedom (see Table 17-7):

$$df_{numerator} = df_{tr} = 2$$
$$df_{denominator} = df_e = 8$$

   b. **Calculating the *p*-value.** Table F (see Appendix 3) gives the critical value for $\alpha = .05$, with $df_{tr} = 2$ and $df_e = 8$, as 4.46 and Table G (see Appendix 3) gives the critical value for $\alpha = .01$ (with the same number of degrees of freedom for each term) as 8.65. Because the calculated test statistic, $F = 5.36$, lies to the right of 4.46, but to the left of 8.65, p is less than .05 and greater than .01. Or, simply, $p < .05$.
   c. **Decision rule.** The decision rule, based on a level of significance $\alpha = .05$, is: **"Reject H_O if $p \le .05$."**
5. **State the conclusion.**
   a. **Statistical conclusion.** Since $p \le .05$, H_O is rejected at the $\alpha = .05$ level of significance.
   b. **Clinical interpretation.** There is at least one statistically significant difference among the mean diastolic blood pressures of the populations of patients receiving each of the three ACE inhibitors ($p < .05$).
   c. **Chance of error.** The probability of error associated with rejecting H_O (the type I error rate) is 5% or less.
   d. As in the CRD, a **multiple comparison procedure** may be used, if the result of the analysis of variance is statistically significant, to determine which pairs of treatment means are different (see IV).

D. **The effect of blocking on experimental error** is demonstrated by Tables 17-4 and 17-7.
   1. Since the 15 values of diastolic blood pressure are the same in both instances, the values for the **total variation** in the readings should be the same for the two designs. Intuitively, the variation due to treatments should be the same as well.

Table 17-7. Analysis of Variance (ANOVA) Table for a Study Comparing the Effect of Three (ACE) Inhibitors on Diastolic Blood Pressure (Matched Subjects)

Source of Variation	*df*	SS	MS	*F*
Among Treatments	2	$SST = 763.33$	$MST = \frac{763}{2} = 381.67$	$F = MST/MSE$
Among Blocks	4	$SSB = 360.0$		$= 5.36$
Error	8	$SSE = 570.0$	$MSE = \frac{570}{8} = 71.25$	
Total	14	$SS_{total} = 1693.33$		

2. Thus, the **difference between the RCB design and the CRD is the loss of the variation attributable to the blocking variable** from the experimental error term.
3. Rearranging the rules for partitioning the total variance in the two designs:

CRD: experimental error = (total variation) − (variation due to treatments)

RCB: experimental error = (total variation) − (variation due to treatments)
− (variation due to blocking)

4. Therefore, the RCB design yields an estimate of experimental error that is free of the systematic effect of the blocking variable and, hence, is smaller than the estimate of experimental error obtained using the CRD (i.e., the **denominator of the test statistic is smaller for the RCB design**). Provided this reduction compensates for the concomitant loss of degrees of freedom, the RCB design increases the power of the statistical test. Furthermore, the treatment groups are **equalized with respect to the blocking variable,** causing its effect as a potentially confounding variable to be virtually eliminated.

IV. MULTIPLE COMPARISON PROCEDURES are those statistical procedures that isolate the specific differences among several population means.

A. Purpose. These procedures try to **determine which members of a set of comparisons are significantly different,** without falling prey to the multiple comparison problem (see I A).

B. Procedures available. Statistical procedures that are designed to accomplish this goal include **Scheffe's test, Tukey's test, Duncan's multiple range test, Dunnett's test, Fisher's least significant difference test,** and **the Bonferroni procedure** (see IV C).

C. The Bonferroni procedure: a specific example.* Table 17-8 summarizes the results of a study comparing the efficacy of four different antihypertensive agents in reducing diastolic blood pressure. Table 17-9 is the completed ANOVA table for this data.

Table 17-8. Reduction in Diastolic Blood Pressure after Treatment with Four Antihypertensive Drugs

	Drug			
	A	**B**	**C**	**D**
Sample Means, $\overline{Y}$:	9.0	13.0	11.0	15.0
Sample Size, n_i:	4	4	4	4
Sample Standard Deviations, s:	1.826	1.826	1.826	1.826

Adapted from Duncan RC, Knapp RG, Miller MC III: *Introductory Biostatistics for the Health Sciences*, 2nd ed. Albany, Delmar, 1983, p 152.

Table 17-9. Analysis of Variance (ANOVA) Table for a Study Comparing the Efficacy of Four Antihypertensive Drugs

Source of Variation	***df***	**SS**	**MS**	***F***
Among Treatments	3	80	26.7	8.1
Error	12	40	3.3	
Total	15	120		

Reprinted from Duncan RC, Knapp RG, Miller MC III: *Introductory Biostatistics for the Health Sciences*, 2nd ed. Albany, Delmar, 1983, p 152.

*This example is adapted from Duncan RC, Knapp RG, Miller MC III: *Introductory Biostatistics for the Health Sciences*, 2nd ed. Albany, Delmar, 1983, pp 152–155.

1. The following five pairs of hypotheses concerning the four drugs are formulated and tested:
   a. $\mathbf{H_O}$: $\mu_A = \mu_B$ versus $\mathbf{H_A}$: $\mu_A \neq \mu_B$
   b. $\mathbf{H_O}$: $\mu_A = \mu_C$ versus $\mathbf{H_A}$: $\mu_A \neq \mu_C$
   c. $\mathbf{H_O}$: $\mu_A = \mu_D$ versus $\mathbf{H_A}$: $\mu_A \neq \mu_D$
   d. $\mathbf{H_O}$: $\mu_B = \mu_C$ versus $\mathbf{H_A}$: $\mu_B \neq \mu_C$
   e. $\mathbf{H_O}$: $\mu_C = \mu_D$ versus $\mathbf{H_A}$: $\mu_C \neq \mu_D$
2. Each of these five individual pairs may be tested using the t statistic, computed from the formula

$$t = \frac{\bar{Y}_i - \bar{Y}_j}{\sqrt{\text{MSE}\left(\frac{1}{n_i} + \frac{1}{n_j}\right)}}$$

   where MSE is the pooled estimate of within-group variation (i.e., the experimental error), derived from the ANOVA table.
   a. The number of degrees of freedom for this test is equal to the degrees of freedom for the error sum of squares (df_e).
   b. Thus, the t statistics corresponding to each of the five hypotheses formulated in IV C 1 are

(1) $$t = \frac{\bar{Y}_A - \bar{Y}_B}{\sqrt{\text{MSE}\left(\frac{1}{n_A} + \frac{1}{n_B}\right)}} = \frac{9.0 - 13.0}{\sqrt{3.3\left(\frac{1}{4} + \frac{1}{4}\right)}} = -3.11$$

(2) $$t = \frac{\bar{Y}_A - \bar{Y}_C}{\sqrt{\text{MSE}\left(\frac{1}{n_A} + \frac{1}{n_C}\right)}} = \frac{9.0 - 11.0}{\sqrt{3.3\left(\frac{1}{4} + \frac{1}{4}\right)}} = -1.56$$

(3) $$t = \frac{\bar{Y}_A - \bar{Y}_D}{\sqrt{\text{MSE}\left(\frac{1}{n_A} + \frac{1}{n_D}\right)}} = \frac{9.0 - 15.0}{\sqrt{3.3\left(\frac{1}{4} + \frac{1}{4}\right)}} = -4.67$$

(4) $$t = \frac{\bar{Y}_B - \bar{Y}_C}{\sqrt{\text{MSE}\left(\frac{1}{n_B} + \frac{1}{n_C}\right)}} = \frac{13.0 - 11.0}{\sqrt{3.3\left(\frac{1}{4} + \frac{1}{4}\right)}} = 1.56$$

(5) $$t = \frac{\bar{Y}_C - \bar{Y}_D}{\sqrt{\text{MSE}\left(\frac{1}{n_C} + \frac{1}{n_D}\right)}} = \frac{11.0 - 15.0}{\sqrt{3.3\left(\frac{1}{4} + \frac{1}{4}\right)}} = -3.11$$

3. To ensure that the **overall error rate per experiment** (the overall error rate for the set of five comparisons) does not exceed the chosen level of significance α, each **individual** test must be carried out at the α/c level of significance, where c = the number of comparisons. For example, to preserve an overall level of significance α = .05 for the five comparisons in this example, each t test must be carried out at the .05/5, or .01, level of significance.
   a. Each calculated test statistic is compared with the tabulated critical value. For a one-tailed test using $\alpha = 1$, $t_{1-\alpha} = t_{.99}$, and $t_{.99}$ with 12 df is 2.681. For a two-tailed test (again, $df = 12$), $t_{(.01/2)} = t_{.005} = 3.0545$.
   b. **Decision rule.** For a two-tailed test (H_A: $\mu_i \neq \mu_j$) and an overall level of significance α = .05, the decision rule is: **If $t > 3.0545$, declare population means μ_i and μ_j significantly different from each other at the α = .05 level of significance.**
   c. Based on this decision rule, the physician can conclude simultaneously that drug A differs from drug B and drug D, and that drug C differs from drug D.
4. The Bonferroni procedure often requires t values not listed in standard t tables.
   a. The necessary values of t may be obtained from a standard t table by linear interpolation.
   b. Alternatively, an approximate value, corresponding to the upper $\alpha/2$ proportion of t with v degrees of freedom, can be calculated from the normal distribution, using the formula

$$t_{(1-\alpha/2)} = z + \frac{z^3 + z}{4(v-2)}$$

where z is the value from the normal distribution corresponding to an area of $(1 - \alpha/2)$ [i.e., $z_{(1-\alpha/2)}$; see Appendix 3, Table B].

c. In the study summarized in Table 17-8, a total of six different comparisons can be carried out between the four population means. Thus, for the set of six comparisons, the critical value of t for a two-tailed test $[t_{1-.05/6(2)}]$ may be estimated as

$$t_{.996} = z_{.996} + \frac{z_{.996}^{\ 3} + z_{.996}}{(4)(10)}$$

$$= 2.65 + \frac{(2.65)^3 + 2.65}{40}$$

$$= 3.18$$

d. The six calculated *t* statistics are compared with the estimated critical value, $t = 3.18$, to determine which pairs of means are significantly different. As before, when a calculated value of t exceeds the critical value, the corresponding population means μ_i and μ_j are declared significantly different at the $\alpha = .05$ level of significance.

5. When the **number of comparisons is large, the Bonferroni procedure is very conservative** (i.e., it has a very low false positive rate but a high false negative rate). As the number of comparisons increases, so does the tabulated critical value. As a result, it is increasingly difficult to detect significant differences between the pairs of means; often, it is more difficult than really necessary. Under these circumstances, a less stringent multiple comparison procedure may be preferred.

V. COMPUTATIONAL FORMULAS AND CALCULATIONS

A. CRD. The calculation of the sums of squares, degrees of freedom, and mean squares (see Table 17-4) for the example given in II B are described below.

1. Total variation and $\mathbf{SS_{total}}$. The total variation is defined as

$$\text{total variation} = \frac{SS_{total}}{df_{total}}$$

and

$$SS_{total} = \sum_{i=1}^{k} \sum_{j=1}^{n_i} Y_{ij}^2 - G^2/N \qquad (17.1)$$

$$df_{total} = N - 1 \qquad (17.2)$$

where $\boldsymbol{k}$ is the number of treatment groups, $\boldsymbol{n_i}$ is the number of observations in the ith treatment group, $\boldsymbol{N} = n_1 + n_2 + \cdots + n_k$, $\Sigma\Sigma\, \boldsymbol{Y_{ij}^2}$ = the sum of squares for all N observations, and $\boldsymbol{G}$ is the sum of all N observations (**grand total**). For the example described in II B,

$$N = n_1 + n_2 + n_3 = 5 + 5 + 5$$
$$= 15$$

$$G = \sum \sum Y_{ij} = 100 + 105 + \cdots + 100 + 95$$
$$= 1390$$

$$SS_{total} = \sum_{i=1}^{k} \sum_{j=1}^{n_i} Y_{ij}^2 - G^2/N = [100^2 + 105^2 + \cdots + 100^2 + 95^2] - (1390)^2/15$$
$$= 1693.3333$$

$$df_{total} = N - 1 = 15 - 1$$
$$= 14$$

2. **Variation due to treatments** (Table 17-4, *row 1*). The sum of squares for the treatments (SST), degrees of freedom (df_{tr}), and the mean square (MST) are computed using the following formulas:

$$\text{SST} = \sum_{i=1}^{k} [T_i^2/(\text{subjects in } i\text{th group})] - G^2/N \qquad (17.3)$$

$$df_{tr} = k - 1 \qquad (17.4)$$

$$\text{MST} = \text{SST}/df_{tr} \qquad (17.5)$$

where $\boldsymbol{k}$ is the number of treatments, $\boldsymbol{n_i}$ is the number of observations in the ith treatment group, $\boldsymbol{N} = n_1 + n_2 + \cdots + n_k$, $\boldsymbol{T_i} = \Sigma Y_{ij}$ for treatment group i, and $\boldsymbol{G} = \Sigma \Sigma Y_{ij}$ = sum of all N observations in the study. For this example,

$$\text{SST} = \sum_{i=1}^{3} [T_i^2/n_i] - G^2/n = \left[\frac{500^2}{5} + \frac{415^2}{5} + \frac{475^2}{5}\right] - \frac{(1390)^2}{15}$$

$$= 763.33$$

$$df_{tr} = k - 1 = 3 - 1$$

$$= 2$$

$$\text{MST} = \text{SST}/df_{tr} = 763.333/2$$

$$= 381.6667$$

3. **Experimental error** (see Table 17-4, *row 2*). The sum of squares and degrees of freedom for experimental error can be calculated as follows:

$$\text{SSE} - \text{SS}_{total} - \text{SST} \qquad (17.6)$$

$$df_e = N - k \qquad (17.7)$$

The equation $df_e = df_{total} - df_{tr}$ can also be used to calculate the degrees of freedom for experimental error.

a. Thus, in this example,

$$\text{SSE} = 1693.333 - 763.333$$

$$= 930$$

$$df_e = 15 - 3$$

$$= 12$$

and the corresponding mean square is

$$\text{MSE} = \text{SSE}/df_e = (\text{SSE})/(N - k) \qquad (17.8)$$

$$= 930/12$$

$$= 77.5$$

b. MSE can also be calculated using an extension of the formula for calculating the pooled variance s_p^2 [see Ch 16 II B 1 c (1) (a)], as follows:

$$\text{MSE} = s_p^2 = \frac{(n_1 - 1)s_1^2 + (n_2 - 1)s_2^2 + \cdots + (n_k - 1)s_k^2}{n_1 + n_2 + \cdots + n_k - k}$$

$$= \frac{\text{SSE}}{N - k}$$

where s_i^2 represents the variance of the measurements in treatment group i.

4. The test statistic, F, is the ratio of MST to MSE; that is,

$$F = \text{MST}/\text{MSE} \tag{17.9}$$
$$= 381.6667/77.5$$
$$= 4.92$$

B. RCB design (see III C; Table 17-7)

1. The sums of squares and degrees of freedom for the total variation and variation due to treatments, as well as the mean square for treatments, are calculated in the same fashion as that described above for the CRD.

2. The computational formulas for the sum of squares, degrees of freedom, and mean square for the variation among the blocks (Table 17-7, *row 2*) are analogous to those used to compute these values for the variation due to treatments. These formulas are:

$$\text{SSB} = \sum_{j=1}^{b} [B_j^2/(\text{number of subjects in block } j)] - G^2/N \tag{17.10}$$

$$df_{\text{bl}} = b - 1 \tag{17.11}$$

$$\text{MSB} = \text{SSB}/(b - 1) \tag{17.12}$$

where $\boldsymbol{B_j}$ is the sum of observations in block j, $\boldsymbol{b}$ is the number of blocks, $\boldsymbol{G}$ is the sum of all observations **(grand total)**, and $\boldsymbol{N} = bk$ = the total number of study participants. For the example in III C,

$$\text{SSB} = \left[\frac{300^2}{3} + \frac{275^2}{3} + \frac{255^2}{3} + \frac{285^2}{3} + \frac{275^2}{3}\right] - \frac{(1390)^2}{15} = 360.0$$

$$df_{\text{bl}} = 5 - 1 = 4$$

$$\text{MSB} = 360/4 = 90$$

Ordinarily, MSB is not computed since it is not used in the calculation of the test statistic F.

3. **Experimental error** (see Table 17-7, *row 3*)
   a. The sum of squares and degrees of freedom for this term are obtained by subtraction, as follows:

$$\text{SSE} = \text{SS}_{\text{total}} - \text{SST} - \text{SSB} \tag{17.13}$$

$$df_{\text{e}} = df_{\text{total}} - df_{\text{tr}} - df_{\text{bl}} \tag{17.14}$$

   Thus, for this example,

$$\text{SSE} = 1693.333 - 763.333 - 360.0 = 570$$

$$df_{\text{e}} = 14 - 2 - 4 = 8$$

   b. As in the CRD, the mean square for experimental error is

$$\text{MSE} = \text{SSE}/df_{\text{e}} = 570/8 = 71.25$$

4. The test statistic F is the ratio of MST and MSE (formula 17.9):

$$F = \text{MST}/\text{MSE} = 381.67/71.25 = 5.36$$

BIBLIOGRAPHY

Duncan RC, Knapp RG, Miller MC III: *Introductory Biostatistics for the Health Sciences*, 2nd ed. Albany, Delmar, 1983.

Elston RC, Johnson WD: *Essentials of Biostatistics*. Philadelphia, FA Davis, 1987, pp 210–232.

Kirk RE: *Procedures for the Behavioral Sciences*. Belmont, CA, Wadsworth, 1968.

Steel RG, Torrie JH: *Principles and Procedures of Statistics*, 2nd ed. New York, McGraw-Hill, 1980, pp 124–125.

PROBLEMS*

17-1. A double blind clinical study was conducted to compare the analgesic efficacy of ibuprofen (400 mg), codeine (60 mg), codeine (30 mg), and a placebo in patients with chronic lower back pain. Twenty participants were randomly assigned to the four treatment groups ($n = 5$ for each group). Two hours after receiving the assigned treatment, patients were asked to rate their degree of pain relief on a scale from 0 to 100. The results are summarized in the table below.

Pain Relief Score			
Ibuprofen (400 mg)	**Codeine (60 mg)**	**Codeine (30 mg)**	**Placebo**
82	80	77	65
89	70	69	75
77	72	67	67
72	90	65	55
92	68	57	63

Adapted from Duncan RC, Knapp RG, Miller MC III: *Introductory Biostatistics for the Health Sciences*, 2nd ed. Albany, Delmar, 1983, p 157.

a. What experimental design was used in this study?

b. State H_O and H_A for the statistical test.

c. Calculate the sums of squares (SS), degrees of freedom (*df*), mean squares (MS), and *F*, and use these values to construct an ANOVA table.

d. Based on the result obtained in problem 17-1c, do the data obtained in this study provide sufficient evidence, at the .05 level of significance, to support the existence of a difference in perceived pain relief among the four treatments? If so, which treatments are different (use an error rate per experiment of 5%)?

17-2. A psychiatrist conducted a study to compare the effects of desipramine, trazodone, and fluoxetine for the treatment of anxious depression in nonpsychotic patients. Twelve such patients, all suffering from a moderate-to-severe major depressive episode accompanied by anxiety, were assigned at random to the three treatment groups. At the end of one month of therapy, a combined anxiety–depression score was recorded for each patient, based on a combination of that patient's responses on the Minnesota Multiphasic Personality Inventory (MMPI) and the Taylor Manifest Anxiety Scale. The results are shown in the table below.

Combined Anxiety–Depression Scores		
Desipramine	**Trazodone**	**Fluoxetine**
25	20	25
15	16	15
20	18	20
14	25	20

Adapted from Duncan RC, Knapp RG, Miller MC III: *Introductory Biostatistics for the Health Sciences*, 2nd ed. Albany, Delmar, 1983, p 158.

a. What experimental design was used for this study?

b. What are H_O and H_A for the statistical test?

c. Calculate the sums of squares, degrees of freedom, mean squares, and *F*. Use these values to construct the ANOVA table.

*The problems in this section are adapted from Duncan RC, Knapp RG, Miller MC III: *Introductory Biostatistics for the Health Sciences*, 2nd ed. Albany, Delmar, 1983, pp 156–158.

d. Based on the result obtained in problem 17-2c, is there a statistically significant difference in mean anxiety–depression scores among the three treatment groups? If so, which drugs are different (use an error rate per experiment of 5%)?

17-3. The psychiatrist described in problem 17-2 replicated this study with a second sample of nonpsychotic patients suffering from anxious depression. She organized these patients into four groups, according to the severity of their illness prior to initiating drug treatment, then randomly assigned the subjects in each of the severity groups to the three drugs (desipramine, trazodone, or fluoxetine). The table below summarizes the combined anxiety–depression scores for the patients at the end of the one-month treatment period.

Block	Combined Anxiety–Depression Score			Block
Initial Level	Desipramine	Trazodone	Fluoxetine	Totals (B_j)
1	35	30	25	90
2	40	25	20	85
3	25	25	20	70
4	30	25	25	80

Adapted from Duncan RC, Knapp RG, Miller MC III: *Introductory Biostatistics for the Health Sciences,* 2nd ed. Albany, Delmar, 1983, p 159.

a. What experimental design was used in this study?

b. State H_O and H_A for the statistical test.

c. Compute the sums of squares, degrees of freedom, mean squares, and *F*, and construct the ANOVA table.

d. Based on the results obtained in problem 17-3c, is there a significant difference in mean anxiety–depression scores for the three drugs (use a 5% level of significance)? If so, which drugs are different, based on an overall error rate per experiment of 5%?

17-4. In a quality control study of clinical chemistry laboratories, samples of four different enzymes were sent to three commercial laboratories. The physician conducting the study wished to compare test results from the three labs. The table below summarizes the data obtained in this study.

Lab	Enzyme Concentration (μg/ml)				Treatment
	1	2	3	4	Totals (T_i)
A	4.2	6.0	3.9	8.3	22.4
B	3.9	7.3	4.0	7.2	22.4
C	5.2	6.5	3.2	6.9	21.8

Adapted from Duncan RC, Knapp RG, Miller MC III: *Introductory Biostatistics for the Health Sciences,* 2nd ed. Albany, Delmar, 1983, p 156.

a. What experimental design was used in this study?

b. What are H_O and H_A for the statistical test?

c. Compute the sums of squares, degrees of freedom, mean squares, and *F*. Use these values to construct the ANOVA table.

d. Based on the answer to problem 17-4c, what conclusion can be drawn regarding the relative performance of the three commercial laboratories (use a 5% level of significance for the test)?

Solutions on p 402.

STUDY QUESTIONS

Directions: Each of the numbered items or incomplete statements in this section is followed by answers or by completions of the statement. Select the **one** lettered answer or completion that is **best** in each case.

Questions 1–4

A researcher wishes to compare the effect of temperature on serum bilirubin levels in the newborn. He first conducts a pilot study to determine the effect of temperature on bilirubin levels of newborn rat pups. Several strains of rats are available; the researcher suspects that these strains vary with respect to baseline bilirubin levels.

1. Which of the following experimental designs should the investigator choose to account for the inherent strain differences in serum bilirubin levels?

(A) CRD
(B) RCB design
(C) Independent sample design
(D) None of the above

2. Based on the experimental design selected in question 1, which of the following procedures would be used to assign subjects randomly to the treatment groups?

(A) Randomly select a convenient number of rat pups from the total population and assign each pup at random to one of the treatments (temperatures)
(B) From each strain, randomly select a group of rat pups equal in size to the number of treatments, then, within each group, randomly assign one rat to each treatment
(C) Assign the first rat pup selected to the first temperature group, the second to the second temperature group, and so on until all available rat pups have been assigned to a comparison group
(D) None of the above

3. If the researcher wishes to use five different temperatures and four strains of rats, how many rat pups does the experiment require?

(A) 5
(B) 4
(C) 9
(D) 20
(E) Cannot be determined from the information given

4. The experiment is carried out with three temperatures and four strains of rats. The F value (test statistic) is calculated as 6.45, while the tabulated critical value for $\alpha = .05$ is 5.14. What is the most appropriate conclusion that can be drawn from this result?

(A) There is a statistically significant difference in mean serum bilirubin levels among the populations of rat pups exposed to the three temperatures ($p < .05$)
(B) No statistically significant difference in mean serum bilirubin levels exists among the populations of rat pups exposed to the three temperatures ($p < .05$)
(C) There is a statistically significant difference in the strain of rats among the three temperatures ($p < .05$)
(D) There is a statistically significant difference in mean bilirubin levels at the three temperatures among the populations of rat pups tested ($p > .05$)
(E) Serum bilirubin levels do not differ among the three temperature settings studied ($p > .05$)

1-B
2-B
3-D
4-A

Questions 5–8

An experiment was conducted to compare the effect of four drugs on serum cholesterol levels in patients with type III familial hypercholesteremia. Three patients were randomly assigned to receive drug A, five to receive drug B, four to receive drug C, and four were randomly assigned to receive drug D. Serum cholesterol levels were recorded after six months of drug therapy.

5. What experimental design was used in this study?

(A) CRD
(B) Paired design
(C) RCB design
(D) Case–control design

6. The investigators report that the *F* value obtained in this study was not statistically significant ($p > .10$). All of the following statements about this conclusion are correct EXCEPT

(A) it is likely that any observed differences in average serum cholesterol levels in response to the four drug treatments occurred by random chance
(B) average serum cholesterol levels do not differ among the four treatment groups
(C) the probability of a type II error (i.e., a failure to detect a difference in population means given that such a difference actually exists) may be high
(D) a clinically important difference in average serum cholesterol levels after the four drug treatments may exist, but the power of the statistical test may have been too low to detect it

7. The data obtained in this experiment were analyzed using an analysis of variance. How many degrees of freedom are associated with the error sum of squares?

(A) 16
(B) 3
(C) 12
(D) 15
(E) None of the above

8. The use of multiple *t* tests (with a 5% level of significance for each test), rather than an analysis of variance, to analyze the data obtained in this study would accomplish which of the following?

(A) A decrease in the chance of a type I error
(B) A decrease in the likelihood of erroneously declaring a difference in means statistically significant
(C) An increase in the overall type I error rate for the entire set of comparisons
(D) A decrease in the probability of a false positive result

9. All of the following statements about the Bonferroni multiple comparison procedure are true EXCEPT

(A) it is used to determine which of a set of means are different while simultaneously controlling for the multiple comparison problem
(B) it is used to control the overall error rate per experiment for the entire set of comparisons
(C) it requires that each of the c individual comparisons be carried out at the α/c level of significance (for a one-tailed test) to preserve the desired overall level of significance α
(D) it represents a modification of the *t* test
(E) it is unnecessary when the *F* value computed in the analysis of variance is statistically significant

---

5-A
6-B
7-C
8-C
9-E

ANSWERS AND EXPLANATIONS

1–4. The answers are: 1-B *[III A 5]*, **2-B** *[III B 3]*, **3-D** *[III B 3 b]*, **4-A** *[III C 4 b, 5 a]*
The additional source of variability in the response variable (i.e., strain differences in baseline serum bilirubin levels) can be controlled by blocking. Thus, the randomized complete block (RCB) design should be chosen to account for the suspected strain differences. The completely random design (CRD), also known as the independent sample design, does not control for strain differences over and above randomization.

When the RCB design is chosen, randomization is carried out within the blocks. Thus, in each block, each treatment is represented by at least one subject. In contrast, when the CRD is selected, a convenient number of subjects is selected from the total population and each subject is randomly assigned to one of the treatment groups. The systematic assignment of subjects to groups may result in a biased comparison of the treatments. It should be avoided in all study designs.

Each treatment must appear at least once within each homogeneous block. Thus, for five treatments (temperatures) and four blocks (strains), the experimenter needs 20 rat pups.

The calculated test statistic, $F = 6.45$, exceeds the tabulated critical value for $\alpha = .05$. The p-value, therefore, is less than .05. Since $p < (\alpha = .05)$, the null hypothesis (H_O) is rejected at the 5% level of significance. There is a statistically significant difference between at least two mean serum bilirubin levels.

5–8. The answers are: 5-A *[II A]*, **6-B** *[II B 5 e; Ch 12 VI C 2]*, **7-C** *[V A 3]*, **8-C** *[I A]*
There were no restrictions on randomization in this study; hence, the completely random design was used. In contrast, the randomized complete block design first groups subjects into homogeneous strata or blocks. Subjects are then randomly assigned to the treatments within each block. A paired study is inappropriate because independent samples were used to compare four treatments. The case–control study design is an observational study design. Subjects are not randomly assigned to comparison groups, but rather "self-select" their group status.

While a failure to achieve statistical significance may, in fact, reflect the absence of differences in serum cholesterol level among the populations of patients receiving the four drugs, it may also indicate low power of the statistical test. Therefore, unless power is known to be high, a failure to reject the null hypothesis (H_O) does not permit the investigator to automatically conclude that no difference exists among the population means. The p-value does not provide the necessary information on β (the probability of a type II error), or its complement, power.

The number of degrees of freedom associated with the error sum of squares is $N - k$, where N = the total number of observations and k = the number of treatment groups. Thus, for this study, $df_e = 16 - 4 = 12$. Since $df_{total} = df_{tr} + df_e$, the degrees of freedom for error can also be calculated by subtraction; that is, $df_e = df_{total} - df_{tr} = (N - 1) - (k - 1) = 15 - 3 = 12$.

The use of multiple t tests increases the probability of a type I error (false positive result) for the entire set of comparisons. That is, the likelihood of detecting a statistically significant difference between at least two means that does not actually exist is greater than when the analysis of variance is used to analyze the data.

9. The answer is E *[II B 5 d; III C 5 d; IV C]*
The Bonferroni procedure is used to carry out multiple comparisons when a significant F value has been obtained in the analysis of variance. This procedure is a modification of the t test, which requires that each individual comparison be conducted at the α/c level of significance (for the one-tailed test; $\alpha/2c$ is used for a two-tailed test). This modification ensures that the overall error rate per experiment for the set of c comparisons is no greater than α.

Comprehensive Exam

Introduction

One of the least attractive aspects of pursuing an education is the necessity of being examined on what has been learned. Instructors do not like to prepare tests, and students do not like to take them.

However, students are required to take many examinations during their learning careers, and little if any time is spent acquainting them with the positive aspects of tests and with systematic and successful methods for approaching them. Students perceive tests as punitive and sometimes feel that they are merely opportunities for the instructor to discover what the student has forgotten or has never learned. Students need to view tests as opportunities to display their knowledge and to use them as tools for developing prescriptions for further study and learning.

A brief history and discussion of the National Board of Medical Examiners (NBME) examinations (i.e., Parts I, II, and III and FLEX) are presented here, along with ideas concerning psychological preparation for the examinations. Also presented are general considerations and test-taking tips, as well as ways to use practice exams as educational tools. (The literature provided by the various examination boards contains detailed information concerning the construction and scoring of specific exams.)

National Board of Medical Examiners Examinations

Before the various NBME exams were developed, each state attempted to license physicians through its own procedures. Differences between the quality and testing procedures of the various state examinations resulted in the refusal of some states to recognize the licensure of physicians licensed in other states. This made it difficult for physicians to move freely from one state to another and produced an uneven quality of medical care in the United States.

To remedy this situation, the various state medical boards decided they would be better served if an outside agency prepared standard exams to be given in all states, allowing each state to meet its own needs and have a common standard by which to judge the educational preparation of individuals applying for licensure.

One misconception concerning these outside agencies is that they are licensing authorities. This is not the case; they are examination boards only. The individual states retain the power to grant and revoke licenses. The examination boards are charged with designing and scoring valid and reliable tests. They are primarily concerned with providing the states with feedback on how examinees have performed and with making suggestions about the

---

The author of this introduction, Michael J. O'Donnell, holds the positions of Assistant Professor of Psychiatry and Director of Biomedical Communications at the University of New Mexico School of Medicine, Albuquerque, New Mexico.

interpretation and usefulness of scores. The states use this information as partial fulfillment of qualifications upon which they grant licenses.

Students should remember that these exams are administered nationwide and, although the general medical information is the same, educational methodologies and faculty areas of expertise differ from institution to institution. It is unrealistic to expect that students will know all the material presented in the exams; they may face questions on the exams in areas that were only superficially covered in their classes. The testing authorities recognize this situation, and their scoring procedures take it into account.

The Exams

The first exam was given in 1916. It was a combination of written, oral, and laboratory tests, and it was administered over a 5-day period. Admission to the exam required proof of completion of medical education and 1 year of internship.

In 1922, the examination was changed to a new format and was divided into three parts. Part I, a 3-day essay exam, was given in the basic sciences after 2 years of medical school. Part II, a 2-day exam, was administered shortly before or after graduation, and Part III was taken at the end of the first postgraduate year. To pass both Part I and Part II, a score equaling 75% of the total points available in each was required.

In 1954, after a 3-year extensive study, the NBME adopted the multiple-choice format. To pass, a statistically computed score of 75 was required, which allowed comparison of test results from year to year. In 1971, this method was changed to one that held the mean constant at a computed score of 500, with a predetermined deviation from the mean to ascertain a passing or failing score. The 1971 changes permitted more sophisticated analysis of test results and allowed schools to compare among individual students within their respective institutions as well as among students nationwide. Feedback to students regarding performance included the reporting of pass or failure along with scores in each of the areas tested.

During the 1980s, the ever-changing field of medicine made it necessary for the NBME to examine once again its evaluation strategies. It was found necessary to develop questions in multidisciplinary areas such as gerontology, health promotion, immunology, and cell and molecular biology. In addition, it was decided that questions should test higher cognitive levels and reasoning skills.

To meet the new goals, many changes have been made in both the form and content of the examination. These changes include reduction in the number of questions to approximately 800 in Part I and Part II to allow students more time on each question, with total testing time reduced on Part I from 13 to 12 hours and on Part II from 12.5 to 12 hours. The basic science disciplines are no longer allotted the same number of questions, which permits flexible weighing of the exam areas. Reporting of scores to schools includes total scores for individuals and group mean scores for separate discipline areas. Only pass/fail designations and total scores are reported to examinees. There is no longer a provision for the reporting of individual subscores to either the examinees or medical schools. Finally, the question format used in the new exams, now referred to as Comprehensive (Comp) I and II, is predominately multiple-choice, best-answer.

The New Format

New questions, designed specifically for Comp I, are constructed in an effort to test the student's grasp of the sciences basic to medicine in an integrated fashion—the questions are designed to be interdisciplinary. Many of these items are presented as vignettes, or case studies, followed by a series of multiple-choice, best-answer questions.

The scoring of this exam is altered. Whereas in the past the exams were scored on a normal curve, the new exam has a predetermined standard, which must be met in order to pass. The exam no longer concentrates on the trivial; therefore, it has been concluded that there is a common base of information that all medical students should know in order to pass. It is anticipated that a major shift in the pass/fail rate for the nation is unlikely. In the past, the average student could only expect to feel comfortable with half the test and eventually would complete approximately 67% of the questions correctly, to achieve a mean score of 500. Although with the standard setting method it is likely that the mean score will change and become higher, it is unlikely that the pass/fail rates will differ significantly from those in the past. During the first testing in 1991, there will not be differential weighing of questions. However, in the future, the NBME will be researching methods of weighing questions based on both the time it takes to answer questions vis-à-vis their difficulty and the perceived importance of the information. In addition, the NBME is attempting to design a method of delivering feedback to the student that will have considerable importance in discovering weaknesses and pinpointing areas for further study in the event that a retake is necessary.

Since many of the proposed changes will be implemented for the first time in June 1991, specific information regarding actual standards, question emphasis, pass/fail rates, and so forth were unavailable at the time of publication. The publisher will update this section as information becomes available and as we attempt to follow the evolution and changes that occur in the area of physician evaluation.

Materials Needed for Test Preparation

In preparation for a test, many students collect far too much study material only to find that they simply do not have the time to go through all of it. They are defeated before they begin because either they leave areas unstudied, or they race through the material so quickly that they cannot benefit from the activity.

It is generally more efficient for the student to use materials already at hand; that is, class notes, one good outline to cover or strengthen areas not locally stressed and to quickly review the whole topic, and one good text as a reference for looking up complex material needing further explanation.

Also, many students attempt to memorize far too much information, rather than learning and understanding less material and then relying on that learned information to determine the answers to questions at the time of the examination. Relying too heavily on memorized material causes anxiety, and the more anxious students become during a test, the less learned knowledge they are likely to use.

Positive Attitude

A positive attitude and a realistic approach are essential to successful test taking. If concentration is placed on the negative aspects of tests or on the potential for failure, anxiety increases and performance decreases. A negative attitude generally develops if the student concentrates on "I must pass" rather than on "I can pass." "What if I fail?" becomes the major factor motivating the student to **run from failure rather than toward success**. This results from placing too much emphasis on scores rather than understanding that scores have only slight relevance to future professional performance.

The score received is only one aspect of test performance. Test performance also indicates the student's ability to use information during evaluation procedures and reveals how this ability might be used in the future. For example, when a patient enters the physician's office with a problem, the physician begins by asking questions, searching for clues, and seeking diagnostic information. Hypotheses are then developed, which will include several potential causes for the problem. Weighing the probabilities, the physician will begin to discard those hypotheses with the least likelihood of being correct. Good differential diagnosis involves the ability to deal with uncertainty, to reduce potential causes to the smallest number, and to use all learned information in arriving at a conclusion.

The same thought process can and should be used in testing situations. It might be termed **paper-and-pencil differential diagnosis**. In each question with five alternatives, of which one is correct, there are four alternatives that are incorrect. If deductive reasoning is used, as in solving a clinical problem, the choices can be viewed as having possibilities of being correct. The elimination of wrong choices increases the odds that a student will be able to recognize the correct choice. Even if the correct choice does not become evident, the probability of guessing correctly increases. Just as differential diagnosis in a clinical setting can result in a correct diagnosis, eliminating choices on a test can result in choosing the correct answer.

Answering questions based on what is incorrect is difficult for many students since they have had nearly 20 years experience taking tests with the implied assertion that knowledge can be displayed only by knowing what is correct. It must be remembered, however, that students can display knowledge by knowing something is wrong, just as they can display it by knowing something is right. **Students should begin to think in the present as they expect themselves to think in the future.**

Paper-and-Pencil Differential Diagnosis

The technique used to arrive at the answer to the following question is an example of the paper-and-pencil differential diagnosis approach.

A recently diagnosed case of hypothyroidism in a 45-year-old man may result in which of the following conditions?

(A) Thyrotoxicosis

(B) Cretinism

(C) Myxedema

(D) Graves' disease

(E) Hashimoto's thyroiditis

It is presumed that all of the choices presented in the question are plausible and partially correct. If the student begins by breaking the question into parts and trying to discover what the question is attempting to measure, it will be possible to answer the question correctly by using more than memorized charts concerning thyroid problems.

- The question may be testing if the student knows the difference between "hypo" and "hyper" conditions.
- The answer choices may include thyroid problems that are not "hypothyroid" problems.
- It is possible that one or more of the choices are "hypo" but are not "thyroid" problems, that they are some other endocrine problems.
- "Recently diagnosed in a 45-year-old man" indicates that the correct answer is not a congenital childhood problem.
- "May result in" as opposed to "resulting from" suggests that the choices might include a problem that **causes** hypothyroidism rather than **results from** hypothyroidism, as stated.

By applying this kind of reasoning, the student can see that choice **A,** thyroid toxicosis, which is a disorder resulting from an overactive thyroid gland ("hyper"), must be eliminated. Another piece of knowledge, that is, Graves' disease is thyroid toxicosis, eliminates choice **D**. Choice **B,** cretinism, is indeed hypothyroidism, but is a childhood disorder. Therefore, **B** is eliminated. Choice **E** is an inflammation of the thyroid gland—here the clue is the suffix "itis." The reasoning is that thyroiditis, being an inflammation, may **cause** a thyroid problem, perhaps even a hypothyroid problem, but there is no reason for the reverse to be true. Myxedema, choice **C,** is the only choice left and the obvious correct answer.

Preparing for Board Examinations

1. Study for yourself. Although some of the material may seem irrelevant, the more you learn now, the less you will have to learn later. Also, do not let the fear of the test rob you of an important part of your education. If you study to learn, the task is less distasteful than studying solely to pass a test.

2. Review all areas. You should not be selective by studying perceived weak areas and ignoring perceived strong areas. This is probably the last time you will have the time and the motivation to review **all** of the basic sciences.

3. Attempt to understand, not just memorize, the material. Ask yourself: To whom does the material apply? Where does it apply? When does it apply? Understanding the connections among these points allows for longer retention and aids in those situations when guessing strategies may be needed.

4. Try to anticipate questions that might appear on the test. Ask yourself how you might construct a question on a specific topic.

5. Give yourself a couple days of rest before the test. Studying up to the last moment will increase your anxiety and cause potential confusion.

Taking Board Examinations

1. In the case of NBME exams, be sure to **pace yourself** to use the time optimally. Each booklet is designed to take 2 hours. You should use all of your allotted time; if you finish too early, you probably did so by moving too quickly through the test.

2. Read each question and all the alternatives carefully before you begin to make decisions. Remember the questions contain clues, as do the answer choices. As a physician, you would not make a clinical decision without a complete examination of all the data: the same holds true for answering test questions.

3. Read the directions for each question set carefully. You would be amazed at how many students make mistakes in tests simply because they have not paid close attention to the directions.

4. It is not advisable to leave blanks with the intention of coming back to answer the questions later. Because of the way Board examinations are constructed, you probably will not pick up any new information that will help you when you come back, and the chances of getting numerically off on your answer sheet are greater than your chances of benefiting by skipping around. If you feel that you must come back to a question, mark the best choice and place a note in the margin. Generally speaking, it is best not to change answers once you have made a decision. Your intuitive reaction and first response are correct more often than changes made out of frustration or anxiety. **Never turn in an answer sheet with**

blanks. Scores are based on the number that you get correct; you are not penalized for incorrect choices.

5. **Do not try to answer the questions on a stimulus-response basis.** It generally will not work. Use all of your learned knowledge.

6. **Do not let anxiety destroy your confidence.** If you have prepared conscientiously, you know enough to pass. Use all that you have learned.

7. **Do not try to determine how well you are doing as you proceed.** You will not be able to make an objective assessment, and your anxiety will increase.

8. **Do not expect a feeling of mastery** or anything close to what you are accustomed to. Remember, this is a nationally administered exam, not a mastery test.

9. **Do not become frustrated or angry** about what appear to be bad or difficult questions. You simply do not know the answers; you cannot know everything.

Specific Test-Taking Strategies

Read the entire question carefully, regardless of format. Test questions have multiple parts. Concentrate on picking out the pertinent key words that might help you begin to problem-solve. Words such as "always," "never," "mostly," "primarily," and so forth play significant roles. In all types of questions, distractors with terms such as "always" or "never" most often are incorrect. Adjectives and adverbs can completely change the meaning of questions—pay close attention to them. Also, medical prefixes and suffixes (e.g., "hypo-," "hyper-," "-ectomy," "-itis") are sometimes at the root of the question. The knowledge and application of everyday English grammar often is the key of dissecting questions.

Multiple-Choice Questions

Read the question and the choices carefully to become familiar with the data as given. Remember, in multiple-choice questions there is one correct answer and there are four distractors, or incorrect answers. (Distractors are plausible and possibly correct or they would not be called distractors.) They are generally correct for part of the question but not for the entire question. Dissecting the question into parts aids in discerning these distractors.

If the correct answer is not immediately evident, begin eliminating the distractors. (Many students feel that they must always start at option A and make a decision before they move to B, thus forcing decisions they are not ready to make.) Your first decisions should be made on those choices you feel the most confident about.

Compare the choices to each part of the question. **To be wrong,** a, choice needs to be **incorrect for only part** of the question. **To be correct,** it must be **totally** correct. If you believe a choice is partially incorrect, tentatively eliminate that choice. Make notes next to the choices regarding tentative decisions. One method is to place a minus sign next to the choices you are certain are incorrect and a plus sign next to those that potentially are correct. Finally, place a zero next to any choice you do not understand or need to come back to for further inspection. Do not feel that you must make final decisions until you have examined all choices carefully.

When you have eliminated as many choices as you can, decide which of those that are left has the highest probability of being correct. Remember to use paper-and-pencil differential diagnosis. Above all, be honest with yourself. If you do not know the answer, eliminate as many choices as possible and choose reasonably.

Vignette-Based Questions

Vignette-based questions are nothing more than normal multiple-choice questions that use the same case, or grouped information, for setting the problem. The NBME has been researching question types that would test the student's grasp of the integrated medical basic sciences in a more cognitively complex fashion than can be accomplished with traditional testing formats. These questions allow the testing of information that is more medically relevant than memorized terminology.

It is important to realize that several questions, although grouped together and referring to one situation or vignette, are independent questions; that is, they are able to stand alone. Your inability to answer one question in a group should have no bearing on your ability to answer other questions in that group.

These are multiple-choice questions, and just as with single best-answer questions, you should use the paper-and-pencil differential diagnosis, as was described earlier.

Single Best-Answer–Matching Sets

Single best-answer–matching sets consist of a list of words or statements followed by several numbered items or statements. Be sure to pay attention to whether the choices can be used more than once, only once, or not at all. Consider each choice individually and carefully. Begin with those with which you are the most familiar. It is important always to break the statements and words into parts, as with all other question formats. **If a choice is only partially correct, then it is incorrect.**

Guessing

Nothing takes the place of a firm knowledge base, but with little information to work with, even after playing paper-and-pencil differential diagnosis, you may find it necessary to guess at the correct answer. A few simple rules can help increase your guessing accuracy. Always guess consistently if you have no idea what is correct; that is, after eliminating all that you can, make the choice that agrees with your intuition or choose the option closest to the top of the list that has not been eliminated as a potential answer.

When guessing at questions that present with choices in numerical form, you will often find the choices listed in ascending or descending order. It is generally not wise to guess the first or last alternative, since these are usually extreme values and are most likely incorrect.

Using the Comprehensive Exam to Learn

All too often, students do not take full advantage of practice exams. There is a tendency to complete the exam, score it, look up the correct answers to those questions missed, and then forget the entire thing.

In fact, great educational benefits can be derived if students would spend more time using practice tests as learning tools. As mentioned earlier, incorrect choices in test questions are plausible and partially correct or they would not fulfill their purpose as distractors. This means that it is just as beneficial to look up the incorrect choices as the correct choices to discover specifically why they are incorrect. In this way, it is possible to learn better test-taking skills as the subtlety of question construction is uncovered.

Additionally, it is advisable to go back and attempt to restructure each question to see if all the choices can be made correct by modifying the question. By doing this, four times as much will be learned. By all means, look up the right answer and explanation. Then, focus on each of the other choices and ask yourself under what conditions they might be correct.

For example, the entire thrust of the sample question concerning hypothyroidism could be altered by changing the first words to read:

"Hyperthyroidism recently discovered in . . ."
"Hypothyroidism prenatally occurring in . . ."
"Hypothyroidism resulting from . . ."

This question can be used to learn and understand thyroid problems in general, not only to memorize answers to specific questions.

In the practice exams that follow, every effort has been made to simulate the types of questions and the degree of question difficulty in the NBME Part I Comprehensive exam. While taking these exams, the student should attempt to create the testing conditions that might be experienced during actual testing situations. Approximately 1 minute should be allowed for each question, and the entire test should be finished before it is scored.

Summary

Ideally, examinations are designed to determine how much information students have learned and how that information is used in the successful completion of the examination. Students will be successful if these suggestions are followed:

- Develop a positive attitude and maintain that attitude.
- Be realistic in determining the amount of material you attempt to master and in the score you hope to attain.
- Read the directions for each type of question and the questions themselves closely and follow the directions carefully.
- Guess intelligently and consistently when guessing strategies must be used.
- Bring the paper-and-pencil differential diagnosis approach to each question in the examination.
- Use the test as an opportunity to display your knowledge and as a tool for developing prescriptions for further study and learning.

National Board examinations are not easy. They may be almost impossible for those who have unrealistic expectations or for those who allow misinformation concerning the exams to produce anxiety out of proportion to the task at hand. They are manageable if they are approached with a positive attitude and with consistent use of all the information that has been learned.

Michael J. O'Donnell

QUESTIONS

Directions: Each of the numbered items or incomplete statements in this section is followed by answers or by completions of the statement. Select the **one** lettered answer or completion that is **best** in each case.

Questions 1–5

In a study designed to test the effect of a new knee brace on running speed, eight college athletes who wore the brace turned in the following times (in minutes) in a 1000-meter speed trial: 4, 2, 5, 2, 4, 5, 5, and 9.

1. The mean finishing time for this group of subjects is

(A) 8.0 minutes
(B) 5.0 minutes
(C) 4.0 minutes
(D) 4.5 minutes
(E) none of the above

2. The median finishing time is

(A) 3.0 minutes
(B) 4.0 minutes
(C) 4.5 minutes
(D) 5.0 minutes
(E) none of the above

3. The modal finishing time is

(A) 2.0 minutes
(B) 4.0 minutes
(C) 5.0 minutes
(D) 9.0 minutes
(E) none of the above

4. The range of times is

(A) 7.0 minutes
(B) 4.0 minutes
(C) 5.0 minutes
(D) 9.0 minutes
(E) none of the above

5. The standard deviation for these data is

(A) 34/7
(B) $\sqrt{34/8}$
(C) $\sqrt{191.5/7}$
(D) $\sqrt{34/7}$
(E) none of the above

6. The following data represent length of hospitalization (in weeks) for five patients who receive no physical therapy after a total hip replacement: 4, 4, 3, 5, 4, and 20. The best measure of central tendency for this set of data is the

(A) mean
(B) range
(C) mode
(D) median
(E) standard deviation

Questions 7–10

In a study of the association between smoking and coronary heart disease, a team of clinical researchers obtained the following frequency distribution of systolic blood pressure readings (in mm Hg) for 37 smokers.

Systolic BP	Frequency	Relative Frequency (%)	Cumulative Frequency (%)
89.5–109.5	5	13.5	13.5
109.5–129.5	15	40.5	54.0
129.5–149.5	*A*	27.0	*C*
149.5–169.5	3	8.1	89.1
169.5–189.5	2	*B*	94.5
189.5–209.5	*D*	5.4	99.9
Totals	37		100.0

7. Value A in the table is

(A) 10
(B) 2
(C) 27
(D) 9
(E) not able to be determined from the given information

8. Value B in the table is

(A) 2.0%
(B) 5.4%
(C) .04%
(D) 27.0%
(E) not able to be determined from the given information

9. Value C in the table is

(A) 27.0%
(B) 81.0%
(C) 54.0%
(D) 10.0%
(E) not able to be determined from the given information

10. The probability that an individual selected at random from this sample of 37 smokers will have a systolic blood pressure between 89.5 mm Hg and 129.5 mm Hg is

(A) .135
(B) .405
(C) .540
(D) .270
(E) none of the above

Questions 11–12

For questions 11–12, assume that the theoretical frequency distribution corresponding to the empirical frequency distribution obtained in the study described for questions 7–10 is the normal distribution. In addition, assume that the population mean μ is 140 mm Hg and the population standard deviation σ is 10 mm Hg.

11. The probability that an individual selected at random from this population will have a systolic blood pressure greater than or equal to 149.5 mm Hg is

(A) .9500
(B) .3289
(C) .1711
(D) .8289
(E) none of the above

12. For this population, 90% of systolic blood pressure readings will fall below which of the following values?

(A) .4000
(B) 152.8
(C) 156.5
(D) 127.2
(E) 123.6

Questions 13–20

A physician wished to investigate the utility of serum creatinine kinase (CK) levels as a diagnostic test for acute myocardial infarction (AMI). She measured CK levels in 360 randomly selected patients under the age of 70 admitted to a particular intensive care unit with a suspected AMI during the preceding 48 hours, selecting a CK level of 80 IU as the positivity criterion. Each patient was also diagnosed by a team of expert cardiologists "blinded" to the CK test results (gold standard). The results are summarized in the table below.

	Acute Myocardial Infarction (AMI)		
CK Test Result	**Present**	**Absent**	**Totals**
Positive (≥80 IU)	215	16	231
Negative (<80 IU)	15	114	129
Totals	230	130	360

Reprinted from Sackett DL, Hayes RB, Tugwell P: *Clinical Epidemiology: A Basic Science for Clinical Medicine.* Boston, Little, Brown, 1985, p 71.

13. The prevalence of AMI in the study population is

(A) 215/231
(B) 230/360
(C) 231/360
(D) 215/230
(E) 215/360

14. The sensitivity of the CK test is

(A) 231/360
(B) 230/360
(C) 215/360
(D) 215/230
(E) 215/231

15. The probability that a patient who has not suffered an AMI will have a negative CK test result is

(A) 114/129
(B) 114/360
(C) 129/360
(D) 114/130
(E) 130/360

16. The probability calculated in question 15 is called the

(A) PVP
(B) FPR
(C) specificity
(D) PVN
(E) sensitivity

17. The probability that a new patient with a CK level equal to 85 IU has had an AMI is

(A) 215/231
(B) 215/230
(C) 230/360
(D) 231/360
(E) 215/360

18. The test characteristic calculated in question 17 is called the

(A) PVP
(B) FPR
(C) specificity
(D) PVN
(E) sensitivity

19. What is the FPR of the CK test?

(A) 114/129
(B) 15/230
(C) 15/129
(D) 114/130
(E) 16/130

20. The probability of a negative CK result among the population of patients who have suffered an AMI is

(A) 114/129
(B) 15/230
(C) 15/129
(D) 114/130
(E) 16/130

21. The PVP of a diagnostic test is defined as

(A) the quantity equal to $1 - \text{specificity}$
(B) the probability of disease given a positive test result, $P(D+|T+)$
(C) the probability that a diseased individual will have a positive test result, $P(T+|D+)$
(D) a function only of the sensitivity and specificity of the test

22. A screening test for Lyme disease has a 7% FPR and an 18% FNR. If the prevalence of Lyme disease in the population is 3%, what is the probability that a patient with a positive test result has this disorder?

(A) $\frac{(.07)(.03)}{(.07)(.03) + (.18)(.97)}$

(B) $\frac{(.18)(.03)}{(.18)(.03) + (.07)(.97)}$

(C) $\frac{(.82)(.03)}{(.82)(.03) + (.07)(.97)}$

(D) $\frac{(.93)(.97)}{(.93)(.97) + (.82)(.03)}$

(E) $\frac{(.93)(.03)}{(.93)(.03) + (.82)(.97)}$

Questions 23–24

A screening test for detecting sexual abuse in children is 95% sensitive and 90% specific. The prevalence of sexual abuse in a particular community is estimated to be approximately .10%.

23. The probability that a child with a positive test result has not been sexually abused is

(A) .950
(B) .009
(C) .991
(D) .900

24. Which of the following statements about the PVP of this test is true?

(A) It is equal to 10%
(B) It decreases as the prevalence of sexual abuse decreases
(C) It can be written as *P*(no abuse|positive test result)
(D) It decreases as the test specificity increases

25. The probability of hepatitis B [$P(D+)$] in a certain patient population is known to be .20 (conversely, $P(D-) = .80$). In a study of a new diagnostic test for hepatitis B, the probability of a positive test result among patients known to have hepatitis, $P(T+|D+)$, is shown to be .90, while the probability that a healthy patient will have a negative test result, $P(T-|D-)$, is shown to be .95. What is the probability that a new patient with a negative test result is truly healthy?

(A) .95

(B) $\frac{(.95)(.80)}{(.95)(.80) + (.10)(.20)}$

(C) $\frac{(.90)(.20)}{(.90)(.20) + (.05)(.80)}$

(D) $\frac{(.95)(.80)}{(.95)(.80) + (.90)(.20)}$

(E) None of the above

Questions 26–28

Total cholesterol values in a random sample of 2500 healthy men are assumed to follow a normal distribution, with a mean of 200 mg/dl and a standard deviation of 40 mg/dl.

26. Using the gaussian method, the normal range for total cholesterol values in this population is

(A) 200 ± 40
(B) $200 \pm (40/\sqrt{2500})$
(C) $200 \pm 2(40/\sqrt{2500})$
(D) $200 \pm 2(40)$
(E) none of the above

27. What proportion of healthy patients in the population will have an abnormal total cholesterol level, if only those values above the upper limit of the normal range are defined as abnormal?

(A) 5%
(B) 95%
(C) 2.5%
(D) .05%
(E) Cannot be determined from the data

28. Factors influencing the validity of the gaussian method for computing the normal range of total cholesterol values include all of the following EXCEPT

(A) the sample size
(B) the frequency distribution of total cholesterol values in the population
(C) the degree to which the sample is representative of the population
(D) the criteria used to designate subjects as healthy
(E) the criteria used to define the total cholesterol level at which therapeutic intervention becomes effective

29. Tay-Sachs disease is an autosomal recessive disorder most prevalent among families of Eastern European Jewish descent. A 28-year-old woman, who had two brothers with Tay-Sachs disease, and her husband, who had a child with the disorder in a previous marriage, seek genetic counseling. What is the probability that their first child will have Tay-Sachs disease (assume that the spontaneous mutation rate is negligible)?

(A) 1/2
(B) 2/3
(C) 1/4
(D) 1/6
(E) None of the above

30. Niemann-Pick disease, a lipid storage disease, is inherited in an autosomal recessive manner. A man and a woman are both identified as carriers for Niemann-Pick disease and informed that their risk of having an affected child is 1/4. If the couple plan to have four children, what is the probability that only one child will have Niemann-Pick disease?

(A) $4(1/4)(3/4)^3$
(B) $(1/4)^3$
(C) 1/4
(D) $4(1/4)^3(3/4)$
(E) None of the above

Questions 31–32

About 1% of the black population in the United States carry the recessive gene for β-thalassemia (Cooley anemia), a severe, progressive hemolytic anemia. A newly married black couple seek genetic counseling because the man is a known carrier for β-thalassemia. The woman's status as a carrier, including her family history, is not known.

31. What is the probability that this couple will have a child with β-thalassemia?

(A) .50
(B) .25
(C) .0025
(D) .01
(E) None of the above

32. If the couple have a healthy child, what is the posterior probability that the woman is a carrier?

(A) $\dfrac{(1/4)(1/100)}{(1/4)(1/100)+(3/4)(99/100)} = .0034$

(B) $\dfrac{(3/4)(1/100)}{(3/4)(1/100)+(1)(99/100)} = .0075$

(C) $\dfrac{(99/100)(3/4)}{(3/4)(99/100)+(1/100)} = .9867$

(D) $\dfrac{(1/4)}{(1/4)(1/100)+(3/4)(99/100)} = .3356$

(E) None of the above

33. A new vinca alkaloid proves highly effective in treating acute granulocytic leukemia. Which of the following rates will be *least* affected by widespread use of this drug?

(A) Five-year survival rate
(B) Prevalence
(C) Incidence
(D) Mortality rate

Questions 34–35

An oncologist determined that 75 of 100 randomly selected leukemia patients had experienced significant exposure to ionizing radiation. One hundred healthy individuals were randomly selected as controls; cases and controls did not differ with respect to age, race, or sex. Sixty of the controls had experienced significant exposure to ionizing radiation.

34. From this study, the relative odds (odds ratio) of leukemia among individuals exposed to ionizing radiation can be calculated as

(A) 1.0
(B) 2.0
(C) 1.4
(D) 4.5
(E) relative odds cannot be directly calculated from this data

35. According to this study, the risk of leukemia after exposure to ionizing radiation is

(A) .56
(B) .50
(C) .75
(D) .60
(E) not able to be calculated from the data

36. As part of a study of risk factors for cardiovascular disease in young adults, a cardiologist measured diastolic blood pressures in high school students in a certain county. All students were then classified according to their place of residence (urban or rural community) and socioeconomic status (SES). The three tables below show the results obtained for white girls between the ages of 17 and 19. The researcher wishes to design a new study to investigate further the possible causes of high blood pressure in this subgroup. Based on the data obtained in the first study, the new study should focus primarily on which of the following strategies?

(A) The identification of the salient factors that distinguish rural and urban lifestyles
(B) The identification of the factors that distinguish among the three socioeconomic groups
(C) The identification of the relevant factors that distinguish urban and rural residents and individuals with a low SES from those with a high SES
(D) The identification of additional risk factors, because neither place of residence nor SES appear to be associated with blood pressure

Mean Systolic Blood Pressure by Place of Residence

Place of Residence	**Mean Systolic Blood Pressure**
Urban	132.0
Rural	118.0
All Places	125.2

Mean Systolic Blood Pressure by Socioeconomic Status (SES)

SES	**Mean Systolic Blood Pressure**
High	107.6
Intermediate	131.4
Low	134.8
All Classes	125.2

Mean Systolic Blood Pressure by Place of Residence and Socioeconomic Status (SES)

	Place of Residence		
SES	**Urban**	**Rural**	**All Places**
High	108.0	106.9	107.6
Intermediate	130.6	132.1	131.4
Low	135.4	134.0	134.8
All Classes	132.0	118.0	125.2

37. A county school system provides lunch to 10,000 school children. During the first week of school, 2500 of these children ate chicken salad later shown to be contaminated with salmonella. The entire population of 10,000 students was subsequently followed for one month to determine whether exposure to salmonella increased the risk of diarrhea. Based on the results (summarized in the table below), the risk of diarrhea in children exposed to salmonella is

	Diarrhea (D+)	No Diarrhea (D−)	Totals
E+	30	2470	2500
E−	60	7440	7500
Totals	90	9910	10,000

E+ = exposed to salmonella; E− = not exposed to salmonella

(A) 3.0 times greater than in children with no such exposure
(B) 1.5 times greater than in children with no such exposure
(C) one-half that of children who were not exposed
(D) one-third that of children who were not exposed
(E) none of the above

Questions 38–40

A prospective cohort study designed to identify risk factors associated with coronary heart disease (CHD), initiated in 1960, is still underway. Of the sample of 1000 white men randomly selected to participate in the study in 1960, 50 had CHD when the study began and 950 were healthy. Two hundred of these healthy subjects could be classified as hypertensive (diastolic blood pressure >90 mm Hg), while 750 were normotensive. At the 20-year recall in 1980, 50 of the 200 hypertensive subjects had suffered an acute myocardial infarction (AMI). Only 75 of the normotensive subjects had suffered an AMI during this period (1960–1980).

38. The prevalence of CHD among white men in 1960 was

(A) 950/1000
(B) 50/950
(C) 50/1000
(D) 200/950
(E) not discernible from the data

39. Assuming that the study can account for all 1000 subjects between 1960 and 1980, what is the incidence of AMI during this period?

(A) 50/200
(B) 125/950
(C) 125/1000
(D) 75/750
(E) The incidence cannot be estimated directly from the data

40. The relative risk (RR) of AMI in hypertensive versus normotensive white men for the period 1960–1980 is

(A) 1.0
(B) 2.0
(C) 2.5
(D) 3.0
(E) 3.5

Questions 41–48

A pediatrician wished to determine the relationship between chronic otitis media in young children and parental history of such infections. From the records of a large pediatric practice, he identified 50 children between one and three years of age who had experienced at least three middle ear infections during the preceding year. Fifty children in the same age group, treated by the same practice for other illnesses, were also identified. The pediatrician interviewed the parents of subjects in both groups to determine their history of chronic otitis media as young children. Of the children with recurrent ear infections, 30 had a family history of chronic otitis media, compared to 20 of the children treated for other illnesses.

41. This study is an example of a

(A) cross-sectional study
(B) prospective cohort study
(C) case–control study
(D) experimental study
(E) randomized controlled clinical trial

42. Which of the following would be *least* likely to threaten the internal validity of this study?

(A) Selection bias
(B) Confounding variables that compete with family history as explanations for chronic otitis media
(C) Improper control for possible age differences
(D) Recall bias

43. The odds ratio (OR) of chronic otitis media in children between the ages of one and three, given a parental history of such infections, is

(A) 2.25
(B) 1.20
(C) 1.50
(D) 0.60
(E) not able to be directly calculated from the data

44. The RR of chronic otitis media in children between the ages of one and three with a parental history of such infections is

(A) 2.25
(B) 1.20
(C) 1.50
(D) 0.60
(E) not able to be directly calculated from the data

45. The absolute risk of chronic otitis media in children between one and three years of age who have a parental history of this disorder is

(A) 2.25
(B) 1.20
(C) 1.50
(D) 0.60
(E) not able to be directly calculated from the data

46. What is the most appropriate statistical test for determining whether a significant association exists between chronic otitis media in children between the ages of one and three and a parental history of otitis media?

(A) Paired *t* test
(B) Chi-square test
(C) Correlation analysis
(D) Analysis of variance
(E) Independent sample (pooled) *t* test

47. The pediatrician conducting this study reports the existence of a statistically significant association between chronic otitis media in children between the ages of one and three and a parental history of such infections ($p < .05$). Which of the following represents the most appropriate interpretation of this finding?

(A) A strong and clinically important association exists between the study variables in the populations from which the samples were drawn
(B) An unbiased comparison was made between the study groups
(C) There is a causal link between chronic otitis media in young children and a family history of this disorder
(D) There is a less than 5% chance that the observed association occurred by random chance

48. All of the following statements about the *p*-value reported by this investigator in question 47 are true EXCEPT

(A) the *p*-value represents the probability that the statistical test will detect a significant association between the study variables when, in fact, such an association does not exist in the populations from which the study samples were drawn
(B) the *p*-value is the probability that samples showing the observed degree of association were drawn by random chance from populations in which the study variables were not associated
(C) the *p*-value defines the probability of rejecting a true null hypothesis (type I error)
(D) the *p*-value measures the strength of the association between the two study variables; the smaller the *p*-value, the stronger the association

Questions 49–52

The Framingham Study is an ongoing large-scale prospective cohort study initiated in 1949 to investigate putative risk factors for coronary heart disease (CHD). Study participants underwent a complete physical examination at the beginning of the study and every two years thereafter. The upper table shown below summarizes the occurrence of CHD at the initial exam, while the lower table depicts the occurrence of CHD over one particular eight-year follow-up period.

Occurrence of Coronary Heart Disease (CHD) at Initial Examination Among 4469 Persons 30–62 Years of Age, Framingham Study

Age (Years)	Males Examined	Males with CHD	Rate per 1000	Females Examined	Females with CHD	Rate per 1000
30–44	1083	5	5	1317	7	5
45–62	941	43	46	1128	21	19
Totals	2024	48		2445	28	

Reprinted from Mausner JS, Kramer S: *Mausner and Bahn's Epidemiology: An Introductory Text,* 2nd ed. Philadelphia, WB Saunders, 1985.

Incidence of Coronary Heart Disease (CHD) Over an Eight-Year Period Among 4995 Persons 30–59 Years of Age Free of CHD at Initial Examination

Age (Years)	Males Examined	Males with CHD	Females Examined	Females with CHD
30–39	825	20	1036	1
40–49	770	51	955	19
50–59	617	81	792	53
Totals	2212	152	2873	73

Reprinted from Mausner JS, Kramer S: *Mausner and Bahn's Epidemiology: An Introductory Text,* 2nd ed. Philadelphia, WB Saunders, 1985.

49. Based on the data in the upper table, the conclusion that the risk of CHD for men between the ages of 30 and 44 is equal to the risk for women in the same age range is

(A) correct
(B) incorrect, because the age distribution for men and women may differ
(C) incorrect, because a rate must be calculated to support this inference
(D) incorrect, because of a failure to distinguish between incidence and prevalence
(E) incorrect, because of the lack of an appropriate control or comparison group

50. Based on the data in the lower table, the absolute risk of developing CHD over the eight-year observation period among men between 30 and 39 years of age is

(A) .00097 (.97 per 1000)
(B) .0687 (68.7 per 1000)
(C) .0242 (24.2 per 1000)
(D) .0262 (26.2 per 1000)
(E) not able to be calculated from the data given

51. The RR of developing CHD for men between the ages of 30 and 39, compared to women in the same age range, is

(A) 20.0
(B) 25.1
(C) 2.42
(D) 2.70
(E) not able to be calculated from the data given

52. Prevalence rates for men and women are approximately equal, while incidence rates are very different. Which of the following statements represents the *least* plausible explanation for this observation?

(A) CHD in young men typically results in myocardial infarction and sudden death, while in women, it typically results in stable angina
(B) Men are at greater risk for developing CHD
(C) CHD follows a more protracted duration in women
(D) The age distribution of men and women differs in the general population

Questions 53–54

As part of a study of suspected risk factors for neural tube defects, a group of infants born with spina bifida during a particular six-month period was identified from the records of a large maternity hospital. The researcher also identified a group of healthy infants born at the same hospital during this time period. Mothers of both groups of infants were questioned about their use of prenatal vitamins. The study found that significantly more mothers of healthy infants took prenatal vitamins than mothers of infants with spina bifida ($p < .001$).

53. The study is an example of a

(A) cohort study
(B) case–control study
(C) experimental study
(D) randomized controlled clinical trial

54. Based on the results of this study, the conclusion that prenatal vitamins prevent birth defects is likely to be

(A) correct
(B) incorrect, because the observed difference in rates can be attributed to random chance
(C) incorrect, because systematic differences between the comparison groups (other than prenatal vitamin use) may account for the observed difference in rates
(D) incorrect, because of the failure to distinguish between incidence and prevalence
(E) incorrect, because of the lack of a suitable control or comparison group

55. A study reports that 65 of 100 patients with multiple sclerosis have cats as household pets. The conclusion that exposure to cats causes multiple sclerosis is

(A) valid for the study population
(B) incorrect, because of the failure to carry out an appropriate statistical analysis of the data
(C) incorrect, because the comparison is not based on exposure rates
(D) incorrect, because the sample size is too small to permit generalization to the entire population of multiple sclerosis patients
(E) incorrect, because of the lack of an appropriate control or comparison group

Questions 56–57

A surgeon wished to compare the surgical and dietary treatment of hemorrhoids. He randomly assigned 100 hemorrhoid patients to the surgery group (rubber band ligation) and 100 to the diet group (high fiber diet). Twelve months later, he questioned both groups about the persistence or recurrence of pain and inflammation. A significantly higher proportion of the surgical patients reported symptomatic relief ($p < .01$).

56. This study is an example of a

(A) prospective cohort study
(B) case–control study
(C) cross-sectional study
(D) randomized controlled clinical trial
(E) historical cohort study

57. The conclusion that surgery is more efficacious than a high fiber diet in alleviating hemorrhoidal symptoms is likely to be

(A) valid for the study population
(B) incorrect, because of the small sample size
(C) incorrect, because selection bias may account for the observed difference between diet and surgery
(D) incorrect, because the observed difference between diet and surgery is likely to have occurred by random chance
(E) incorrect, because subjects assigned to the diet group may have failed to comply with the prescribed high fiber diet

58. In a double-blind study of the efficacy of anticoagulant therapy in preventing thromboembolic disease and reducing mortality after a myocardial infarction, patients admitted to a certain intensive care unit after suffering an acute myocardial infarction (AMI) were randomly assigned to receive warfarin or a placebo. Patients in the warfarin group were also given prophylactic antiarrhythmic agents and were permitted early ambulation. There was a statistically significant reduction in both the occurrence of thromboembolic disease and mortality among the patients receiving warfarin therapy compared to the placebo group ($p < .05$). The conclusion that warfarin is the treatment of choice for patients suffering an AMI is likely to be

(A) correct
(B) incorrect, because the observed differences between warfarin and placebo can be attributed to random chance
(C) incorrect, because the observed differences between warfarin and placebo can be attributed to selection bias
(D) incorrect, because the additional therapeutic interventions (antiarrhythmic drugs and early ambulation) received by the warfarin group provide a competing explanation for the observed differences in morbidity and mortality
(E) incorrect, because surveillance bias can account for the observed results; patients receiving warfarin may have been more carefully monitored than those receiving the placebo

59. An outbreak of food poisoning occurred among 400 people who ate Thanksgiving dinner at a certain homeless shelter. The tables below depict the attack rates among those who ate turkey or dressing and those who did not eat these foods. Based on these data, which food or combination of foods was probably responsible for the outbreak?

	Ate Dressing		
	All Diners	**Ill**	**Attack Rate (%)**
Ate Turkey	300	225	75
Did Not Eat Turkey	100	84	84

	Did not Eat Dressing		
	All Diners	**Ill**	**Attack Rate (%)**
Ate Turkey	150	3	2
Did Not Eat Turkey	50	0	0

(A) Turkey alone (without dressing)
(B) Dressing alone (without turkey)
(C) The combination of turkey and dressing
(D) Cannot be determined

60. Death rates from coronary heart disease (CHD) are compared for two populations. The crude (unadjusted) mortality rates for the populations are approximately equal. However, the age-adjusted mortality rate for population 1 is much higher than the corresponding age-adjusted mortality rate for population 2. What is the most likely explanation for this disparity?

(A) The two populations have the same age distribution
(B) Population 1 is younger, on the average, than population 2
(C) Population 1 is older, on the average, than population 2
(D) The disparity cannot be explained from the data because no information is provided on the relative age distributions of the two populations

61. Scores on the final exam in gross anatomy at a certain medical school are assumed to follow a normal distribution. The class average is 65 and the standard deviation is 10. Which of the following statements accurately describes the distribution of exam scores?

(A) Five percent of the class will have scores above 84.6
(B) The 99th percentile value is 90.8
(C) Five percent of the class will have grades below 45.4
(D) The 50th percentile value is 65

62. To assess the health care needs of a low-income community, a team of health care planners surveyed a random sample of residents selected from the local phone directory. Fewer than 10% of those contacted reported visiting a physician in the preceding five years, even though they had experienced at least one illness during this period. In response to the results of this survey, free physical examinations were offered at a walk-in clinic in the community. A second survey of patients who took advantage of this new examination program found that over 40% reported seeing a physician at least once for a physical exam during the preceding five years. Which of the following statements best explains the apparent discrepancy between the two surveys?

(A) The first survey sample was biased toward the most affluent residents (only those who could afford telephones were contacted); hence, this survey underestimated the true proportion of individuals who saw a physician during the preceding five years
(B) The first survey sample was biased because of recall error (subjects were unlikely to remember all visits to a physician); hence, this survey underestimated the true proportion of individuals who saw a physician during the preceding five years
(C) The first survey lacked an appropriate control or comparison group; hence, the results are invalid
(D) The second survey sample was biased because individuals requesting the free exam were likely to be more health conscious than the average community resident; hence, this survey overestimated the true proportion of residents who saw a physician during the preceding five years

63. A study of the effect of the serotonin antagonist ketanserin on blood pressure reported* that, "in the racially mixed patient population there was a statistically significant positive correlation (r) between pretreatment diastolic blood pressure (DBP) and the change in blood pressure (ΔBP) following treatment with ketanserin ($p <$ .05)." What is the most appropriate interpretation of this finding?

(A) The population correlation coefficient ρ is much larger than zero
(B) The relationship between DBP and ΔBP is described by a straight line with a positive slope
(C) The observed association between DBP and ΔBP is likely to be due to random chance
(D) There is a strong relationship between DBP and ΔBP

64. A measure of the strength of the association between a putative risk factor and a disease is provided by which of the following?

(A) Magnitude of the absolute risk incurred by those exposed to the risk factor
(B) Magnitude of the p-value
(C) Attributable risk
(D) RR or OR
(E) None of the above

65. A statistical test fails to reject the hypothesis that the correlation between blood pressure and serum cholesterol level is equal to zero at the 5% level of significance. What is the correct interpretation of this result?

(A) The correlation between blood pressure and serum cholesterol levels in the population from which the sample was drawn is equal to zero
(B) The correlation between blood pressure and serum cholesterol level is statistically significant ($p < .05$)
(C) The results are not statistically significant; that is, there is not enough evidence to substantiate the claim of an association between blood pressure and serum cholesterol level
(D) Higher blood pressure values tend to be associated with higher serum cholesterol levels
(E) The observed correlation is unlikely to have occurred as a result of random chance

*Kosoglou T, et al: Antihypertensive response to ketanserin: influence of race and weight. *J Clin Pharmacol* 28:1017–1022, 1988.

Questions 66–69

In a clinical trial of a novel antihypertensive agent, each of four patients received the drug for 14 days and a placebo for 14 days; the treatments were assigned in a random order. Changes in diastolic blood pressure from baseline (ΔBP) during the treatment period were recorded for all subjects. The investigator wished to determine if the mean change from baseline in diastolic blood pressure was greater for the new drug than for the placebo.

66. What is H_A for the statistical test?

(A) The mean change in diastolic blood pressure produced by the drug does not differ from that produced by the placebo
(B) The mean change in diastolic blood pressure produced by the drug is different from that produced by the placebo
(C) The mean change in diastolic blood pressure is greater after the drug than after the placebo
(D) The drug is not efficacious
(E) None of the above

67. Which of the following statistical tests is most appropriate for analyzing this data?

(A) Independent sample (pooled) *t* test
(B) Correlation analysis
(C) Paired *t* test
(D) Chi-square test
(E) Analysis of variance using the completely random design

68. All of the following statements about this statistical test are true EXCEPT

(A) it is a one-tailed test
(B) it must be carried out at the 5% level of significance
(C) it is more likely to reject H_O if $\alpha = .05$ than if $\alpha = .01$
(D) it compares the calculated test statistic to a tabulated critical value with 3 degrees of freedom

69. The investigator computes a test statistic of 3.0; the tabulated critical value for $\alpha = .05$ is 2.35. What is the most appropriate conclusion that can be drawn from this result?

(A) Reject H_O at the 5% level of significance; there is not a statistically significant difference between the mean change in diastolic blood pressure elicited by the drug and that elicited by the placebo
(B) Do not reject H_O at the 5% level of significance; there is insufficient evidence to demonstrate a greater mean change in diastolic blood pressure for the drug compared to the placebo
(C) Reject H_O at the 5% level of significance; the mean change in diastolic blood pressure elicited by the drug is greater than that elicited by the placebo
(D) Do not reject H_O at the 5% level of significance; there is a statistically significant difference between the mean change in diastolic blood pressure elicited by the drug and that elicited by the placebo

70. If a statistical test is carried out at the $\alpha = .01$ level of significance, which of the following statements most accurately describes the probable validity of the conclusion?

(A) The probability of rejecting a true H_O is .01
(B) The probability of accepting H_A when it is, in fact, true is .01
(C) The probability of rejecting H_O when it is indeed false is .01
(D) the probability of rejecting H_O is greater than it would be if α were equal to .05

71. A clinical researcher conducts a study to compare the effect of a new histamine receptor antagonist and placebo on peptic ulcer healing times. She reports that the difference in healing times between the two treatments is statistically significant at the $\alpha = .05$ level of significance ($p < .05$). The correct interpretation of the level of uncertainty associated with this conclusion is

(A) the probability of obtaining a difference greater than or equal to that observed, given that the two treatments are actually similarly efficacious, is greater than .05
(B) the probability of accepting H_O, given that there is no difference between the two treatments, is .05
(C) the probability that the healing time observed after treatment with the new drug truly does differ from the healing time observed after placebo is .95
(D) the probability that the statistical test will detect a difference in healing times when, in fact, no such difference exists, is less than or equal to .05
(E) the probability that a particular peptic ulcer patient will benefit from receiving the new drug is .95

Questions 72–74

A group of stroke patients participated in a study to compare the degree of functional recovery resulting from three different physical therapy regimens. Because the investigator felt that the initial level of disability would affect functional recovery, he grouped the subjects into four disability categories (where category 1 represented the most severe initial disability level and category 4, the least severe). He then randomly assigned three subjects in each disability category to receive each of the three therapy regimens. At the end of six months, functional recovery was measured on a quantitative scale by an independent physical therapist. The investigator wished to compare these functional recovery scores for the three regimens.

72. What is the most appropriate statistical test for analyzing this data?

(A) Analysis of variance, using the CRD
(B) Analysis of variance, using the RCB design
(C) Chi-square test
(D) Pooled *t* test
(E) Correlation analysis

73. The investigator reports that the results of this study were statistically significant at the 5% level of significance. The most appropriate conclusion that can be drawn from this result is

(A) there is no difference in the degree of functional recovery produced by the three physical therapy regimens
(B) patients who receive physical therapy have a greater degree of functional recovery than those who do not
(C) all three of the therapy regimens result in different degrees of functional recovery
(D) at least two of the therapy regimens result in differing degrees of functional recovery

74. The investigator wishes to use the Bonferroni procedure for pairwise comparisons of the three treatment means. What correction factor must he use (for a two-tailed comparison) to insure that the overall level of significance for the set of comparisons does not exceed 5%?

(A) .05/4
(B) .05/3
(C) $\frac{.05/3}{2}$
(D) $\frac{.05/4}{2}$
(E) .05

Questions 75–78

A neurologist postulates that the extent of physical disability following a cerebrovascular accident (CVA) is related to personality type. Five hundred such patients were classified according to the severity of their physical deficit (mild or severe) and simultaneously assigned to one of four personality groups (1 = most prone to depression and 4 = least prone to depression), based on a personality assessment questionnaire developed by the investigator. The table below depicts the number of subjects in each category.

Personality Type	Severity of Condition		
	Severe	Mild	Totals
1	60	40	100
2	60	40	100
3	132	68	200
4	48	52	100
Totals	300	200	500

75. This study is an example of a

(A) prospective cohort study
(B) case–control study
(C) cross-sectional study
(D) experimental study
(E) randomized controlled clinical trial

76. What is the most appropriate statistical test for determining whether a significant association exists between the extent of physical disability and personality type?

(A) Analysis of variance
(B) Independent sample (pooled) *t* test
(C) Paired *t* test
(D) Chi-square test
(E) Correlation / regression analysis

77. The neurologist computes a test statistic of 9.00; the tabulated critical value for $\alpha = .05$ is 7.81. What is the most appropriate conclusion she can draw from this result?

(A) Reject H_O at the 5% level of significance; conclude that the severity of the disability is related to personality type
(B) Reject H_O at the 5% level of significance; conclude that no association exists between the extent of disability and personality type
(C) Do not reject H_O at the 5% level of significance; conclude that there is no association between the severity of the physical deficit and personality type
(D) Do not reject H_O at the 5% level of significance; conclude that there is an association between the extent of the disability and personality type
(E) None of the above

78. In a similar study, the investigators report that there is a statistically significant association between the severity of the physical deficit observed after a CVA and personality type ($p < .001$). What is the most appropriate interpretation of this finding?

(A) There is a strong, clinically important association between personality type and the degree of physical disability after a CVA
(B) Those patients who are most prone to depression are least likely to make a full recovery after suffering a CVA
(C) All competing explanations for the observed association have been eliminated (i.e., all potential sources of bias have been controlled)
(D) It is unlikely that a sample giving rise to an association as large as that observed would have been drawn by random chance from a population in which no association exists between the two study variables

79. In a study reported by Koshy and coworkers, pregnant women with sickle cell anemia were randomly assigned to one of two groups. The first group received a prophylactic transfusion of frozen red blood cells, while the second group received such a transfusion only in the event of a medical or obstetric emergency. The authors report, "There were no statistically significant differences in perinatal outcome between the offspring of mothers with sickle cell disease who were assigned to the treatment group with prophylactic transfusions and those who were not."* The most appropriate interpretation of this finding is

(A) prophylactic transfusion is ineffective in reducing adverse perinatal outcomes among the offspring of women with sickle cell anemia
(B) a clinically important difference in perinatal outcomes may not have been detected due to low power of the statistical test
(C) a biased comparison has been made
(D) the observed difference in perinatal outcome between children born to the two groups of mothers is unlikely to be due to random chance

80. Two physicians carry out a study to determine the efficacy of a particular diet and exercise program in chronically obese women. They determined that the regression line for predicting weight loss at the end of one year on the program (Y), based on initial weight (X), is $\hat{Y} = -25 + .3X$. What is the most appropriate interpretation of this finding?

(A) There is a statistically significant positive linear relationship between initial weight and weight loss after one year on the diet and exercise program
(B) All study participants lost at least 25 pounds during the one-year observation period
(C) The slope of the line is positive; its magnitude indicates the presence of a strong association between initial weight and subsequent weight loss while on the program
(D) Women who weighed more initially lost more weight, on the average, during the one-year study period

Questions 81–83

A double-blind clinical study was conducted to compare the efficacy of ibuprofen (400 mg), codeine (60 mg) and a placebo in providing pain relief after minor dental surgery. Fifteen participants were randomly assigned in groups of five to the three treatment groups. Two hours after receiving the drug or placebo, patients were asked to rate their degree of pain relief on a scale from 0 to 100, and the average score for each treatment was calculated. The results were not statistically significant at the 5% level of significance.

81. The chance of error associated with this conclusion is

(A) the type II error rate (β), which is .95
(B) the type II error rate (β), which is .05
(C) the type II error rate, which is unknown
(D) the power of the statistical test, which is unknown
(E) the level of significance α, which is .05

82. Which of the following statements best defines the power of the statistical test?

(A) Power is the probability that the test will detect a difference between the three drugs when, in fact, no such difference exists
(B) Power is the probability that the test will fail to detect a difference in the level of pain relief provided by the three treatments when they actually do differ
(C) Power is the probability that the three treatments provide different levels of pain relief
(D) Power is the probability that there is no difference in the level of pain relief provided by the three treatments
(E) Power is the probability that the test will correctly detect a difference among the three treatments which does, in fact, exist

---

*Koshy M, Burd L, Wallace D, et al: Prophylactic red cell transfusions in pregnant patients with sickle cell disease: a randomized cooperative study. *N Engl J Med* 319:1447–1452, 1988.

83. What is the most appropriate conclusion that can be drawn from the results of this study?

(A) There is no difference in the level of pain relief provided by the three treatments
(B) There is a high probability ($>.95$) that no difference exists among the three treatments
(C) Fewer than 5% of patients can expect satisfactory pain relief from ibuprofen (400 mg) or codeine (60 mg)
(D) The observed difference in perceived pain relief is unlikely to be due to random chance
(E) A clinically important difference between the three treatments may actually exist, but the power of the statistical test may have been too low to detect it

Questions 84–87

A pharmaceutical company conducts a randomized controlled clinical trial to determine the efficacy of a newly developed topical solution for inducing hair growth in men with male pattern baldness. The researchers decide that average hair count must increase by at least 20% in the drug-treated group, compared to the placebo, to be considered a clinically important difference. They select a sample size that will ensure 80% power for the statistical test, and carry out this test at the 5% level of significance.

84. The probability that this statistical test will fail to detect a clinically important difference in average hair growth between the comparison groups, given that such a difference actually exists between the populations from which the study samples were drawn

(A) increases as the specified effect size decreases
(B) increases as the sample size increases
(C) increases as α, the level of significance increases
(D) increases as the power of the test increases

85. The probability that this statistical test will detect the specified clinically important difference in average hair growth, given that such a difference does indeed exist between the populations represented by the study samples, is

(A) .95
(B) .05
(C) .80
(D) .20
(E) impossible to determine from the data

86. Which of the following statements best describes α, the level of significance?

(A) The chance of detecting a statistically significant difference between the comparison groups is 95%
(B) If average hair growth does not differ between the population of patients treated with the new medication and those treated with the placebo, the chance of detecting a statistically significant difference between the samples drawn from these populations is 5% or less
(C) If average hair growth differs between the populations of patients receiving the two treatments (drug and placebo), the chance that the statistical test will detect a significant difference between the samples drawn from these populations is at least 95%
(D) A statistically significant result will be obtained 5% of the time that a true difference exists between the populations represented by the study samples
(E) None of the above

87. The investigators report a highly significant difference in average hair count between men receiving the new medication and men treated with the placebo solution ($p < .0005$). What is the most appropriate conclusion they can draw from this result?

(A) Fewer than .05% of patients treated with the new medication will fail to experience hair growth
(B) The probability that a particular patient treated with the new medication will experience significant hair growth is high
(C) If the new topical solution is truly efficacious, the probability of observing the reported increase in hair growth is less than .05%
(D) If the new medication is not efficacious, the probability of observing the reported difference in hair growth is less than .05%
(E) None of the above

Questions 88–90

The decision tree on the facing page was constructed to select the best course of treatment for a 24-year-old woman hospitalized with fever (104.2° F), nausea, and vomiting. Physical examination of this patient, a kidney transplant recipient, also revealed rales in the left base, but no significant abdominal signs. The possibility of a subdiaphragmatic abscess is raised, and the utility of surgery debated.

If surgery is elected, the following factors must be considered.

An abscess may or may not be present.
Complete excision of the abscess may or may not be possible.
If the abscess is completely excised, the prognosis is excellent. If the abscess cannot be completely removed, it may spontaneously resolve or result in sepsis and ultimately, death.
Recovery from the surgery may be uneventful, or the procedure may result in serious complications or even death (regardless of whether or not an abscess is actually present).

If, however, surgery is not performed, the following factors must be considered.

The patient may or may not recover if her symptoms are not due to a subdiaphragmatic abscess (e.g., if they are caused by acute pancreatitis).
If the patient does indeed have a surgically correctable subdiaphragmatic abscess, it may resolve spontaneously or lead to sepsis and death.

The utilities listed in the last column of the figure represent the expected survival rates associated with each outcome.

88. From this decision tree, the proportion of patients with a comparable clinical profile who have a surgically correctable lesion can be estimated as

(A) .25
(B) .55
(C) .30
(D) .70
(E) .75

89. The expected utility at node B is

(A) 50.0
(B) 63.8
(C) 62.5
(D) 75.0
(E) none of the above

90. According to the decision tree, what is the most appropriate course of action?

(A) Forego surgery, because those patients who are not treated surgically have a lower expected mortality than those who do
(B) Perform the surgery, because the expected survival associated with the surgery branch of the tree is higher than that associated with the no surgery branch
(C) Perform the surgery, because the expected utility associated with the surgery option is lower than that associated with the no surgery option
(D) Forego surgery, because there is insufficient data to arrive at a satisfactory decision

91. If the positivity criterion for a diagnostic test is shifted to a more stringent value (i.e., to the right along the horizontal axis of the frequency distribution), all of the following statements about the performance of the test are true EXCEPT

(A) sensitivity decreases
(B) specificity increases
(C) the number of true positives increases
(D) the number of false negatives increases

92. All of the following are disadvantages of the gaussian method of determining the normal range for a diagnostic test EXCEPT

(A) the prevalence of all diseases is at least 5%, if clinical values outside both the upper and lower limits are taken as indicative of disease
(B) multiple testing increases the likelihood that a patient will be diagnosed with the disease in question
(C) this method requires physicians to keep track of test performance characteristics in their own practices
(D) this method ignores the fact that diagnostic test results rarely follow a normal distribution
(E) the method does not take into account the frequency distribution of the test variable in the diseased population

93. A screening test for adenosine deaminase deficiency, a rare genetic immunodeficiency disorder, is both highly sensitive and highly specific. All of the following statements accurately describe the performance characteristics of this test EXCEPT

(A) the false positive rate is low
(B) the predictive value positive is high
(C) the true negative rate is high
(D) the false negative rate is low
(E) the true positive rate is high

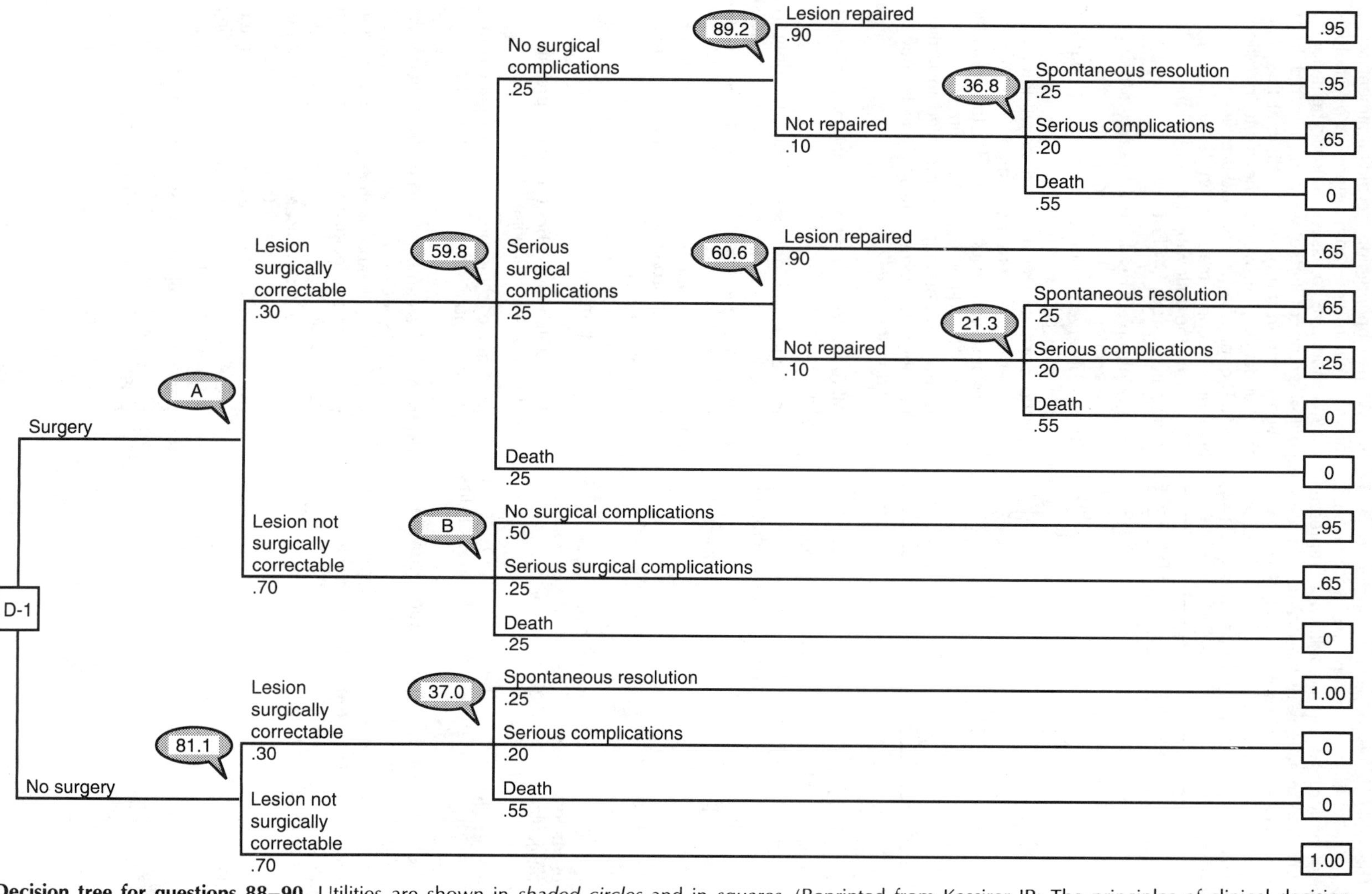

Decision tree for questions 88–90. Utilities are shown in *shaded circles* and in *squares.* (Reprinted from Kassirer JP: The principles of clinical decision making: an introduction to decision analysis. *Yale J Biol Med* 49:149–164, 1976.)

94. The table below depicts the incidence and prevalence of measles in a particular city over a 10-year period. All of the following represent plausible interpretations of the changes in these frequency measures EXCEPT

Cases of Measles Among Children Ages 1–4, 1976–1985

Year	Incidence	Prevalence
1976	24.5	41.8
1977	24.9	41.2
1978	23.8	40.9
1979	24.6	40.1
1980	24.1	38.4
1981	24.7	37.9
1982	24.2	35.3
1983	23.9	33.2
1984	25.1	29.8
1985	24.5	27.2

(A) average recovery time declined over the 10-year period
(B) public health measures to prevent new cases of measles were successful
(C) the mortality rate from measles increased over the 10-year period
(D) the duration of the disease decreased between 1976 and 1985

95. All of the following statements about a cohort study are true EXCEPT

(A) the results of this type of study provide an estimate of the absolute risk of developing a particular disease following exposure to a suspected risk factor
(B) the results of this type of study provide an unbiased comparison between cases of disease and controls
(C) the results of this type of study provide an estimate of relative risk
(D) the results of this type of study provide an estimate of the probability of developing a particular disease among those exposed to a suspected risk factor

96. Women undergoing steroid treatment at an infertility clinic were administered a psychological questionnaire to assess neurotic behavior. A control group of fertile women, selected from patients seen in the obstetrics clinic at the same medical facility, answered the same questionnaire. The infertile women had significantly higher neurosis scores than the fertile women ($p < .01$). All of the following interpretations of this finding are correct EXCEPT

(A) the observed result could occur if infertility causes neurotic behavior
(B) the observed result could occur if neurotic behavior causes infertility
(C) the observed result could be attributed to the effects of steroid treatment
(D) the observed results could be the result of systematic differences between the comparison groups that are unrelated to fertility
(E) the observed results are likely to be due to random chance

97. All of the following statements about disease incidence are true EXCEPT

(A) incidence will decrease if a new drug is effective in reducing deaths from the disease
(B) incidence measures the absolute risk of developing the disease
(C) incidence is the probability that a healthy individual will develop the disease during a specified time period
(D) incidence will decrease if a particular prevention program is effective

98. True statements about disease prevalence include all of the following EXCEPT

(A) prevalence is the proportion of people in a given population who currently have the disease
(B) prevalence can be estimated from the results of a cross-sectional study
(C) prevalence is approximately equal to the product of the incidence of the disease and its duration
(D) prevalence is a measure of existing cases of the disease at a particular point in time (or over a specified time period), rather than the proportion of new cases that develop during a specified period of time
(E) prevalence will decrease if a new drug reduces mortality associated with the disease, even if it does not result in a cure

99. The mean survival time following chemotherapy for 2500 patients suffering from rhabdomyosarcoma is 48 months, with a standard deviation of 10 months; the median survival time is 30 months. All of the following statements about patients undergoing chemotherapy for this malignancy are true EXCEPT

(A) ninety-five percent of the patients in the population will survive between 28.4 and 67.6 months
(B) approximately half of the patients in the population will survive more than 30 months
(C) a physician can be 95% confident that the mean survival time for the population of patients falls between 47.6 and 48.4 months
(D) approximately half of the patients in the population survive less than 30 months

100. A randomized controlled clinical trial was conducted to evaluate the relative efficacy of leg elevation and an antidiuretic in reducing edema in patients with stasis ulcers of the leg. Patients were randomly assigned to the two treatments. After a two-week treatment period, the average reduction in leg volume, as determined by water displacement, was recorded for all subjects. A 90% confidence interval estimate of the difference in the average reduction in leg volume produced by the two treatments was calculated as 19.1 to 40.9 cm^3. All of the following represent correct interpretations of this confidence interval EXCEPT

(A) the difference in mean reduction in leg volume between the population of patients treated with leg elevation and the population treated with the antidiuretic drug is between 19.1 and 40.9 cm^3
(B) the physician conducting this study can be 90% confident that the average difference in the reduction in leg volume experienced by the two populations of patients is between 19.1 and 40.9 cm^3
(C) this 90% confidence interval is narrower than the corresponding 95% confidence interval
(D) in repeated sampling of the populations in question, 90 out of every 100 confidence intervals would be expected to contain the true difference between the population means (i.e., $\mu_1 - \mu_2$, where μ_i represents the average reduction in leg volume for population i)

101. All of the following statements about a case–control study are true EXCEPT

(A) this study design is the design of choice for studying rare diseases
(B) both absolute risk and RR can be directly calculated from the results of this type of study
(C) a case–control study is less expensive and time-consuming than a prospective cohort study
(D) a case–control study is more vulnerable to bias than a prospective cohort study

ANSWERS AND EXPLANATIONS

1. The answer is D *[Ch 1 III B 1 a]*
The mean finishing time ($\bar{Y}$) for this group of eight subjects is the arithmetic average of their individual times:

$$\bar{Y} = \sum Y_i / n$$
$$= 36/8$$
$$= 4.5 \text{ minutes}$$

2. The answer is C *[Ch 1 III B 1 b]*
For an even number of observations, the median is the arithmetic average of the two middle values when the observations are arranged in order of ascending magnitude. Here, the eight running times, arranged in order of magnitude, are 2, 2, 4, 4, 5, 5, 5 and 9. Since n is even, the median is the average of the two middle values, 4 and 5: $(4 + 5)/2 = 4.5$ minutes.

3. The answer is C *[Ch 1 III B 1 c]*
The mode is the value in a set of observations that occurs most frequently. For this data, the modal running time is 5.0 minutes.

4. The answer is A *[Ch 1 III B 2 a]*
The range is the difference between the slowest and the fastest finishing time. Thus,

$$\text{range} = 9 - 2$$
$$= 7.0 \text{ minutes}$$

5. The answer is D *[Ch 1 III B 2 b (2)]*
The standard deviation of the eight running times (s) is calculated from the formula

$$s = \sqrt{\frac{\sum Y_i^2 - \left(\sum Y_i\right)/n}{n-1}}$$
$$= \sqrt{\frac{196 - (36)^2/8}{8-1}}$$
$$= \sqrt{34/7}, \text{ or } 2.2 \text{ minutes}$$

6. The answer is D *[Ch 1 IV C 1 b (2), 2]*
When a data set contains either a very large or a very small value (i.e., when the distribution is skewed), the mean is pulled in the direction of the extreme value and may not adequately represent the center of the frequency distribution. For highly skewed distributions, the median (midpoint of a set of values arranged in order of ascending magnitude) is a more representative measure of the central tendency of the frequency distribution. In this example, the median (4 weeks) is more representative of the center than is the mean (6.67 weeks). Note that only one value (20) falls above the mean, whereas half fall above and half below the median.

7–10. The answers are: 7-A *[Ch 1 III A 1 c (1)–(2); Table 1-2]*, **8-B** *[Ch 1 III A 1 c (2); Table 1-2]*, **9-B** *[Ch 1 III A 1 c (3); Table 1-2]*, **10-C** *[(Ch 2 I A, D 1, 2]*
Value A in the table represents the frequency of systolic blood pressure readings falling between 129.5 and 149.5 mm Hg. Absolute frequency is the product of the relative frequency of the values encompassed by this interval and the total number of values. For this study, 27% of the 37 systolic blood pressure readings fall between 129.5 and 149.5 mm Hg. Therefore,

$$A = (\text{relative frequency}) \times (\text{total number of observations})$$
$$= (.27)(37)$$
$$= 10$$

Value B represents the relative frequency of systolic blood pressure readings between 169.5 and

189.5 mm Hg; that is, the proportion of readings encompassed by this interval:

$$\begin{aligned} B &= \text{(absolute frequency)/(total number of observations)} \\ &= 2/37 \\ &= .054, \text{ or } 5.4\% \end{aligned}$$

Cumulative frequency is the proportion of values falling below the upper limit of each interval. Thus, value C represents the proportion of systolic blood pressure readings below 149.5 mm Hg. This cumulative frequency is the sum of the relative frequencies of all intervals up to and including the interval 129.5 to 149.5 mm Hg:

$$\begin{aligned} C &= \text{(relative frequency, 89.5} - \text{109.5 mm Hg)} \\ &\quad + \text{(relative frequency, 109.5} - \text{129.5 mm Hg)} \\ &\quad + \text{(relative frequency, 129.5} - \text{149.5 mm Hg)} \\ &= 13.5\% + 40.5\% + 27.0\% \\ &= 81.0\% \end{aligned}$$

In other words, 81% of the blood pressure readings are less than or equal to 149.5 mm Hg.

The probability that an individual selected at random from this sample of 37 smokers will have a systolic blood pressure between 89.5 and 129.5 mm Hg is computed by dividing the number of values in this interval by the total number of measurements:

$$\begin{aligned} P(89.5 \leq Y \leq 129.5) &= (5 + 15)/37 \\ &= .540 \end{aligned}$$

This probability can also be obtained from the relative frequency column (13.5% + 40.5% = 54%, or .054) or from the cumulative frequency column, as the proportion of values below 129.5 mm Hg (54% or .054).

11–12. The answers are: 11-C *[Ch 10 II C 2, 5, 6]*, **12-B** *[Ch 10 II C 7, 8]*
The raw score, $Y = 149.5$, is first converted to a z score:

$$\begin{aligned} z &= \frac{Y - \mu}{\sigma} \\ &= \frac{149.5 - 140.0}{10} \\ &= .95 \end{aligned}$$

The probability that an individual selected at random from a population in which $\mu = 140$ mm Hg and $\sigma = 10$ mm Hg will have a systolic blood pressure greater than or equal to 149.5 mm Hg is the area under the standard normal curve (Table B, Appendix 3) to the right of this z value:

$$\begin{aligned} P(Y \geq 149.5) &= P(z \geq .95) \\ &= 1 - .8289 \\ &= .1711 \end{aligned}$$

The systolic blood pressure below which 90% of blood pressure readings in the population fall is the 90th percentile of the frequency distribution. To obtain this value, it is first necessary to determine the z score that has an area of .90 below (to the left of) it. This z value is given in the standard normal table (Table B, Appendix 3) as 1.28. The systolic blood pressure measurement corresponding to $z = 1.28$ is calculated using the formula

$$z = \frac{Y - \mu}{\sigma}$$

Thus,

$$\begin{aligned} 1.28 &= \frac{Y - 140}{10} \\ Y &= 140 + 10(1.28) \\ &= 152.8 \text{ mm Hg} \end{aligned}$$

13–20. The answers are: 13-B *[Ch 3 II C 5]*, **14-D** *[Ch 3 II C 1]*, **15-D** *[Ch 3 II C 2]*, **16-C** *[Ch 3 II C 2]*, **17-A** *[Ch 3 I A 2; II C 6]*, **18-A** *[Ch 3 II C 6]*, **19-E** *[Ch 3 II C 4]*, **20-B** *[Ch 3 II C 3]*
The prevalence of acute myocardial infarction (AMI) in the population is the proportion of patients in the sample with a confirmed AMI, as defined by the gold standard:

$$\begin{aligned}\text{prevalence} &= P(\text{D+}) \\ &= 230/360\end{aligned}$$

This estimate of prevalence is valid only when the numbers of diseased and healthy patients (the totals of columns 1 and 2 in the table) are *not* fixed by the investigator.

Sensitivity is the probability that a diseased individual will have a positive test result. It is estimated as the number of patients with a positive CK test among the total number of patients with a proven AMI, as defined by the gold standard:

$$\begin{aligned}\text{sensitivity} &= P(\text{T+}|\text{D+}) \\ &= \frac{\text{diseased with positive test}}{\text{all diseased patients}} \\ &= 215/230\end{aligned}$$

The probability of a negative test result in a patient who has not suffered an AMI, $P(\text{T}-|\text{D}-)$, is the proportion of negative tests among the healthy population:

$$\begin{aligned}P(\text{T}-|\text{D}-) &= \frac{\text{disease-free with negative test}}{\text{all disease-free patients}} \\ &= 114/130\end{aligned}$$

This quantity is the specificity, or true negative rate (TNR), of the CK test.

A CK value greater than the positivity criterion, 80 IU, represents a positive test result, while a value below 80 IU represents a negative test result. Hence, the new patient, with a CK level of 85 IU, has a positive CK test result. The probability that this patient has actually had an AMI is $P(\text{D+}|\text{T+})$:

$$\begin{aligned}P(\text{D+}|\text{T+}) &= \frac{\text{diseased with positive test}}{\text{all patients with positive test}} \\ &= 215/231\end{aligned}$$

This probability is the predictive value positive (PVP) of the diagnostic test. From a diagnostic point of view, PVP is more informative than sensitivity $[P(\text{T+}|\text{D+})]$, the probability that a diseased individual will have a positive test result.

The false positive rate (FPR) of the CK test is the probability that a healthy individual will have a positive test result, $P(\text{T+}|\text{D}-)$:

$$\begin{aligned}\text{FPR} &= P(\text{T+}|\text{D}-) \\ &= \frac{\text{disease-free with positive test}}{\text{all disease-free patients}} \\ &= 16/130\end{aligned}$$

Other definitions of FPR exist in the medical literature. For example, FPR can be estimated as the proportion of false positives among the total population:

$$\begin{aligned}\text{FPR} &= P(\text{T+ and D}-) \\ &= P(\text{T+}|\text{D}-)P(\text{D}-) \\ &= \frac{\text{disease-free with positive test}}{\text{total number of patients screened}} \\ &= 16/360\end{aligned}$$

FPR can also be estimated as the proportion of false positives among those patients with positive test results, that is, as 1 − PVP:

$$\begin{aligned}\text{FPR} &= P(\text{D}-|\text{T+}) \\ &= \frac{\text{healthy patients with positive test}}{\text{all patients with positive test}} \\ &= 16/231\end{aligned}$$

The probability of a negative CK test result in the diseased population is the false negative rate (FNR), $P(\text{T}-|\text{D}+)$, which is the complement of sensitivity:

$$\begin{aligned}\text{FNR} &= P(\text{T}-|\text{D}+)\\ &= 1 - \text{sensitivity}\\ &= 1 - P(\text{T}+|\text{D}+)\\ &= 1 - (215/230)\\ &= 15/230\end{aligned}$$

21. The answer is B *[Ch 3 II C 6; III B 2, C 4]*
The predictive value positive (PVP) is the probability that a patient who has a positive test result actually has the disease in question, or $P(\text{D}+|\text{T}+)$. The quantity 1 − specificity is the false positive rate (FPR), $P(\text{T}+|\text{D}-)$. The PVP is a function not only of sensitivity and specificity, but also of the prevalence of the disease in the population.

22. The answer is C *[Ch 3 III C 2]*
The false positive rate (FPR) for the screening test, $P(\text{T}+|\text{D}-)$, is known to be .07. Test specificity $[P(\text{T}-|\text{D}-)]$, the probability that a healthy patient will have a negative test result, is the complement of the FPR: $1 - P(\text{T}+|\text{D}-) = .93$. The proportion of negative test results among those patients with Lyme disease, the false negative rate (FNR), or $P(\text{T}-|\text{D}+)$, is known to be .18. Test sensitivity is the complement of FNR, that is, $P(\text{T}+|\text{D}+) = 1 - P(\text{T}-|\text{D}+) = .82$. The prevalence of Lyme disease, $P(\text{D}+)$, is 3% and $P(\text{D}-) = 1 - P(\text{D}+) = .97$. The probability that a patient with a positive test result has Lyme disease [the predictive value positive (PVP), or $P(\text{D}+|\text{T}+)$] can be computed from these known conditional probabilities using Bayes' rule:

$$\begin{aligned}P(\text{D}+|\text{T}+) &= \frac{P(\text{T}+|\text{D}+)P(\text{D}+)}{P(\text{T}+|\text{D}+)P(\text{D}+) + P(\text{T}+|\text{D}-)P(\text{D}-)}\\ &= \frac{(.82)(.03)}{(.82)(.03) + (.07)(.97)}\end{aligned}$$

23–24. The answers are: 23-C *[Ch 3 II C 1–4; III C 1]*, **24-B** *[Ch 3 III B 2, C 4 c; Figure 3-4]*
The test sensitivity (i.e., the probability that a child who has been sexually abused will have a positive test result) is given as .95. The complement of sensitivity, the false negative rate [FNR, or $P(\text{T}-|\text{D}+)$], is therefore $1 - .95 = .05$. Test specificity is known to be .90. Thus, its complement, the false positive rate [FPR, or $P(\text{T}+|\text{D}-)$] is $1 - .90 = .10$. The prevalence of sexual abuse, $P(\text{D}+)$, is .001 and $P(\text{D}-) = 1 - .001 = .999$. Using Bayes' rule, the probability that a child who has a positive test result has not been sexually abused is

$$\begin{aligned}P(\text{D}-|\text{T}+) &= \frac{P(\text{T}+|\text{D}-)P(\text{D}-)}{P(\text{T}+|\text{D}-)\,P(\text{D}-) + P(\text{T}+|\text{D}+)\,P(\text{D}+)}\\ &= \frac{(.10)(.999)}{(.10)(.999) + (.95)(.001)}\\ &= .991\end{aligned}$$

The predictive value positive (PVP), the probability that a child with a positive test result has been sexually abused, decreases as the prevalence of sexual abuse in the community decreases. At the extreme, when the prevalence of sexual abuse in the population is equal to zero, all positive test results represent false positives and PVP = 0. In this instance, $\text{PVP} = 1 - P(\text{D}-|\text{T}+) = 1 - .991 = .009$. The greater the specificity, or true negative rate (TNR), of the test, the lower the false positive rate (FPR). Therefore, as specificity increases, PVP also increases.

25. The answer is B *[Ch 3 III C 3]*
Given that $P(\text{D}+) = .20$, $P(\text{D}-) = .80$, $P(\text{T}+|\text{D}+) = .90$, and $P(\text{T}-|\text{D}-) = .95$, the probability that a patient with a negative test result does not have hepatitis B, $P(\text{D}-|\text{T}-)$, is

$$P(D-|T-) = \frac{P(T-|D-)P(D-)}{P(T-|D-)P(D-) + P(T-|D+)\ P(D+)}$$

$$= \frac{(.95)(.80)}{(.95)(.80) + (1 - .90)(.20)}$$

26–28. The answers are: 26-D *[Ch 3 I B 1 b (4)]*, **27-C** *[Ch 3 I B 1 b (5), d (2)]*, **28-E** *[Ch 3 I B 1, 4]*
The gaussian method defines the normal range for values of a diagnostic variable as the mean ± two standard deviations. Hence, for this example, the normal range = 200 ± 2(40) = 120 to 280 mg/dl.

Ninety-five percent of the total cholesterol values for the healthy population fall between 120 and 280 mg/dl, while 5% of the values fall outside this range. That is, 2.5% of the values are greater than 280 mg/dl and 2.5% are less than 120 mg/dl. If only the upper limit is used to define the presence of disease, 2.5% of healthy individuals will be erroneously classified as diseased.

The criteria used to define the cholesterol level at which therapeutic intervention becomes effective affects the validity of a positivity criterion that is selected by the therapeutic method, not by the gaussian method. The validity of the normal range, as defined by the gaussian method, is influenced by the sample size (the larger the sample size, the more accurate the estimate of the population mean μ and standard deviation σ), the frequency distribution of the diagnostic variable (deviations from normality invalidate the definition of the normal range), the representativeness of the sample (a biased sample will lead to an inaccurate estimate of μ and σ), and the accuracy with which individuals are classified as healthy or diseased.

29. The answer is D *[Ch 6 I A–B]*
Because the patient had two siblings with Tay-Sachs disease, her parents are both obligate carriers. The probability that she herself is a carrier, *P*(patient carrier), is therefore 2/3. Her husband, who has already fathered one child with the disorder, is also a proven carrier. The probability that this couple's first child will have Tay-Sachs disease is

$$\begin{aligned} P(\text{child affected}) &= P(\text{child affected and parents carriers}) \\ &\quad + P(\text{child affected and parents not carriers}) \\ &= P(\text{child affected}|\text{parents carriers}) \\ &\quad \times P(\text{parents carriers}) + 0 \\ &= P(\text{child affected}|\text{parents carriers}) \\ &\quad \times P(\text{patient carrier}) \times P(\text{husband carrier}) \\ &= (1/4)(2/3)(1) \\ &= 1/6 \end{aligned}$$

30. The answer is A *[Ch 6 II A–B]*
Given that the number of affected children $k = 1$, the number of trials (i.e., number of children) $n = 4$, and the probability that any child born to these two carriers will have Niemann-Pick disease = 1/4, the probability that only one of the four children will have the disorder can be computed using the binomial formula, where p = the probability that an individual selected at random from the population will have the disease, and q = the probability that such a randomly selected individual will not have the disease. Thus,

$$\begin{aligned} P(k = 1) &= \binom{n}{k} p^k (1 - p)^{n-k} \\ &= \binom{4}{1} (1/4)(3/4)^3 \\ &= \frac{4!}{1!\ 3!} (1/4)(3/4)^3 \\ &= 4(1/4)(3/4)^3 \end{aligned}$$

31–32. The answers are: 31-C *[Ch 6 I B]*, **32-B** *[Ch 6 III B 3–4]*
The man is a known carrier for β-thalassemia; that is, *P*(father carrier) = 1. His wife's family history is not known. Therefore, the probability that she is a carrier is assumed to equal that of the black

population in general: *P*(mother carrier) = .01, or 1/100. The probability that a child born to this couple will have β-thalassemia is

$$\begin{aligned} P(\text{child affected}) &= P(\text{child affected and parents carriers}) \\ &\quad + P(\text{child affected and parents not carriers}) \\ &= P(\text{child affected}|\text{parents carriers}) \\ &\quad \times P(\text{mother carrier}) \times P(\text{father carrier}) + 0 \\ &= (1/4)(1/100)(1) \\ &= .0025 \end{aligned}$$

Following the birth of a healthy child (C), the posterior probability that the patient's wife (M) is a carrier can be computed using Bayes' rule:

$P(\text{M carrier}|\text{C healthy})$

$$= \frac{P(\text{C healthy}|\text{M carrier})\ P(\text{M carrier})}{P(\text{C healthy}|\text{M carrier})P(\text{M carrier}) + P(\text{C healthy}|\text{M not carrier})P(\text{M not carrier})}$$

$$= \frac{(3/4)(1/100)}{(3/4)(1/100) + (1)(99/100)} = .0075$$

33. The answer is C *[Ch 7 I B, C 1 a, 2, D 2]*
The number of new cases of acute granulocytic leukemia (incidence) will be least affected by a new drug therapy. Because of the efficacy of the drug, the five-year survival rate (the probability that a patient will survive at least five years after the initial diagnosis) will increase and the mortality rate (the probability of dying from the disease) will decrease. If more patients survive, the number of existing cases (i.e., the prevalence) will increase.

34–35. The answers are: 34-B *[Ch 14 II B 4 b–c]*, **35-E** *[Ch 8 II B 2 c (5) (b); Ch 14 II B 4 b (1)]*
The data first should be summarized in a 2 × 2 table, as shown below:

	Leukemia (D+)			No Leukemia (D−)	Totals
Exposed to Radiation (E+)	75	*a*	*b*	60	135
Not Exposed to Radiation (E−)	25	*c*	*d*	40	65
Totals	100			100	200

The odds ratio (OR) can be calculated from this data, using the formula

$$\begin{aligned} \text{OR} &= ad/bc \\ &= (75)(40)/(60)(25) \\ &= 2.0 \end{aligned}$$

Thus, the odds that leukemia patients have been exposed to ionizing radiation are twice as great as those of the controls. Since leukemia is a rare condition, the OR approximates relative risk (RR). In other words, an individual who is exposed to ionizing radiation is twice as likely to develop leukemia as one who is not.

This study utilized a case–control study design (i.e., leukemia patients and healthy controls were identified at the onset and followed backward in time to determine exposure). Absolute risk [$P(\text{D}+|\text{E}+)$], the probability of a disease among exposed individuals, cannot be calculated directly in a case–control study. An investigator who wishes to calculate absolute risk must use a cohort study, in which exposed and nonexposed individuals are followed forward in time to determine the incidence of the disease.

36. The answer is B *[Ch 8 II D 3; Tables 8-2, 8-3, and 8-4]*
The results of the first study suggest a difference in systolic blood pressure between students residing in urban communities and those residing in rural communities, as well as differences among the three

SES groups (statistical tests would be needed to demonstrate that these differences are not due to random chance). However, when the place of residence is cross-classified by SES, the apparent difference between urban and rural students disappears, while mean systolic blood pressure continues to differ between the high and low SES groups. Thus, focusing on factors that distinguish among the three socioeconomic groups would be the best strategy for the researcher's new study.

37. The answer is B *[Ch 14 II A 2, 3; Table 14-1]*
The risk of diarrhea following salmonella exposure is $P(D+|E+) = 30/2500 = .012$, while the risk for students with no exposure is $P(D+|E-) = 60/7500 = .008$. Thus, the relative risk (RR) is calculated as

$$\begin{aligned} RR &= (.012)/(.008) \\ &= 1.5 \end{aligned}$$

That is, a child exposed to the contaminated chicken salad was 1.5 times more likely to contract diarrhea than a child who was not.

38–40. The answers are: 38-C *[Ch 7 I A 2]*, **39-B** *[Ch 7 I B 1 c]*, **40-C** *[Ch 14 II A 2]*
The prevalence of coronary heart disease (CHD) among white men in 1960 is calculated as the proportion of CHD cases in the original sample of 1000 men:

$$\begin{aligned} \text{prevalence} &= \frac{\text{number with CHD}}{\text{total number in sample}} \\ &= 50/1000 \end{aligned}$$

The incidence of acute myocardial infarction (AMI) between 1960 and 1980 is the number of new cases that developed during this time among the 950 subjects who were healthy at the outset of the study. Over the 20-year follow-up period, 50 + 75, or 125 of these subjects suffered an AMI. Thus, the incidence is

$$\begin{aligned} \text{incidence} &= \frac{\text{number of new cases}}{\text{total number at risk}} \\ &= 125/950 \end{aligned}$$

The risk (incidence) of AMI among hypertensive subjects is $P(\text{AMI}|\text{hypertension}) = 50/200$. The risk of AMI among normotensive subjects is $P(\text{AMI}|\text{normal blood pressure}) = 75/750$. Therefore, the relative risk (RR) of AMI for hypertensive versus normotensive subjects is

$$\begin{aligned} RR &= \frac{P(\text{AMI}|\text{hypertension})}{P(\text{AMI}|\text{normal blood pressure})} \\ &= \frac{P(D+|E+)}{P(D+|E-)} \\ &= \frac{50/200}{75/750} \\ &= 2.5 \end{aligned}$$

41–48. The answers are: 41-C *[Ch 8 II B 2 c]*, **42-C** *[Ch 8 II C 1, 2 b, 3 b, D 1]*, **43-A** *[Ch 14 II B 4 b–c]*, **44-E** *[Ch 14 II A 4, B 4 d (2)]*, **45-E** *[Ch 8 II B 2 c (5) (b); Ch 14 II A 4, B 4 b (1)]*, **46-B** *[Ch 13 II B, C 3; Figure 13-5]*, **47-D** *[Ch 12 V A; VI C 1; Ch 13 II C 3 b (5)]*, **48-D** *[Ch 12 II E 3; VI C 1 b; Ch 13 II D 4]*
This study utilizes a case–control study design; that is, it begins with the selection of cases of the disease in question (chronic otitis media) and disease-free controls. Both groups are then followed backward in time to determine exposure to the putative risk factor (parental history of ear infections).

Restriction is used in this study to equalize the comparison groups with respect to age (cases and controls are limited to children between the ages of one and three). However, the comparison groups may differ with respect to other factors. If these extraneous systematic differences affect the response variable (parental history of chronic otitis media) as well as the outcome (chronic otitis media in children between one and three years of age), a biased comparison will occur. For example, low socioeconomic status (leading to a reduced access to health care, poor nutrition, etc.) may be the critical factor mediating the occurrence of chronic otitis media in both parents and their children. Case–control studies are particularly prone to recall bias. That is, the parents of children suffering from recurrent otitis media may be far more likely to remember their own history of this disorder.

The results of this study can be summarized in the following 2 × 2 table:

Parental History	Ear Infections; Cases (D+)			No Ear Infections; Controls (D−)	Totals
Childhood Ear Infections (E+)	30	*a*	*b*	20	50
No Childhood Ear Infections (E−)	20	*c*	*d*	30	50
Totals	50			50	100

The odds ratio (OR) can be calculated from the values in this table, using the formula

$$\begin{aligned} OR &= (ad)/(bc) \\ &= (30)(30)/(20)(20) \\ &= 2.25 \end{aligned}$$

That is, the odds that children with chronic otitis media have at least one parent who also suffered from this disorder are 2.25 times greater than the odds of parental history of infection for controls.

Neither the absolute risk of chronic otitis media in children with a parental history of this disorder $[P(D+|E+)]$ nor the absolute risk for children with no such family history $[P(D+|E-)]$ can be calculated directly from the results of a case–control study. Thus, relative risk (RR), $P(D+|E+)/P(D+|E-)$, is also impossible to calculate directly. If the prevalence of otitis media in the general population were low, and if the control group is representative of the healthy population with respect to family history of chronic otitis media, then the OR would be an estimate of RR. Since otitis media is *common* rather than rare, no such estimate can be made.

The absolute risk of chronic otitis media, given a parental history of the disorder $[P(D+|E+)]$ cannot be directly calculated from the results of this case–control study. Estimates of absolute risk can be obtained only in a prospective cohort study, in which exposed and nonexposed subjects are followed forward in time to determine the incidence of the disease in question.

When an investigator wishes to detect an association between two study variables (or differences among proportions) and the data are in the form of counts (such as number of children infected), the chi-square test is the appropriate statistical procedure. (For matched samples, McNemar's test, a variant of the basic chi-square test, is most appropriate.) The paired *t* test is used to detect differences between two means when study subjects have been matched and the data are continuous, while the independent sample (pooled) *t* test is selected for two independent samples. Correlation analysis allows the investigator to determine the magnitude of the association between two variables measured on an interval/ratio scale. Analysis of variance is used in studies involving interval/ratio data to compare the equality of several population means.

Statistical significance implies only that it is unlikely that an association at least as large as that observed would be obtained by random chance in two samples drawn from populations in which no such association actually exists. A statistically significant association is not necessarily either strong or clinically important; even a weak association may prove significant when the sample size is large. Furthermore, a statistically significant association is not proof of causality. Other confounding variables related to both exposure and the outcome may cause the two to "travel together." Statistical significance does not guarantee an absence of bias. Only careful attention to proper study design and the control of potential sources of bias can ensure internal validity.

The p-value does not measure the strength of an association between two study variables, but is rather the actual probability of committing a type I error (concluding that an association exists, when, in fact, it does not). If $p \leq \alpha$, the preselected risk of a type I error the investigator is willing to tolerate, the null hypothesis is rejected.

49–52. The answers are: 49-D *[Ch 7 I A 1, B]*, **50-C** *[Ch 7 I B 1 c]*, **51-B** *[Ch 14 II A 2, 3]*, **52-D** *[Ch 7 I C; Figure 7-4]*

Incidence, not prevalence, measures the risk of developing a given disease. The upper table accompanying the question presents information gathered in the initial examination. Five of the 1083 men between the ages of 30 and 44 were shown to have coronary heart disease (CHD) at the initial exam, compared to seven of the 1317 women in this age range. Therefore, the data in this table represent prevalence, rather than incidence, rates (see *last column* in upper table accompanying the question). The age distribution (30–44 years) is the same for both comparison groups (men and women), and each group acts as a control for the other.

Absolute risk is the number of new cases of CHD that develop over the specified time period divided by the number of individuals at risk (i.e., the number who were healthy at the beginning of the observation period). That is,

$$\begin{aligned}\text{absolute risk for men, ages 30–39} &= P(\text{D}+|\text{E}+)\\ &= 20/825\\ &= .0242 \text{ (24.2 per 1000)}\end{aligned}$$

The absolute risk of developing CHD for women between the ages of 30 and 39, $P(\text{D}+|\text{E}-) = 1/1036$. Thus, the relative risk (RR) for men compared to women is

$$\begin{aligned}\text{RR} &= \frac{P(\text{D}+|\text{E}+)}{P(\text{D}+|\text{E}-)}\\ &= \frac{(20)/(825)}{(1)/(1036)}\\ &= 25.1\end{aligned}$$

Men between the ages of 30 and 39 are 25.1 times more likely to develop CHD than women in the same age range.

Mausner and Kramer report, "The explanation for this discrepancy [between incidence and prevalence] lies in the different course of disease in young men and women. In the men, the disease manifests itself as myocardial infarction and sudden death. In women, the disease is more likely to present as anginal attacks, that is, attacks of chest pain of brief duration that do not generally endanger life. With the longer duration of the disease in women, prevalence can actually be equal in the two sexes despite the much greater incidence in men."* In other words, while the number of *new* cases (i.e., the incidence) of men with CHD may be high because men are more likely than women to get the disease, the prevalence (number of *existing* cases) in men may still be low because the disease is more short-lived and deadly in men. Thus, because the disease occurs more often but is of shorter duration in men and occurs less often but lasts longer in women, the prevalence of the disease in the two populations can be equal.

53–54. The answers are: 53-B *[Ch 8 II B 2 c]*, **54-C** *[Ch 8 II B 2 c (5) (a), C]*
This study began with the identification of two groups of subjects (infants with spina bifida and healthy infants) who were followed backward in time to determine prenatal exposure to vitamins. Thus, it is an example of a case–control study. In both cohort and experimental studies, subjects are followed forward in time to determine the incidence of disease as a function of exposure to a suspected risk factor.

Uncontrolled systematic differences between the comparison groups may explain the observed difference in prenatal vitamin use. For example, mothers of healthy infants may have been more health conscious than mothers of infants born with spina bifida. Thus, in addition to using prenatal vitamins, they also may have been more likely than other mothers to obtain good prenatal care, refrain from smoking, and avoid exposure to radiation and other environmental risk factors. These extraneous factors may have accounted for the reduced incidence of neural tube defects. The faulty inference of causality is not related to random chance ($p < .001$), failure to distinguish incidence and prevalence, or the use of rates of disease rather than rates of exposure.

55. The answer is E *[Ch 8 II B 1 a, b; Ch 9 I A 2]*
This study lacks a control or comparison group. Therefore, the rate of exposure to cats among healthy people is unknown (it may be higher than that among the multiple sclerosis patients) and the inference of causality is incorrect. Sample size and statistical analysis are irrelevant; neither can correct the fatal flaw in this study.

56–57. The answers are: 56-D *[Ch 8 II B 1 a, b, 2; Ch 9 I A]*, **57-E** *[Ch 9 II C 1, 3 b; Ch 12 VI C 1 b]*
This study is an example of a randomized controlled clinical trial, in which subjects are randomly assigned to the comparison groups. A prospective cohort study is the observational design most like

---

*Mausner J, Kramer S: *Mausner and Bahn's Epidemiology: An Introductory Text*, 2nd. ed. Philadelphia, WB Saunders, 1985, p 54.

the randomized controlled clinical trial. In this design, however, subjects are not randomly assigned to comparison groups but rather "self-select" their group status.

An unappealing diet, like a medication with unpleasant side effects, is likely to result in poor patient compliance. A failure of subjects to comply with the prescribed high fiber diet may therefore explain its lower success rate. Because subjects were randomly assigned to comparison groups, selection bias is an unlikely explanation for the observed difference in efficacy. The small, statistically significant *p*-value indicates that the sample size was sufficiently large and that the observed difference in efficacy is unlikely to be due to random chance.

58. The answer is D *[Ch 9 II C 1 b, D 1, 2 b; Ch 12 VI C 1 b]*
The additional interventions provided to the warfarin group may well explain the reduction in thromboembolic disease and subsequent mortality among these patients. Thus, the results of this study do not support the conclusion that warfarin therapy is the best treatment for patients suffering an acute myocardial infarction (AMI). The small *p*-value ($p < .05$) indicates that the observed differences between the two groups are unlikely to be the result of random chance. Because subjects were randomly assigned to comparison groups, selection bias is unlikely. Blinding was used to control surveillance bias.

59. The answer is B *[Ch 7 I D 4]*
The attack rate for those who ate both turkey and dressing is 75%. The rate for those who ate only the dressing is 84%, compared to 2% for those who ate only the turkey. No cases of food poisoning occurred among those who ate neither food. Based on this information, the dressing is most likely to have been the contaminated food.

60. The answer is B *[Ch 7 I D 2; Ch 8 II D 4; IV B–D]*
If crude mortality rates are the same for the two populations, but population 1 has a higher age-adjusted rate, then population 1 must be younger, on the average, than population 2. The same conclusion can be drawn if the two populations have equal age-adjusted mortality rates, but population 1 has a lower crude mortality rate.

61. The answer is D *[Ch 3 I B 1 a; Ch 10 II B, C 2, 7, 8; Figures 10-4 and 10-6]*
For symmetric distributions, such as the normal distribution, the 50th percentile value is the mean ($\mu = 65$); 50% of the values lie above the mean and 50% lie below it. Ninety-five percent of the values of a normally distributed variable fall within ± 1.96 (often rounded to 2) standard deviations of the mean. Thus, 95% of the exam scores will fall within $65 \pm 1.96(10) = 45.4$ to 84.6, while 2.5% fall above 84.6 and 2.5% fall below 45.4. The 99th percentile value is that value below which 99% of the test scores will fall. The standard normal table (Appendix 3, Table B) gives $z = 2.33$ as the z value with an area $= .99$ below it. The raw score corresponding to this z value is computed using the formula

$$z = \frac{Y - \mu}{\sigma}$$

$$2.33 = \frac{Y - 65}{10}$$

$$Y = 65 + 10(2.33) = 88.3$$

62. The answer is D *[Ch 8 II C 1, 2 a (1), 3 b (2), 4 c]*
Those who took advantage of the free physical examination are likely to represent a more health conscious segment of the population. Thus, the biased sample used in the second survey apparently overestimated the true proportion who had seen a physician in the preceding five years. Telephone ownership in a low-income community may indeed be restricted to the least poverty-stricken residents. However, this bias would be likely to overestimate, not underestimate, the proportion of the population who saw a physician during the preceding five years. Recall bias is equally likely to have compromised both surveys. If the purpose of a study is primarily descriptive, as in this case, a comparison or control group is unnecessary.

63. The answer is B. *[Ch 15 II A 1 a (2), B 2 e, D 2]*
The correlation coefficient r is a measure of the degree of linear association between two study variables. When r is positive, this relationship is described by a straight line with a positive slope. A statistically significant correlation indicates that it is unlikely that a value of r equal to or greater than

that observed could be obtained by random chance from a sample drawn from a population in which the study variables are unrelated (i.e., in which $\rho = 0$). Even a weak correlation (i.e., small deviations from $\rho = 0$) may prove statistically significant if the sample size is large.

64. The answer is D *[Ch 14 I B 2 b; II A 3 b, B 3 c; III A 1]*
Relative risk (RR) and the odds ratio (OR) are two measures of the strength of the association between a suspected risk factor and a disease. Both indicate how much more likely the disease is to occur in those exposed to the risk factor than those with no such exposure. Absolute risk is not a measure of the strength of the association; RR may be high even when absolute risk is low. Attributable risk measures the excess risk of the disease among those exposed to the risk factor, over and above that experienced by individuals who are not exposed. The size of the *p*-value indicates the likelihood that the observed association occurred by random chance but does not measure the strength of the association.

65. The answer is C *[Ch 12 II D 1–2; VI C 2; Ch 15 II B]*
When the null hypothesis (H_O) is not rejected, the results are reported as "not statistically significant" (i.e., the observed correlation is likely to have occurred by random chance). This failure to reject H_O may indicate that no such correlation exists in the population from which the sample was drawn ($\rho = 0$) or it may reflect low power of the statistical test. In the absence of information on power, it is impossible to discriminate between these two alternatives.

66–69. The answers are: 66-C *[Ch 12 II A 1; III B]*, **67-C** *[Ch 15 II A; Ch 16 I A, B; II A; III A; Figure 16-3 Ch 17 I B 1]*, **68-B** *[Ch 12 II E 1 b; V A 3; Ch 13 II D 1 b; Ch 16 III A 2 b (2) (b)]*, **69-C** *[Ch 12 III E 3, F; IV D, E; Ch 16 III B 1 d–e]*
The investigator wishes to know if the new drug has a *greater effect* on diastolic blood pressure than the placebo (not just whether it causes a change in diastolic blood pressure as compared to the placebo). Thus, the alternative hypothesis (H_A), which generally states that a predicted effect exists is: "The mean change in diastolic blood pressure elicited by the drug is greater than that elicited by the placebo."

A paired *t* test is appropriate because the investigator wishes to compare two population means (ΔBP), using matched samples (the subjects serve as their own controls). The independent sample (pooled *t* test) is used to compare the equality of two means obtained from independent samples, while the analysis of variance is used when more than two groups are to be compared. The chi-square test is used to evaluate associations involving frequency data. Correlation analysis defines the degree of association between two quantitative study variables.

Because H_A specifies a direction (i.e., that the change in blood pressure will be greater after treatment), a one-tailed statistical test must be used. Although the 5% level of significance is most common, the investigator may select any value for α. A statistical test using this level of significance is more likely to reject the null hypothesis, H_O, (i.e., is more powerful) than a test based on a level of significance $\alpha = .01$ (as α increases, β decreases and power increases). The paired *t* test is associated with $n - 1$ degrees of freedom; in this instance, $df = 4 - 1 = 3$.

Since the calculated test statistic is greater than the tabulated critical value, the H_O is rejected. In other words, the area under the frequency distribution to the right of the test statistic is less than 5% ($p < .05$). The mean change in diastolic blood pressure elicited by the new drug is significantly greater than that elicited by the placebo.

70. The answer is A *[Ch 12 II E 1; V A; VI B 1]*
By definition, α, the level of significance, is the probability of rejecting the null hypothesis (H_O) when it is actually true. Conversely, the probability of accepting a true H_O is $1 - \alpha$. The probability of rejecting H_O, given that it is indeed false, is the power of the statistical test. The probability of rejecting H_O when $\alpha = .01$ is less than when $\alpha = .05$, since as α decreases, β increases and power decreases.

71. The answer is D *[Ch 12 II E 1, 3; V A; VI B]*
The level of significance (or the *p*-value) provides information on the probability of a type I error. That is, it defines the probability of detecting a difference in the average healing time elicited by the new drug and that elicited by the placebo when no such difference actually exists in the populations from which the samples were drawn. The probability of accepting a true null hypothesis (H_O) is the complement of the level of significance (i.e., $1 - \alpha$). The level of significance does not define the likelihood that either H_O or H_A are true, nor does it measure the probability that a given patient will benefit from the new drug.

72–74. The answers are: 72-B *[Ch 15 I B; Ch 16 I A, B; II A; Ch 17 II A; III A, B 3]*, **73-D** *[Ch 17 III C 5]*, **74-C** *[Ch 17 IV C 3]*
The analysis of variance is used with continuous data to compare the equality of three or more means. When subjects can be grouped into homogeneous blocks prior to treatment, on the basis of a factor thought to affect the response variable, the randomized complete block (RCB) design is more appropriate than the completely random design (CRD). Here, the blocking variable is the initial level of disability. When an investigator wishes to compare only two treatment means obtained from independent samples, the pooled t test should be chosen. Chi-square procedures are applicable to frequency, not interval/ratio, data.

When the null hypothesis (H_O) is rejected on the basis of an analysis of variance (i.e., the results are statistically significant), the correct conclusion is that at least two of the treatment means differ. The analysis of variance procedure itself provides no further information on which treatment means differ. Multiple comparison tests, such as the Bonferroni procedure, must be used to answer this question.

For three treatments (A, B, and C), there are three possible pairwise comparisons: A versus B, A versus C, and B versus C. The Bonferroni procedure requires that each individual comparison be carried out at the α/c level of significance, where $c =$ the number of comparisons $= 3$. Because the investigator desires a two-tailed comparison, the level of significance $\alpha/3$ must be divided by 2, resulting in a $(.05/3)/2$, or .0083, level of significance.

75–78. The answers are: 75-C *[Ch 8 I B 1, 2 a–d]*, **76-D** *[Ch 13 II B; Ch 15 I B; III A 1; Ch 16 I; II A; III A; Ch 17 I B 1]*, **77-A** *[Ch 12 III F 1; Ch 13 II C 1 b (5)]*, **78-D** *[Ch 13 II C 1 c (5), D 4]*
The patients participating in this study were simultaneously classified according to the severity of their disability and their personality type. Hence, this is an example of a cross-sectional study.

When the data are in the form of counts (i.e., the number of people in each combination of categories), the chi-square test is used to test for the presence of a significant association between the two study variables. Both the paired and pooled t tests, as well as the analysis of variance, use continuous, not frequency, data to test for the equality of means. Correlation analysis estimates the magnitude of the association between two quantitative variables, while regression analysis is used to derive an equation for predicting the value of one variable based on values of the other.

The calculated test statistic exceeds the tabulated critical value; therefore, the p-value is less than $\alpha = .05$. The null hypothesis (H_O) is rejected at the 5% level of significance, indicating there is a statistically significant association between the severity of the physical deficit and personality type.

A statistically significant result indicates that it is unlikely (here, a less than .1% chance) that an association equal to or greater than that observed could be obtained from a sample drawn from a population in which no association exists between the two study variables. Statistical significance, however, does not provide an estimate of the strength of the association, nor does it imply the existence of a cause-and-effect relationship. Because a cross-sectional design was used, this study is particularly vulnerable to antecedent–consequence uncertainty. No statistical test can overcome flaws in the study design that lead to a biased comparison.

79. The answer is B *[Ch 12 VI C 2 b]*
A failure to achieve statistical significance may occur if there is no difference between the comparison groups or if the power of the statistical test is too low to detect such a difference. A result that is not statistically significant is likely to have occurred by random chance. Statistical tests provide no information on internal or external validity.

80. The answer is D *[Ch 15 III A 1–2]*
The regression equation ($\hat{Y} = -25 + .3X$) describes a positive linear relationship between initial weight (X) and weight loss after one year on the prescribed diet and exercise regimen (Y). Thus, women who weighed more initially lost more weight. Both the statistical significance of the association between initial weight and ultimate weight loss, as well as the actual magnitude of the association, are unknown (the size of the slope is a function of the unit of measurement, not a measure of the strength of the association). Not all patients lost at least 25 pounds. For example, a patient whose initial weight was 100 pounds lost, on the average, $-25 + (.3)(100) = 5$ pounds.

81–83. The answers are: 81-C *[Ch 12 II E 2; V B]*, **82-E** *[Ch 12 V C]*, **83-E** *[Ch 12 VI B, C 2]*
The chance of error associated with the failure to reject the null hypothesis (H_O)—that is, with a result that is not statistically significant—is β, the probability of a type II error. In conditional probability notation, β is written: $P(H_O \text{ not rejected} | H_O \text{ false})$. It may also be thought of as the probability that the statistical test will fail to detect a clinically important difference among the three treatments, given that such a difference actually does exist in the populations from which the study samples were drawn. The value of β is unknown in this example.

Power, the complement of β, is the probability that the statistical test will detect a clinically important difference in the level of pain relief afforded by the three treatments when such a difference truly exists in the populations from which the study samples were drawn [P(reject $H_O|H_O$ false)]. Like β, the value of power is unknown.

The failure of the statistical test to detect a difference in the level of pain relief afforded by the three treatments may reflect low power of the test, rather than the true absence of such a difference in the populations from which the study samples were drawn. The values of α, β, and power represent probabilities associated with test performance, not statements of probability regarding the validity of H_O or H_A. These probabilities also provide no information on the likelihood that any individual patient will benefit from a given treatment; indeed, a highly significant difference among the treatments can occur even if none are efficacious. A negative conclusion indicates that the observed difference among the three treatments is likely to have occurred by random chance.

84–87. The answers are: 84-A *[Ch 12 II E 2; V B; VI A 1–3]*, **85-C** *[Ch 12 V C]*, **86-B** *[Ch 12 II E 1; V A]*, **87-D** *[Ch 12 VI B, C 1]*

The probability that the statistical test will fail to detect a clinically important difference, given that this difference actually exists, is β, the chance of a type II error. As the sample size or level of significance α increases, β decreases. Because it is the complement of power, β also decreases if power increases. As the clinically important difference (effect size) decreases, β increases.

The probability that the statistical test will detect a true difference of a specified effect size is the power of that test. Here, power is equal to .80.

In conditional probability notation, the level of significance α may be written in terms of the null hypothesis (H_O): P(reject $H_O|H_O$ true) = .05. That is, if there truly is no difference in average hair count elicited by the new drug and that elicited by the placebo, a statistical test carried out on samples drawn from these populations will erroneously detect a difference between the treatment means 5% of the time.

The highly significant result of the statistical test indicates that the observed difference between the drug and placebo is unlikely to have occurred (a less than .05% chance) by random chance (i.e., without the drug being truly efficacious). The values for α, β, and power do not define the proportion of patients who will benefit from the new drug, nor do they predict the response of any individual patient.

88–90. The answers are: 88-C *[Ch 5 II B; Figure 5-5]*, **89-B** *[Ch 5 III B 4; Figure 5-7]*, **90-A** *[Ch 5 III B 5]*

The decision tree lists the probability that a surgically correctable abscess is present (for both the "surgery" and "no surgery" branches) as .30. The complement of this probability, that is, the likelihood that a surgically correctable abscess is absent, is 1 − .30 = .70. The probability of death from surgical complications is .25.

The expected utility at node A is obtained by averaging out the utilities and probabilities associated with the branches emanating from this node. For the branch labeled no surgical complications, the utility is .95 and the corresponding probability is .50. The utility associated with the branch labeled serious surgical complications is .65; the probability associated with this alternative is .25. Death has a utility of 0 and the probability of this outcome following surgery is .25. Thus, the expected utility at A is

$$\text{expected utility at node A} = (95)(.50) + (65)(.25) + (0)(.25)$$

$$= .638, \text{ or } 63.8\%$$

The utility values represent the expected survival rates associated with each outcome on the tree. The no surgery branch has the highest expected survival rate. Hence, this is the preferred course of action.

91. The answer is C *[Ch 4 I B 2; Figure 4-1]*

As the positivity criterion becomes more stringent, fewer positive test results will be reported, even among those who have the disease. Thus, sensitivity (the proportion of positive test results among the diseased population) decreases, while the false negative rate increases. Specificity, the proportion of negative test results among the healthy population, also increases as the positivity criterion is shifted to the right.

92. The answer is C *[Ch 3 I B 1 d; Ch 4 I D]*

The predictive value method, not the gaussian method, requires physicians to keep track of current values for test performance characteristics in their practices. The gaussian method defines as normal that range of values for a diagnostic test variable which encompass 95% of the values in the healthy population; it has several disadvantages. Assuming that values of the diagnostic variable follow a normal distribution, the range encompassing 95% of the values is equal to the mean ± two standard deviations. Thus, when both values above the upper limit and values below the lower limit are taken to indicate the presence of disease, at least 5% of the healthy population will be labeled diseased. A second disadvantage of the gaussian method is that multiple testing increases the chance of a positive test result. For example, if five independent tests are performed, the probability of a positive result on at least one is equal to $1 - (.95)^n = 1 - (.95)^5 = .23$. Still another disadvantage of this method is the fact that many diagnostic variables have skewed or bimodal frequency distributions. Finally, because the method is based on the frequency distribution of the test variable in the healthy population, it has no biologic basis for defining the presence or absence of the disease.

93. The answer is B *[Ch 3 II C 1–4, 6, D; III B 2 b, C 4]*
Sensitivity, $P(T+|D+)$, is the true positive rate; its complement, $P(T-|D+)$, is the false negative rate. If the sensitivity of the screening test is high, the true positive rate is high and the false negative rate is low. Specificity, $P(T-|D-)$, is the true negative rate; its complement, $P(T+|D-)$, is the false positive rate. If the specificity of the screening test is high, the true negative rate is high, while the false positive rate is low. The predictive value positive (PVP) is a function of sensitivity, specificity, and prevalence. For rare diseases (i.e., prevalence is low), PVP may be low, even if test sensitivity and specificity are high.

94. The answer is B *[Ch 7 I C 1–2]*
Over the 10-year observation period, the incidence of measles remained approximately the same, while the prevalence steadily declined. Had efforts to prevent new cases of measles been successful, incidence would have declined as well. The reduction in prevalence (existing cases at a given point in time) could have occurred as a result of a decrease in recovery time or an increase in mortality; both reduce the pool of existing cases.

95. The answer is B *[Ch 8 II B 2 a; Ch 14 II A 2, 4–5]*
Because a cohort study begins with the identification of exposed (E+) and nonexposed (E−) people, exposure rates cannot be estimated from the results of this type of study. By following both the exposed and nonexposed groups forward in time, the investigator can determine the proportion of individuals who develop the disease in each comparison group. These proportions estimate the absolute risk (incidence) of the disease among those exposed to the suspected risk factor and those with no such exposure. Relative risk, the ratio of these two incidence rates [i.e., $P(D+|E+)/P(D+|E-)$], can then be determined.

96. The answer is E *[Ch 8 II B 2 d, C 1, 2; Ch 12 II D 1; VI C 1]*
A statistically significant result is unlikely to be attributable to random chance. The higher neurosis scores in the infertile women could be due to a causal link between infertility and this sort of psychiatric disturbance. However, because a cross-sectional study design was used, it is unclear which factor is the antecedent and which the consequence. Steroid use by the infertile women could also explain the observed difference in neurosis scores (cointervention bias), as could systematic differences between the two comparison groups that are unrelated to fertility. In particular, of the observational study designs, cross-sectional studies are vulnerable to selection bias.

97. The answer is A *[Ch 7 I B, C 2]*
Incidence, the probability that a healthy individual will develop a particular disease during a specified time period, is calculated by dividing the number of new cases occurring over that period by the number of people at risk. A new drug that reduces disease mortality will not alter the number of new cases (incidence), although it can affect prevalence (the proportion of existing cases). In contrast, a prevention program that is successful in preventing the disease will lead to a decrease in incidence.

98. The answer is E *[Ch 7 I A, C 1 a, 2 b; Ch 8 II B 2 d]*
Prevalence, the probability that a member of a particular population will have the disease in question at a specified point in time, is calculated as the proportion of existing cases of a disease among the total number in the sample (both cases and healthy subjects). It can be estimated from the results of a cross-sectional study. Assuming that the study variables are relatively stable over time, prevalence is approximately equal to the product of incidence and the average duration of the disease. Any factor that increases disease duration (including a drug that prolongs life without curing the disease) will increase the pool of existing cases and, hence, increase the prevalence.

99. The answer is A *[Ch 1 IV C 1 b; Ch 11 I A 1 b, c B 1; II B; Figure 11-1]*
The interval described by the mean ± 1.96 standard deviations (i.e., 28.4 months to 67.6 months) will encompass 95% of the survival times in the population only if survival times are normally distributed. Because a normal distribution is symmetric, the mean and the median are approximately equal. Here, however, the median (30) is less than the mean (48), indicating a skewed frequency distribution. Hence, the interval from 28.4 to 67.6 does not necessarily include 95% of the survival times in the population. The 95% confidence interval estimate of the mean survival time is calculated as

$$\begin{aligned}\text{endpoints of interval} &= \text{mean} \pm (\text{confidence coefficient} \times \text{standard error of estimate}) \\ &= \overline{Y} \pm (1.96)(s/\sqrt{n}) \\ &= 48 \pm (1.96)(10/\sqrt{2500}) \\ &= 47.6 \text{ to } 48.4\end{aligned}$$

A physician can be 95% confident that the mean survival time of patients in the population falls between these two values. Fifty percent of survival times are greater than the median (30 months), while 50% of survival times are less than 30 months.

100. The answer is A *[Ch 11 I B 2, C 1]*
A confidence interval can be interpreted in two ways. Over the long run, if 100 samples were drawn from the population in question and 100 confidence intervals constructed, 90 of these intervals can be expected to include the unknown difference between the two population means. Or, in terms of a single sample, the investigator can be 90% confident that the calculated interval encompasses the unknown difference in mean leg volume between the patients treated with leg elevation and those treated with the antidiuretic drug. A 95% confidence interval is wider than a 90% confidence interval. That is, the price of greater confidence is an increase in the width of the confidence interval.

101. The answer is B *[Ch 8 II B 2 c (3) (b), (4)–(5); Ch 14 II A 4]*
In contrast to a prospective cohort study, a case–control study begins with the identification of healthy individuals and those suffering from the disease in question. Thus, neither absolute risk nor relative risk (RR) can be directly calculated from the results of a case–control study. This study design is especially useful for studying rare diseases. Because it requires few study subjects and utilizes existing data, a case–control study is less expensive and less time-consuming than a prospective cohort study. However, it is more vulnerable to potential sources of bias (because both the outcome and exposure have already occurred).

Solutions

CHAPTER 1

1-1. a. The mean, median, and mode of the phenacetin levels for each group of study participants are shown in the table below. The mean and median of the levels taken from the nonsmoking participants are nearly equal, which suggests that the frequency distribution for this group will be symmetric (or nearly symmetric). On the other hand, the mean of the phenacetin levels taken from the smoking participants is larger than the median, which suggests that the frequency distribution for this group will be skewed to the right (i.e., have a tail on the left).

Mean, Median, and Mode of Plasma Phenacetin Levels (μg/ml) for Both Study Groups in Problem 1-1

	Nonsmoking Group ($n = 14$)	**Smoking Group ($n = 12$)**
$\sum Y$	28.63	11.21
$\sum Y^2$	69.0989	26.1715
Mean	2.05	.93
Median	2.03	.57
Mode	(no mode)	.01

b. The range, variance, and standard deviation of the phenacetin levels for each group of study participants are calculated as

Range = highest value − lowest value
nonsmokers = 3.55 − .45 = 3.10
smokers = 3.80 − .01 = 3.79

$$\textbf{Variance } (s^2) = \frac{\sum Y^2 - \frac{\left(\sum Y\right)^2}{n}}{n-1}$$

$$\text{nonsmokers} = \frac{69.0989 - \frac{(28.63)^2}{14}}{13} = .81$$

$$\text{smokers} = \frac{26.1715 - \frac{(11.21)^2}{12}}{11} = 14.3$$

Standard deviation (s) $= \sqrt{s^2}$
nonsmokers $= \sqrt{.81} = .90$
smokers $= \sqrt{1.43} = 1.2$

c. The frequency distribution for each group of study participants is presented in the following table and shown in the following histograms. The frequency distribution of plasma phenacetin levels for the nonsmoking group is symmetric; for the smoking group it is skewed.

Grouped Frequency Distributions for Data in Problem 1-1

Plasma Phenacetin (μg/ml)	Frequency		Relative Frequency	
	Nonsmokers	Smokers	Nonsmokers	Smokers
0.005–0.505	1	5	.07	.42
0.505–1.005	1	2	.07	.17
1.005–1.505	2	2	.14	.17
1.505–2.005	3	1	.21	.08
2.005–2.505	3	0	.21	0
2.505–3.005	2	1	.14	.08
3.005–3.505	1	0	.07	0
3.505–4.005	1	1	.07	.08
Total	14	12	.98	1.00

Adapted from Duncan RC, Knapp RG, Miller MC III: *Introductory Biostatistics for the Health Sciences*, 2nd ed. Albany, Delmar, 1983, p 209.

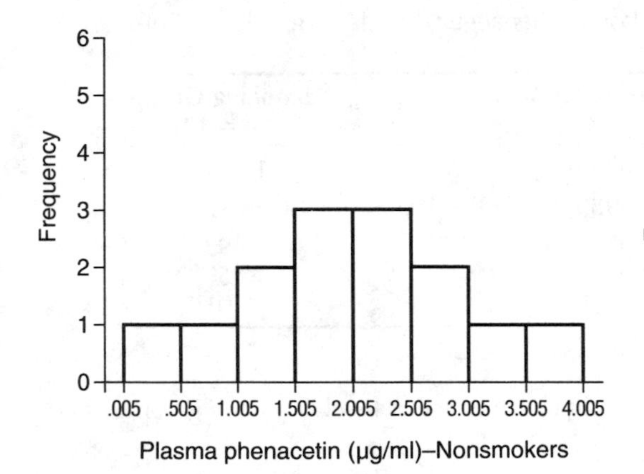

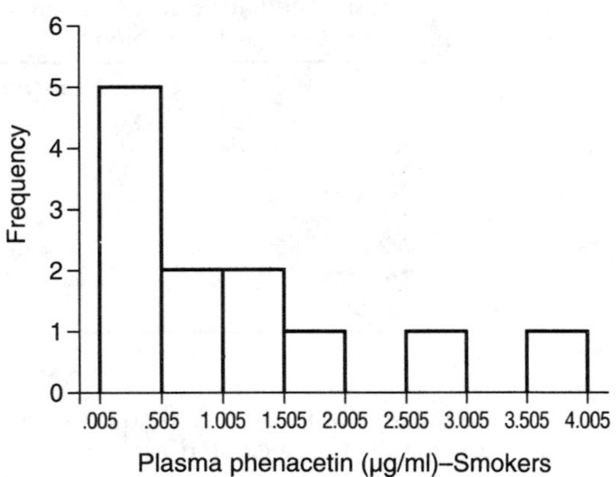

d. From the statistical results of the study, it appears that cigarette smokers have lower plasma phenacetin concentrations than nonsmokers.

e. The mean phenacetin level for the smoking participants of the study gives a misleading indication of the central value, since the frequency distribution for this group is skewed. In a skewed distribution, the mean is shifted toward the tail and away from the center of the distribution.

f. If phenacetin metabolism increases with age, then plasma phenacetin levels would be expected to be higher in older individuals than in younger ones. If average age of smokers is compared with average age of nonsmokers in this study (see table below), it appears that, in general, nonsmokers are older than smokers. Thus, if age affects phenacetin metabolism, it may be that age—not smoking—accounts for the lower levels of phenacetin among smokers in this study.

Comparison of Study Groups in Problem 1-1 According to Age (in years)

Measure	Nonsmoking Group	Smoking Group
Mean	47.2	35.4
Median	46.5	30.0
Standard deviation	15.1	14.6

1-2. a. The mean and standard deviation for each of the three groups of measurements are calculated as

$$\textbf{Mean} = \sum_{i=1}^{n} \frac{Y_i}{n}$$

$$\text{Instrument \#1} = \frac{100}{10} = 10$$

$$\text{Instrument \#2} = \frac{90}{10} = 9$$

$$\text{Instrument \#3} = \frac{100}{10} = 10$$

$$\textbf{Standard deviation } (s) = \sqrt{\frac{\sum Y_i^2 - \frac{\left(\sum Y_i\right)^2}{n}}{n-1}}$$

$$\text{Instrument \#1} = \sqrt{\frac{1148 - \frac{(100)^2}{10}}{9}} = 4.055$$

$$\text{Instrument \#2} = \sqrt{\frac{822 - \frac{(90)^2}{10}}{9}} = 1.155$$

$$\text{Instrument \#3} = \sqrt{\frac{1012 - \frac{(100)^2}{10}}{9}} = 1.155$$

b. Instrument 1 is accurate but imprecise (i.e., it produces measurements that average around the true value of 10 mg/ml but that vary widely from it), instrument 2 is inaccurate but precise (i.e., it produces measurements that are systematically shifted from the true value of 10 mg/ml but that vary little from the central value of 9 mg/ml), and instrument 3 is both accurate and precise.

CHAPTER 2

2-1. a. The probability that a man who visits the clinic has gonorrhea can be determined by referring directly to the marginal total for gonorrhea (D+ by culture result) in the table for problem 2-1,

$$P(\text{D+}) = 183/240 = .7625$$

Thus, an estimated 76.25% of men who visit the clinic have gonorrhea.

b. The probability that a man who visits the clinic has a positive gonodectin (Gd) test can be determined by referring to the marginal total for positive Gd results (T+),

$$P(\text{T+}) = 184/240 = .7667$$

c. The probability that a man has both a positive Gd test and gonorrhea is

$$P(\text{T+ and D+}) = 175/240 = .7292$$

To verify this figure using $P(\text{T+}|\text{D+})P(\text{D+})$,

$$P(\text{T+}|\text{D+})P(\text{D+}) = (175/183)(183/240) = 175/240$$

This could also be calculated using $P(\text{D+}|\text{T+})P(\text{T+})$.

d. The probability that a man tests negative and does not have gonorrhea is

$P(\text{T}- \text{ and } \text{D}-) = 48/240 = .20$

To verify this figure using $P(\text{T}+ \text{ and } \text{D}-)/P(\text{D}-)$,

$P(\text{T}+ \text{ and } \text{D}-)/P(\text{D}-) = (9/240)/(57/240) = 9/57$

e. The probability that a man with gonorrhea tests positive is

$P(\text{T}+|\text{D}+) = 175/183 = .9563$

To verify this figure using $P(\text{T}+ \text{ and } \text{D}+)/P(\text{D}+)$,

$P(\text{T}+ \text{ and } \text{D}+)/P(\text{D}+) = (175/240)/(183/240) = 175/183$

f. The probability that a man who does not have gonorrhea tests negative is

$P(\text{T}-|\text{D}-) = 48/57 = .8421$

To verify this figure using $P(\text{T}- \text{ and } \text{D}-)/P(\text{D}-)$,

$P(\text{T}- \text{ and } \text{D}-)/P(\text{D}-) = (48/240)/(57/240) = 48/57$

g. The probability that a man without gonorrhea tests positive is

$P(\text{T}+|\text{D}-) = 9/57 = .1579$

To verify this figure using $P(\text{T}+ \text{ and } \text{D}-)/P(\text{D}-)$,

$P(\text{T}+ \text{ and } \text{D}-)/P(\text{D}-) = (9/240)/(57/240) = 9/57$

h. The probability that a man with gonorrhea tests negative is

$P(\text{T}-|\text{D}+) = 8/183 = .0437$

To verify this figure using $P(\text{T}- \text{ and } \text{D}+)/P(\text{D}+)$,

$P(\text{T}- \text{ and } \text{D}+)/P(\text{D}+) = (8/240)/(183/240) = 8/183$

i. The probability that a man who tests positive has gonorrhea is

$P(\text{D}+|\text{T}+) = 175/184 = .9511$

To verify this figure using $P(\text{D}+ \text{ and } \text{T}+)/P(\text{T}+)$,

$P(\text{D}+ \text{ and } \text{T}+)/P(\text{T}+) = (175/240)/(184/240) = 175/184$

j. The probability that this man has gonorrhea given that the Gd test is negative is

$P(\text{D}+|\text{T}-) = 8/56 = .1429$

The probability that the man does not have gonorrhea is

$P(\text{D}-|\text{T}-) = 48/56 = .8571$

k. Physicians frequently must interpret positive or negative test results in an attempt to judge the likelihood of disease in a given patient. Therefore, the probability of most interest (from a diagnostic point of view) is the probability of disease given a positive test result, or $P(\text{D}+|\text{T}+)$. $P(\text{T}+|\text{D}+)$ is the probability of a positive test result in patients known to be diseased.

l. The probability that a man who comes to the clinic has a positive Gd test result or gonorrhea is

$$\begin{aligned} P(\text{T}+ \text{ or } \text{D}+) &= P(\text{T}+) + P(\text{D}+) - P(\text{D}+ \text{ and } \text{T}+) \\ &= 184/240 + 183/240 - 175/240 \\ &= 192/240 = .80 \end{aligned}$$

m. Using data from the table for problem 2-1 to confirm the summation principle for joint probabilities

$P(\text{D}+) = P(\text{D}+ \text{ and } \text{T}+) + P(\text{D}+ \text{ and } \text{T}-)$

$183/240 = 175/240 + 8/240$

$183/240 = 183/240$

2-2. a. From the table in problem 2-2, the probability that a student becomes ill after eating barbecue is

$$P(\text{ill}|\text{ate barbecue}) = 90/120 = .75$$

b. From the table in problem 2-2, the probability that a student becomes ill if no barbecue is eaten is

$$P(\text{ill}|\text{no barbecue}) = 20/80 = .25$$

c. From the table in problem 2-2, the probability that a student does not become ill after eating barbecue is

$$P(\text{not ill}|\text{ate barbecue}) = 30/120 = .25$$

d. From the table in problem 2-2, the probability that a student who attended the party becomes ill is

$$P(\text{ill}) = 110/200 = .55$$

e. From the table in problem 2-2, the probability that a student with food poisoning ate barbecue is

$$P(\text{ate barbecue}|\text{ill}) = 90/110 = .82$$

f. From the table in problem 2-2, the probability that a student who attended the party did not eat barbecue is

$$P(\text{no barbecue}) = 80/200 = .40$$

g. The ratio of the probability of being ill after eating barbecue to the probability of being ill after not eating barbecue is

$$\frac{P(\text{ill}|\text{ate barbecue})}{P(\text{ill}|\text{no barbecue})} = \frac{90/120}{20/80} = 3$$

This means that a student who ate barbecue at the party is 3 times more likely to be ill than a student who did not eat barbecue.

2-3. a. Using relative frequencies from column three of the table in problem 2-3, the probability that the pesticide-exposed worker's red blood cell (RBC) cholinesterase level is between 7.95 and 11.95 is

$$(7.95 \leq Y \leq 11.95) = .229 + .400 = .629,$$

the probability that the RBC cholinesterase level is greater than 15.95 is

$$P(Y > 15.95) = .029,$$

and the probability that the level is less than 9.95 is

$$P(Y < 9.95) = .229 + .029 = .258$$

b. Using frequencies from column two of the table in problem 2-3, the number of pesticide workers with RBC cholinesterase values greater than 11.95 is $12(9 + 2 + 1)$. Of these, 1 worker has an RBC cholinesterase value greater than 15.95. The probability that a pesticide-exposed worker's RBC cholinesterase level is greater than 15.95, if known to be greater than 11.95, therefore, is

$$P(Y \geq 15.95|Y \geq 11.95) = 1/12 = .083$$

2-4. The probability of a child being a boy or a girl is independent of the sex of other children previously born to a couple. Thus, assuming a .5 probability of the birth of a girl, the probability of the successive birth of four girls is

$$P(4 \text{ girls}) = (.5)^4 = .0625$$

Likewise, using formula 2.3, the probability that the fourth child is a boy given that the first three children are girls is

$$P(\text{boy}|3 \text{ girls}) = P(\text{boy}) = .5$$

Note. $P(A|B) = P(A)$ if A and B are independent.

2-5. Given a .001 failure rate for a certain cardiac arrest alarm, the probability that a cardiac arrest will not be signaled while two of these alarms (A1 and A2) are in service is

$$P(\text{A1 fails and A2 fails}) = P(\text{A2 fails}|\text{A1 fails})P(\text{A1 fails})$$

Failure of A1 is independent of the performance of A2. Therefore,

$$P(\text{A2 fails}|\text{A1 fails}) = P(\text{A2 fails})$$

and,

$$P(\text{A1 and A2 fail}) = P(\text{A2 fails})P(\text{A1 fails})$$
$$= (.001)(.001)$$
$$= .000001$$

CHAPTER 3

3-1. Given that cholesterol values in this population follow a gaussian distribution, the normal range is calculated as

$$L = \text{mean} \pm 2 \text{ standard deviations}$$
$$= 150 \pm 2(15)$$
$$= 120 \text{ mg/dl to } 180 \text{ mg/dl}$$

3-2. Cholesterol levels below 92 mg/dl fall in the bottom 2.5% of values, while values above 205 mg/dl are in the upper 2.5%. The normal range, therefore, consists of values between 92 mg/dl and 205 mg/dl.

3-3. a. The prevalence of major depression can be determined by referring to the marginal total (D+) in the table for problem 3-3

$$P(\text{D}+) = 215/368 = .58$$

b. The sensitivity of the dexamethasone suppression test (DST) is

$$P(\text{T}+|\text{D}+) = 84/215 = .39$$

The specificity of the test is

$$P(\text{T}-|\text{D}-) = 148/153 = .97$$

The false positive rate (FPR) of the DST is

$$P(\text{T}+|\text{D}-) = 5/153 = .03$$

The false negative rate (FNR) of the DST is

$$P(\text{T}-|\text{D}+) = 131/215 = .61$$

Predictive value positive (PVP) for the DST is

$$P(\text{D}+|\text{T}+) = 84/89 = .94$$

Predictive value negative (PVN) for the DST is

$$P(\text{D}-|\text{T}-) = 148/279 = .53$$

c. The probability that a patient who does not have major depression will have a negative DST result is the specificity, calculated as

$$P(\text{T}-|\text{D}-) = 148/153 = .97$$

d. The probability that a patient with a negative DST result does not have major depression is the PVN, calculated as

$$P(\text{D}-|\text{T}-) = 148/279 = .53$$

3-4. a. An arbitrary sample size of 1000 is chosen. If $P(\text{D}+)$ is .58, there are 580 patients with depression and 420 who are not depressed. Since sensitivity, $P(\text{T}+|\text{D}+)$, is .43, 43% of the 580 depressed patients—or 249.4 (retain the decimals to improve accuracy)—have a positive DST result and 330.6 have a negative DST result. Since specificity, $P(\text{T}-|\text{D}-)$, is .96, 96% of the 420 patients who are not depressed, or 403.2, have a negative DST result and 16.8 have a positive result. These calculations are summarized in the following table.

A 2 × 2 Table Showing Results of the Back Calculation Method for Revised Estimate of Dexamethasone Suppression Test (DST) Performance

	Depression (D+)	No Depression (D−)	Totals
DST Result (T+)	249.4	16.8	266.2
DST Result (T−)	330.6	403.2	733.8
Totals	580	420	1000

Using the information in the table, the predictive value positive (PVP) and predictive value negative (PVN) are determined as

$$\text{PVP} = P(\text{D}+|\text{T}+) = 249.4/266.2 = .94$$

$$\text{PVN} = P(\text{D}-|\text{T}-) = 403.2/733.8 = .55$$

b. Given sensitivity, $P(\text{T}+|\text{D}+)$, is .43,

$$P(\text{T}-|\text{D}+) = 1 - .43 = .57$$

Given specificity, $P(\text{T}-|\text{D}-)$, is .96,

$$P(\text{T}+|\text{D}-) = 1 - .96 = .04$$

Given prevalence, $P(\text{D}+)$, is .58, calculate its complement, $P(\text{D}-)$, as $1 - .58 = .42$. Then, using Bayes' rule,

$$\text{PVP} = P(\text{D}+|\text{T}+) = \frac{P(\text{T}+|\text{D}+)P(\text{D}+)}{P(\text{T}+|\text{D}+)P(\text{D}+) + P(\text{T}+|\text{D}-)P(\text{D}-)}$$

$$= \frac{(.43)(.58)}{(.43)(.58) + (.04)(.42)}$$

$$= .94$$

Similarly,

$$\text{PVN} = P(\text{D}-|\text{T}-) = \frac{P(\text{T}-|\text{D}-)P(\text{D}-)}{P(\text{T}-|\text{D}-)P(\text{D}-) + P(\text{T}-|\text{D}+)P(\text{D}+)}$$

$$= \frac{(.96)(.42)}{(.96)(.42) + (.57)(.58)}$$

$$= .55$$

3-5. a. The probability that an individual with a positive test result is depressed is $P(\text{D}+|\text{T}+)$, the predictive value positive (PVP). Given sensitivity = .43, specificity = .96, and prevalence = .05, PVP may be calculated using Bayes' rule:

$$\text{PVP} = P(\text{D}+|\text{T}+) = \frac{P(\text{T}+|\text{D}+)P(\text{D}+)}{P(\text{T}+|\text{D}+)P(\text{D}+) + P(\text{T}+|\text{D}-)P(\text{D}-)}$$

$$= \frac{(.43)(.05)}{(.43)(.05) + (.04)(.95)}$$

$$= .36$$

b. The probability that an individual with a negative test result is not depressed is $P(\text{D}-|\text{T}-)$, the predictive value negative (PVN). Given sensitivity = .43, specificity = .96, prevalence $[P(\text{D}+)] = .05$, and $P(\text{D}-) = 1 - .05 = .95$, PVN may be calculated using Bayes' rule:

$$\text{PVN} = P(\text{D}-|\text{T}-) = \frac{P(\text{T}-|\text{D}-)P(\text{D}-)}{P(\text{T}-|\text{D}-)P(\text{D}-) + P(\text{T}-|\text{D}+)P(\text{D}+)}$$

$$= \frac{(.96)(.95)}{(.96)(.95) + (.57)(.05)}$$

$$= .97$$

c. The prevalence in this instance is .50; therefore, $P(\text{D}-)$ is $1 - .50 = .50$. The probability that this individual is depressed is the PVP, which may be calculated using Bayes' rule and the new value for prevalence:

$$\text{PVP} = P(\text{D}+|\text{T}+) = \frac{P(\text{T}+|\text{D}+)P(\text{D}+)}{P(\text{T}+|\text{D}+)P(\text{D}+) + P(\text{T}+|\text{D}+)P(\text{D}-)}$$

$$= \frac{(.43)(.50)}{(.43)(.50) + (.04)(.50)}$$

$$= .91$$

3-6. a. The probability that a patient with a positive test has polysplenia is $P(\text{T}-|\text{D}+)$, the predictive value positive (PVP). Since the false negative rate (FNR) and sensitivity are complements,

$\text{FNR} = P(\text{T}-|\text{D}+) = .18$ and sensitivity $= P(\text{T}+|\text{D}+) = .82$

Similarly, prevalence, $P(\text{D}+) = .003$ and $P(\text{D}-) = 1 - .003 = .997$. Using Bayes' rule,

$$\text{PVP} = P(\text{D}+|\text{T}+) = \frac{P(\text{T}+|\text{D}+)P(\text{D}+)}{P(\text{T}+|\text{D}+)P(\text{D}+) + P(\text{T}+|\text{D}-)P(\text{D}-)}$$

$$= \frac{(.82)(.003)}{(.82)(.003) + (.07)(.997)}$$

$$= .03$$

b. Since the false positive rate (FPR) and specificity are complements,

$\text{FPR} = P(\text{T}+|\text{D}-) = .07$ and specificity $= P(\text{T}-|\text{D}-) = .93$

Using Bayes' rule,

$$P(\text{D}+|\text{T}-) = \frac{P(\text{T}-|\text{D}+)P(\text{D}+)}{P(\text{T}-|\text{D}+)P(\text{D}+) + P(\text{T}-|\text{D}-)P(\text{D}-)}$$

$$= \frac{(.18)(.003)}{(.18)(.003) + (.93)(.997)}$$

$$= .0006$$

c. The probability that a person with a negative test result does not have polysplenia is the predictive value negative (PVN), $P(\text{D}-|\text{T}-)$. This is the complement of the value calculated in problem 3-6b,

$P(\text{D}-|\text{T}-) = 1 - .0006 = .9994$

3-7. Prevalence, $P(\text{D}+)$, in this population $= .30$. Therefore, $P(\text{D}-) = 1 - .30 = .70$. Using this information in conjunction with the value for sensitivity computed in problem 3-6a, Bayes' rule may be used to compute the predictive value positive (PVP):

$$\text{PVP} = P(\text{D}+|\text{T}+) = \frac{P(\text{T}+|\text{D}+)P(\text{D}+)}{P(\text{T}+|\text{D}+)P(\text{D}+) + P(\text{T}+|\text{D}-)P(\text{D}-)}$$

$$= \frac{(.82)(.30)}{(.82)(.30) + (.07)(.70)}$$

$$= .83$$

CHAPTER 4

4-1. a. The value closest to the upper left corner of the ROC curve, 195.6, represents the best choice for the positivity criterion when the consequences of false positive and false negative test results are equally detrimental to the patient.

b. The test is maximally accurate at the point closest to the upper left corner of the ROC curve, 195.6.

c. When it is more harmful to label the patient diseased even though he is not than to miss a diagnosis, the most leftward point, 221.6, should be selected as the positivity criterion. At this point, the false positive rate is minimized.

d. When the consequences of missing a diagnosis are grave, the value farthest to the right of the ROC curve, 175.2, should be selected as the positivity criterion. At this point, the false negative rate is minimized.

CHAPTER 5

5-1. a. The decision tree below was constructed to evaluate the relative merit of performing an appendectomy versus observing the patient for 6 hours. This decision is represented on the tree by decision node D-1. The possible outcomes of this decision are represented by chance nodes C-1 through C-8.

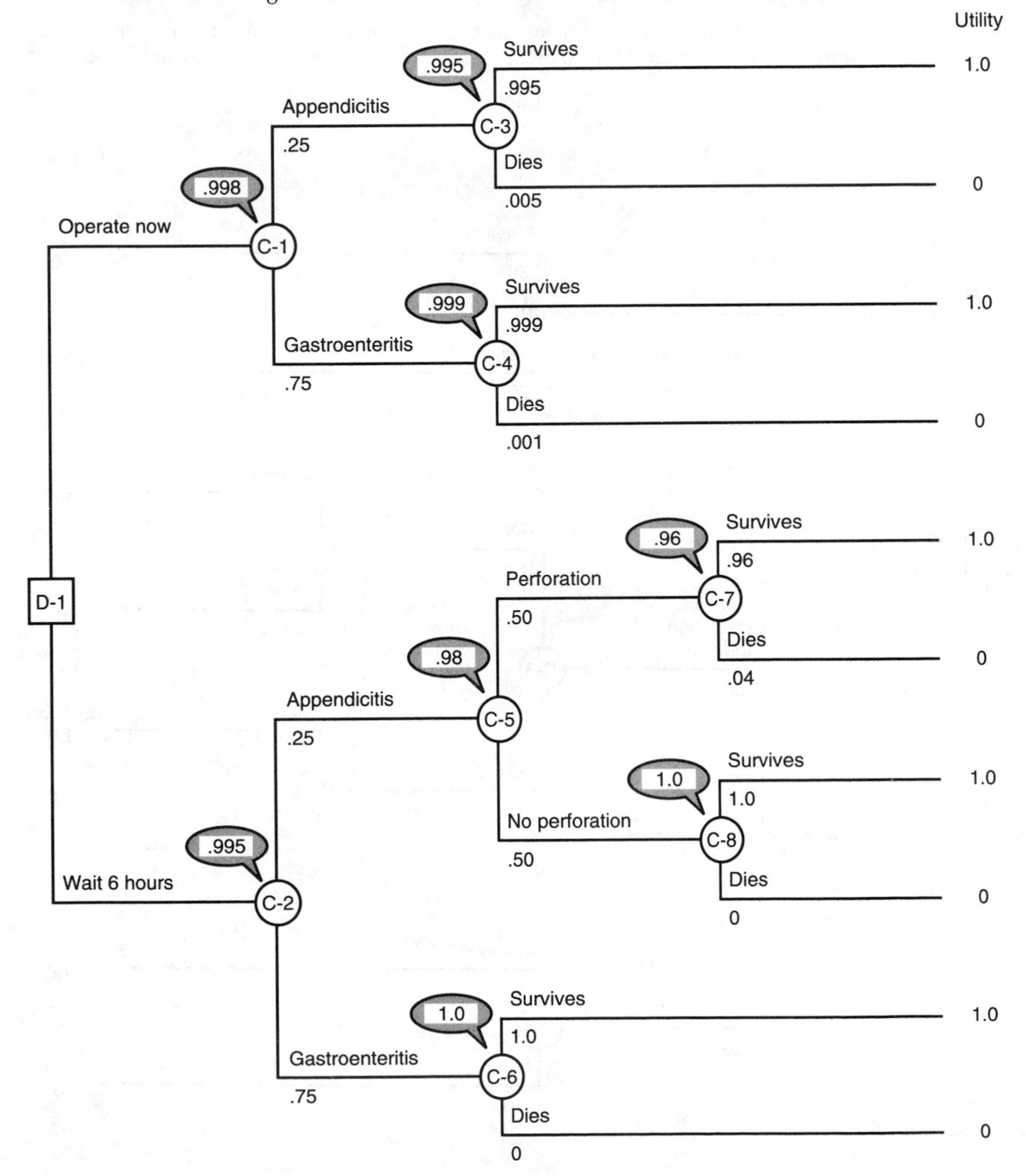

The expected utility associated with each chance node is obtained by multiplying the probability and the utility corresponding to each branch emanating from the node, then summing the products for all branches:

C-8: (0)(0) + (1.0)(1.0) = 1.0

C-7: (.04)(0) + (1.0)(.96) = .96

C-6: (0)(0) + (1.0)(1.0) = 1.0

C-5: (.50)(1.0) + (.50)(.96) = .98

C-4: (.001)(0) + (1.0)(.999) = .999

C-3: (.005)(0) + (.995)(1.0) = .995

C-2: (.75)(1.0) + (.25)(.98) = .995

C-1: (.75)(.999) + (.25)(.995) = .998

Thus, the expected utility of immediate surgery, .998, is greater than that of keeping the patient under observation for 6 hours, .995, so appendectomy is the best course of action for this patient.

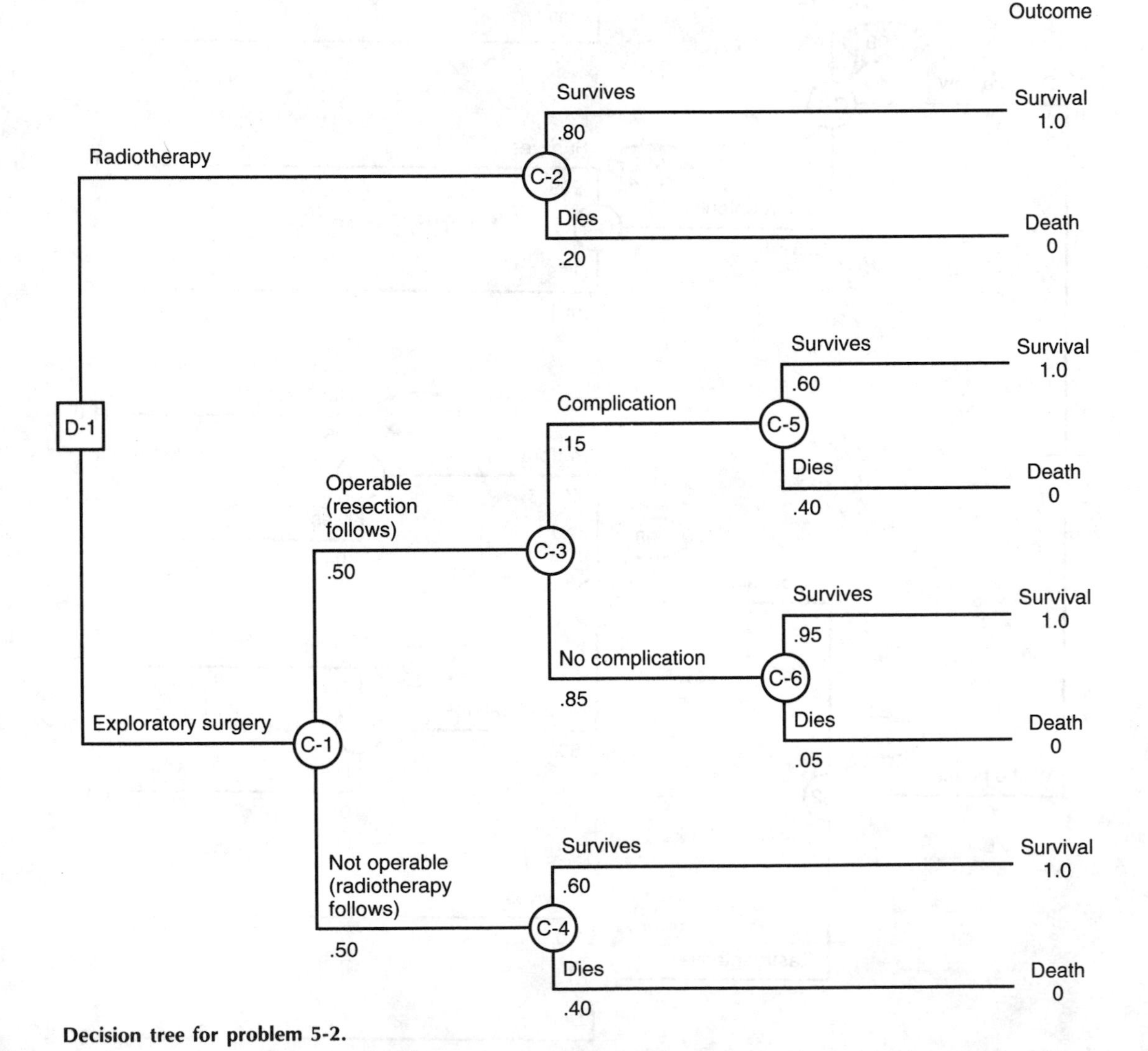

Decision tree for problem 5-2.

b. The validity of this decision can be evaluated by performing a sensitivity and threshold analysis, that is, by calculating the expected utilities for surgery and observation over a range of values for the probability of appendicitis in patients with similar symptoms, for the probability of perforation, for the probability of surviving surgery, and for the probability of surviving perforation.

c. The threshold value, derived from such an analysis, is that value for the probability of appendicitis which changes the preferred course of action. For values above the threshold, the decision "operate now" has the highest utility, while for values below the threshold, the strategy "wait 6 hours" is optimal.

5-2. In the decision tree on the facing page (constructed according to the information given in problem 5-2), the decision to perform exploratory surgery or initiate radiotherapy is represented by decision node D-1 and the outcomes of this decision by chance nodes C-1 through C-6. The expected utilities associated with the chance nodes are:

C-6: (.05)(0) + (.95)(1.0) = .95

C-5: (.40)(0) + (.60)(1.0) = .60

C-4: (.40)(0) + (.60)(1.0) = .60

C-3: (.85)(.95) + (.15)(.60) = .8975

C-2: (.20)(0) + (.80)(1.0) = .80

C-1: (.50)(.60) + (.8975)(.50) = .7488

The option "forego exploratory surgery and initiate radiotherapy" is associated with the highest expected utility and, hence, represents the optimal course of action.

CHAPTER 6

6-1. a. The condition "at least two" is satisfied if two, three, or four of the children have sickle-cell anemia (SCA). Therefore, the probability that the couple will have at least two affected children can be written $P(Y=2 \text{ or } Y=3 \text{ or } Y=4)$. Using the binomial formula and the addition rule of probabilities (given that $k=2$, the number of trials $n=4$, and p, the probability that any child born to these two carriers will have SCA = .25),

$$\begin{aligned} P(Y \geq 2) &= P(Y=2) + P(Y=3) + P(Y=4) \\ &= \binom{4}{2}(1/4)^2(3/4)^2 + \binom{4}{3}(1/4)^3(3/4) + \binom{4}{4}(1/4)^4(3/4)^0 \\ &= 6(1/4)^2(3/4)^2 + 4(1/4)^3(3/4) + 1(1/4)^4 \\ &= .21 + .05 + .004 \\ &= .26 \end{aligned}$$

b. The condition "no more than one" is satisfied when none or only one of the four children have SCA. That is, the probability that the couple will have no more than one affected child can be written $P(Y=0 \text{ or } Y=1)$. Using the binomial formula and the addition rule (given that $k=1$, $n=4$, and $p=.25$),

$$\begin{aligned} P(Y \leq 1) &= P(Y=0) + P(Y=1) \\ &= \binom{4}{0}(1/4)^0(3/4)^4 + \binom{4}{1}(1/4)^1(3/4)^3 \\ &= 1(3/4)^4 + 4(1/4)(3/4)^3 \\ &= .32 + .42 \\ &= .74 \end{aligned}$$

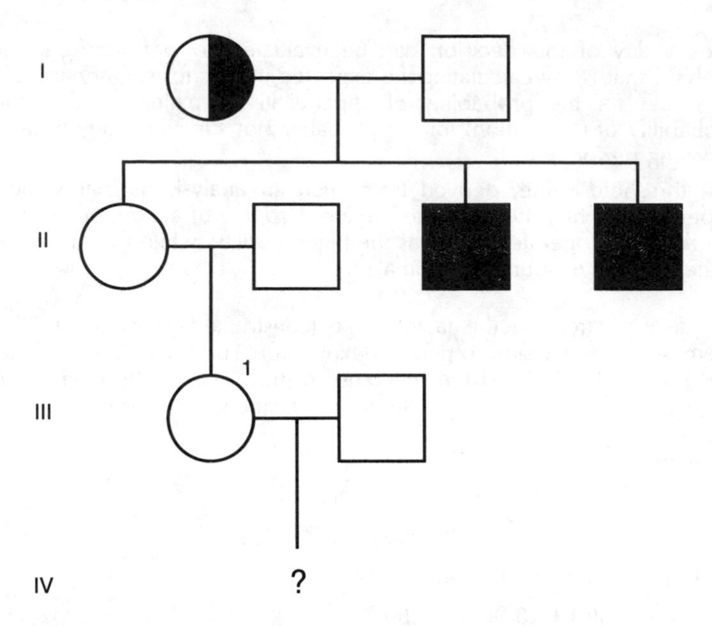

6-2. a. From the information given, the pedigree above can be constructed.

Because the patient has two uncles with Lesch-Nyhan syndrome, her grandmother must be a carrier; hence, the probability that her mother is also a carrier is 1/2. If her mother is a carrier, the probability that the patient herself is a carrier is 1/2. Using the summation principle for joint probabilities and the multiplication rule,

$$\begin{aligned} P(\text{patient carrier}) &= P(\text{patient carrier and mother carrier}) + P(\text{patient carrier and mother not carrier}) \\ &= P(\text{patient carrier}|\text{mother carrier})P(\text{mother carrier}) + 0 \\ &= (1/2)(1/2) \\ &= .25 \end{aligned}$$

Similarly, the probability that the patient will have a son with Lesch-Nyhan syndrome is calculated as

$$\begin{aligned} P(\text{son affected}) &= P(\text{son affected and patient carrier}) + P(\text{son affected and patient not carrier}) \\ &= P(\text{son affected}|\text{patient carrier})P(\text{patient carrier}) + 0 \\ &= (1/2)(1/4) \\ &= .125 \end{aligned}$$

b. The existence of four normal siblings alters the probability that the patient and her mother are carriers (see the pedigree on the facing page). To determine the probability that the patient's mother is a carrier, first define M = mother a carrier; $\bar{\text{M}}$ = mother **not** a carrier; and N = no affected sons. Then, using Bayes' Rule,

$$\begin{aligned} P(\text{mother carrier}|\text{no affected sons}) &= P(\text{M}|\text{N}) \\ &= \frac{P(\text{N}|\text{M})P(\text{M})}{P(\text{N}|\text{M})P(\text{M}) + P(\text{N}|\bar{\text{M}})P(\bar{\text{M}})} \end{aligned}$$

$P(\text{N}|\text{M})$, the probability that the patient's mother will have no affected sons if she is a carrier, is calculated using the binomial formula (n, the number of sons, =4, and p, the probability that any given son is affected when the patient's mother is a carrier, = 1/2) as follows:

$$\begin{aligned} P(\text{N}|\text{M}) = P(Y = 0) &= \binom{4}{0}(1/2)^0(1/2)^4 \\ &= 1/16 \end{aligned}$$

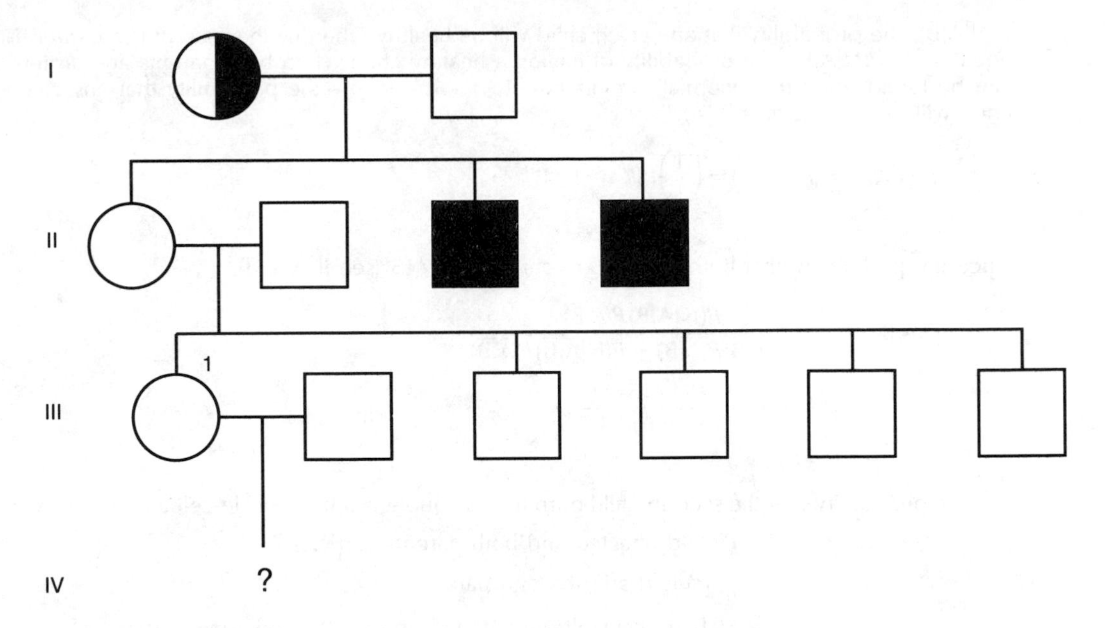

$P(N|\bar{M})$, the probability that the patient's mother has no affected sons if she is not a carrier, is equal to 1. Substituting these probabilities in Bayes' Rule gives

$$P(M|N) = \frac{P(N|M)P(M)}{P(N|M)P(M) + P(N|\bar{M})P(\bar{M})}$$

$$= \frac{(1/16)(1/2)}{(1/16)(1/2) + (1)(1/2)}$$

$$= 1/17$$

The revised probability that the patient is a carrier is

$$\begin{aligned} P(\text{patient carrier}) &= P(\text{patient carrier and mother carrier}) + \\ &\quad P(\text{patient carrier and mother not a carrier}) \\ &= P(\text{patient carrier}|\text{mother carrier})P(\text{mother carrier}) + 0 \\ &= (1/2)(1/17) \\ &= 1/34 \end{aligned}$$

The revised probability that she will have a son with Lesch-Nyhan syndrome is

$$\begin{aligned} P(\text{son affected}) &= P(\text{son affected and patient carrier}) + \\ &\quad P(\text{son affected and patient not a carrier}) \\ &= P(\text{son affected}|\text{patient carrier})P(\text{patient carrier}) + 0 \\ &= (1/2)(1/34) \\ &= .015 \end{aligned}$$

6-3. The probability that both husband and wife are carriers, given that they already have one unaffected child, may be written $P(\text{husband and wife carriers}|\text{1 healthy child}) = P(A,B|N)$. Using Bayes' Rule,

$$P(A,B|N) = \frac{P(N|A,B)P(A,B)}{P(N|A,B)P(A,B) + P(N|\overline{A,B})P(\overline{A,B})}$$

$P(N|\overline{A,B})$, the probability that any given child will be healthy when neither parent is a carrier, is equal to 1. $P(N|A,B)$, the probability of having a healthy child when both parents are carriers, can be found using the binomial formula ($k = 1$, $n = 1$, and p = the probability that any given child will be unaffected = 3/4):

$$P(N|A,B) = P(Y = 1) = \binom{1}{1}(3/4)^1(1/4)^0$$
$$= 3/4$$

Since the probability that both parents are carriers is 1/15 (see II A 2 d),

$$P(A,B|N) = \frac{P(N|A,B)P(A,B)}{P(N|A,B)P(A,B) + P(N|\overline{A,B})P(\overline{A,B})}$$
$$= \frac{(3/4)(1/15)}{(3/4)(1/15) + (1)(14/15)}$$
$$= 3/59$$

and the probability that the second child born to this couple will have sickle-cell anemia (SCA) is

$$P(\text{child affected}) = P(\text{child affected and both parents carriers}) + P(\text{child affected and parents not carriers})$$
$$= P(\text{child affected}|\text{both parents carriers})P(\text{both parents carriers}) + 0$$
$$= (1/4)(3/59)$$
$$= .013$$

6-4. a. The probability that this couple will have three healthy children can be calculated using the binomial formula (given that p, the probability that any given child will be healthy, is 3/4).

$$P(Y = 3) = \binom{3}{3}(3/4)^3(1/4)^0$$
$$= 27/64 = .42$$

b. The probability that at least one of the three children will have phenylketonuria (PKU) is the complement of the probability that none will have the disorder. That is,

$$P(Y \geq 1) = P(Y = 1) + P(Y = 2) + P(Y = 3)$$
$$= 1 - P(Y = 0)$$
$$= 1 - (27/64)$$
$$= .58$$

CHAPTER 7

7-1. a. Among the 120 students who ate barbecue, 90 became ill. The attack rate for this group is

$$P(\text{ill}|\text{ate barbecue}) = 90/120 = .75$$

b. Twenty cases of food poisoning occurred among the 80 students who did not eat barbecue. The attack rate for this group is

$$P(\text{ill}|\text{no barbecue}) = 20/80 = .25$$

c. Of the 100 students who ate coleslaw, 67 contracted food poisoning. The attack rate for this group is

$$P(\text{ill}|\text{ate coleslaw}) = 67/100 = .67$$

d. Forty-three of the 100 students who did not eat coleslaw contracted food poisoning. The attack rate for this group is

$$P(\text{ill}|\text{no coleslaw}) = 43/100 = .43$$

7-2. The difference in attack rates for those students who ate barbecue and those who did not is the difference between the conditional probabilities calculated in problems 7-1a and 7-1b:

$$P(\text{ill}|\text{ate barbecue}) - P(\text{ill}|\text{no barbecue}) = .75 - .25 = .50$$

Similarly, the difference in attack rates for those students who ate coleslaw and those who did not is the difference between the conditional probabilities calculated in problems 7-1c and 7-1d:

$$P(\text{ill}|\text{ate coleslaw}) - P(\text{ill}|\text{ate no coleslaw}) = .67 - .43 = .24$$

In other words, the excess risk of food poisoning attributable to eating barbecue is 50%, while that attributable to eating coleslaw is 24%.

7-3. a. Of the 90 students who ate both barbecue and coleslaw, 65 contracted food poisoning. The attack rate for those who ate both foods is therefore

$$P(\text{ill}|\text{ate barbecue and coleslaw}) = 65/90 = .72$$

b. Only 2 of the 10 students who ate coleslaw, but not barbecue, became ill. The attack rate for this group is

$$P(\text{ill}|\text{ate coleslaw but not barbecue}) = 2/10 = .20$$

c. Twenty-five of the 30 students who ate barbecue but did not eat coleslaw contracted food poisoning. The attack rate for this group is

$$P(\text{ill}|\text{ate barbecue but not coleslaw}) = 25/30 = .83$$

d. Eighteen cases of food poisoning were reported among the 70 students who ate neither food. The attack rate for this group is

$$P(\text{ill}|\text{ate neither barbecue nor coleslaw}) = 18/70 = .26$$

e. Because the highest attack rate occurred in those students who ate barbecue but not coleslaw, the barbecue was most likely to have been contaminated.

7-4. The ratio of the two attack rates (i.e., the conditional probabilities) is

$$\frac{P(\text{ill}|\text{ate barbecue})}{P(\text{ill}|\text{no barbecue})} = \frac{90/120}{20/80} = 3$$

This means that a student who ate barbecue at the party is three times more likely to be ill than a student who did not eat barbecue.

7-5. a. Of the 1000 men sampled in 1960, 100 had CHD. The prevalence of CHD in 1960 when the study was initiated is therefore 100/1000 = .1.

b. When the study began in 1960, 900 of the men sampled were initially healthy and at risk for CHD. During the follow-up period (i.e., between 1960 and 1985), 110 of these men developed CHD. Therefore, the cumulative incidence of CHD over the 15-year observation period is 110/900 = .12.

7-6. a. On January 1, 1989, the practice reported 2 existing cases of sinusitis among the population of 10 patients. That is, the prevalence of sinusitis on this date is 2/10 = .2.

b. The two patients who had sinusitis at the beginning of the follow-up period must be excluded from the population at risk. Of the eight patients at risk, four contracted sinusitis during the one-year observation period. Thus, the CI of sinusitis for 1989 is 4/8 = .5.

7-7. The prevalence of a disease is approximately equal to the product of the incidence and average duration:

$$\text{prevalence} \approx \text{incidence} \times \text{average duration}$$

By rewriting this relationship and substituting the values given for the prevalence and incidence of Huntington's disease, the average duration of the disorder is calculated as

$$\text{average duration} \approx \text{prevalence}/\text{incidence}$$
$$\approx .003/.0003$$
$$\approx 10 \text{ years}$$

7-8. The most likely explanation for this data is that COPD is more rapidly fatal in patients over 65. The shorter survival time in elderly patients decreases the "prevalence pool" (i.e., the number of existing cases), in spite of the fact that more new cases of COPD occur in older than in younger patients.

CHAPTER 8

8-1. a. Except for the occurrence of rheumatoid arthritis, controls should resemble cases as closely as possible. In particular, cases and controls should be equally likely to have been prescribed oral contraceptives (i.e., they should have visited a physician at comparable rates; see also the answer to problem 8-1c).

Cases and controls should be drawn from the same population (e.g., a study utilizing hospitalized cases should employ a control group of hospitalized patients). In a hospital-based study, control patients should not be undergoing treatment for a condition that may have affected their exposure to the putative risk factor (e.g., breast cancer patients would be less likely to have been prescribed oral contraceptives).

In this study, a random sample of women undergoing treatment for other inflammatory joint conditions (e.g., systemic lupus erythematosus) at the same five outpatient rheumatology clinics, selected in the same fashion as the cases, could serve as an appropriate control group.

b. The response variable is the proportion of cases and controls who used oral contraceptives. Prior oral contraceptive use could be determined in both comparison groups from a questionnaire, personal interview, or evaluation of existing medical records. To avoid antecedent–consequence uncertainty, the date when rheumatoid arthritis was diagnosed should also be ascertained.

c. Bias may result from any of the following situations.

Cases and controls were not equally likely to have been prescribed oral contraceptives. For example, if one of the comparison groups included more patients with a history of cardiovascular disease (a contraindication for oral contraceptive use), this group would have a spuriously low proportion of oral contraceptive users.

Controls were not prescribed oral contraceptives because of mild arthritic symptoms, even though these symptoms did not lead to a diagnosis of rheumatoid arthritis. Such patients may subsequently forget their earlier symptoms and report only the date of an actual diagnosis of rheumatoid arthritis. In this case, oral contraceptive use may have occurred prior to the diagnosis, but after the true onset of the disease.

Cases and controls differed systematically with respect to known risk factors for rheumatoid arthritis, such as a family history of the disorder (**prognostic susceptibility** bias).

Participating physicians were more likely to enroll a patient with rheumatoid arthritis in the study if she was an oral contraceptive user, leading to a spuriously high proportion of oral contraceptive users among the cases.

Cases recalled oral contraceptive use more accurately than controls (recall bias), or if oral contraceptive use was recorded more accurately for these patients.

Patients with rheumatoid arthritis made more visits to a physician, and hence, had more opportunities to have been prescribed oral contraceptives.

Cases who were also oral contraceptive users were more likely to be diagnosed correctly. In this instance, the cases would have a higher rate of oral contraceptive use than the general population of patients with rheumatoid arthritis.

The false negative rate differs systematically for cases and controls.

Rheumatoid arthritis patients who also used oral contraceptives were more likely to participate in the study than patients who did not use oral contraceptives.

d. The comparison groups should be equalized to limit the effect of potentially confounding variables, such as age, family history of rheumatoid arthritis, date of diagnosis, smoking status, marital status, age at menopause, age at menarche, number of pregnancies, miscarriages, or live births. Methods to control these confounding variables include matching, restriction, stratification, and statistical adjustment.

8-2. a. A cohort of women who use oral contraceptives would be enrolled in the study as the exposed group. For each oral contraceptive user, another woman of the same age who has never taken oral contraceptives would be recruited as part of the nonexposed cohort. "Cross-overs"—women who enter the study as oral contraceptive users and subsequently stop taking them or vice versa—may be handled as suggested by the Royal College of General Practitioners of the U.K.* In this study, women who began taking oral contraceptives after the study began were included in the exposed group, while those who discontinued oral contraceptive use after enrollment were designated ex-users. All three comparison groups—users (i.e., exposed), non-users (i.e., nonexposed), and ex-users—were followed forward in time to determine the proportion of subjects in each group who developed rheumatoid arthritis.

b. Bias may result from any of the following situations.

Undetected cases of rheumatoid arthritis are preferentially enrolled in one of the comparison groups, resulting in a spuriously high incidence in that group.

Early symptoms of undiagnosed rheumatoid arthritis systematically alter the way study subjects are treated. For example, if mild arthritis symptoms prompt oral contraceptive users to stop taking them, the incidence of rheumatoid arthritis among the exposed subjects will be spuriously low.

Comparison groups differ systematically with respect to the proportion of fertile women or the proportion of women with medical conditions (e.g., prior history of breast cancer) that contraindicate the use of oral contraceptives.

The comparison groups differ systematically with respect to prognostic factors known to be associated with rheumatoid arthritis (prognostic susceptibility bias).

Previous oral contraceptive use leads to arthritic symptoms prior to study enrollment and thus systematically "screens out" patients who ultimately develop rheumatoid arthritis.

Oral contraceptive users are followed more rigorously than the nonexposed group, providing a greater opportunity for the diagnosis of rheumatoid arthritis (surveillance bias).

Subjects are preferentially lost from one of the comparison groups and this loss is associated with the subsequent occurrence of rheumatoid arthritis. For example, if early symptoms of rheumatoid arthritis increase the number of dropouts among oral contraceptive users, an erroneously low incidence of arthritis will be observed in this group.

8-3. a. The target population in this study consists of all men who have committed rape. By restricting participation in the study to those men convicted of the crime, the investigator excludes rapists who were not reported by their victims, not caught, not convicted if apprehended, or not sentenced to prison. This sample, therefore, is not representative of the target population.

b. The response variable is illegal drug use. Exposure might be measured in terms of the type and amount of a particular drug abused or the length of time since such abuse began. One of the many difficulties an investigator measuring a "sensitive" response variable such as drug abuse might expect to encounter is questionable validity of the response. For example, cases may overestimate their illegal drug use as an excuse for their actions, while controls may be reluctant to admit to drug abuse. Survey research techniques for improving the validity of responses to sensitive questions exist, but do not entirely erase doubts about response validity.

c. An appropriate control group would be a random sample of men who have never committed rape, drawn from residents of the state in which the prison is located. Some of the difficulties the investigator might expect to encounter during the sampling process include: obtaining an accurate list, or **sampling frame,** of potential subjects (i.e., all men residing in the state in question); contacting those selected for inclusion in the study, which may prove both costly and time-consuming; and accounting for potential subjects who decline to participate in the study and thus increase the likelihood of sampling bias.

---

*Royal College of General Practitioners' Oral Contraception Study: Reduction in incidence of rheumatoid arthritis associated with oral contraceptives. *Lancet* 1:569–571, 1978.

d. Potentially confounding variables include age, race, and socioeconomic status. Methods for limiting their influence include matching, restriction (e.g., select only middle-class individuals as subjects), stratification, and statistical adjustment.

e. There are several advantages of using the case–control study design to examine this question.

Fewer ethical concerns are raised with this study design than with the use of a prospective cohort study, in which potential rapists would merely be monitored without intervention.

The case–control study design is faster and more economical than a cohort study.

If rape is a rare event, the case–control study design generates more study subjects and hence is the most practical.

Disadvantages of this study design include such difficulties as selecting a suitable control group, resolving antecedent–consequence uncertainty, and controlling for the numerous sources of potential bias.

8-4. A group of confirmed drug abusers with no previous rape convictions (ideally, abusers who have never committed rape) and a control group of men who are not drug abusers and who have no previous rape convictions would be followed forward in time to determine the incidence of rape (see also the solution to problem 8-3e).

8-5. a. The cross-sectional study design is the most appropriate for estimating prevalence.

b. A random sample of individuals is selected from the community in question and tested to determine the number who have AIDS.

c. Prevalence is estimated as the proportion of study subjects who test positive for AIDS.

d. Information on the prevalence of AIDS in this community can be used to plan and evaluate health care needs and services, to assess the health care "burden" imposed on the community, and to determine the need for prevention programs.

8-6. a. A cohort study is used to estimate incidence.

b. Healthy residents from the sample described in the solution to problem 8-5b are followed forward in time for a specified period to determine the rate at which new cases of AIDS occur.

c. Cumulative incidence (CI) is estimated as the proportion of new cases of AIDS occurring during the specified time period among initially healthy individuals at risk. Incidence density (ID) is estimated by dividing the number of new AIDS cases by the total number of person-years at risk accrued by the initially healthy cohort.

d. Cumulative incidence (CI) provides a measure of the risk that an individual who is currently healthy will contract AIDS. A comparison of the incidence rates in two groups of subjects exposed or not exposed to a putative risk factor may be used to test the strength of the association between the factor and the occurrence of AIDS. Measures of incidence are also useful in evaluating the effectiveness of prevention programs.

8-7. a. The crude death rate (unadjusted rate) is a weighted sum of the age-specific death rates where the weighting constant is the proportion of the population in each age interval, that is,

$$\text{crude death rate} = \text{age-specific death rate} \times \text{proportion of the population in age interval}$$

Therefore,

$$\begin{aligned}\text{crude death rate (blacks)} &= (40)(.8)+(65)(.2)\\ &= 45 \text{ deaths per 1000 people}\end{aligned}$$

b. Similarly,

$$\begin{aligned}\text{crude death rate (whites)} &= (28)(.2)+(60)(.8)\\ &= 53.6 \text{ deaths per 1000 people}\end{aligned}$$

c. The crude death rate for whites (see the solution to problem 8-7b) is greater than that for blacks (see the solution to problem 8-7a), even though age-specific death rates for blacks are higher in each age-group, because the *proportion* of the population in each age-group affects the crude death rates. Thus, a large proportion of the white community (.8) is in an age-group with a high death rate (over 65, where the death rate is 60/1000), and a small proportion of the black community is in the over-65 age-group, where the death rate for blacks is 65/1000. Although the proportion of blacks who are under 65 (an age group with a high death rate, 40/1000) is high (.8), it is not high enough to compensate for the proportion of whites over 65.

d. Because age-specific rates are rates *within* particular age ranges (in this case, the specific rates for both blacks and whites who are under 65), a comparison of two age-specific rates would not be affected by differences in the overall age distributions of the two groups. Thus, the fact that the proportion of people under 65 among blacks differs from the proportion of people under 65 among whites does not explain why the death rates for the two groups differ.

e. The data in Table 8-6 can be age-adjusted by replacing the proportion of the population comprising each age interval in the race-specific calculations (the second term in the product calculated in problems 8-7a and 8-7b) by the corresponding age proportion in a standard population (e.g., the proportion of those under and over 65 years of age in the U.S. population at the last census).

$$\text{age-adjusted rate (blacks)} = (40)(.4)+(65)(.6)$$
$$= 55 \text{ deaths per } 1000 \text{ people}$$
$$\text{age-adjusted rate (whites)} = (28)(.4)+(60)(.6)$$
$$= 47.2 \text{ deaths per } 1000 \text{ people}$$

8-8. In developing countries, age-specific death rates are typically elevated during childhood (i.e., a large proportion of the population dies at a young age). Therefore, although age-specific death rates in the developing country may be higher in all age categories than those in the U.S., the crude (overall) death rate is depressed by the low number of elderly people.

CHAPTER 9

9-1. In conducting a randomized controlled clinical trial of their hypothesis that immediate physical therapy reduces permanent impairment after a cerebrovascular accident (CVA), the two physicians must perform five sequential operations (see II).

Formulate a hypothesis. The hypothesis must clearly define the research question within the constraints of available resources and identify the response variable that will be measured. In this study, the neurologists decide to measure physical impairment (the response variable) using an ordinal rating scale devised by a physical therapist at their teaching hospital. Their research hypothesis may be stated as, "Stroke patients who receive immediate physical therapy (i.e., as soon as their condition stabilizes) have significantly lower impairment scores than stroke patients who do not receive immediate physical therapy."

Select participants. Volunteers for the study are recruited from the population of stroke patients treated in the CVA rehabilitation unit of the teaching hospital.

Allocate subjects to comparison groups. The neurologists employ a computer-based random number generator to randomly allocate subjects to receive either the experimental treatment (immediate physical therapy) or the control treatment (delayed physical therapy).

Administer treatment and measure outcome. Both groups of patients are managed identically throughout the study, except for the timing of physical therapy (immediate or delayed). Patients and participating physicians are asked to rate the perceived level of impairment three months, six months, and one year after the CVA; these rankings are averaged to obtain an "impairment score" for each patient at the three time intervals.

Both patients and participating physicians will obviously know which treatment a given patient has received (immediate or delayed physical therapy). For this reason, blinding would prove difficult, if not impossible, in this study.

Analyze data. Appropriate statistical procedures are used to compare impairment scores in those patients receiving immediate and delayed physical therapy at the three time intervals.

Because study participants are not randomly selected from the target population of all stroke patients, the study results are unlikely to be generalizable to the target population. Not only does volunteerism represent a potential threat to external validity, but the study population (stroke patients treated in the rehabilitation unit) may represent the most seriously afflicted patients. Routine delays in admitting patients to this unit may also impede the administration of "immediate" physical therapy, diluting any treatment effect.

9-2. Random allocation equalizes comparison groups with respect to potentially confounding variables. As a result, the conclusions drawn from experimental studies, including randomized controlled clinical trials, are more likely to be valid (i.e., free of bias) than those drawn from observational studies.

CHAPTER 10

10-1. Because serum cholesterol values in the population follow a normal distribution, the probabilities to be determined in problems 10-1a through 10-1e can be derived from the standard normal table (Appendix 3, Table B). In each problem, the specified Y values must first be converted to z scores, using the formula

$$z = \frac{\text{observation} - \text{its population mean}}{\text{its standard deviation}}$$
$$= \frac{Y_i - \mu}{\sigma}$$

The areas corresponding to the calculated z scores are then obtained from the standard normal table.

a. The z scores corresponding to the endpoints of the specified interval (i.e., $Y_1 = 219.5$ mg/dl and $Y_2 = 259.5$ mg/dl) are calculated using the formula given above:

$$z_1 = \frac{219.5 - 242.2}{45.4} = -.5 \qquad z_2 = \frac{259.5 - 242.2}{45.4} = .38$$

The probability that a particular cholesterol value will fall within this interval is

$$P(219.5 \le Y \le 259.5) = P(-.5 \le z \le .38)$$
$$= .6480 - .3085$$
$$= .3395$$

b. First, convert the endpoints of the specified interval to their corresponding z scores:

$$z_1 = \frac{139.5 - 242.2}{45.4} = -2.26 \qquad z_2 = \frac{219.5 - 242.2}{45.4} = -.5$$

The proportion of values included in this interval can then be calculated as

$$P(139.5 \le Y \le 219.5) = P(-2.26 \le z \le -.5)$$
$$= .3085 - .0119$$
$$= .2966$$

c. To convert the specified endpoints to z scores:

$$z_1 = \frac{159.5 - 242.2}{45.4} = -1.82 \qquad z_2 = \frac{179.5 - 242.2}{45.4} = -1.38$$

The probability that a person selected at random from the population will have a cholesterol value in this interval is

$$P(159.5 \le Y \le 179.5) = P(-1.82 \le z \le -1.38)$$
$$= .0838 - .0344$$
$$= .0494$$

d. The population mean, 242.2 mg/dl, corresponds to a z score of 0. Hence, the probability that a cholesterol value will exceed the population mean is $P(Y \ge 242.2) = P(z \ge 0) =$

1 − .5 = .5. In other words, half of the population has a serum cholesterol value greater than or equal to 242.2 mg/dl.

e. The z score corresponding to 219.5 mg/dl was calculated in problems 10-1a and 10-1b as −.5. Thus, the proportion of cholesterol values less than 219.5 mg/dl, $P(Y \leq 219.5)$, is equal to $P(z \leq -.5) = .3085$. That is, 30.85% of the individuals in this population have serum cholesterol levels less than 219.5 mg/dl.

10-2. a. Using the standard normal table (Appendix 3, Table B), the area lying to the left of a point 1.96 standard deviations below the population mean (i.e., to the left of $z = -1.96$) is .0250. The area lying to the left of a point 1.96 standard deviations above the population mean (i.e., the area to the left of $z = 1.96$) is .9750. Therefore, the area within ±1.96 standard deviations of the mean, $P(-1.96 \leq z \leq 1.96) = .9750 - .0250 = .95$. That is, 95% of the values in a population following a normal distribution fall within 1.96 standard deviations of the mean.

b. Similarly, the area within ± 2.58 standard deviations of the mean, $P(-2.58 \leq z \leq 2.58) = .9950 - .0050 = .99$. In other words, 99% of the values in a normally distributed population lie within 2.58 standard deviations of the mean.

10-3. Given that values for systolic blood pressure are normally distributed, with a true population mean $\mu = 120$ mm Hg and standard deviation $\sigma = 20$ mm Hg, the probabilities to be computed in problems 10-3a through 10-3e can be obtained from the standard normal table (Appendix 3, Table B), after converting the specified Y values to z scores as described in problem 10-1.

a. The z score corresponding to 150 mm Hg is

$$z = \frac{150 - 120}{20} = 1.5$$

The probability that a randomly selected individual will have a systolic pressure reading less than or equal to 150 mm Hg is

$$P(Y \leq 150) = P(z \leq 1.5) = .9332$$

Similarly, the z score corresponding to 110 mm Hg is

$$z = \frac{110 - 120}{20} = -.5$$

The probability of obtaining a systolic pressure reading less than or equal to this value is

$$P(Y \leq 110) = P(z \leq -.5) = .3085$$

b. The z score corresponding to 160 mm Hg measures the distance between this observation and the population mean:

$$z = \frac{160 - 120}{20} = 2$$

The value in question lies 2 standard deviations above the mean.

c. Ninety-five percent of the observations in a normally distributed population fall within 1.96 standard deviations of the population mean (i.e., between $z = -1.96$ and $z = +1.96$). The blood pressure measurements corresponding to these z scores are found by solving for Y in the formula given in the solution to problem 10-1:

$$1.96 = \frac{Y_1 - 120}{20} \qquad -1.96 = \frac{Y_2 - 120}{20}$$

$$= 159.2 \qquad = 80.8$$

In other words, 95% of systolic blood pressure readings in this population lie between 80.8 and 159.2 mm Hg.

Similarly, 90% of the observations comprising a normally distributed population lie within 1.65 standard deviations (i.e., between $z = -1.65$ and $z = +1.65$). The corresponding Y values are given by

$$1.65 = \frac{Y_1 - 120}{20} \qquad -1.65 = \frac{Y_2 - 120}{20}$$

$$= 153 \qquad\qquad = 87$$

Ninety percent of systolic blood pressure readings in this population are between 87 mm Hg and 153 mm Hg.

d. The endpoints of the specified interval are first converted to z scores:

$$z_1 = \frac{100 - 120}{20} \qquad z_2 = \frac{140 - 120}{20}$$

$$= -1 \qquad\qquad = 1$$

The proportion of readings that fall within this interval is

$$P(100 \leq Y \leq 140) = P(-1 \leq z \leq 1)$$
$$= .8413 - .1587$$
$$= .6826, \text{ or } 68.3\%$$

e. The proportion of readings that fall within the specified interval is obtained by converting the endpoints to z scores and computing the area between the z scores using the standard normal table (Appendix 3, Table B).

$$z_1 = \frac{60 - 120}{20} \qquad z_2 = \frac{180 - 120}{20}$$

$$= -3 \qquad\qquad = 3$$

And

$$P(60 \leq Y \leq 180) = P(-3 \leq z \leq 3)$$
$$= .9987 - .0013$$
$$= .9974, \text{ or } 99.7\%$$

Since 99.7% of systolic pressure readings for the population fall between 60 and 180 mm Hg, 1 − .997, or .3% of readings, lie outside this interval.

10-4. The central limit theorem states that the mean of the population of sample means ($\mu_{\bar{y}}$) equals the mean of the parent population (in this instance, $\mu_y = \mu_{\bar{y}} = 120$ mm Hg). The standard deviation of the population of sample means, $\sigma_{\bar{y}} = \sigma/\sqrt{n} = 20/\sqrt{100} = 2$.

To obtain the required probabilities in problems 10-4a through 10-4f, sample mean values are first converted to the corresponding z scores, using the formula

$$z = \frac{\text{distance between } \bar{Y} \text{ and } \mu}{\text{standard deviation of the population of } \bar{Y}\text{s}}$$
$$= \frac{\bar{Y} - \mu}{\sigma/\sqrt{n}}$$

The probabilities may then be obtained from the standard normal table (Appendix 3, Table B).

a. First, convert $\bar{Y} = 126$ mm Hg to the corresponding z score:

$$z = \frac{126 - 120}{20/\sqrt{100}}$$
$$= 3$$

The probability that a given sample mean will exceed 126 mm Hg is thus

$$P(\bar{Y} \geq 126) = P(z \geq 3)$$
$$= 1 - .9987$$
$$= .0013$$

b. The z score corresponding to this sample mean (calculated in problem 10-1a) measures the distance between it and the true population mean. That is, a sample mean equal to 126 mm Hg lies three standard deviations above the population mean.

c. Ninety-five percent of all sample means fall within $\pm$ 1.96 standard deviations of the mean (i.e., between $z = -1.96$ and $z = 1.96$). The corresponding $\bar{Y}$ values are calculated as

$$-1.96 = \frac{\bar{Y}_1 - 120}{20/\sqrt{100}} \qquad 1.96 = \frac{\bar{Y}_2 - 120}{20/\sqrt{100}}$$
$$\bar{Y}_1 = 116.08 \qquad \bar{Y}_2 = 123.92$$

That is, 95% of the sample means lie between 116.08 and 123.92 mm Hg.

d. The observed sample mean ($\bar{Y} = 126$ mm Hg) falls outside the range expected to comprise 95% of the sample means. Apparently, the mean of the population of systolic pressure readings for male medical students exceeds the mean systolic pressure for men between the ages of 20 and 24.

e. Ten percent of the sample means lie more than 1.28 standard deviations above the mean (i.e., above $z = 1.28$). The $\bar{Y}$ value corresponding to this z score is

$$1.28 = \frac{\bar{Y} - 120}{20/\sqrt{100}}$$
$$\bar{Y} = 122.56$$

In other words, 10% of the sample means are greater than 122.56 mm Hg.

f. The z score corresponding to $\bar{Y} = 115$ mm Hg is

$$z = \frac{115 - 120}{20/\sqrt{100}}$$
$$= -2.5$$

The probability of selecting a sample with a mean systolic pressure reading less than or equal to 115 mm Hg is

$$P(\bar{Y} \leq 115) = P(z \leq -2.5)$$
$$= .0062$$

CHAPTER 11

11-1. a. The unknown population parameter to be estimated is μ, the mean urinary calcium excretion rate for the population of men receiving bumetanide. The point estimate of μ is the sample mean, $\bar{Y}$, which is 7.5 mg/hr. Because the population sample standard deviation σ is unknown, the standard error $\sigma_{\bar{Y}}$ is estimated by $s/\sqrt{n}$; the appropriate confidence coefficient, $t_{.975}$, obtained from the t distribution (Table C, Appendix 3) for $n - 1 = 8$ *df*, is 2.306.

The general formula for calculating a confidence interval is

endpoints of interval = estimate $\pm$ (confidence coefficient $\times$ standard error of estimate)

Thus, in this instance, the 95% confidence interval (CI_μ) is

$$CI_\mu = \bar{Y} \pm t_{.975}(s/\sqrt{n})$$
$$= 7.5 \pm 2.306(6/\sqrt{9})$$
$$= 2.9 \text{ to } 12.1$$

The physician conducting this study may therefore be 95% confident that the unknown mean urinary calcium excretion rate for the population of men taking bumetanide lies between 2.9 and 12.1 mg/hr.

b. The parameter to be estimated is $\mu_D - \mu_{ND}$, the difference in mean excretion rate between the population receiving the drug and the drug-free population. The pooled estimate of the common unknown standard deviation σ is s_p; the corresponding pooled variance is computed using the formula

$$s_p^2 = \frac{s_D^2(n_D - 1) + s_{ND}^2(n_{ND} - 1)}{n_D + n_{ND} - 2}$$

$$= \frac{36(9-1)+4(16-1)}{9+16-2}$$

$$= 15.1304$$

The number of degrees of freedom is

$$df = n_D + n_{ND} - 2$$

$$= 9 + 16 - 2$$

$$= 23$$

The 95% confidence coefficient, $t_{(1-\alpha/2)} = t_{.975}$ (given $df = 23$), is 2.0687 (see Table C, Appendix 3). The 95% confidence interval $[\mathrm{CI}_{(\mu_D - \mu_{ND})}]$ is calculated using the formula

$$\mathrm{CI}_{(\mu_D-\mu_{ND})} = (\overline{Y}_D - \overline{Y}_{ND}) \pm t_{.975}\sqrt{\frac{s_p^2}{n_D} + \frac{s_p^2}{n_{ND}}}$$

$$= (7.5 - 6.5) \pm (2.0687)(1.6207)$$

$$= -2.4 \text{ to } 4.4$$

In other words, the physician conducting the study can be 95% confident that the difference in mean excretion rate between the population of men receiving bumetanide and the population of men who did not receive the drug falls within the interval −2.4 to 4.4 mg/hr.

11-2. a. The unknown population parameter to be estimated is μ, the mean serum cholesterol level in the population of men undergoing coronary bypass surgery. The point estimate of the population mean is $\overline{Y}$, the serum cholesterol level in the sample of 100 men, which is 260 mg/dl. The standard error of this estimate, $\sigma_{\overline{Y}}$, is $\sigma/\sqrt{n}$.

The standard deviation of serum cholesterol measurements in the population of men who have undergone coronary bypass surgery is assumed to be equal to that of the population of healthy men (i.e., $\sigma = 40$ mg/dl). Because the population standard deviation is known, the standard normal table (Table B, Appendix 3) can be used to derive the 95% confidence coefficient. This value, $z_{(1-\alpha/2)} = z_{.975}$, is 1.96.

Using the general formula for a confidence interval estimate, the 95% confidence interval (CI_μ) is calculated as

endpoints of interval = estimate ± (confidence coefficient × standard error of estimate)

$$\mathrm{CI}_\mu = \overline{Y} \pm z(\sigma/\sqrt{n})$$

$$= 260 \pm (1.96)(40/\sqrt{100})$$

$$= 252.2 \text{ to } 267.8$$

Thus, the researcher conducting this study can be 95% confident that the true mean serum cholesterol level for men undergoing coronary bypass surgery lies between 252.2 and 267.8 mg/dl.

b. Since this mean serum cholesterol level for the population of normal healthy men is not contained in the interval computed in problem 11-2a, the researcher can conclude that the mean serum cholesterol levels for the two populations are different.

11-3. a. This investigation is an example of a paired study design. The parameter to be estimated is μ_D, the unknown mean of the population of differences between the serum digoxin concentration four hours after injection and eight hours after injection. The point estimate of this difference is the sample mean difference, $\overline{Y}_d$, which is

$$\overline{Y}_d = \sum_{i=1}^{n} d_i / n = -.6/9 = -.067$$

and the standard error $\sigma_{\overline{Y}_d}$ is estimated by $s_{\overline{Y}_d} = s_d/\sqrt{n}$. Since σ_D, the standard deviation of the population of differences is unknown, the confidence coefficient is derived from the t distribution (Table C, Appendix 3). The 95% confidence coefficient, $t_{(1-\alpha/2)} = t_{.975}$ (with $9 - 1 = 8$ df), is 2.306. The 95% confidence interval (CI_{μ_D}) is

$$\begin{aligned} CI_{\mu_D} &= \overline{Y}_d \pm t_{.975}(s_d/\sqrt{n}) \\ &= -.067 \pm 2.306(.024) \\ &= -.01 \text{ to } -.12 \end{aligned}$$

The investigator conducting the study may be 95% confident that this interval (−.01 to −.12 ng/ml) contains the mean difference in serum digoxin concentration between the population of four-hour measurements and the population of eight-hour measurements.

b. The comparison between intravenous injection and continuous infusion is an independent sample study design. The parameter to be estimated in this instance is $\mu_{injection} - \mu_{infusion}$ $(\mu_1 - \mu_2)$, the difference in mean serum digoxin concentration between the population receiving the injection and the population receiving the infusion. This difference is estimated by the difference between the sample means, $\overline{Y}_1 - \overline{Y}_2 = .989 - .950 = .04$. The common unknown standard deviation σ is estimated by s_p; the corresponding pooled variance s_p^2 is given by the formula

$$\begin{aligned} s_p^2 &= \frac{s_1^2(n_1 - 1) + s_2^2(n_2 - 1)}{n_1 + n_2 - 2} \\ &= \frac{.033(9-1) + .04(11-1)}{9 + 11 - 2} \\ &= .037 \end{aligned}$$

Given that the number of degrees of freedom $df = 9 + 11 - 2 = 18$, the 99% confidence coefficient, $t_{(1-\alpha/2)} = t_{.995}$, can be derived from the t distribution; Table C (see Appendix 3) lists this value as 2.878. The 99% confidence interval $[CI_{(\mu_1 - \mu_2)}]$ is computed using the formula

$$\begin{aligned} CI_{(\mu_1-\mu_2)} &= (\overline{Y}_1 - \overline{Y}_2) \pm t_{.995}\sqrt{\frac{s_p^2}{n_1} + \frac{s_p^2}{n_2}} \\ &= (.989 - .950) \pm (2.878)(.087) \\ &= -.21 \text{ to } .29 \text{ ng/ml} \end{aligned}$$

The investigator conducting the study can be 99% confident that this interval encompasses the population mean difference between the serum digoxin concentration eight hours after rapid injection and the concentration observed eight hours after intravenous infusion.

11-4. a. If the success rate at the hospital in question is equal to the reported national success rate, the investigator should identify 10 unsuccessful operations among the 100 patient records.

b. The parameter to be estimated is P, the population proportion of successful operations employing the new procedure at this ophthalmologic hospital. The point estimate of P is $\hat{p}$, the proportion of successful operations among the 100 cases surveyed:

$$\hat{p} = 85/100 = .85$$

The 95% confidence coefficient, $z_{(1-\alpha/2)} = z_{.975}$, is obtained from the standard normal table (Table B, Appendix 3) and is equal to 1.96. Using the general formula for computing a confidence interval, the 95% confidence interval (CI_p) is endpoints of interval = estimate ± (confidence coefficient × standard error of estimate)

$$\text{CI}_P = \hat{p} \pm 1.96\sqrt{\frac{\hat{p}(1-\hat{p})}{n}}$$

$$= .85 \pm 1.96\sqrt{\frac{(.15)(.85)}{100}}$$

$$= .78 \text{ to } .92$$

Thus, the investigator conducting this study can be 95% confident that the true proportion of successful operations for chronic angle-closure glaucoma at this particular hospital is between 78% and 92%.

11-5. The parameter to be estimated is $P_O - P_C$, the difference between the proportion of hypertensive women among those using oral contraceptives and the proportion of hypertensive women among those using other methods of contraception. The point estimate of this difference is $\hat{p}_O - \hat{p}_C$, the difference between the two proportions in the sample, that is, $(8/40) - (15/60) = .20 - .25 = -.05$. The 99% confidence coefficient, $z_{(1-\alpha/2)} = z_{.995} = 2.58$, is obtained from Table B (see Appendix 3). Substituting in the general formula, the required confidence interval $[\text{CI}_{(P_O-P_C)}]$ is computed as

endpoints of interval = estimate ± (confidence coefficient × standard error of estimate)

$$\text{CI}_{(P_O-P_C)} = (\hat{p}_O - \hat{p}_C) \pm 2.58\sqrt{\frac{\hat{p}_O(1-\hat{p}_O)}{n_O} + \frac{\hat{p}_C(1-\hat{p}_C)}{n_C}}$$

$$= -.05 \pm 2.58\sqrt{\frac{(.20)(.80)}{40} + \frac{(.25)(.75)}{60}}$$

$$= -.27 \text{ to } .17$$

The research team conducting this study can be 99% confident that the difference in the proportion of hypertensive women among the two populations lies between −27% and 17%.

CHAPTER 12

12-1. The figure below illustrates the calculation of power for this statistical test.

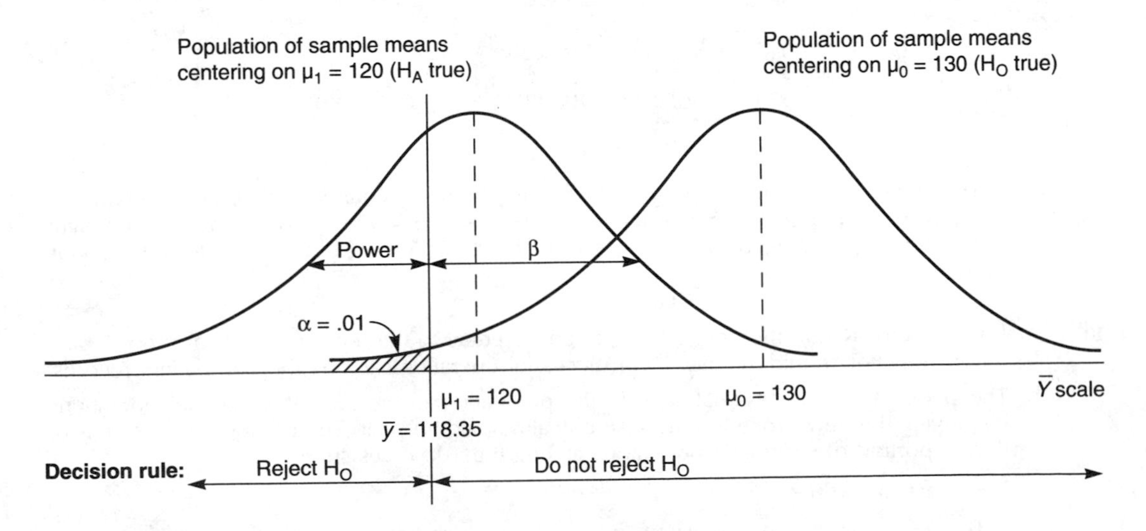

The critical value associated with the desired level of significance, $\alpha = .01$ (*shaded area*), is given in Appendix 3, Table B as $z = -2.33$. The $\bar{Y}$ value corresponding to the positivity criterion ($z = -2.33$) is calculated using the formula

$$z = \frac{\overline{Y} - \mu_0}{\sigma/\sqrt{n}}$$

$$-2.33 = \frac{\overline{Y} - 130}{20/\sqrt{16}}$$

$$\overline{Y} = 118.35$$

Power is defined as the probability that the statistical test will reject the null hypothesis (H_O), given that it is indeed false. That is,

$$\text{power} = P(\text{reject } H_O | H_O \text{ false and } H_A \text{ true})$$

$$= P(\overline{Y} \leq 118.35 | \mu_1 = 120 \text{ mm Hg})$$

$$= P\left(z \leq \frac{\overline{Y} - \mu_1}{\sigma/\sqrt{n}}\right)$$

$$= P\left(z \leq \frac{118.35 - 120}{5}\right)$$

$$= P(z \leq -.33)$$

$$= .3707$$

The chosen statistical test has a 37% chance of rejecting the H_O when this hypothesis is truly false. That is, there is a 37% chance that the statistical test will detect a 10 mm Hg difference between the mean blood pressures of those receiving the new drug treatment and those receiving the standard drug treatment, if such a difference exists.

12-2. a. The investigator can evaluate the hypothesis that the new drug is more effective at lowering systolic blood pressure by applying the five steps of statistical hypothesis testing summarized in Ch 12 IV.

State the hypothesis to be tested. The two opposing hypotheses (the null hypothesis, H_O, and the alternative hypothesis, H_A) are:

$\mathbf{H_O}$: $\mu_{new} = 130$ mm Hg

$\mathbf{H_A}$: $\mu_{new} < 130$ mm Hg

Select a sample and collect data. In this instance, the average systolic blood pressure in the sample of 16 patients is 119.5 mm Hg, with a standard deviation of 20 mm Hg.

Calculate the test statistic. Using the general formula for calculating the test statistic,

$$\text{test statistic} = \frac{\text{sample estimate} - \text{hypothesized population value}}{\text{standard deviation}}$$

$$z = \frac{\overline{Y} - \mu_0}{\sigma/\sqrt{n}}$$

$$= \frac{119.5 - 130}{20/\sqrt{16}}$$

$$= -2.1$$

Evaluate the evidence against H_O. The test statistic is compared with its frequency distribution (shown in the figure on next page) to determine the probability that this value was obtained by random chance (i.e., the p-value), based on the chosen level of significance α. In this case, the area to the left of the test statistic, $z = -2.1$ (see the figure on next page *hatched and cross-hatched areas*), that is, the p-value, is greater than the level of significance ($\alpha = .01$; see the figure on next page, *cross-hatched area*).

State the conclusion. Since the calculated test statistic does not fall in the area to the left of the positivity criterion ($z = -2.33$; see the figure on next page, *cross-hatched area*), that is, $p > \alpha$, H_O is not rejected at the .01 level of significance. The cardiologist conducting this study concludes that the data do not provide sufficient evidence, at this level of significance, to support the hypothesis that patients receiving the new drug have a lower mean systolic blood pressure than patients receiving the standard drug.

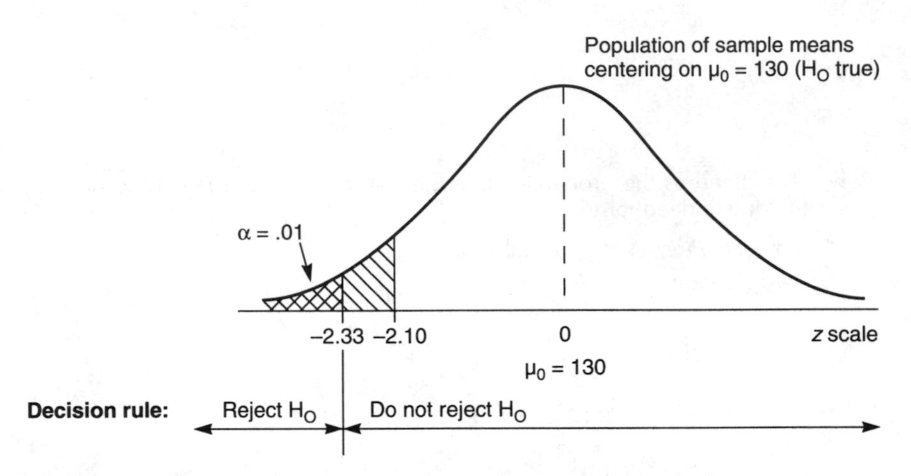

b. The p-value is the probability of obtaining a test statistic less than or equal to -2.1 by random chance from a population with a true mean $\mu = 130$ mm Hg (i.e., for which H_O is true).

$$p = P(z \leq -2.1)$$
$$= .0179 \text{ or } .02 \text{ (from Appendix 3, Table B)}$$

c. The level of significance, $\alpha = .01$, represents the maximum probability of a type I error (i.e., rejecting H_O when it is, in fact, true) that the investigator is willing to tolerate. The value of β, the probability of committing a type II error (i.e., failing to reject H_O when it is false) is the complement of power: $\beta = 1 - \text{power} = 1 - .37 = .63$. In other words, the investigator runs a 63% risk of failing to detect a decrease in systolic blood pressure of at least 10 mm Hg in patients taking the new drug compared to patients taking the standard drug, given that such a difference actually exists in the relevant populations.

In light of the low power and correspondingly high risk of a type II error, the cardiologist realizes that the result she has obtained from the 16-patient sample is inconclusive. Her failure to detect the clinically important 10 mm Hg difference in mean systolic blood pressure may well be attributable to the low power of the statistical test, rather than a valid indicator that this difference does not exist.

12-3. The critical value (positivity criterion) associated with $\alpha = .10$ is -1.28 and the corresponding $\overline{Y}$ value can be derived as follows:

$$z = \frac{\overline{Y} - \mu_O}{\sigma/\sqrt{n}}$$

$$-1.28 = \frac{\overline{Y} - 130}{20/\sqrt{16}}$$

$$\overline{Y} = 130 - 1.28(5)$$

$$= 123.60 \text{ mm Hg}$$

The revised value of power is computed as

$$\text{power} = P(\text{reject } H_O \mid H_O \text{ false and } H_A \text{ true})$$
$$= P(\overline{Y} \leq 123.6 \mid \mu_1 = 120)$$
$$= P\left(z \leq \frac{123.6 - 120}{5}\right)$$
$$= P(z \leq .72)$$
$$= .7642, \text{ or } .76$$

The new value of β is

$$\begin{aligned}\beta &= 1 - \text{power}\\ &= 1 - .76\\ &= .24\end{aligned}$$

Therefore, increasing the value of α from .01 to .10 decreases the value of β (from .63 to .24) and increases the power of the statistical test (from .37 to .76).

12-4. In this instance, the population mean proposed by the alternative hypothesis H_A is $\mu_1 = 125$ mm Hg. The $\bar{Y}$ value (positivity criterion) corresponding to $\alpha = .01$ (i.e., $z = -2.33$) was calculated in problem 12-1 as 118.35 mm Hg. The revised value of power is calculated as

$$\begin{aligned}\text{power} &= P(\text{reject } H_O \mid H_O \text{ false and } H_A \text{ true})\\ &= P(\bar{Y} \leq 118.35 \mid \mu_1 = 125)\\ &= P\left(z \leq \frac{118.35 - 125}{5}\right)\\ &= P(z \leq -1.33)\\ &= .0918, \text{ or } .09\end{aligned}$$

The complement of this value is $\beta = 1 - .09 = .91$. Thus, as the effect size decreases, the power of the statistical test decreases and the corresponding probability of a type II error (β) increases.

12-5. The *p*-value estimates the probability that an observed (sample) effect can be explained by random chance. However, the magnitude of the *p*-value does not reflect the corresponding magnitude of this effect; modest effect sizes may be associated with small *p*-values when the sample size *n* is large. Therefore, it is not possible to draw a conclusion about the relative effectiveness of the two drugs from the reported *p*-values alone.

CHAPTER 13

13-1. a. Because subjects were randomly allocated to comparison groups, this is an example of an experimental study. The response variable is the number of new cavities at the end of the six-month observation period.

b. Testing the hypotheses involves five steps.

State the hypotheses. H_O in this case is: "There is no association between dental hygiene instruction and the number of new cavities six months later. That is, the proportion of children in each of the three cavity frequency categories is constant for the experimental (instruction) and control (no instruction) groups," and **H_A** is: "Dental hygiene instruction is related to the number of new cavities occurring during the six-month observation period. That is, the proportion of children in each of the three cavity frequency categories differs for the experimental and control groups."

Select a sample and collect data. The data from the study appear in the table in problem 13-1. Note that the row totals are fixed, while the column totals are random.

Calculate the test statistic. The expected frequencies for each cell are given in parentheses in the table for problem 13-1 and are calculated using the simplified formula:

$$E_{ij} = \frac{(\text{total row } i)(\text{total column } j)}{\text{grand total}}$$

The test statistic is

$$\begin{aligned}\chi_c^2 &= \sum_{\text{all } ij} \frac{(O_{ij} - E_{ij})^2}{E_{ij}}\\ &= \frac{(30-25)^2}{25} + \frac{(15-15)^2}{15} + \frac{(5-10)^2}{10} + \frac{(20-25)^2}{25} + \frac{(15-15)^2}{15} + \frac{(15-10)^2}{10}\\ &= 7.0\end{aligned}$$

Evaluate the evidence against H_O. H_O should be rejected if the p-value is less than the level of significance $\alpha = .05$. The p-value (the probability that the calculated test statistic was obtained by random chance from populations in which no association exists between dental hygiene instruction and the occurrence of new cavities) is derived from Table A (see Appendix 3). Given that the number of degrees of freedom $df = (r-1)(c-1) = 2$, the tabulated χ^2 value corresponding to $\alpha = .05$ is 5.991. Since the calculated test statistic $\chi_c^2 = 7.0$ falls to the right of 5.991 on the horizontal axis of the chi square frequency distribution, $p < .05$.

State the conclusion. Because $p < .05$, H_O is rejected at the 5% level of significance. That is, there is a statistically significant association between dental hygiene instruction and the occurrence of new cavities during the succeeding six months.

13-2. a. Because study subjects self-selected their comparison group status (oral contraceptive or other method of contraception) and both variables were measured concurrently at a single point in time, this investigation is an example of a cross-sectional study.

b. Testing the hypothesis involves five steps.

State the hypothesis. H_O in this case is: "The group of women who use oral contraceptives contains the same proportion of hypertensive subjects as the group of women who use other methods of contraception," and H_A is: "The proportion of hypertensive women among those who use oral contraceptives differs from the proportion of hypertensive women among those using another method of contraception."

Select a sample and collect data. The results of the study are summarized in the table for this problem. Note that neither the row nor the column totals are fixed.

Calculate the test statistic. The expected frequencies for each cell are given in parentheses in the table for problem 13-2 and are calculated using the simplified formula:

$$E_{ij} = \frac{(\text{total row } i)(\text{total column } j)}{\text{grand total}}$$

The test statistic is

$$\chi_c^2 = \sum_{\text{all } ij} \frac{(O_{ij} - E_{ij})^2}{E_{ij}}$$

$$= \frac{(8.0 - 9.2)^2}{9.2} + \frac{(32.0 - 30.8)^2}{30.8} + \frac{(15.0 - 13.8)^2}{13.8} + \frac{(45.0 - 46.2)^2}{46.2}$$

$$= .34$$

Evaluate the evidence against H_O. The probability of computing a test statistic equal to or greater than $\chi_c^2 = .34$ when H_O is true (i.e., the p-value) is derived from Table A (see Appendix 3). Given that the number of degrees of freedom associated with the test is $df = (r-1)(c-1) = 1$, the value of the tabulated χ^2 for $\alpha = .10$ is listed in Table A (Appendix 3) as 2.706. Because the calculated test statistic, $\chi_c^2 = .34$, falls to the left of 2.706 (i.e., $\alpha = .10$) on the horizontal axis of the chi-square frequency distribution, the area to the right of .34 (the p-value) is greater than .10 ($p > .10$). H_O should be rejected if $p < \alpha$, that is, if $p < .01$.

State the conclusion. Since the p-value is not less than $\alpha = .01$ ($p > .10$), H_O is not rejected at the 1% level of significance. The results of this study provide insufficient evidence to demonstrate that the proportion of hypertensive women among those using oral contraceptives differs from the proportion of hypertensive women among those using other methods of contraception.

c. The chance of error associated with a failure to reject H_O is β, the probability of a type II error. In this study, β, the probability that the statistical test will fail to detect a difference in the proportion of hypertensive patients between the two comparison groups when such a difference actually exists, is unknown. Without additional information on the value of β, or its complement, power, it is not possible to determine whether this failure to achieve statistical significance is due to the absence of such a difference or to a low value of power for the test.

13-3. a. Study subjects were randomly assigned to the two comparison groups (desipramine alone or desipramine and alprazolam); hence, this was an experimental study.

b. Testing the hypothesis involves five steps.

State the hypothesis. $\mathbf{H_O}$ in this case is: "The proportion of patients with satisfactory response times is the same in both treatment groups, that is, $P_1 = P_2$," and $\mathbf{H_A}$ is: "The proportion of patients with satisfactory response times differs for the two treatment groups, that is, $P_1 \neq P_2$."

Select a sample and collect data. The results of the study are summarized in the table below. Note that the row totals (drug treatment) are fixed, while the column totals (response time) are random.

	Number of Patients		
	Satisfactory Response Time	**Unsatisfactory Response Time**	**Totals**
Desipramine Alone	12 (10)*	18 (20)	30
Desipramine + Alprazolam	8 (10)	22 (20)	30
Totals	20	40	60

*Numbers in parentheses represent the calculated expected frequencies.

Calculate the test statistic. The expected frequencies for each cell (above table, *parentheses*) are calculated using the formula:

$$E_{ij} = \frac{(\text{total row } i)(\text{total column } j)}{\text{grand total}}$$

The test statistic is

$$\chi_c^2 = \sum_{\text{all } ij} \frac{(O_{ij} - E_{ij})^2}{E_{ij}}$$

$$= \frac{(12-10)^2}{10} + \frac{(18-20)^2}{20} + \frac{(8-10)^2}{10} + \frac{(22-20)^2}{20}$$

$$= 1.2$$

Evaluate the evidence against $\mathbf{H_O}$. H_O is rejected when the probability of computing a test statistic equal to or greater than $\chi_c^2 = 1.2$ when H_O is true (the *p*-value) is less than $\alpha = .05$. For $df = (r-1)(c-1) = 1$, the value of the tabulated χ^2 for $\alpha = .10$ is listed in Table A (see Appendix 3) as 2.706. Because the calculated test statistic, $\chi_c^2 = 1.2$, lies to the left of 2.706 (i.e., $\alpha = .10$) on the horizontal axis of the chi-square frequency distribution, $p > .10$.

State the conclusion. Since p is not less than $\alpha (p > .10)$, H_O is not rejected at the 5% level of significance. The results of this study are inconclusive; there is not sufficient evidence to demonstrate a difference in the number of patients with satisfactory response times between the two treatment groups.

c. The chance of error associated with this conclusion is β, the probability of a type II error. The magnitude of β is unknown for this study.

CHAPTER 14

14-1. a. Two groups, one with the characteristic in question (i.e., a major ECG abnormality) and one without (exposed and nonexposed, respectively) were followed forward in time to determine the number of deaths resulting from CHD. Therefore, this is an example of a prospective cohort study.

b. The absolute risk of dying from CHD for black men between the ages of 40 and 64 is the probability that a black man in this age group will die of CHD during the 20-year study period, given that he was known to have a major ECG abnormality at the start of the study. From the table in Problem 14-1,

$$P(\text{death from CHD}|\text{major ECG abnormality}) = 8/47, \text{ or } .17$$

c. Relative risk (RR) is the ratio of the incidence of the outcome (i.e., D+ = death from CHD) among those who are exposed (i.e., E+ = presence of a major ECG abnormality) and the incidence among those who are not exposed (i.e., E− = absence of a major ECG abnormality):

$$\text{RR} = \frac{P(\text{D}+|\text{E}+)}{P(\text{D}+|\text{E}-)}$$

$$= \frac{P(\text{death}|\text{major ECG abnormality})}{P(\text{death}|\text{no major ECG abnormality})}$$

$$= \frac{8/47}{10/144}$$

$$= 2.45$$

Thus, a black man in this population who has a major ECG abnormality is 2.45 times more likely to die from CHD over the specified time period than a black man with no such abnormality.

d. The odds ratio (OR) is the ratio of the odds of dying from CHD if a major ECG abnormality is present to the odds of dying from CHD in the absence of an ECG abnormality. That is,

$$O_{\text{D}+|\text{E}+} = \text{odds of dying from CHD if a major ECG abnormality is present}$$

$$= \frac{P(\text{D}+|\text{E}+)}{P(\text{D}-|\text{E}+)}$$

$$= \frac{P(\text{death}|\text{major ECG abnormality})}{P(\text{survival}|\text{major ECG abnormality})}$$

$$= \frac{8/47}{39/47}$$

$$= 8/39$$

$$O_{\text{D}+|\text{E}-} = \text{odds of dying from CHD if no major ECG abnormality is present}$$

$$= \frac{P(\text{D}+|\text{E}-)}{P(\text{D}-|\text{E}-)}$$

$$= \frac{P(\text{death}|\text{no major ECG abnormality})}{P(\text{survival}|\text{no major ECG abnormality})}$$

$$= \frac{10/144}{134/144}$$

$$= 10/134$$

$$\text{OR} = \frac{O_{\text{D}+|\text{E}+}}{O_{\text{D}+|\text{E}-}}$$

$$= \frac{8/39}{10/134}$$

$$= \frac{(8)(134)}{(39)(10)}$$

$$= 2.75$$

Alternatively, OR can be calculated using formula 14.3 (OR $= ad/bc$), the simplified rule for independent samples. Either way, the odds that a black man with a major ECG abnormality will die from CHD are 2.75 times greater than the odds that a black man with no ECG abnormality will die from CHD. If death from CHD is a rare event in this population, OR approximates RR, that is, the risk of death from CHD is 2.75 times higher for a black man with a major ECG abnormality than for a black man with no such abnormality.

e. Using Woolf's method, the 95% confidence interval estimate of *ln* OR ($CI_{ln\,OR}$) is

$$CI_{ln\,OR} = ln\ OR \pm 1.96\sqrt{\frac{1}{a} + \frac{1}{b} + \frac{1}{c} + \frac{1}{d}}$$

$$= ln(2.75) \pm 1.96\sqrt{\frac{1}{8} + \frac{1}{39} + \frac{1}{10} + \frac{1}{134}}$$

$$= .02 \text{ to } 2.01$$

and the 95% confidence interval for the OR is

$$CI_{OR} = e^{.02} \text{ to } e^{2.01}$$

$$= 1.02 \text{ to } 7.44$$

Thus, the research team conducting this study can be 95% confident that the OR for the population represented by the study sample falls between 1.02 and 7.44.

f. The two opposing hypotheses for the statistical test are the **null hypothesis (H_O):** "There is no association between death from CHD and the presence of a major ECG abnormality in the study population (i.e., OR for the population = 1)," and the **alternative hypothesis (H_A):** "There is a major association between death from CHD and the presence of a major ECG abnormality in the study population (i.e., OR for the population $\neq 1$)."

If the 95% confidence interval for the OR includes 1, H_A is not rejected. However, from the answer to problem 14-1e, the investigators can be 95% certain that the OR for the study population falls within the interval 1.02 to 7.44. The confidence interval does not contain 1; therefore, a statistically significant association exists between death from CHD and the presence of a major ECG abnormality ($p \leq .05$).

g. Attributable risk (AR) measures the excess risk of death from CHD that can be attributed to the presence of a major ECG abnormality. It is calculated using formula 14.6:

$$AR = P(D+|E+) - P(D+|E-)$$

$$= P(\text{death}|\text{major ECG abnormality}) - P(\text{death}|\text{no major ECG abnormality})$$

$$= (8/47) - (10/144)$$

$$= .10$$

The excess risk of death associated with the presence of a major ECG abnormality is .10.

14-2. a. Four groups of subjects with different levels of exposure to the putative risk factor (alcohol) are followed forward in time to determine the occurrence of the relevant outcome (i.e., death from all causes). Thus, this is an example of a prospective cohort study.

b. The absolute risk of death for a heavy drinker is the probability that an individual who consumes 91 or more drinks per month will die during the 10-year study period. Using the data in the table in problem 14-2,

$$P(\text{death}|\text{heavy drinker}) = 175/350, \text{ or } .50$$

c. The relative risk (RR) for a heavy drinker compared to a light drinker is the ratio of the incidence of death among heavy drinkers [i.e., $P(\text{death}|\text{heavy drinker}) = .50$] to the incidence of death among light drinkers, [$P(\text{death}|\text{light drinker}) = 500/2500 = .20$], that is,

$$\text{RR} = \frac{P(\text{D+}|\text{E+})}{P(\text{D+}|\text{E}-)}$$

$$= \frac{P(\text{death}|\text{heavy drinker})}{P(\text{death}|\text{light drinker})}$$

$$= .50/.20$$

$$= 2.50$$

RR alone does not provide information on absolute risk for heavy drinkers, which is the incidence of death among these individuals [i.e., $P(\text{D+}|\text{E+})$].

d. The RR of death from all causes for a heavy drinker compared to an abstainer is

$$\text{RR} = \frac{P(\text{death}|\text{heavy drinker})}{P(\text{death}|\text{abstainer})}$$

$$= .50/.25$$

$$= 2.0$$

e. The RR of death from all causes for a moderate drinker compared to a light drinker is

$$\text{RR} = \frac{P(\text{death}|\text{moderate drinker})}{P(\text{death}|\text{light drinker})}$$

$$= .24/.20$$

$$= 1.2$$

f. The odds of death from all causes for a heavy drinker, $O_{\text{D+}|\text{E+}}$, are

$$O_{\text{D+}|\text{E+}} = \frac{P(\text{death}|\text{heavy drinker})}{P(\text{survival}|\text{heavy drinker})}$$

$$= \frac{175/350}{175/350}$$

$$= 175/175$$

Similarly, the odds of death for a light drinker are

$$O_{\text{D+}|\text{E}-} = \frac{P(\text{death}|\text{light drinker})}{P(\text{survival}|\text{light drinker})}$$

$$= \frac{500/2500}{2000/2500}$$

$$= 500/2000$$

The odds ratio (OR) for heavy compared to light drinking is

$$\text{OR} = \frac{175/175}{500/2000}$$

$$= 4.0$$

The odds of dying from any cause during the 10-year study period are four times greater for heavy drinkers than light drinkers.

g. The 95% confidence interval estimate of the OR is calculated using Woolf's method. Thus, the 95% confidence interval for *ln* OR is

$$CI_{ln\,OR} = ln\ OR \pm 1.96\sqrt{\frac{1}{a}+\frac{1}{b}+\frac{1}{c}+\frac{1}{d}}$$

$$= ln(4) \pm 1.96\sqrt{\frac{1}{175}+\frac{1}{175}+\frac{1}{500}+\frac{1}{2000}}$$

$$= 1.15 \text{ to } 1.62$$

while the 95% confidence interval estimate of OR is

$$CI_{OR} = e^{1.15} \text{ to } e^{1.62}$$

$$= 3.2 \text{ to } 5.0$$

The investigator can be 95% confident that the population OR falls between 3.2 and 5.0.

h. The opposing hypotheses to be tested are the **null hypothesis (H_O):** the population OR = 1, and the **alternative hypothesis (H_A):** the population OR ≠ 1.

Since 1 is not included in the 95% confidence interval computed in the answer to problem 14-2g, H_O is rejected at the $\alpha = .05$ level of significance. There is a significant association between heavy alcohol consumption and all-cause mortality ($p \leq .05$). The chance of error associated with this conclusion (i.e., the type I error rate) is less than 5%.

i. Using formula 14.6, the attributable risk (AR) associated with heavy drinking compared to abstinence is

$$AR = P(\text{death}|\text{heavy drinker}) - P(\text{death}|\text{abstainer})$$

$$= (175/350) - (300/1200) = .25$$

Thus, the excess risk of death conferred by heavy drinking as compared to abstinence is .25.

14-3. a. Two groups of subjects, one with rheumatoid arthritis (RA) and the other without, were followed backward in time to determine prior oral contraceptive use. Thus, this is an example of a case–control study.

b. For a case–control study, the odds ratio (OR) is

$$OR = \frac{\text{odds that a diseased individual has been exposed}}{\text{odds that a healthy individual has been exposed}}$$

$$= \frac{O_{E+|D+}}{O_{E+|D-}}$$

[Formula 14.3 ($OR = ad/bc$) can also be used.]

In this instance,

$$O_{E+|D+} = \text{odds that a woman with RA has been exposed to oral contraceptives}$$

$$= \frac{P(\text{oral contraceptive use}|\text{RA})}{P(\text{no oral contraceptive use}|\text{RA})}$$

$$= \frac{40/100}{60/100} = 40/60$$

$O_{E+|D-}$ = Odds that a woman who does not have RA has been exposed to oral contraceptives

$$= \frac{P(\text{oral contraceptive use}|\text{no RA})}{P(\text{no oral contraceptive use}|\text{no RA})}$$

$$= \frac{120/200}{80/200} = 120/80$$

Thus, the OR is

$$OR = \frac{40/60}{120/80} = .44$$

The odds of developing RA are .44 times as great for women who use oral contraceptives as for women who do not use these drugs. Thus, rather than being a risk factor for RA, oral contraceptive use appears to protect a woman against this disorder.

Incidence cannot be directly determined from the results of a case–control study. Therefore, relative risk (RR), which is a ratio of incidence rates, cannot be directly calculated either when this study design is employed. If RA is a rare disease, the OR provides an estimate of RR. That is, the risk of developing RA incurred by a woman who uses oral contraceptives is .44 times that of a woman who does not.

c. Using Woolf's method of computing a confidence interval for OR, the 95% confidence interval estimate of *ln* OR is calculated as

$$\begin{aligned} CI_{ln\,OR} &= ln\ OR \pm 1.96\sqrt{\frac{1}{a}+\frac{1}{b}+\frac{1}{c}+\frac{1}{d}} \\ &= ln(.44) \pm 1.96\sqrt{\frac{1}{40}+\frac{1}{120}+\frac{1}{60}+\frac{1}{80}} \\ &= -1.30 \text{ to } -.32 \end{aligned}$$

The 95% confidence interval estimate of OR is then computed as

$$\begin{aligned} CI_{OR} &= e^{-1.30} \text{ to } e^{-.32} \\ &= .27 \text{ to } .73 \end{aligned}$$

Thus, the researcher conducting this study can be 95% confident that the OR for the population falls between .27 and .73.

d. The hypotheses to be tested are the **null hypothesis (H_O):** "The occurrence of RA is unrelated to oral contraceptive use (population OR = 1)," and the **alternative hypothesis (H_A):** "There is an association between the use of oral contraceptives and RA (population OR ≠ 1)."

The 95% confidence interval estimate of OR, calculated in the answer to problem 14-3c, does not include 1. Therefore, H_O is rejected at the $\alpha = .05$ level of significance. Oral contraceptive use exerts a statistically significant protective effect against the occurrence of RA ($p \leq .05$).

14-4. a. Formula 14.4 can be used to calculate the odds ratio (OR):

$$\begin{aligned} OR &= b/c \\ &= 20/50 = .40 \end{aligned}$$

The OR is interpreted as in problem 14-3b [i.e., the odds of developing rheumatoid arthritis (RA) are .44 times as great for women who use oral contraceptives as for those who do not]. Again, if RA is a rare condition, OR estimates relative risk (RR).

b. Matching was used in this study to equalize the comparison groups (cases and controls) with respect to the potentially confounding variables, age and year of registry.

CHAPTER 15

15-1. a. S_{xx}, S_{yy}, and S_{xy} are computed prior to calculating the slope and *Y* intercept of the sample regression line, using formula 15.4, formula 15.5, and formula 15.3 respectively:

$$S_{xx} = \sum X^2 - \frac{\left(\sum X\right)^2}{n} = 1002 - \frac{(84)^2}{9} = 218$$

$$S_{yy} = \sum Y^2 - \frac{\left(\sum Y\right)^2}{n} = 642 - \frac{(72)^2}{9} = 66$$

$$S_{xy} = \sum XY - \frac{\left(\sum X \sum Y\right)}{n} = 780 - \frac{(84)(72)}{9} = 108$$

The slope of the regression line, b_1, is then obtained using formula 15.7:

$$b_1 = \frac{S_{xy}}{S_{xx}} = \frac{108}{218} = .4954$$

The Y intercept, b_0, is computed using formula 15.8:

$$b_0 = \overline{Y} - b_1\overline{X} = 8.0 - (.4954)(9.3333) = 3.3762$$

Hence, the sample regression equation is

$$\hat{Y} = 3.3762 + .4954X$$

b. The five steps of statistical hypothesis testing outlined in Chapter 12 IV may be used to test the hypothesis that sleeping time and drug dose are linearly related.

State the hypothesis. The null hypothesis, $\mathbf{H_O}$, is: "Sleeping time and drug dose are not linearly related; that is, the slope of the population regression line is 0 ($\beta_1 = 0$)," and the alternative hypothesis, $\mathbf{H_A}$, is: "Sleeping time is linearly related to dose; that is, the slope of the population regression line differs significantly from 0 ($\beta_1 \neq 0$)."

Select a sample and collect data (see the table in problem 15-1). The sample slope, $b_1 = .4954$ (calculated in the answer to problem 15-1a), provides an estimate of the population slope β_1.

Calculate the test statistic. To calculate the test statistic, it is first necessary to compute the SSE, the error, or residual, sum of squares, and $s_{y.x}$, the standard error of the estimate. These values are obtained from formulas 15.10 and 15.9, respectively.

$$\text{SSE} = S_{yy} - b_1 S_{xy} = 66 - (.4954)(108) = 12.4968$$

$$s_{y.x} = \sqrt{\frac{\text{SSE}}{n-2}} = \sqrt{\frac{12.4968}{7}} = 1.3361$$

Formula 15.11 is then used to compute the test statistic:

$$t = \frac{b_1 - \beta_1}{s_{y.x}/\sqrt{S_{xx}}} = \frac{.4954 - 0}{.0905} = 5.5$$

Evaluate the evidence against $\mathbf{H_O}$. The tabulated critical value (see Table C, Appendix 3) for $\alpha = .005$ and $df = n - 2 = 7$ is 3.4995. Since the calculated test statistic ($t = 5.5$) lies to the right of 3.4995 on the horizontal axis of the frequency distribution, the area to the right of the test statistic is less than .5%. This value is doubled to obtain the p-value (H_A does not specify a particular direction; hence, this is a two-tailed test). Thus, the chance that the observed results came from a population in which the slope of the regression line relating sleeping time and dose is equal to zero is less than 1% ($p < .01$).

State the conclusion. Because the p-value is less than the preselected level of significance, $\alpha = .05$, H_O is rejected at the 5% level of significance. The results of the statistical test support the conclusion that there is a statistically significant linear dose–response relationship in the population from which the study sample was drawn ($p < .01$).

c. The sleeping time for a patient receiving a 12 mM/kg dose of the new drug (i.e., $X = 12$) can be predicted from the regression equation:

$$\hat{Y} = 3.3762 + .4954(12) = 9.32 \text{ hours}$$

15-2. a. As in problem 15-1a, S_{xx}, S_{yy}, and S_{xy} are calculated using formula 15.4, formula 15.5, and formula 15.3:

$$S_{xx} = \sum X^2 - \frac{\left(\sum X\right)^2}{n} = 24{,}900 - \frac{(450)^2}{9} = 2400$$

$$S_{yy} = \sum Y^2 - \frac{\left(\sum Y\right)^2}{n} = 211{,}475 - \frac{(1365)^2}{9} = 4450$$

$$S_{xy} = \sum XY - \frac{\sum X \sum Y}{n} = 71{,}450 - \frac{614{,}250}{9} = 3200$$

The slope of the sample regression line is

$$b_1 = \frac{S_{xy}}{S_{xx}} = \frac{3200}{2400} = 1.3333$$

The Y intercept of the sample regression line is

$$b_0 = \overline{Y} - b_1\overline{X} = 151.6667 - (1.3333)(50) = 85$$

Hence, the sample regression equation relating shock intensity and diastolic blood pressure is

$$\hat{Y} = 85 + 1.3333X$$

b. The standard five-step statistical testing procedure should be used

State the hypothesis. The null hypotheses, $\mathbf{H_O}$, is: "Shock intensity and diastolic blood pressure are not linearly related; the slope of the population regression line is 0 ($\beta_1 = 0$)," and the alternative hypothesis, $\mathbf{H_A}$, is: "There is a linear relationship between shock intensity and diastolic blood pressure; the slope of the population regression line differs significantly from 0 ($\beta_1 \neq 0$)."

Select a sample and collect data (see the table in problem 15-2). The sample slope ($b_1 = 1.3333$; see answer to problem 15-2a) is an estimate of the population slope β_1.

Calculate the test statistic. SSE (the residual sum of squares) and $s_{y.x}$ (the standard error of the estimate) are computed first, using formulas 15.10 and 15.9:

$$SSE = S_{yy} - b_1 S_{xy} = 4450 - (1.3333)(3200) = 183.3344$$

$$s_{y.x} = \sqrt{\frac{SSE}{n-2}} = \sqrt{\frac{183.3344}{7}} = 5.1178$$

Substituting in formula 15.11,

$$t = \frac{b_1 - \beta_1}{s_{y.x}/\sqrt{S_{xx}}} = \frac{1.3333 - 0}{.1045} = 12.8$$

Evaluate the evidence against $\mathbf{H_O}$. In the t distribution table (Appendix 3, Table C), the critical value for $\alpha = .005$ and $df = n - 2 = 7$ is 3.4995. The calculated test statistic, $t = 12.8$, lies to the right of 3.4995 on the horizontal axis of the frequency distribution; hence, the area to the right of the test statistic is less than .5%. Because H_A does not specify direction, the test is two-tailed and this value is doubled to obtain the p-value (i.e., $p < .01$). The probability that the observed results were obtained from a population in which shock intensity and blood pressure are unrelated (the slope of the population regression line equals 0) is less than 1%.

State the conclusion. Because the p-value is less than the chosen level of significance ($\alpha = .05$), H_O is rejected. It may be concluded that a significant linear relationship between shock intensity and blood pressure exists in the study population of monkeys ($p < .01$).

c. The coefficient of determination, r^2, is the proportion (SSR) of the total variation (SST) in the response variable (i.e., in diastolic blood pressure) that can be attributed to its linear relationship to shock intensity. Thus,

$$SSR = b_1 S_{xy} = (1.3333)(3200) = 4266.56$$

$$SST = S_{yy} = 4450$$

$$r^2 = \frac{SSR}{SST} = \frac{4266.56}{4450} = .96$$

Ninety-six percent of the total variation in blood pressure can be attributed to variation in shock intensity.

d. The diastolic blood pressure reading corresponding to a 60 μvolt shock can be predicted from the regression equation, setting X equal to 60:

$$\hat{Y} = 85 + (1.3333)(60) = 164.998 \text{ mm Hg}$$

15-3. a. Prior to calculating the correlation coefficient, S_{xx}, S_{yy}, and S_{xy} must be computed, using formulas 15.4, 15.5, and 15.3:

$$S_{xx} = \sum X^2 - \frac{\left(\sum X\right)^2}{n} = 85.1323 - \frac{(31.41)^2}{12} = 2.9166$$

$$S_{yy} = \sum Y^2 - \frac{\left(\sum Y\right)^2}{n} = 2{,}816{,}050 - \frac{(5720)^2}{12} = 89{,}516.6667$$

$$S_{xy} = \sum XY - \frac{\left(\sum X \sum Y\right)}{n} = 15{,}357.55 - \frac{(31.41)(5720)}{12} = 385.45$$

The sample correlation coefficient (r) is then calculated using formula 15.2:

$$r = \frac{S_{xy}}{\sqrt{(S_{xx})(S_{yy})}} = \frac{385.45}{\sqrt{(2.9166)(89{,}516.6667)}} = .7544$$

b. The coefficient of determination r^2 is the proportion of the total variation in the response variable (National Board Examination score) that can be explained by its linear relationship with X (GPA).

$$r^2 = (.7544)^2 = .57$$

Thus, 57% of the variation in National Board scores can be attributed to variation in GPA.

c. The general five-step procedure for statistical hypothesis testing should be used.

State the hypothesis. The null hypothesis, $\mathbf{H_O}$, is: "There is no linear relationship between examination score and GPA; that is, the population correlation coefficient is zero ($\rho = 0$)," and the alternative hypothesis, $\mathbf{H_A}$, is: "Examination scores and GPA are linearly related; the population correlation coefficient differs significantly from zero ($\rho \neq 0$)."

Select a sample and collect data (see the table in problem 15-3). The sample correlation coefficient $r = .7544$ (calculated in the answer to problem 15-3a) is an estimate of the population correlation coefficient ρ.

Calculate the test statistic. The sample correlation coefficient ($r = .7544$) serves as the test statistic.

Evaluate the evidence against $\mathbf{H_O}$. The tabulated critical value (see Appendix 3, Table H) for $\alpha = .05$ and $df = n - 2 = 10$ is .5760. Since the calculated test statistic, $r = .7544$, is greater than this tabulated value, the p-value is less than .05.

State the conclusion. Since $p < \alpha = .05$, H_O is rejected at the 5% level of significance. There is a statistically significant correlation between GPA and National Board score in the study population ($p < .05$).

d. The slope (b_1) and Y intercept (b_0) of the regression equation relating GPA (X) and National Board score (Y) are obtained from formula 15.7 and formula 15.8:

$$b_1 = \frac{S_{xy}}{S_{xx}} = \frac{385.45}{2.9166} = 132.1562$$

$$b_0 = \bar{Y} - b_1\bar{X} = 476.6667 - (132.1562)(2.6175) = 130.7479$$

Therefore, the sample regression equation is

$$\hat{Y} = 130.7479 + 132.1562X$$

e. The National Board Examination score for a student with a GPA of 3.0 can be predicted by substituting $X = 3.0$ in the regression equation:

$$\hat{Y} = 130.7479 + (132.1562)(3.0) = 527.2$$

CHAPTER 16

16-1. a. The hypothesis that the mean urinary calcium excretion rate differs in the bumetanide-treated and control groups is tested using the five-step procedure outlined in Chapter 12 IV.

State the hypothesis. H_O" is: "The mean urinary calcium excretion rate is the same for the drug-treated and the control groups (i.e., $\mu_B - \mu_C = 0$, where μ_B represents the mean urinary calcium excretion rate for the population of patients receiving bumetanide and μ_C represents the mean urinary calcium excretion rate for the control population)," and H_A is: "Mean urinary calcium excretion rates differ in the two groups (i.e., $\mu_B - \mu_C \neq 0$)."

Select a sample and collect data. The table on the next page summarizes the results of this study.

Calculate the test statistic. Because this study employed independent samples, the independent sample (pooled) t test is appropriate. The estimate of the common unknown variance s_p^2 is derived from formula 16.3:

$$s_p^2 = \frac{(n_1 - 1)s_1^2 + (n_2 - 1)s_2^2}{n_1 + n_2 - 2} = \frac{8(6)^2 + 15(2)^2}{9 + 16 - 2} = 15.1304$$

and the test statistic t is computed using formula 16.2:

$$t = \frac{(\bar{Y}_1 - \bar{Y}_2)}{\sqrt{\frac{s_p^2}{n_1} + \frac{s_p^2}{n_2}}} = \frac{7.5 - 6.5}{\sqrt{(15.1304/9) + (15.1304/16)}} = .62$$

Evaluate the evidence against H_O. The probability of obtaining a test statistic as large as that calculated above ($t = .62$) when H_O is true (i.e., the p-value) is the area to the right of the test statistic under the frequency distribution tabulated in Table C (see Appendix 3). The number of degrees of freedom (df) for the test is $n_1 + n_2 - 2 = 23$. The test statistic, $t = .62$, lies to the left of the tabulated value for $\alpha = .10$, $df = 23$ ($t = 1.319$) on the horizontal axis of the frequency distribution and the area to the right of the test statistic is therefore greater than 10% ($p > .10$). Since this is a two-tailed test (i.e., H_A does not specify direction) this p-value is doubled ($p > .20$). H_O is rejected if $p \leq \alpha$.

State the conclusion. Because $p > \alpha$, H_O cannot be rejected. The data obtained in this study provide insufficient evidence to conclude that mean urinary calcium excretion differs in the population of patients receiving bumetanide and the control population.

b. Using the general formula for computing a confidence interval (see Ch 11 I B 1), the 95% confidence interval estimate of the difference in mean urinary calcium excretion between the population of patients receiving bumetanide and the control population [$CI_{(\mu_B - \mu_C)}$] is

endpoints of interval = estimate ± (confidence coefficient × standard error of estimate)

$$CI_{(\mu_B - \mu_C)} = (\bar{Y}_1 - \bar{Y}_2) \pm t_{(.975,\ df=23)} \sqrt{\frac{s_p^2}{n_1} + \frac{s_p^2}{n_2}}$$

$$= (7.5 - 6.5) \pm (2.069)(1.62)$$

$$= -2.35 \text{ to } 4.35\ \mu g/dl$$

Thus, the investigator conducting this study can be 95% confident that the difference in mean urinary calcium excretion rate between the two populations lies between −2.35 and 4.35 μg/dl.

16-2. a. The hypothesis can be tested using the five steps of statistical hypothesis testing.

State the hypotheses. H_O is: "Mean cumulative weight loss is the same for the population of patients receiving propranolol and the control population (i.e., $\mu_1 = \mu_2$; or $\mu_1 - \mu_2 = 0$, where μ_1 represents the mean weight loss for the population treated with propranolol and μ_2 represents the mean weight loss for the control population)," and H_A is: "Mean cumulative weight loss differs in the population of patients receiving propranolol and the control population (i.e., $\mu_1 \neq \mu_2$; or $\mu_1 - \mu_2 \neq 0$)."

Select a sample and collect data (shown in the table for problem 16-2).

	Bumetanide	Control
Sample Size	$n_1 = 9$	$n_2 = 16$
Sample Mean	$\overline{Y}_1 = 7.5$ mg/hr	$\overline{Y}_2 = 6.5$ mg/hr
Sample Standard Deviation	$s_1 = 6.0$	$s_2 = 2.0$

Calculate the test statistic. Because subjects were not matched, the pooled t test is the appropriate statistical procedure. Formula 16.3 is used to calculate the pooled variance:

$$s_p{}^2 = \frac{(n_1 - 1)s_1{}^2 + (n_2 - 1)s_2{}^2}{n_1 + n_2 - 2} = \frac{11(10)^2 + 10(8)^2}{21} = 82.857$$

Formula 16.2 is used to compute the test statistic:

$$t = \frac{(\overline{Y}_1 - \overline{Y}_2)}{\sqrt{\frac{s_p{}^2}{n_1} + \frac{s_p{}^2}{n_2}}} = \frac{120 - 70}{\sqrt{(82.857/12) + (82.857/11)}} t = 13.2$$

Evaluate the evidence against H_O. According to Table C (see Appendix 3), the tabulated value for $\alpha = .005$, with $n_1 + n_2 - 2 = 21$ degrees of freedom, is 2.8314. The calculated test statistic, $t = 13.2$, lies to the right of this value on the horizontal axis of the frequency distribution. Therefore, the area to the right of the test statistic is less than .005. Since this is a two-tailed test (H_A does not specify direction), this value is doubled and $p < .01$. The null hypothesis is rejected if $p \leq \alpha$.

State the conclusion. Because $p < \alpha = .05$, H_O is rejected at the 5% level of significance. Mean cumulative weight loss differs significantly between the population of patients receiving propranolol and the control population ($p < .01$).

b. The 95% confidence interval estimate of the difference in mean cumulative weight loss between the drug-treated population and the control population [$CI_{(\mu_1 - \mu_2)}$] is

endpoints of interval = estimate ± (confidence coefficient × standard error of estimate)

$$CI_{(\mu_1 - \mu_2)} = (\overline{Y}_1 - \overline{Y}_2) \pm t_{(.975,\ df\ =21)} \sqrt{\frac{s_p{}^2}{n_1} + \frac{s_p{}^2}{n_2}}$$

$$= (120 - 70) \pm (2.0796)(3.799)$$

$$= 42.1 \text{ to } 57.9 \text{ g}$$

The investigator conducting this study can be 95% confident that the difference in mean cumulative weight loss during insulin-induced hypoglycemia for the two populations lies between 42.1 g and 57.9 g.

16-3. a. The hypothesis that mean serum digoxin concentration four hours after a bolus injection differs significantly from mean digoxin concentration eight hours after injection can be tested using the five steps of hypothesis testing.

State the hypothesis. H_O is: "Mean serum digoxin levels are the same four hours and eight hours after injection; that is, $\mu_D = 0$, where μ_D represents the mean of the population of differences between the four hour and eight hour measurements [$d_i = (Y_{1j} - Y_{2j})$]," and **H_A** is: "The mean serum digoxin level four hours after a bolus injection differs from the mean serum digoxin level eight hours after injection; that is, $\mu_D \neq 0$."

Select a sample and collect data (shown in the table for problem 16-3 through 16-6).

Calculate the test statistic. Subjects in this study were self-paired (i.e., serum digoxin levels were determined for the same subject four and eight hours after injection). Therefore, the paired t test is the most appropriate statistical procedure. The test statistic for the paired t test is calculated using formula 16.4:

$$t = \frac{\overline{Y}_d - \mu_0}{s_d/\sqrt{n}} = \frac{-.067 - 0}{.071/\sqrt{9}} = -2.839$$

Evaluate the evidence against H$_O$. The probability of obtaining a test statistic as large as −2.8, given that H$_O$ is true, is the area to the left of the calculated test statistic (since the test statistic has a negative value). However, because the t distribution is symmetric, the absolute value of the test statistic can be used to determine the p-value (i.e., the area to the left of a negative test statistic equals the area to the right of a positive test statistic). The tabulated value (see Appendix 3, Table C) corresponding to $\alpha = .025$, with $df = n - 1 = 8$, is 2.306. The calculated test statistic, $t = |-2.8| = 2.8$, lies to the right of this value on the horizontal axis of the frequency distribution. Therefore, the area to the right of the test statistic is less than 2.5% ($p < .025$). This is a two-tailed test; hence, the p-value is doubled ($p < .05$). In other words, the chance that a difference in sample means as extreme as .067 (i.e., $t = -2.8$) could have been obtained by random chance when the difference between the population means ($\mu_D = \mu_1 - \mu_2$) is zero is less than 5%. The null hypothesis is rejected if the p-value is sufficiently small (i.e., if $p \leq \alpha$).

State the conclusion. Because $p \leq .05$, H$_O$ is rejected at the 5% level of significance. There is a statistically significant difference between mean serum digoxin concentration four hours after a rapid intravenous injection and eight hours after such an injection ($p < .05$).

b. The 95% confidence interval estimate of the difference between the mean serum digoxin concentration four hours and eight hours after injection (CI$_D$) is

endpoints of interval = estimate ± (confidence coefficient × standard error of estimate)

$$CI_D = \overline{Y}_d \pm (t_{.975,\ df=8})(s_d/\sqrt{n})$$
$$= -.067 \pm (2.306)(.024)$$
$$= -.01 \text{ to } -.12 \text{ ng/ml}$$

The researchers conducting this study can be 95% confident that this interval includes the mean difference in serum digoxin concentrations for the populations of four-hour and eight-hour measurements.

16-4. First, the two physicians specify the level of significance, power, standard deviation, and the effect size.

The level of significance, $\alpha = .05$. A two-tailed test will again be used to compare mean serum digoxin concentrations at the two time points.

Power = .80. Thus, β, the complement of power, is $1 - .80 = .20$.

The standard deviation, σ_D, can be estimated by s_d, the sample standard deviation obtained in the pilot study ($s_d = .071$).

The effect size in units of standard deviation (Δ) can be calculated based on the absolute effect size (.04 ng/ml) and the estimated value of σ_D:

$$\Delta = \frac{|\text{absolute effect size}|}{\sigma_D} = \frac{.04}{.071} = .563$$

The correct sample size, $n = 28$, is located in Table D (see Appendix 3) in the panel corresponding to $\alpha = .05$ (two-tailed test), at the intersection of the column labeled "$\beta = .20$" and the row labeled "$\Delta = .55$."

16-5. a. The five steps of statistical hypothesis testing are used to evaluate this hypothesis.

State the hypothesis. H$_O$ is: "The mean serum digoxin concentration eight hours after rapid injection (μ_1) equals the mean serum digoxin concentration (μ_2) eight hours after intravenous infusion; that is, $\mu_1 = \mu_2$ (or $\mu_1 - \mu_2 = 0$)," and **H$_A$** is: "The mean serum digoxin concentration eight hours after rapid injection is greater than that eight hours after intravenous infusion; that is, $\mu_1 > \mu_2$ (or $\mu_1 - \mu_2 > 0$)."

Select a sample and collect data (shown in the table for problem 16-3; data for intravenous infusion appear in problem 16-5).

Calculate the test statistic. Because the two samples are independent, the pooled t test is appropriate. Formula 16.3 is used to calculate the pooled variance s_p^2:

$$s_p^2 = \frac{(n_1 - 1)s_1^2 + (n_2 - 1)s_2^2}{n_1 + n_2 - 2} = \frac{8(.183)^2 + 10(.2)^2}{9 + 11 - 2} = .037$$

while t, the test statistic, is computed using formula 16.2:

$$t = \frac{(\bar{Y}_1 - \bar{Y}_2)}{\sqrt{\frac{s_p^2}{n_1} + \frac{s_p^2}{n_2}}} = \frac{.989 - .950}{\sqrt{(.037/9) + (.037/11)}} = .451$$

Evaluate the evidence against H_O. The probability of obtaining a test statistic as large as $t = .45$ when H_O is true is the area to the right of this value under the frequency distribution given in Appendix 3, Table C. The calculated test statistic, $t = .45$, lies to the left of 1.330, the tabulated value for $\alpha = .10$, $df = n_1 + n_2 - 2 = 18$, on the horizontal axis of the frequency distribution. Therefore, the area to the right of the test statistic is greater than 10% ($p > .10$ for this one-tailed test). H_O is rejected if the p-value is sufficiently small (i.e., if $p \leq \alpha$).

State the conclusion. Since the p-value is greater than $\alpha = .01$, H_O is not rejected. There is insufficient evidence to conclude that the mean serum digoxin concentration is higher eight hours after rapid injection than eight hours after intravenous infusion ($p > .10$).

This failure to detect a statistically significant difference between the population means may indicate that a difference of the specified magnitude may truly not exist (H_O true). Alternatively, such a difference may exist, but the power of the statistical test was too low to detect it. Thus, the conclusion, "No difference exists between the population means (H_O true)" cannot be drawn from a negative result unless power is known to be sufficiently high.

b. The 99% confidence interval estimate of the population mean difference in serum digoxin concentration eight hours after rapid injection and eight hours after intravenous infusion [$CI_{(\mu_1 - \mu_2)}$] is

endpoints of interval = estimate ± (confidence coefficient × standard error of estimate)

$$CI_{(\mu_1 - \mu_2)} = (\bar{Y}_1 - \bar{Y}_2) \pm t_{(.995,\ df\ =18)} \sqrt{\frac{s_p^2}{n_1} + \frac{s_p^2}{n_2}}$$
$$= (.989 - .950) \pm (2.878)(.087)$$
$$= -.21 \text{ to } .29 \text{ ng/ml}$$

The two physicians can be 99% confident that this interval includes the true difference in serum digoxin concentrations eight hours after rapid injection and eight hours after intravenous infusion.

16-6. a. The level of significance, power, standard deviation, and effect size are first specified by the investigators.

The level of significance, α, equals .01. A one-tailed test (H_A specifies a direction) will be used to analyze the data.

Power = .90, and β, the complement of power, is therefore $1 - .90 = .10$.

The standard deviation, σ, is estimated by $s_p = .193$, the pooled standard deviation obtained in the pilot study.

The effect size in units of standard deviation (Δ) can be calculated based on the absolute effect size (.1 ng/ml) and the estimated value of σ:

$$\Delta = \frac{|\text{absolute effect size}|}{\sigma} = .10/.193 = .51$$

The correct sample size, $n = 106$, is located in Table D (see Appendix 3) in the panel corresponding to $\alpha = .01$ (one-tailed test), at the intersection of the column labeled "$\beta = .10$" and the row labeled "$\Delta = .50$."

b. Given a sample size $n_1 = n_2 = 11$, an effect size $\Delta = .50$, and $\alpha = .01$, the power of the test is much less than 50% ($n_1 = n_2 = 45$ yields a value for power of 50%). Thus, the failure to observe a clinically important difference of at least .1 ng/ml is likely to be due to the low power of the statistical test. The investigators cannot conclude, therefore, that this difference does not exist (i.e., that H_O is true).

CHAPTER 17

17-1. a. No restrictions have been placed on the randomization of subjects to the four comparison groups; hence, this study utilizes a completely random design (CRD).

b. The null and alternative hypotheses (H_O and H_A) are:

H_O: "The average pain relief scores are the same for all four drugs (i.e., $\mu_1 = \mu_2 = \mu_3 = \mu_4$, where μ_1 = the average pain relief score of the patients receiving ibuprofen, μ_2 = the average pain relief score of patients receiving 60 mg of codeine, μ_3 = the average pain relief score of those patients receiving 30 mg of codeine, and μ_4 = the average pain relief score of the placebo group)."

H_A: "There is a difference in the analgesic efficacy of the four treatments (at least two of the population means are unequal)."

c. The entries in the analysis of variance (ANOVA) table (shown below) are computed from the data in the table in problem 17-1.

Source of Variation	*df*	SS	MS	*F*
Treatments	3	983.6	327.867	5.16
Error	16	1017.2	63.575	
Total	19	2000.8		

To calculate the total variation, the sum of squares and degrees of freedom are computed using formulas 17.1 and 17.2 (see V A):

$$SS_{total} = \sum_{i=1}^{k} \sum_{j=1}^{n_i} Y_{ij}^2 - G^2/n$$

$$= \sum_{i=1}^{4} \sum_{j=1}^{5} Y_{ij}^2 - (1452)^2/20 = 107{,}416 - 105{,}415.2 = 2000.8$$

$$df_{total} = N - 1 = 20 - 1 = 19$$

To calculate the variation due to treatments, the sum of squares, degrees of freedom, and mean square for the treatments are calculated using formulas 17.3, 17.4, and 17.5:

$$SST = \sum_{i=1}^{k} [T_i^2/(\text{subjects in } i\text{th group})] - G^2/N$$

$$= \sum_{i=1}^{4} [T_i^2/n_i] - (1452)^2/20$$

$$= \left[\frac{(412)^2 + (380)^2 + (335)^2 + (325)^2}{5}\right] - 105{,}415.2 = 983.6$$

$$df_{tr} = k - 1 = 4 - 1 = 3$$

$$MST = SST/df_{tr} = 983.6/3 = 327.867$$

To calculate the experimental error, the sum of squares, degrees of freedom, and mean square for error are derived from formulas 17.6, 17.7, and 17.8:

$$SSE = SS_{total} - SST = 2000.8 - 983.6 = 1017.2$$

$$df_e = N - k = 20 - 4 = 16$$

$$MSE = SSE/df_e = 1017.2/16 = 63.575$$

The test statistic, *F*, is computed using formula 17.9:

$$F = MST/MSE = 327.867/63.575 = 5.16$$

d. Table F (see Appendix 3) gives the critical value for $F_{(.05,\ df=3,16)}$ as 3.24. The calculated test statistic, $F = 5.16$, exceeds this critical value; therefore, the area to the right of the test statistic (the p-value) is less than .05. Since $p < \alpha = .05$, H_O is rejected at the 5% level of significance. At least one pair of population treatment means differ (i.e., the data support the existence of a difference in perceived pain relief among the four treatments; $p < .05$).

The Bonferroni procedure can be used to identify which of the population means are different. This procedure uses the formula

$$t = \frac{\overline{Y}_i - \overline{Y}_j}{\sqrt{\text{MSE}\left(\frac{1}{n_i} + \frac{1}{n_j}\right)}}$$

to calculate the t statistic for each of the six possible pairwise comparisons. Thus, the following values of t can be calculated.

For the hypothesis $\mu_1 = \mu_2$,

$$t = \frac{82.4 - 76}{\sqrt{63.575\left(\frac{1}{5} + \frac{1}{5}\right)}} = 1.27$$

For the hypothesis $\mu_1 = \mu_3$,

$$t = \frac{82.4 - 67}{\sqrt{63.575\left(\frac{1}{5} + \frac{1}{5}\right)}} = 3.05$$

For the hypothesis $\mu_1 = \mu_4$,

$$t = \frac{82.4 - 65}{\sqrt{63.575\left(\frac{1}{5} + \frac{1}{5}\right)}} = 3.45$$

For the hypothesis $\mu_2 = \mu_3$,

$$t = \frac{67 - 67}{\sqrt{63.575\left(\frac{1}{5} + \frac{1}{5}\right)}} = 1.78$$

For the hypothesis $\mu_2 = \mu_4$,

$$t = \frac{76 - 65}{\sqrt{63.575\left(\frac{1}{5} + \frac{1}{5}\right)}} = 2.18$$

For the hypothesis $\mu_3 = \mu_4$,

$$t = \frac{67 - 65}{\sqrt{63.575\left(\frac{1}{5} + \frac{1}{5}\right)}} = .40$$

To preserve the overall error rate per experiment of $\alpha = .05$, each of the t statistics is compared with $t_{(1-.05/6)} = t_{.992}$ for a one-tailed test, and $t_{[1-(.05/6)/2]} = t_{.996}$ for the two-tailed alternative ($\mu_i \neq \mu_j$). This critical value of t can be derived from the normal distribution (Appendix 3, Table B), using the formula

$$t_{.996} = z_{.996} + \frac{z_{.996}{}^3 + z_{.996}}{4(v-2)} = 2.65 + \frac{(2.65)^3 + 2.65}{(4)(14)} = 3.03$$

Two means are significantly different if the calculated t value exceeds 3.03 (for a two-tailed test). Thus, ibuprofen simultaneously differs from codeine (30 mg) and the placebo.

17-2. a. Subjects were randomly assigned to the three treatment groups without restrictions. Therefore, this study is an example of the completely random design (CRD).

b. The **null hypothesis (H_O)** is: "Mean anxiety–depression scores are equivalent for the populations of patients receiving desipramine, trazodone, and fluoxetine (i.e., $\mu_A = \mu_B = \mu_C$, where μ_A = the mean anxiety–depression score for the population receiving desipramine, μ_B = the mean anxiety–depression score for the population receiving trazodone, and μ_C = the mean anxiety–depression score for the population receiving fluoxetine)." The **alternative hypothesis (H_A)** is: "Mean anxiety–depression scores differ among the three treatment groups (i.e., at least two of the population means are unequal)."

c. The entries in the analysis of variance (ANOVA) table shown below are computed from the data in the table for problem 17-2.

Source of Variation	*df*	SS	MS	*F*
Treatments	2	5.167	2.583	.135
Error	9	171.750	19.083	
Total	11	176.917		

To calculate total variation, formulas 17.1 and 17.2 are used to compute the sum of squares and degrees of freedom for this row of the table. Thus,

$$SS_{total} = \sum_{i=1}^{k} \sum_{j=1}^{n_i} Y_{ij}^2 - G^2/n$$

$$= \sum_{i=1}^{3} \sum_{j=1}^{4} Y_{ij}^2 - (233)^2/12 = 4701 - 4524.083 = 176.917$$

$$df_{total} = N - 1 = 12 - 1 = 11$$

To calculate the variation due to treatments, formulas 17.3, 17.4, and 17.5 are used to derive the sum of squares, degrees of freedom, and mean square for treatments. Thus,

$$SST = \sum_{i=1}^{k} [T_i^2/(\text{subjects in } i\text{th group}) - G^2/N$$

$$= \sum_{i=1}^{3} [T_i^2/n_i] - (233)^2/12 = \left[\frac{(74)^2 + (79)^2 + (80)^2}{4}\right] - 4524.083 = 5.167$$

$$df_{tr} = k - 1 = 3 - 1 = 2$$

$$MST = SST/df_{tr} = 5.167/2 = 2.583$$

To calculate experimental error, formulas 17.6, 17.7, and 17.8 are used to derive the sum of squares, degrees of freedom, and mean square for this row of the ANOVA table. Thus,

$$SSE = SS_{total} - SST = 176.917 - 5.167 = 171.75$$

$$df_e = N - k = 12 - 3 = 9$$

$$MSE = SSE/df_e = 171.75/9 = 19.083$$

Formula 17.9 is used to calculate the test statistic as follows:

$$F = MST/MSE = 2.583/19.083 = .135$$

d. Table F gives the critical value for $F_{(.05,\ df=2,9)}$ as 4.26. Since the calculated test statistic, $F = .135$, lies to the left of this critical value, the p-value is greater than .05. Thus, $p > (\alpha = .05)$, and H_O is not rejected at the 5% level of significance. The data provide insufficient evidence to conclude that the average anxiety–depression scores differ among the populations of patients treated with desipramine, trazodone, and fluoxetine. This failure to detect a statistically significant difference between the population means may reflect the true absence of such a difference (i.e., H_O true) or low power of the statistical test. A negative result cannot be equated with the conclusion "there is no difference between the population means" unless the power of the statistical test is known to be sufficiently high.

17-3. a. Subjects in this study were grouped into homogeneous strata, corresponding to the severity of their illness prior to treatment, then randomly assigned to the treatments within these blocks. Therefore, this study is an example of the randomized complete block (RCB) design.

b. The opposing hypotheses for the statistical test are the same as those formulated for problem 17-2b (see the solution to problem 17-2b).

c. The table below is the analysis of variance (ANOVA) table for the data presented in the table in problem 17-3.

Source of Variation	*df*	SS	MS	*F*
Treatments	2	204.167	102.083	6.39
Blocks	3	72.917	24.306	
Error	6	95.833	15.972	
Total	11	372.917		

To calculate total variation, the sum of squares and degrees of freedom are calculated using formulas 17.1 and 17.2. Thus,

$$SS_{total} = \sum_{i=1}^{k} \sum_{j=1}^{b} Y_{ij}^2 - G^2/n$$

$$= \sum_{i=1}^{3} \sum_{j=1}^{4} Y_{ij}^2 - (325)^2/12 = 9175 - 8802.083 = 372.917$$

$$df_{total} = N - 1 = 12 - 1 = 11$$

To calculate variation due to treatments, the sum of squares, degrees of freedom, and mean square for this row of the ANOVA table are obtained using formulas 17.3, 17.4, and 17.5. Thus,

$$SST = \sum_{i=1}^{k} [T_i^2/(\text{number of subjects in } i\text{th group})] - G^2/N$$

$$= \sum_{i=1}^{3} [T_i^2/b] - (325)^2/12 = \left[\frac{(130)^2 + (105)^2 + (90)^2}{4}\right] - 8802.083 = 204.167$$

$$df_{tr} = k - 1 = 3 - 1 = 2$$

$$MST = SST/df_{tr=} 204.167/2 = 102.083$$

To calculate variation within blocks, formulas 17.10, 17.11, and 17.12 are used to derive the sum of squares, degrees of freedom, and mean square for the blocks. Thus

$$SSB = \sum_{j=1}^{b} [B_j^2/(\text{number of subjects in block } j)] - G^2/N$$

$$= \sum_{i=1}^{4} [B_j^2/n] - (325)^2/12$$

$$= \left[\frac{(90)^2 + (85)^2 + (70)^2 + (80)^2}{3}\right] - 8802.083 = 8875 - 8802.083 = 72.917$$

$$df_{bl} = b - 1 = 4 - 1 = 3$$

$$MSB = SSB/df_{bl} = 72.917/3 = 24.306$$

Because the RCB design was used in this study, the variation in the response variable resulting from systematic variation in initial severity can be subtracted from the experimental error term. Thus, formulas 17.13 and 17.14 are used to compute the sum of squares and degrees of freedom as follows:

$$\text{SSE} = \text{SS}_{\text{total}} - \text{SST} - \text{SSB} = 372.917 - 204.167 - 72.917 = 95.833$$

$$df_{\text{e}} = df_{\text{total}} - df_{\text{tr}} - df_{\text{bl}} = 11 - 2 - 3 = 6$$

The mean square is computed using formula 17.8:

$$\text{MSE} = \text{SSE}/df_{\text{e}} = 95.833/6 = 15.972$$

Formula 17.9 is used to calculate the test statistic as follows:

$$F = \text{MST}/\text{MSE} = 102.083/15.972 = 6.39$$

d. Table H (see Appendix 3) gives the critical value for $F_{(.05,\ df=2,6)}$ as 5.14. The calculated test statistic, $F = 6.39$, exceeds this critical value; therefore, the area to the right of the test statistic is less than .05 ($p < .05$). Since $p < (\alpha = .05)$, H_O is rejected at the 5% level of significance. There is sufficient evidence to conclude that a difference in mean anxiety–depression score exists among the three drug treatments.

The Bonferroni procedure (used with an overall error rate per experiment of 5%) identifies those population means that differ. The t statistics for the three possible pairwise comparisons are calculated using the formula

$$t = \frac{\overline{Y}_i - \overline{Y}_j}{\sqrt{\text{MSE}\left(\frac{1}{n_i} + \frac{1}{n_j}\right)}}$$

For the hypothesis $\mu_1 = \mu_2$,

$$t = \frac{32.5 - 26.25}{\sqrt{15.972\left(\frac{1}{4} + \frac{1}{4}\right)}} = 2.21$$

For the hypothesis $\mu_1 = \mu_3$,

$$t = \frac{32.5 - 22.5}{\sqrt{15.972\left(\frac{1}{4} + \frac{1}{4}\right)}} = 3.54$$

For the hypothesis $\mu_2 = \mu_3$,

$$t = \frac{26.25 - 22.5}{\sqrt{15.972\left(\frac{1}{4} + \frac{1}{4}\right)}} = 1.33$$

Each t statistic is compared to the critical value $t_{[1-(\alpha/c)/2]} = t_{[1-(.05/3)/2]} = t_{.992}$ for the two-tailed alternative ($\mu_i \neq \mu_j$). This critical value is not given in Table C (see Appendix 3) and therefore must be estimated from the normal distribution, using the formula

$$t_{.992} = z_{.992} + \frac{z_{.992}^{\ 3} + z_{.992}}{4(v-2)}$$

$$= 2.41 + \frac{(2.41)^3 + 2.41}{(4)(4)} = 3.44$$

Thus, only desipramine and fluoxetine differ significantly at the overall error rate $\alpha = .05$.

17-4. a. This study uses a randomized complete block (RCB) design. The research question involves a comparison among three commercial laboratories (the treatments), controlling for systematic variation in enzyme concentration by testing samples of four different enzymes (the blocks). Within each block, enzyme samples are randomly assigned to the three labs.

b. The **null hypothesis (H_O)** is: "The three laboratories obtain similar results ($\mu_A = \mu_B = \mu_C$, where μ_A = the mean enzyme concentration obtained by laboratory A; μ_B = the mean concentration obtained by laboratory B, and μ_C = the mean concentration obtained by laboratory C). The **alternative hypothesis (H_A)** is: "The mean enzyme concentrations obtained by the three laboratories differ (at least one pair of means is unequal)."

c. The table below is the completed analysis of variance (ANOVA) table for this study.

Source of Variation	*df*	SS	MS	*F*
Blocks	3	28.34	9.44	
Treatments	2	.06	.03	.06
Error	6	3.19	.53	
Total	11			

To calculate total variation, formula 17.1 is used to compute the sum of squares for this row of the table, while formula 17.2 is used to compute the number of degrees of freedom. Thus,

$$SS_{total} = \sum_{i=1}^{k} \sum_{j=1}^{b} Y_{ij}^2 - G^2/n$$

$$= \sum_{i=1}^{3} \sum_{j=1}^{4} Y_{ij}^2 - (66.6)^2/12 = 401.22 - 369.63 = 31.59$$

$$df_{total} = N - 1 = 12 - 1 = 11$$

To calculate variation due to treatments, formulas 17.3, 17.4, and 17.5 are used to calculate the sum of squares, degrees of freedom, and mean square for this row of the ANOVA table. Thus,

$$SST = \sum_{i=1}^{k} [T_i^2/(\text{number of subjects in } i\text{th group})] - G^2/N$$

$$= \sum_{i=1}^{3} [T_i^2/b] - (66.6)^2/12$$

$$= \left[\frac{(22.4)^2 + (22.4)^2 + (21.8)^2}{4}\right] - 369.63 = .06$$

$$df_{tr} = k - 1 = 3 - 1 = 2$$

$$MST = SST/df_{tr} = .06/2 = .03$$

To calculate the variation within blocks, the sum of squares, degrees of freedom, and mean square for the blocks are computed using formulas 17.10, 17.11, and 17.12. Thus,

$$SSB = \sum_{j=1}^{b} [B_j^2/(\text{number of subjects in block } j)] - G^2/N$$

$$= \sum_{i=1}^{4} [B_j^2/n] - (66.6)^2/12$$

$$= \left[\frac{(13.3)^2 + (19.8)^2 + (11.1)^2 + (22.4)^2}{3}\right] - 369.63 = 28.32$$

$$df_{bl} = b - 1 = 4 - 1 = 3$$

$$MSB = SSB/df_{bl} = 28.32/3 = 9.44$$

To calculate experimental error, formulas 17.13, 17.14, and 17.8 are used to compute the sum of squares, degrees of freedom, and mean square for the RCB design:

$$SSE = SS_{total} - SST - SSB = 31.59 - .06 - 28.34 = 3.19$$

$$df_e = df_{total} - df_{tr} - df_{bl} = 11 - 2 - 3 = 6$$

$$MSE = SSE/df_e = 3.19/6 = .53$$

Formula 17.9 is used to calculate the test statistic:

$$F = MST/MSE = .03/.53 = .06$$

d. Table F (see Appendix 3) gives the critical value for $F_{(.05,\ df=2,6)}$ as 5.14. The calculated test statistic, $F = .06$, lies to the left of this value on the horizontal axis of the frequency distribution of the test statistic and the area to the right of the test statistic (i.e., the p-value) is greater than .05. Since $p > (\alpha = .05)$, H_O is not rejected at the 5% level of significance. The data do not provide sufficient evidence to conclude that enzyme determinations differ among the three commercial laboratories. However, in the absence of information on power, this negative result cannot be equated with the conclusion that the population means are equal (i.e., H_O true), since the failure of the statistical test to detect a difference in means may be due to low power of the test.

Appendixes

APPENDIX 1: SUMMATION NOTATION

A. Definition. The mathematical symbol Σ (capital sigma) is a frequently used symbol in statistics; its presence in a formula indicates that a group of numbers must be added. In general, the summation operation may be expressed as

$$\sum_{i=1}^{n} Y_i = Y_1 + Y_2 + \cdots + Y_n$$

where Σ = the summation operator, Y_i = the variable to be summed, i = the index of summation, and 1 and n designate the range of summation. Thus, the symbol means "add all the observed values of a variable whose subscripts are between 1 and n, inclusive."

B. Application. The following examples use the set of numbers $Y_1 = 1$, $Y_2 = 3$, $Y_3 = 2$, $Y_4 = 4$, and $Y_5 = 5$.

1. The summation operation is illustrated as follows

   a. **Example 1**

$$\sum_{i=1}^{5} Y_i = Y_1 + Y_2 + Y_3 + Y_4 + Y_5$$
$$= 1 + 3 + 2 + 4 + 5 = 15$$

   b. **Example 2**

$$\sum_{i=2}^{4} Y_i = Y_2 + Y_3 + Y_4$$
$$= 3 + 2 + 4 = 9$$

   c. **Example 3**

$$\sum_{i=1}^{5} Y_i^2 = Y_1{}^2 + Y_2{}^2 + Y_3{}^2 + Y_4{}^2 + Y_5{}^2$$
$$= 1^2 + 3^2 + 2^2 + 4^2 + 5^2 = 55$$

   d. **Example 4**

$$\left(\sum_{i=1}^{5} Y_i\right)^2 = (Y_1 + Y_2 + Y_3 + Y_4 + Y_5)^2$$
$$= (1 + 3 + 2 + 4 + 5)^2$$
$$= 15^2 = 225$$

2. Note that $\sum Y_i{}^2 \neq \left(\sum Y_i\right)^2$.

   a. **Example 5**

$$\sum_{i=1}^{5} (Y_i - 3) = (1 - 3) + (3 - 3) + (2 - 3) + (4 - 3) + (5 - 3)$$
$$= (-2) + 0 + (-1) + 1 + 2 = 0$$

   b. **Example 6**

$$\sum_{i=1}^{5} (Y_i - 3)^2 = (1 - 3)^2 + (3 - 3)^2 + (2 - 3)^2 + (4 - 3)^2 + (5 - 3)^2$$
$$= (-2)^2 + 0^2 + (-1)^2 + 1^2 + 2^2 = 10$$

APPENDIX 2: DERIVATION OF BAYES' RULE

A. Rules of probability – a review

1. Definition of conditional probability in terms of joint probability:

$$P(\text{A}|\text{B}) = \frac{P(\text{A and B})}{P(\text{B})}$$

2. Definition of joint probability in terms of conditional probability:

$$P(\text{A } \textit{and} \text{ B}) = P(\text{B}|\text{A})\ P(\text{A})$$
$$\text{or } P(\text{A}|\text{B})\ P(\text{B})$$

3. Summation principle (mutually exclusive events):

$$P(\text{B}) = P(\text{B } \textit{and} \text{ A}_1) + P(\text{B } \textit{and} \text{ A}_2)$$

B. Derivation of Bayes' Rule. The following example, the derivation of predictive value positive (PVP), illustrates the derivation of Bayes' Rule.

1. Using the definition of conditional probability in terms of joint probability:

$$P(\text{D+}|\text{T+}) = \frac{P(\text{D+ and T+})}{P(\text{T+})}$$

2. Using the definition of joint probability in terms of conditional probability, the numerator can be rewritten:

$$P(\text{D+}|\text{T+}) = \frac{P(\text{T+}|\text{D+})\ P(\text{D+})}{P(\text{T+})}$$

3. Because D+ and D− are **mutually exclusive** and **exhaustive** events, the summation principle may be used to rewrite the denominator:

$$P(\text{T+}) = P(\text{T+ and D+}) + P(\text{T+ and D−})$$

Thus,

$$P(\text{D+}|\text{T+}) = \frac{P(\text{T+}|\text{D+})\ P(\text{D+})}{P(\text{T+ and D+}) + P(\text{T+ and D−})}$$

4. Using the definition of joint probability in terms of conditional probability, the denominator can now be rewritten:

$$P(\text{D+}|\text{T+}) = \frac{P(\text{T+}|\text{D+})\ P(\text{D+})}{P(\text{T+}|\text{D+})\ P(\text{D+}) + P(\text{T+}|\text{D−})\ P(\text{D−})} = \text{PVP}$$

APPENDIX 3: TABLES OF VALUES

Table A. Percentiles of the Chi-Square Distribution

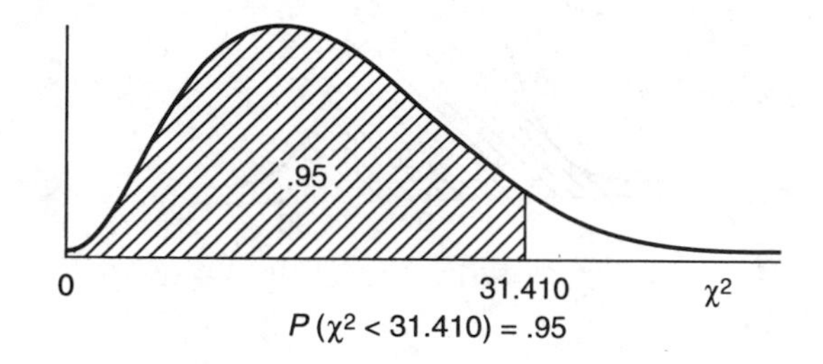

df	$\chi^2_{.005}$	$\chi^2_{.025}$	$\chi^2_{.05}$	$\chi^2_{.90}$	$\chi^2_{.95}$	$\chi^2_{.975}$	$\chi^2_{.99}$	$\chi^2_{.995}$
1	.0000393	.000982	.00393	2.706	3.841	5.024	6.635	7.879
2	.0100	.0506	.103	4.605	5.991	7.378	9.210	10.597
3	.717	.216	.352	6.251	7.815	9.348	11.345	12.838
4	.207	.484	.711	7.779	9.488	11.143	13.277	14.860
5	.412	.831	1.145	9.236	11.070	12.832	15.086	16.750
6	.676	1.237	1.635	10.645	12.592	14.449	16.812	18.548
7	.989	1.690	2.167	12.017	14.067	16.013	18.475	20.278
8	1.344	2.180	2.733	13.362	15.507	17.535	20.090	21.955
9	1.735	2.700	3.325	14.684	16.919	19.023	21.666	23.589
10	2.156	3.247	3.940	15.987	18.307	20.483	23.209	25.188
11	2.603	3.816	4.575	17.275	19.675	21.920	24.725	26.757
12	3.074	4.404	5.226	18.549	21.026	23.336	26.217	28.300
13	3.565	5.009	5.892	19.812	22.362	24.736	27.688	29.819
14	4.075	5.629	6.571	21.064	23.685	26.119	29.141	31.319
15	4.601	6.262	7.261	22.307	24.996	27.488	30.578	32.801
16	5.142	6.908	7.962	23.542	26.296	28.845	32.000	34.267
17	5.697	7.564	8.672	24.769	27.587	30.191	33.409	35.718
18	6.265	8.231	9.390	25.989	28.869	31.526	34.805	37.156
19	6.844	8.907	10.117	27.204	30.144	32.852	36.191	38.582
20	7.434	9.591	10.851	28.412	31.410	34.170	37.566	39.997
21	8.034	10.283	11.591	29.615	32.671	35.479	38.932	41.401
22	8.643	10.982	12.338	30.813	33.924	36.781	40.289	42.796
23	9.260	11.688	13.091	32.007	35.172	38.076	41.638	44.181
24	9.886	12.401	13.848	33.196	36.415	39.364	42.980	45.558
25	10.520	13.120	14.611	34.382	37.652	40.646	44.314	46.928
26	11.160	13.844	15.379	35.563	38.885	41.923	45.642	48.290
27	11.808	14.573	16.151	36.741	40.113	43.194	46.963	49.645
28	12.461	15.308	16.928	37.916	41.337	44.461	48.278	50.993
29	13.121	16.047	17.708	39.087	42.557	45.722	49.588	52.336
30	13.787	16.791	18.493	40.256	43.773	46.979	50.892	53.672
35	17.192	20.569	22.465	46.059	49.802	53.203	57.342	60.275
40	20.707	24.433	26.509	51.805	55.758	59.342	63.691	66.766
45	24.311	28.366	30.612	57.505	61.656	65.410	69.957	73.166
50	27.991	32.357	34.764	63.167	67.505	71.420	76.154	79.490
60	35.535	40.482	43.188	74.397	79.082	83.298	88.379	91.952
70	43.275	48.758	51.739	85.527	90.531	95.023	100.425	104.215
80	51.172	57.153	60.391	96.578	101.879	106.629	112.329	116.321
90	59.196	65.647	69.126	107.565	113.145	118.136	124.116	128.299
100	67.328	74.222	77.929	118.498	124.342	129.561	135.807	140.169

Reprinted from Hald A, Sinkbaek SA: A table of percentage points of the χ distribution. *Skandinavisk Aktuarietidskrift* 33:168–175, 1950.

Table B. Normal Curve Areas $P(z \leq z_0)$. Entries in the Body of the Table are Areas Between $-\infty$ and z

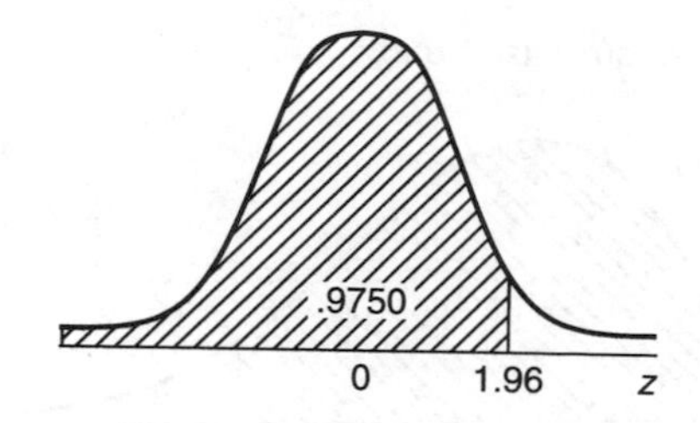

z	**−0.09**	**−0.08**	**−0.07**	**−0.06**	**−0.05**	**−0.04**	**−0.03**	**−0.02**	**−0.01**	**0.00**	z
−3.80	.0001	.0001	.0001	.0001	.0001	.0001	.0001	.0001	.0001	.0001	−3.80
−3.70	.0001	.0001	.0001	.0001	.0001	.0001	.0001	.0001	.0001	.0001	−3.70
−3.60	.0001	.0001	.0001	.0001	.0001	.0001	.0001	.0001	.0002	.0002	−3.60
−3.50	.0002	.0002	.0002	.0002	.0002	.0002	.0002	.0002	.0002	.0002	−3.50
−3.40	.0002	.0003	.0003	.0003	.0003	.0003	.0003	.0003	.0003	.0003	−3.40
−3.30	.0003	.0004	.0004	.0004	.0004	.0004	.0004	.0005	.0005	.0005	−3.30
−3.20	.0005	.0005	.0005	.0006	.0006	.0006	.0006	.0006	.0007	.0007	−3.20
−3.10	.0007	.0007	.0008	.0008	.0008	.0008	.0009	.0009	.0009	.0010	−3.10
−3.00	.0010	.0010	.0011	.0011	.0011	.0012	.0012	.0013	.0013	.0013	−3.00
−2.90	.0014	.0014	.0015	.0015	.0016	.0016	.0017	.0018	.0018	.0019	−2.90
−2.80	.0019	.0020	.0021	.0021	.0022	.0023	.0023	.0024	.0025	.0026	−2.80
−2.70	.0026	.0027	.0028	.0029	.0030	.0031	.0032	.0033	.0034	.0035	−2.70
−2.60	.0036	.0037	.0038	.0039	.0040	.0041	.0043	.0044	.0045	.0047	−2.60
−2.50	.0048	.0049	.0051	.0052	.0054	.0055	.0057	.0059	.0060	.0062	−2.50
−2.40	.0064	.0066	.0068	.0069	.0071	.0073	.0075	.0078	.0080	.0082	−2.40
−2.30	.0084	.0087	.0089	.0091	.0094	.0096	.0099	.0102	.0104	.0107	−2.30
−2.20	.0110	.0113	.0116	.0119	.0122	.0125	.0129	.0132	.0136	.0139	−2.20
−2.10	.0143	.0146	.0150	.0154	.0158	.0162	.0166	.0170	.0174	.0179	−2.10
−2.00	.0183	.0188	.0192	.0197	.0202	.0207	.0212	.0217	.0222	.0228	−2.00
−1.90	.0233	.0239	.0244	.0250	.0256	.0262	.0268	.0274	.0281	.0287	−1.90
−1.80	.0294	.0301	.0307	.0314	.0322	.0329	.0336	.0344	.0351	.0359	−1.80
−1.70	.0367	.0375	.0384	.0392	.0401	.0409	.0418	.0427	.0436	.0446	−1.70
−1.60	.0455	.0465	.0475	.0485	.0495	.0505	.0516	.0526	.0537	.0548	−1.60
−1.50	.0559	.0571	.0582	.0594	.0606	.0618	.0630	.0643	.0655	.0668	−1.50
−1.40	.0681	.0694	.0708	.0721	.0735	.0749	.0764	.0778	.0793	.0808	−1.40
−1.30	.0823	.0838	.0853	.0869	.0885	.0901	.0918	.0934	.0951	.0968	−1.30
−1.20	.0985	.1003	.1020	.1038	.1056	.1075	.1093	.1112	.1131	.1151	−1.20
−1.10	.1170	.1190	.1210	.1230	.1251	.1271	.1292	.1314	.1335	.1357	−1.10
−1.00	.1379	.1401	.1423	.1446	.1469	.1492	.1515	.1539	.1562	.1587	−1.00
−0.90	.1611	.1635	.1660	.1685	.1711	.1736	.1762	.1788	.1814	.1841	−0.90
−0.80	.1867	.1894	.1922	.1949	.1977	.2005	.2033	.2061	.2090	.2119	−0.80
−0.70	.2148	.2177	.2206	.2236	.2266	.2296	.2327	.2358	.2389	.2420	−0.70
−0.60	.2451	.2483	.2514	.2546	.2578	.2611	.2643	.2676	.2709	.2743	−0.60
−0.50	.2776	.2810	.2843	.2877	.2912	.2946	.2981	.3015	.3050	.3085	−0.50
−0.40	.3121	.3156	.3192	.3228	.3264	.3300	.3336	.3372	.3409	.3446	−0.40
−0.30	.3483	.3520	.3557	.3594	.3632	.3669	.3707	.3745	.3783	.3821	−0.30
−0.20	.3859	.3897	.3936	.3974	.4013	.4052	.4090	.4129	.4168	.4207	−0.20
−0.10	.4247	.4286	.4325	.4364	.4404	.4443	.4483	.4522	.4562	.4602	−0.10
0.00	.4641	.4681	.4721	.4761	.4801	.4840	.4880	.4920	.4960	.5000	0.00

Table B (*continued*)

z	**0.00**	**0.01**	**0.02**	**0.03**	**0.04**	**0.05**	**0.06**	**0.07**	**0.08**	**0.09**	*z*
0.00	.5000	.5040	.5080	.5120	.5160	.5199	.5239	.5279	.5319	.5359	0.00
0.10	.5398	.5438	.5478	.5517	.5557	.5596	.5636	.5675	.5714	.5753	0.10
0.20	.5793	.5832	.5871	.5910	.5948	.5987	.6026	.6064	.6103	.6141	0.20
0.30	.6179	.6217	.6255	.6293	.6331	.6368	.6406	.6443	.6480	.6517	0.30
0.40	.6554	.6591	.6628	.6664	.6700	.6736	.6772	.6808	.6844	.6879	0.40
0.50	.6915	.6950	.6985	.7019	.7054	.7088	.7123	.7157	.7190	.7224	0.50
0.60	.7257	.7291	.7324	.7357	.7389	.7422	.7454	.7486	.7517	.7549	0.60
0.70	.7580	.7611	.7642	.7673	.7704	.7734	.7764	.7794	.7823	.7852	0.70
0.80	.7881	.7910	.7939	.7967	.7995	.8023	.8051	.8078	.8106	.8133	0.80
0.90	.8159	.8186	.8212	.8238	.8264	.8289	.8315	.8340	.8365	.8389	0.90
1.00	.8413	.8438	.8461	.8485	.8508	.8531	.8554	.8577	.8599	.8621	1.00
1.10	.8643	.8665	.8686	.8708	.8729	.8749	.8770	.8790	.8810	.8830	1.10
1.20	.8849	.8869	.8888	.8907	.8925	.8944	.8962	.8980	.8997	.9015	1.20
1.30	.9032	.9049	.9066	.9082	.9099	.9115	.9131	.9147	.9162	.9177	1.30
1.40	.9192	.9207	.9222	.9236	.9251	.9265	.9279	.9292	.9306	.9319	1.40
1.50	.9332	.9345	.9357	.9370	.9382	.9394	.9406	.9418	.9429	.9441	1.50
1.60	.9452	.9463	.9474	.9484	.9495	.9505	.9515	.9525	.9535	.9545	1.60
1.70	.9554	.9564	.9573	.9582	.9591	.9599	.9608	.9616	.9625	.9633	1.70
1.80	.9641	.9649	.9656	.9664	.9671	.9678	.9686	.9693	.9699	.9706	1.80
1.90	.9713	.9719	.9726	.9732	.9738	.9744	.9750	.9756	.9761	.9767	1.90
2.00	.9772	.9778	.9783	.9788	.9793	.9798	.9803	.9808	.9812	.9817	2.00
2.10	.9821	.9826	.9830	.9834	.9838	.9842	.9846	.9850	.9854	.9857	2.10
2.20	.9861	.9864	.9868	.9871	.9875	.9878	.9881	.9884	.9887	.9890	2.20
2.30	.9893	.9896	.9898	.9901	.9904	.9906	.9909	.9911	.9913	.9916	2.30
2.40	.9918	.9920	.9922	.9925	.9927	.9929	.9931	.9932	.9934	.9936	2.40
2.50	.9938	.9940	.9941	.9943	.9945	.9946	.9948	.9949	.9951	.9952	2.50
2.60	.9953	.9955	.9956	.9957	.9959	.9960	.9961	.9962	.9963	.9964	2.60
2.70	.9965	.9966	.9967	.9968	.9969	.9970	.9971	.9972	.9973	.9974	2.70
2.80	.9974	.9975	.9976	.9977	.9977	.9978	.9979	.9979	.9980	.9981	2.80
2.90	.9981	.9982	.9982	.9983	.9984	.9984	.9985	.9985	.9986	.9986	2.90
3.00	.9987	.9987	.9987	.9988	.9988	.9989	.9989	.9989	.9990	.9990	3.00
3.10	.9990	.9991	.9991	.9991	.9992	.9992	.9992	.9992	.9993	.9993	3.10
3.20	.9993	.9993	.9994	.9994	.9994	.9994	.9994	.9995	.9995	.9995	3.20
3.30	.9995	.9995	.9995	.9996	.9996	.9996	.9996	.9996	.9996	.9997	3.30
3.40	.9997	.9997	.9997	.9997	.9997	.9997	.9997	.9997	.9997	.9998	3.40
3.50	.9998	.9998	.9998	.9998	.9998	.9998	.9998	.9998	.9998	.9998	3.50
3.60	.9998	.9998	.9999	.9999	.9999	.9999	.9999	.9999	.9999	.9999	3.60
3.70	.9999	.9999	.9999	.9999	.9999	.9999	.9999	.9999	.9999	.9999	3.70
3.80	.9999	.9999	.9999	.9999	.9999	.9999	.9999	.9999	.9999	.9999	3.80

Reprinted from Daniel W: *Biostatistics: A Foundation for Analysis in the Health Sciences*, 4th ed. New York, John Wiley, 1989, pp 687–688.

Table C. The t Distribution

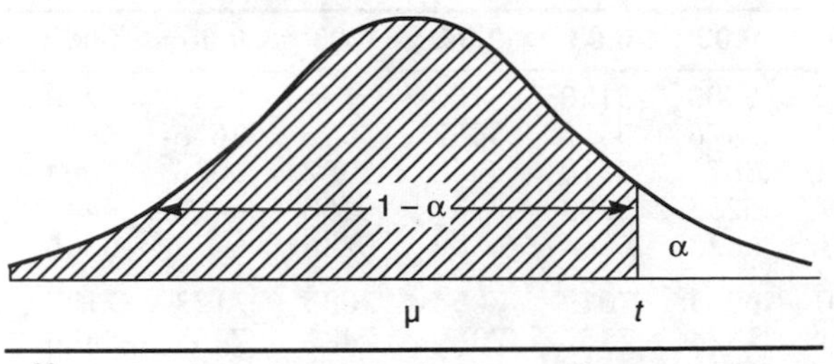

df	$t_{.90}$	$t_{.95}$	$t_{.975}$	$t_{.99}$	$t_{.995}$
1	3.078	6.3138	12.706	31.821	63.657
2	1.886	2.9200	4.3027	6.965	9.9248
3	1.638	2.3534	3.1825	4.541	5.8409
4	1.533	2.1318	2.7764	3.747	4.6041
5	1.476	2.0150	2.5706	3.365	4.0321
6	1.440	1.9432	2.4469	3.143	3.7074
7	1.415	1.8946	2.3646	2.998	3.4995
8	1.397	1.8595	2.3060	2.896	3.3554
9	1.383	1.8331	2.2622	2.821	3.2498
10	1.372	1.8125	2.2281	2.764	3.1693
11	1.363	1.7959	2.2010	2.718	3.1058
12	1.356	1.7823	2.1788	2.681	3.0545
13	1.350	1.7709	2.1604	2.650	3.0123
14	1.345	1.7613	2.1448	2.624	2.9768
15	1.341	1.7530	2.1315	2.602	2.9467
16	1.337	1.7459	2.1199	2.583	2.9208
17	1.333	1.7396	2.1098	2.567	2.8982
18	1.330	1.7341	2.1009	2.552	2.8784
19	1.328	1.7291	2.0930	2.539	2.8609
20	1.325	1.7247	2.0860	2.528	2.8453
21	1.323	1.7207	2.0796	2.518	2.8314
22	1.321	1.7171	2.0739	2.508	2.8188
23	1.319	1.7139	2.0687	2.500	2.8073
24	1.318	1.7109	2.0639	2.492	2.7969
25	1.316	1.7081	2.0595	2.485	2.7874
26	1.315	1.7056	2.0555	2.479	2.7787
27	1.314	1.7033	2.0518	2.473	2.7707
28	1.313	1.7011	2.0484	2.467	2.7633
29	1.311	1.6991	2.0452	2.462	2.7564
30	1.310	1.6973	2.0423	2.457	2.7500
35	1.3062	1.6896	2.0301	2.438	2.7239
40	1.3031	1.6839	2.0211	2.423	2.7045
45	1.3007	1.6794	2.0141	2.412	2.6896
50	1.2987	1.6759	2.0086	2.403	2.6778
60	1.2959	1.6707	2.0003	2.390	2.6603
70	1.2938	1.6669	1.9945	2.381	2.6480
80	1.2922	1.6641	1.9901	2.374	2.6388
90	1.2910	1.6620	1.9867	2.368	2.6316
100	1.2901	1.6602	1.9840	2.364	2.6260
120	1.2887	1.6577	1.9799	2.358	2.6175
140	1.2876	1.6558	1.9771	2.353	2.6114
160	1.2869	1.6545	1.9749	2.350	2.6070
180	1.2863	1.6534	1.9733	2.347	2.6035
200	1.2858	1.6525	1.9719	2.345	2.6006
∞	1.282	1.645	1.96	2.326	2.576

Adapted from Lentner C (ed): *Geigy Scientific Tables*, 8th ed, volume 2. Basle, Switzerland, Ciba-Geigy, 1982, pp 30–33.

Table D. Student's t Distribution: Number of Observations for Single-sample t Test and Paired t Test

	Level of t Test																			
Single-sided Test	$\alpha = 0.005$					$\alpha = 0.01$					$\alpha = 0.025$					$\alpha = 0.05$				
Double-sided Test	$\alpha = 0.01$					$\alpha = 0.02$					$\alpha = 0.05$					$\alpha = 0.1$				
$\beta =$	0.01	0.05	0.1	0.2	0.5	0.01	0.05	0.1	0.2	0.5	0.01	0.05	0.1	0.2	0.5	0.01	0.05	0.1	0.2	0.5
Value of $\Delta = \frac{\mu - \mu_0}{\sigma}$																				
0.05																				
0.10																				
0.15																				122
0.20										139					99					70
0.25					110					90				128	64			139	101	45
0.30				134	78				115	63			119	90	45		122	97	71	32
0.35			125	99	58			109	85	47		109	88	67	34		90	72	52	24
0.40		115	97	77	45		101	85	66	37	117	84	68	51	26	101	70	55	40	19
0.45		92	77	62	37	110	81	68	53	30	93	67	54	41	21	80	55	44	33	15
0.50	100	75	63	51	30	90	66	55	43	25	76	54	44	34	18	65	45	36	27	13
0.55	83	63	53	42	26	75	55	46	36	21	63	45	37	28	15	54	38	30	22	11
0.60	71	53	45	36	22	63	47	39	31	18	53	38	32	24	13	46	32	26	19	9
0.65	61	46	39	31	20	55	41	34	27	16	46	33	27	21	12	39	28	22	17	8
0.70	53	40	34	28	17	47	35	30	24	14	40	29	24	19	10	34	24	19	15	8
0.75	47	36	30	25	16	42	31	27	21	13	35	26	21	16	9	30	21	17	13	7
0.80	41	32	27	22	14	37	28	24	19	12	31	22	19	15	9	27	19	15	12	6
0.85	37	29	24	20	13	33	25	21	17	11	28	21	17	13	8	24	17	14	11	6
0.90	34	26	22	18	12	29	23	19	16	10	25	19	16	12	7	21	15	13	10	5
0.95	31	24	20	17	11	27	21	18	14	9	23	17	14	11	7	19	14	11	9	5
1.00	28	22	19	16	10	25	19	16	13	9	21	16	13	10	6	18	13	11	8	5

Table D *(continued)*

	Level of *t* Test																			
Single-sided Test	$\alpha = 0.005$					$\alpha = 0.01$					$\alpha = 0.025$					$\alpha = 0.05$				
Double-sided Test	$\alpha = 0.01$					$\alpha = 0.02$					$\alpha = 0.05$					$\alpha = 0.1$				
$\beta =$	0.01	0.05	0.1	0.2	0.5	0.01	0.05	0.1	0.2	0.5	0.01	0.05	0.1	0.2	0.5	0.01	0.05	0.1	0.2	0.5
Value of $\Delta = \frac{\mu - \mu_0}{\sigma}$																				
1.10	24	19	16	14	9	21	16	14	12	8	18	13	11	9	6	15	11	9	7	
1.20	21	16	14	12	8	18	14	12	10	7	15	12	10	8	5	13	10	8	6	
1.30	18	15	13	11	8	16	13	11	9	6	14	10	9	7		11	8	7	6	
1.40	16	13	12	10	7	14	11	10	9	6	12	9	8	7		10	8	7	5	
1.50	15	12	11	9	7	13	10	9	8	6	11	8	7	6		9	7	6		
1.60	13	11	10	8	6	12	10	9	7	5	10	8	7	6		8	6	6		
1.70	12	10	9	8	6	11	9	8	7		9	7	6	5		8	6	5		
1.80	12	10	9	8	6	10	8	7	7		8	7	6			7	6			
1.90	11	9	8	7	6	10	8	7	6		8	6	6			7	5			
2.00	10	8	8	7	5	9	7	7	6		7	6	5			6				
2.10	10	8	7	7		8	7	6	6		7	6				6				
2.20	9	8	7	6		8	7	6	5		7	6				6				
2.30	9	7	7	6		8	6	6			6	5				5				
2.40	8	7	7	6		7	6	6			6									
2.50	8	7	6	6		7	6	6			6									
3.00	7	6	6	5		6	5	5			5									
3.50	6	5	5			5														
4.00	6																			

Reprinted from Weast RC (ed): *CRC Handbook of Tables for Probability and Statistics*, 2nd ed. Boca Raton, FL, CRC Press, 1968, p 287.

Table E. Student's t Distribution: Number of Observations for Pooled t Test

	Level of t Test																			
Single-sided Test	$\alpha = 0.005$					$\alpha = 0.01$					$\alpha = 0.025$					$\alpha = 0.05$				
Double-sided Test	$\alpha = 0.01$					$\alpha = 0.02$					$\alpha = 0.05$					$\alpha = 0.1$				
$\beta =$	0.01	0.05	0.1	0.2	0.5	0.01	0.05	0.1	0.2	0.5	0.01	0.05	0.1	0.2	0.5	0.01	0.05	0.1	0.2	0.5
Value of $\Delta = \frac{\mu_0 - \mu_1}{\sigma}$																				
0.05																				
0.10																				
0.15																				
0.20																				137
0.25															124					88
0.30										123					87					61
0.35					110					90					64				102	45
0.40					85					70				100	50			108	78	35
0.45				118	68				101	55			105	79	39		108	86	62	28
0.50				96	55			106	82	45		106	86	64	32		88	70	51	23
0.55			101	79	46		106	88	68	38		87	71	53	27	112	73	58	42	19
0.60		101	85	67	39		90	74	58	32	104	74	60	45	23	89	61	49	36	16
0.65		87	73	57	34	104	77	64	49	27	88	63	51	39	20	76	52	42	30	14
0.70	100	75	63	50	29	90	66	55	43	24	76	55	44	34	17	66	45	36	26	12
0.75	88	66	55	44	26	79	58	48	38	21	67	48	39	29	15	57	40	32	23	11
0.80	77	58	49	39	23	70	51	43	33	19	59	42	34	26	14	50	35	28	21	10
0.85	69	51	43	35	21	62	46	38	30	17	52	37	31	23	12	45	31	25	18	9
0.90	62	46	39	31	19	55	41	34	27	15	47	34	27	21	11	40	28	22	16	8
0.95	55	42	35	28	17	50	37	31	24	14	42	30	25	19	10	36	25	20	15	7
1.00	50	38	32	26	15	45	33	28	22	13	38	27	23	17	9	33	23	18	14	7

Table E *(continued)*

	Level of *t* Test																			
Single-sided Test	$\alpha = 0.005$					$\alpha = 0.01$					$\alpha = 0.025$					$\alpha = 0.05$				
Double-sided Test	$\alpha = 0.01$					$\alpha = 0.02$					$\alpha = 0.05$					$\alpha = 0.1$				
$\beta =$	0.01	0.05	0.1	0.2	0.5	0.01	0.05	0.1	0.2	0.5	0.01	0.05	0.1	0.2	0.5	0.01	0.05	0.1	0.2	0.5
Value of $\Delta = \frac{\mu_0 - \mu_1}{\sigma}$																				
1.10	42	32	27	22	13	38	28	23	19	11	32	23	19	14	8	27	19	15	12	6
1.20	36	27	23	18	11	32	24	20	16	9	27	20	16	12	7	23	16	13	10	5
1.30	31	23	20	16	10	28	21	17	14	8	23	17	14	11	6	20	14	11	9	5
1.40	27	20	17	14	9	24	18	15	12	8	20	15	12	10	6	17	12	10	8	4
1.50	24	18	15	13	8	21	16	14	11	7	18	13	11	9	5	15	11	9	7	4
1.60	21	16	14	11	7	19	14	12	10	6	16	12	10	8	5	14	10	8	6	4
1.70	19	15	13	10	7	17	13	11	9	6	14	11	9	7	4	12	9	7	6	3
1.80	17	13	11	10	6	15	12	10	8	5	13	10	8	6	4	11	8	7	5	
1.90	16	12	11	9	6	14	11	9	8	5	12	9	7	6	4	10	7	6	5	
2.00	14	11	10	8	6	13	10	9	7	5	11	8	7	6	4	9	7	6	4	
2.10	13	10	9	8	5	12	9	8	7	5	10	8	6	5	3	8	6	5	4	
2.20	12	10	8	7	5	11	9	7	6	4	9	7	6	5		8	6	5	4	
2.30	11	9	8	7	5	10	8	7	6	4	9	7	6	5		7	5	5	4	
2.40	11	9	8	6	5	10	8	7	6	4	8	6	5	4		7	5	4	4	
2.50	10	8	7	6	4	9	7	6	5	4	8	6	5	4		6	5	4	3	
3.00	8	6	6	5	4	7	6	5	4	3	6	5	4	4		5	4	3		
3.50	6	5	5	4	3	6	5	4	4		5	4	4	3		4	3			
4.00	6	5	4	4		5	4	4	3		4	4	3			4				

Reprinted from Weast RC (ed): *CRC Handbook of Tables for Probability and Statistics*, 2nd ed. Boca Raton, FL, CRC Press, 1968, p 289.

Table F. *F* Distribution ($\alpha = 0.05$)

Denominator Degrees of Freedom (*df*)	Numerator Degrees of Freedom (*df*) 1	2	3	4	5	6	7	8	9	10	12	15	20	24	30	40	60	120	∞
1	161.4	199.5	215.7	224.6	230.2	234.0	236.8	238.9	240.5	241.9	243.9	245.9	248.0	249.1	250.1	251.1	252.2	253.3	254.3
2	18.51	19.00	19.16	19.25	19.30	19.33	19.35	19.37	19.38	19.40	19.41	19.43	19.45	19.45	19.46	19.47	19.48	19.49	19.50
3	10.13	9.55	9.28	9.12	9.01	8.94	8.89	8.85	8.81	8.79	8.74	8.70	8.66	8.64	8.62	8.59	8.57	8.55	8.53
4	7.71	6.94	6.59	6.39	6.26	6.16	6.09	6.04	6.00	5.96	5.91	5.86	5.80	5.77	5.75	5.72	5.69	5.66	5.63
5	6.61	5.79	5.41	5.19	5.05	4.95	4.88	4.82	4.77	4.74	4.68	4.62	4.56	4.53	4.50	4.46	4.43	4.40	4.36
6	5.99	5.14	4.76	4.53	4.39	4.28	4.21	4.15	4.10	4.06	4.00	3.94	3.87	3.84	3.81	3.77	3.74	3.70	3.67
7	5.59	4.74	4.35	4.12	3.97	3.87	3.79	3.73	3.68	3.64	3.57	3.51	3.44	3.41	3.38	3.34	3.30	3.27	3.23
8	5.32	4.46	4.07	3.84	3.69	3.58	3.50	3.44	3.39	3.35	3.28	3.22	3.15	3.12	3.08	3.04	3.01	2.97	2.93
9	5.12	4.26	3.86	3.63	3.48	3.37	3.29	3.23	3.18	3.14	3.07	3.01	2.94	2.90	2.86	2.83	2.79	2.75	2.71
10	4.96	4.10	3.71	3.48	3.33	3.22	3.14	3.07	3.02	2.98	2.91	2.85	2.77	2.74	2.70	2.66	2.62	2.58	2.54
11	4.84	3.98	3.59	3.36	3.20	3.09	3.01	2.95	2.90	2.85	2.79	2.72	2.65	2.61	2.57	2.53	2.49	2.45	2.40
12	4.75	3.89	3.49	3.26	3.11	3.00	2.91	2.85	2.80	2.75	2.69	2.62	2.54	2.51	2.47	2.43	2.38	2.34	2.30
13	4.67	3.81	3.41	3.18	3.03	2.92	2.83	2.77	2.71	2.67	2.60	2.53	2.46	2.42	2.38	2.34	2.30	2.25	2.21
14	4.60	3.74	3.34	3.11	2.96	2.85	2.76	2.70	2.65	2.60	2.53	2.46	2.39	2.35	2.31	2.27	2.22	2.18	2.13
15	4.54	3.68	3.29	3.06	2.90	2.79	2.71	2.64	2.59	2.54	2.48	2.40	2.33	2.29	2.25	2.20	2.16	2.11	2.07
16	4.49	3.63	3.24	3.01	2.85	2.74	2.66	2.59	2.54	2.49	2.42	2.35	2.28	2.24	2.19	2.15	2.11	2.06	2.01
17	4.45	3.59	3.20	2.96	2.81	2.70	2.61	2.55	2.49	2.45	2.38	2.31	2.23	2.19	2.15	2.10	2.06	2.01	1.96
18	4.41	3.55	3.16	2.93	2.77	2.66	2.58	2.51	2.46	2.41	2.34	2.27	2.19	2.15	2.11	2.06	2.02	1.97	1.92
19	4.38	3.52	3.13	2.90	2.74	2.63	2.54	2.48	2.42	2.38	2.31	2.23	2.16	2.11	2.07	2.03	1.98	1.93	1.88
20	4.35	3.49	3.10	2.87	2.71	2.60	2.51	2.45	2.39	2.35	2.28	2.20	2.12	2.08	2.04	1.99	1.95	1.90	1.84
21	4.32	3.47	3.07	2.84	2.68	2.57	2.49	2.42	2.37	2.32	2.25	2.18	2.10	2.05	2.01	1.96	1.92	1.87	1.81
22	4.30	3.44	3.05	2.82	2.66	2.55	2.46	2.40	2.34	2.30	2.23	2.15	2.07	2.03	1.98	1.94	1.89	1.84	1.78
23	4.28	3.42	3.03	2.80	2.64	2.53	2.44	2.37	2.32	2.27	2.20	2.13	2.05	2.01	1.96	1.91	1.86	1.81	1.76
24	4.26	3.40	3.01	2.78	2.62	2.51	2.42	2.36	2.30	2.25	2.18	2.11	2.03	1.98	1.94	1.89	1.84	1.79	1.73
25	4.24	3.39	2.99	2.76	2.60	2.49	2.40	2.34	2.28	2.24	2.16	2.09	2.01	1.96	1.92	1.87	1.82	1.77	1.71
26	4.23	3.37	2.98	2.74	2.59	2.47	2.39	2.32	2.27	2.22	2.15	2.07	1.99	1.95	1.90	1.85	1.80	1.75	1.69
27	4.21	3.35	2.96	2.73	2.57	2.46	2.37	2.31	2.25	2.20	2.13	2.06	1.97	1.93	1.88	1.84	1.79	1.73	1.67
28	4.20	3.34	2.95	2.71	2.56	2.45	2.36	2.29	2.24	2.19	2.12	2.04	1.96	1.91	1.87	1.82	1.77	1.71	1.65
29	4.18	3.33	2.93	2.70	2.55	2.43	2.35	2.28	2.22	2.18	2.10	2.03	1.94	1.90	1.85	1.81	1.75	1.70	1.64
30	4.17	3.32	2.92	2.69	2.53	2.42	2.33	2.27	2.21	2.16	2.09	2.01	1.93	1.89	1.84	1.79	1.74	1.68	1.62
40	4.08	3.23	2.84	2.61	2.45	2.34	2.25	2.18	2.12	2.08	2.00	1.92	1.84	1.79	1.74	1.69	1.64	1.58	1.51
60	4.00	3.15	2.76	2.53	2.37	2.25	2.17	2.10	2.04	1.99	1.92	1.84	1.75	1.70	1.65	1.59	1.53	1.47	1.39
120	3.92	3.07	2.68	2.45	2.29	2.17	2.09	2.02	1.96	1.91	1.83	1.75	1.66	1.61	1.55	1.50	1.43	1.35	1.25
∞	3.84	3.00	2.60	2.37	2.21	2.10	2.01	1.94	1.88	1.83	1.75	1.67	1.57	1.52	1.46	1.39	1.32	1.22	1.00

Reprinted from Pearson ES, Hartley HO (eds): *Biometrika Tables for Statisticians*, 3rd ed, volume I, London, Cambridge University Press, 1966, pp 171, 173.

Table G. *F* Distribution ($\alpha = 0.01$)

Denominator Degrees of Freedom (*df*)	Numerator Degrees of Freedom (*df*) 1	2	3	4	5	6	7	8	9	10	12	15	20	24	30	40	60	120	∞
1	4052	4999.5	5403	5625	5764	5859	5928	5981	6022	6056	6106	6157	6209	6235	6261	6287	6313	6339	6366
2	98.5	99	99.17	99.25	99.3	99.33	99.36	99.37	99.39	99.4	99.42	99.43	99.45	99.46	99.47	99.47	99.48	99.49	99.5
3	34.12	30.82	29.46	28.71	28.24	27.91	27.67	27.49	27.35	27.23	27.05	26.87	26.69	26.60	26.50	26.41	26.32	26.22	26.13
4	21.20	18.00	16.69	15.98	15.52	15.21	14.98	14.80	14.66	14.55	14.37	14.20	14.02	13.93	13.84	13.75	13.65	13.56	13.46
5	16.26	13.27	12.06	11.39	10.97	10.67	10.46	10.29	10.16	10.05	9.89	9.72	9.55	9.47	9.38	9.29	9.20	9.11	9.02
6	13.75	10.92	9.78	9.15	8.75	8.47	8.26	8.10	7.98	7.87	7.72	7.56	7.40	7.31	7.23	7.14	7.06	6.97	6.88
7	12.25	9.55	8.45	7.85	7.46	7.19	6.99	6.84	6.72	6.62	6.47	6.31	6.16	6.07	5.99	5.91	5.82	5.74	5.65
8	11.26	8.65	7.59	7.01	6.63	6.37	6.18	6.03	5.91	5.81	5.67	5.52	5.36	5.28	5.20	5.12	5.03	4.95	4.86
9	10.56	8.02	6.99	6.42	6.06	5.80	5.61	5.47	5.35	5.26	5.11	4.96	4.81	4.73	4.65	4.57	4.48	4.40	4.31
10	10.04	7.56	6.55	5.99	5.64	5.39	5.20	5.06	4.94	4.85	4.71	4.56	4.41	4.33	4.25	4.17	4.08	4.00	3.91
11	9.65	7.21	6.22	5.67	5.32	5.07	4.89	4.74	4.63	4.54	4.40	4.25	4.10	4.02	3.94	3.86	3.78	3.69	3.60
12	9.33	6.93	5.95	5.41	5.06	4.82	4.64	4.50	4.39	4.30	4.16	4.01	3.86	3.78	3.70	3.62	3.54	3.45	3.36
13	9.07	6.70	5.74	5.21	4.86	4.62	4.44	4.30	4.19	4.10	3.96	3.82	3.66	3.59	3.51	3.43	3.34	3.25	3.17
14	8.86	6.51	5.56	5.04	4.69	4.46	4.28	4.14	4.03	3.94	3.80	3.66	3.51	3.43	3.35	3.27	3.18	3.09	3.00
15	8.68	6.36	5.42	4.89	4.56	4.32	4.14	4.00	3.89	3.80	3.67	3.52	3.37	3.29	3.21	3.13	3.05	2.96	2.87
16	8.53	6.23	5.29	4.77	4.44	4.20	4.03	3.89	3.78	3.69	3.55	3.41	3.26	3.18	3.10	3.02	2.93	2.84	2.75
17	8.40	6.11	5.18	4.67	4.34	4.10	3.93	3.79	3.68	3.59	3.46	3.31	3.16	3.08	3.00	2.92	2.83	2.75	2.65
18	8.29	6.01	5.09	4.58	4.25	4.01	3.84	3.71	3.60	3.51	3.37	3.23	3.08	3.00	2.92	2.84	2.75	2.66	2.57
19	8.18	5.93	5.01	4.50	4.17	3.94	3.77	3.63	3.52	3.43	3.30	3.15	3.00	2.92	2.84	2.76	2.67	2.58	2.49
20	8.10	5.85	4.94	4.43	4.10	3.87	3.70	3.56	3.46	3.37	3.23	3.09	2.94	2.86	2.78	2.69	2.61	2.52	2.42
21	8.02	5.78	4.87	4.37	4.04	3.81	3.64	3.51	3.40	3.31	3.17	3.03	2.88	2.80	2.72	2.64	2.55	2.46	2.36
22	7.95	5.72	4.82	4.31	3.99	3.76	3.59	3.45	3.35	3.26	3.12	2.98	2.83	2.75	2.67	2.58	2.50	2.40	2.31
23	7.88	5.66	4.76	4.26	3.94	3.71	3.54	3.41	3.30	3.21	3.07	2.93	2.78	2.70	2.62	2.54	2.45	2.35	2.26
24	7.82	5.61	4.72	4.22	3.90	3.67	3.50	3.36	3.26	3.17	3.03	2.89	2.74	2.66	2.58	2.49	2.40	2.31	2.21
25	7.77	5.57	4.68	4.18	3.85	3.63	3.46	3.32	3.22	3.13	2.99	2.85	2.70	2.62	2.54	2.45	2.36	2.27	2.17
26	7.72	5.53	4.64	4.14	3.82	3.59	3.42	3.29	3.18	3.09	2.96	2.81	2.66	2.58	2.50	2.42	2.33	2.23	2.13
27	7.68	5.49	4.60	4.11	3.78	3.56	3.39	3.26	3.15	3.06	2.93	2.78	2.63	2.55	2.47	2.38	2.29	2.20	2.10
28	7.64	5.45	4.57	4.07	3.75	3.53	3.36	3.23	3.12	3.03	2.90	2.75	2.60	2.52	2.44	2.35	2.26	2.17	2.06
29	7.60	5.42	4.54	4.04	3.73	3.50	3.33	3.20	3.09	3.00	2.87	2.73	2.57	2.49	2.41	2.33	2.23	2.14	2.03
30	7.56	5.39	4.51	4.02	3.70	3.47	3.30	3.17	3.07	2.98	2.84	2.70	2.55	2.47	2.39	2.30	2.21	2.11	2.01
40	7.31	5.18	4.31	3.83	3.51	3.29	3.12	2.99	2.89	2.80	2.66	2.52	2.37	2.29	2.20	2.11	2.02	1.92	1.80
60	7.08	4.98	4.13	3.65	3.34	3.12	2.95	2.82	2.72	2.63	2.50	2.35	2.20	2.12	2.03	1.94	1.84	1.73	1.60
120	6.85	4.79	3.95	3.48	3.17	2.96	2.79	2.66	2.56	2.47	2.34	2.19	2.03	1.95	1.86	1.76	1.66	1.53	1.38
∞	6.63	4.61	3.78	3.32	3.02	2.80	2.64	2.51	2.41	2.32	2.18	2.04	1.88	1.79	1.70	1.59	1.47	1.32	1.00

Reprinted from Pearson ES, Hartley HO (eds): *Biometrika Tables for Statisticians*, 3rd ed, volume I, London, Cambridge University Press, 1966, pp 171, 173.

Table H. Critical Values of the Correlation Coefficient for Different Levels of Significance

*df**	.05	.01	*df*	.05	.01	*df*	.05	.01
1	.996917	.9998766	11	.5529	.6835	25	.3809	.4869
2	.95000	.990000	12	.5324	.6614	30	.3494	.4487
3	.8783	.95873	13	.5139	.6411	35	.3246	.4182
4	.8114	.91720	14	.4973	.6226	40	.3044	.3932
5	.7545	.8745	15	.4821	.6055	45	.2875	.3721
6	.7067	.8343	16	.4683	.5897	50	.2732	.3541
7	.6664	.7977	17	.4555	.5751	60	.2500	.3248
8	.6319	.7646	18	.4438	.5614	70	.2319	.3017
9	.6021	.7348	19	.4329	.5487	80	.2172	.2830
10	.5760	.7079	20	.4227	.5368	90	.2050	.2673
						100	.1946	.2540

*The degrees of freedom (*df*) = (the number of pairs in the sample − 2).
Reprinted from Fisher RA: *Statistical Methods for Research Workers*, 14th ed. New York, Hafner Press, 1970, p 209.

Index

Note: Page numbers in *italics* denote illustrations; those followed by a *t* denote tables; those followed by Q denote questions; and those followed by E denote explanations.

A

Absolute frequency, calculation and application of, 324Q, 344E
Absolute risk
case–control studies, 128Q, 130E, 249Q, 253E, 330Q, 350E
characteristics of, 248Q, 252E
cohort studies, 331Q, 351E
comparative risk measures and, 233
measures of frequency, 233
Accuracy
defined, 4, 15Q, 16E
diagnostic testing evaluation, 42
vs. precision and, 5
Addition rule of probability, 22–23
calculation formula for, 24
Age-adjusted mortality rate, 99
applications for, 333Q, 353E
calculation formula, 122–123
Age-specific mortality rate, 99
example of, 107Q, 108E
Aggregate data, coefficient of determination (r^2) and, 263
Alternative hypothesis (H_A)
analysis of variance testing, completely random design (CRD), 296
chi-square testing
case–control study design, 221
cross-sectional study design, 212–213
evaluating evidence, 188, 190–191
odds ratio (OR), 241
statistical tests, 187
clinical trial applications, 335Q, 354E
α. *See* Level of significance
Analogy, and causality, 121
Analysis of variance (*F* test)
completely random design (CRD), 295–298
calculation formulas, 304–306
characteristics, 293
hypothesis testing, 295–296
null hypothesis (H_O), evidence against, 298
test statistic calculation, 296–297
computational forms and calculations, 304–306
multiple comparison groups, 293–294
Bonferroni procedure, 302–304
procedures, 302–304
purpose of, 293
randomized complete block (RCB) design, 298–302, 336Q, 355E
artificial matching, 300
calculation formulas, 306
characteristics, 298–299
experimental errors, 301–302
hypothesis testing, 300
natural matching, 300
null hypothesis (H_O) evaluation, 301
self-matching, 299–300
test statistic calculation, 300–301
study design and, 293–294
test statistic calculation, 294
variations in, 294
Analysis of variance (ANOVA) table
completely random design (CRD), 296–297
randomized complete block (RCB) design, 301
Analytic studies, risk assessment, 111
Antecedent–consequence uncertainty
cases and controls, 127Q, 129E
risk assessment, 115
Artificial matching
analysis of variance testing, randomized complete block (RCB) design, 300
matched sampling, population means testing, 278
Associations
causality, 120–121
chi-square test, 212
case–control study design, 219–221
cohort study design, 216–219
cross-sectional studies, 212–216
experimental study design, 216–219
matched sample studies, 222–223
negative results, 225
one-tailed vs. two-tailed hypothesis testing, 223–225
statistical significance and, 229Q, 230E
validity, 225
clinical medicine, 109
confidence interval estimation, 181Q, 184E
frequency (count) data, 212–225
magnitude of, 233–234
matched sampling, population means testing, 279
measures of, 233
risk assessment, case-control study, 114–115
regression analysis, 272Q, 274E
risk assessment
historical cohort study, 112–113
prospective cohort study, 111–112
statistical significance of, correlation analysis, 261–262
Attack rate, defined, 100
Attributable fraction in exposed (AF_E), 242–243
Attributable fraction in total population (AF_T), 243–244
Attributable risk, 112
characteristics of, 248Q, 252E
incidence, 251Q, 254E
measures of potential impact, 241–242
Attributable risk percent in the exposed. *See* Attributable fraction in exposed (AF_E)
Averages,
coefficient of determination (r^2), 263
decision analysis, diagnostic strategy, 70

B

Back calculation, diagnostic testing, 39–40
Basic risk statement, measures of frequency, 233
Bayes' rule
derivation of, 412
diagnostic testing, 41–42
heritability assessment, 84–87
odds ratio (OR) calculation, 244
rules of probability, 412
Behavioral (life-habit) risk factors, 109
Berkson's bias (Berkson's fallacy), 118
β (type II error)
analysis of variance testing, 298
hypothesis testing, 189
effect size, 195–197
false negative rate (FNR), 195
interpretation guidelines, 201
power, 195
probability, 205Q, 206Q, 207E, 208E
uncertainty, 198–199
Bias
antecedent–consequence uncertainty, 127Q, 129E
Berkson's bias (Berkson's fallacy), 118
chance and, 4
cross-sectional studies, 342Q, 357E
defined, 4, 15Q, 16E
diagnostic suspicion bias, 117
information bias
risk assessment, 117
measurement errors and, 4, 14Q, 16E
migration bias, risk assessment, 116–117
in observational studies, 110
post-randomization, 141Q, 144E
random sampling and, 334Q, 353E
randomized controlled clinical trials, 137
examples of, 140Q, 143E
risk assessment, 115–118
sampling bias, 4, 117–118
selection bias, risk assessment, 116
surveillance bias, 117

Binomial theorem, heritability assessment, 82–84
Biologic variation, defined, 3
Blind studies, randomized controlled clinical trials, 136
See also Double-blind studies
Blocking techniques, analysis of variance testing, 299–300
Blocking variables, analysis of variance testing
CRD vs. RCB, 302
randomized complete block (RCB) design, 300
Blood type measurements, nominal scale for, 1
Bonferroni procedure
analysis of variance testing, 302–304
characteristics of, 310Q, 311E
correction factors, 336Q, 355E
Bruton's agammaglobulinemia, heritability assessment case study for, 84–86

C

Cancer staging, ordinal scale measurement, 1
Carrier status
heritability assessment, 81–82, 327Q, 348E
Bayes' rule, 85–86
genetic counseling and, 89Q, 90E
Case–control studies
absolute risk, 249Q, 253E
attributable fraction in total population (AF_T), 243–244
causality, 121
characteristics of, 343Q, 358E
chi-square testing, 219–222
matched samples, 222–223
design, 329Q, 350E
hypothesis testing, 127Q, 129E
information bias, 117
odds ratio (OR), 249Q, 253E
derivation techniques, 244–245
prevalence, 249Q, 253E
relative risk (RR), 249Q, 253E
risk assessment with, 113–114, 128Q, 130E, 330Q, 350E, 351E
advantages of, 126Q, 129E
information bias, 117
sampling bias, 117–118
selection bias, 117
risk factors, 332Q, 352E
sampling bias, 117–118
validity, 329Q, 350E
Case-fatality rate
defined, 100
example of, 107Q, 108E
Case-referent study, risk assessment, 113–114
Catagorical data, defined, 275
Catchment area, risk assessment, 114
Categorical data, decision tree, 209
Category-specific mortality rate, 99
Causal factors, risk factors as, 109
Causality, 120–122
analogy, 121
association, 120–121
coefficient of determination (r^2), 263
correlation analysis, 271Q, 274E
dose–response relationship, 121
plausibility, 121
risk factors, case–control studies, 332Q, 352E
specificity, 121
study design, 120–121
temporal relationship, 121
Cause-specific mortality rate, 99
Central limit theorem
characteristics, 161Q, 163E
population of sample means, 155
sampling distribution, 152–160
Chance
bias and, 4
defined, 15Q, 16E
Chance nodes
decision tree construction, 61
diagnostic strategy, 63–64
probability assignment, 67–68
expected utility and, 76–77Q, 79E
Charleston Heart Study
cumulative incidence studies in, 95
frequency of disease, prevalence calculations, 93, 95
Chi-square distribution, percentile table, 413
Chi-square testing
applications, 337Q, 355E
case–control studies, 330Q, 350E
categorical data, 275
critical value, 228Q, 230E
cross-sectional study, 115
defined, 212
random sampling and, 228Q, 230E
Class intervals
endpoints of, 5
frequency table and, 5–6
number of, 5
Classic regression model, 256
Clinical data, 5–10
defined, 5–11
disease occurrence and, 100
confidence interval estimation, 182Q, 185E
hypothesis testing
α, β, and power interpretation, 201
statistical significance and insignificance, 201–202
uncertainty, 198–200
numerical summaries, 7–10
measures of central tendency, 7–8
measures of spread, 8–10
pictorial description, 5–7
frequency table, 5–6
histogram, 6–7
population of sample means, 153
variation in, 3–5
accuracy, 4
bias, 4
precision, 4–5
Clinical starting point, decision tree construction, 63–64
Clinically important difference, hypothesis testing, 195
Coefficient of determination (r^2)
characteristics of, 262
definition, 262
regression analysis, 267–268, 272Q, 274E
Cohort studies
characteristics of, 342Q, 357E
chi-square testing, 216–219
matched samples, 222–223
incidence, 329Q, 350E
mortality rates, 128Q, 130E
odds ratio (OR) calculation, 244
prevalence and, 329Q, 350E
relative risk (RR), 329Q, 350E
risk assessment with, 107Q, 108E, 331Q, 351E
selection bias, 128Q, 130E
Comparative frequency, measures of association, 233
Comparative risk, 234–241
odds ratio (OR), 236–241
calculation, 236–237
case–control study
independent (unmatched) samples, 237–239
matched samples, 239–240
confidence intervals, 240–241
definition, 236
hypothesis testing, 241
interpretation, 237
study design, 237–239
relative risk (RR), 234–236
statements, 233
Comparison groups, risk assessment, 332Q, 352E
Completely random design (CRD)
analysis of variance (*F* test), 295–298
applications, 310Q, 311E
calculation formulas, 304–306
characteristics, 295
null hypothesis (H_O, 295–296
evidence against, 298
test statistic calculation, 296–297
randomized controlled clinical trials, 134
subject-to-subject variation, 298
Concurrent control group, randomized controlled clinical trials, 133
Conditional probability
calculation formula for, 21–24
in terms of joint probability, 21
summation principle, 23
defined, 20
heritability assessment, 81
measures of frequency, 233
notation for, 20, 27Q, 29E
randomized controlled clinical trial, 339Q, 356E
vs. unconditional probability, 20
Confidence coefficient, confidence intervals, 168
computational formula, 168–169
interval width, 169
matched sampling, population means, 281
means estimation, 174
single population mean estimation
known standard deviation, 169–170
unknown standard deviation, 171
population proportion, 175
two population means, 172–173
proportions, 176
Confidence intervals
applications, 169–176
mean of differences, paired study, 173–174
normally distributed variable, 180Q, 183E
single population mean estimation
known standard deviation, 169–170
proportion estimation, 174–175
unknown standard deviation, 170–171
two population means, difference estimation
proportions estimation, 175–176
unknown standard deviation, 171–173
defined, 167–169
formulas, 182Q, 185E
interval width, 180Q, 183E
matched sampling, 279

mathematical derivation of, 176–177
notation and terminology, 168–169
odds ratio (OR), 240–241
pooled *t* test, 278, 289Q, 291E
randomized controlled clinical trials, 343Q, 358E
standard deviation, 181Q, 184E
width factors, 169
Confounding variables, risk assessment, 115–118
controls for, 118–120
in observational studies, 110
Consistency, and causality, 121
Continuous data
correlation analysis, 255
data representation and organization, 255–258
decision trees, 209
t test, 275
See also Interval/ratio data
Continuous variables
random variables
characteristics, 162Q, 165E
theoretical frequency distribution, 146
risk assessment, 112
prospective cohort studies, 112
Control groups, randomized controlled clinical trial, 131
Control tables, risk assessment studies, 119
Correlation analysis, 259–263
coefficient of determination (r^2), 262–263
correlation coefficient (*r*)
calculations for, 258–260
interpretation of, 263
defined, 255
dependent and independent variables, 255–256
statistical significance of associations, 261–262
Correlation coefficient (*r*)
applications of, 334Q, 353E
association and, 273Q, 274E
characteristics of, 259, 271Q, 274E
critical values table, 423
Correlation model, 256
Count data
categorical data, 275
chi-square testing and, 212
risk factor–disease association, 111
Covariates, risk assessment studies, 120
Critical value
analysis of variance testing
Bonferroni procedure, 304
completely random design (CRD), 298
chi-square testing, 228Q, 230E
correlation coefficient, table, 423
probability theory and, 19
quantitative diagnostic testing, 31
test statistic, 335Q, 354E
t test, 289Q, 291E
Cross-sectional studies
bias, 342Q, 357E
causality, 121
characteristics of, 337Q, 355E
chi-square testing, 212–216
incidence in, 250Q, 253E
odds ratio (OR), 250Q, 253E
prevalence, 250Q, 253E
relative risk (RR), 250Q, 253E
risk assessment, 115
characteristics of, 126Q, 129E
Crude mortality rate, 99
age-adjusted rates, 122–123
example of, 107Q, 108E
risk assessment studies, 120
"Culturally desirable" method of diagnostic testing, 33
Cumulative frequency
calculation and application of, 324Q, 344E
defined, 6
Cumulative incidence (CI)
calculation formula, 94–95
defined, 93–94
disease frequency measurement and, 106Q, 108E
incidence density (ID) and, 97
Cutoff value, probability theory and, 19

D

Data analysis, randomized controlled clinical trials, 136–137
Data representation
interval/ratio data, 255–258
nonlinear data, 258
notation, 255
scatter diagram, 256–258
study design, 255–256
tabular representation, 256
Decision analysis
advantages of, 61
decision tree construction, 61–62
defined, 61
expected utility determination, 62
optimal diagnostic strategy selection, 63–73
clinical summary, 63
decision tree construction, 63–64
decision tree pruning, 64–66
probability selection and assignment, 64–69
sensitivity analysis, 73
threshold analysis, 73
2 × 2 table construction, 66
utility assignment, 67–69
probability assignment, 62
properties, 61
sensitivity analysis and, 62
threshold analysis and, 63
utility assignment, 62
Decision rule
analysis of variance testing
Bonferroni procedure, 303
completely random design (CRD), 298
randomized complete block (RCB) design, 301
chi-square testing
cohort study design, 218–219
cross-sectional study design, 215
matched (paired) samples, 223
hypothesis testing, 188
effect size, 197
null hypothesis (H_O), 192
positivity criterion, 194
p-value, 189
level of significance (α level), 189
null hypothesis (H_O), 192
correlation analysis, 261–262
regression analysis, 266–267
test statistic
population means, 281
t distribution, 277
Decision tree
analysis of variance testing, completely random design (CRD), 295
association measurement, 233–234
chi-square testing
case–control study design, 219–220
cohort study design, 217–218
cross-sectional study design, 213–214
matched (paired) samples, 224
construction, 61–62, 340Q–341Q, 356E
diagnostic strategy and, 63–64
probability assignment, 76–77Q, 79E
correlation or regression analysis, 256
decision analysis, expected utilities, 71–72
pruning, 62–65
statistical test selection, 209–212
Degrees of freedom
analysis of variance testing
Bonferroni procedure, 303
completely random design (CRD), 297, 310Q, 311E
chi-square testing
cross-sectional study design, 215
test statistic and, 228Q, 229Q, 230E
confidence interval estimation, 170–171
Denominator
analysis of variance testing, 296
measures of frequency, 233
Dependent (response) variable
correlation or regression analysis, 255–256
randomized controlled clinical trials, 132
Descriptive studies
risk assessment, 111
selection criteria, 10–11
Diagnostic and Statistical Manual of Mental Disorders (DSM-III-R), 1
Diagnostic hypothesis, decision analysis and, 63
Diagnostic method. *See* Predictive value method
Diagnostic strategy
decision analysis and, 63–73
clinical summary, 63
decision tree construction, 63–64
probability assignment, 64–69
sensitivity analysis, 73
threshold analysis, 73
utility assignment, 69–72
Diagnostic suspicion bias
prospective cohort studies, 117
risk assessment, 117
Diagnostic testing
performance evaluation, 340Q, 357E
accuracy of, 2 × 2 table technique and, 18–19
decision analysis, 64, 66
false negative rate (FNR), 36–38
false positive rate (FPR), 36–37
gold standard for, 34
graphic and tabular representation of results, 34–36
multiple testing, 43–45
new test procedures, 42–43
normal vs. abnormal criteria, 31–33
predictive value negative (PVN), 38
predictive value positive (PVP), 37–38
prevalence, 37
revision of estimates, 39–42
back calculation method, 39–40
Bayes' rule method, 41–42
prevalence estimation, 39
prevalence, sensitivity and predictive ability, 40
sensitivity analysis, 39

sensitivity, 36, 38
specificity, 36–39
positivity criterion, 340Q, 356E
predictive value positive (PVP), 103, 326Q, 347E
probability theory and, 28Q, 29E
qualitative tests, defined, 31
quantitative tests, defined, 31
vs. statistical tests, 193
Dichotomous variables
chi-square testing, matched (paired) samples, 223
risk assessment, 111–112
prospective cohort studies, 111
Dimensionless value, correlation coefficient as, 259
Direct estimate, risk assessment, 112
Disease occurrence
defined, 100
frequency measures of, 109
probability characteristics and, 20
Disease-free population
gaussian distribution with, 32
gold standard procedure for, 28Q, 29E
Dispersion. *See* Measures of spread
Distribution
skewed, histograms, 6
symmetric, histograms, 6
Dose–response relationship, causality, 121
Double-blind studies
error rates, 338Q, 355E
placebo effect, 141Q, 144E
morbidity and mortality rates with, 333Q, 353E
randomized controlled clinical trials, 136
Duchenne muscular dystrophy, heritability assessment, 89Q, 90E

E

Ecological fallacy, 263
Effect size
hypothesis testing, 195–197, 205Q, 207E
uncertainty manipulation, 199–200
power of test, 339Q, 355E
t tests, 282
power of test, 287Q, 291E
Efficacious treatment, randomized controlled clinical trials, 132
Empirical distribution, characteristics, 162Q, 165E
Empirical frequency distribution, 145–146
Empirical probabilities, gaussian distribution, 152
Environmental risk factors, 109
Epidemiological studies, dependent and independent variables, 255
Error probability
analysis of variance (*F* test), 294
completely random design (CRD), 298
randomized complete block (RCB) design, 301
chi-square testing, 229Q, 230E
cohort study design, 219
cross-sectional study design, 216
correlation analysis, 261–262
double-blind studies, 338Q, 355E
Estimates, in decision analysis, 62
Etiologic fraction in exposed, *see* Attributable fraction in exposed (AF_E)
Excess risk. *See* Attributable risk
Expected frequency, chi-square testing
cohort study design, 218
cross-sectional study design, 213
Expected utility
decision analysis, 62
diagnostic strategies, 70–72
decision tree construction, 76–77Q, 79E, 340Q–341Q, 356E
Experimental error, analysis of variance testing
calculation formulas, 306
randomized complete block (RCB) design, 298–299
Experimental studies
analysis of variance (*F* test), 293–294
randomized complete block (RCB) design, 299
chi-square testing, 216–219
matched samples, 222–223
decision trees, 209
dependent and independent variables, 255
matched sampling, population means testing, 279
odds ratio (OR) calculation, 244
vs. prospective cohort studies, 140Q, 143E
randomized block design, 142Q, 144E
risk assessment in, 109–110
Experimental units, analysis of variance (*F* test), 294
Explanatory trials, randomized controlled clinical trials, 136
Exposure determination
attributable fraction in total population (AF_T), 243
risk assessment, case–control study, 114
External validity, risk assessment, 117–118
Eye color, nominal scale measurement, 1

F

False negative rate (FNR)
calculation formulas, 36, 325Q, 346E
defined, 36
hypothesis testing, 194–195
notation for, 36
positivity criterion selection, 53
prevalence, 326Q, 347E
receiver operator characteristic (ROC) curve, 55–56
sensitivity and, 38
False negatives
defined, 35
diagnostic testing, gold standard procedure, 34
False positive rate (FPR)
alternative definitions, 37
calculation formula, 37, 325Q, 346E
defined, 37
hypothesis testing, 192–194
notation, 37
predictive value method, 53
prevalence, 326Q, 347E
receiver operator characteristic (ROC) curve, 55–56
specificity and, 38–39
False positives
defined, 35
diagnostic testing, gold standard procedure, 34
F distribution
analysis of variance testing
completely random design (CRD), 298
randomized complete block (RCB) design, 301
tables, 421–422
Fixed cohort, cumulative incidence (CI) and, 93–94
Fixed numbers, risk assessment, 111–112
Fluctuating disease severity, 131
Folding back, decision analysis, 70
Force of morbidity or mortality. *See* Incidence density (ID)
Frequency
cumulative, defined, 6
defined, 6
expressions of, 6
relative, defined, 6
Frequency (count) data
associations involving, 212–225
chi-square tests, 212
case–control studies, 219–222
cohort or experimental studies, 216–219
cross-sectional studies, 212–216
matched sample studies, 222–223
negative results, 225
one-tailed vs. two-tailed hypothesis testing, 223–225
validity, 225
correlation analysis, 255
decision tree, 209
ordinal data, 10–11
Frequency distribution
analysis of variance testing
completely random design (CRD), 298
randomized complete block (RCB) design, 301
calculation and application of, 324Q, 344E
characteristics, 162Q, 165E
chi-square testing, 215
confidence intervals, 167–168
diagnostic testing, 43, 49Q, 50E
empirical, 145
frequency table and, 5
null hypothesis (H_O), in regression analysis, 266–267
population of sample means, 153–154
probability representation, 157–158
precision and, 5
probability and, 17–18
skewed distributions, 14Q, 16E
statistics, 145–146
test statistic
correlation analysis, 261–262
population means, 281
t distribution, 277
theoretical, 145, 324Q, 345E
Frequency of disease, 93–103. *See also* Measures of frequency
Frequency of observation, descriptive statistics, 10
Frequency table, class intervals, 5–6
F statistic, analysis of variance testing, 297
F value, analysis of variance testing
completely random design (CRD), 310Q, 311E
randomized complete block (RCB) design, 309Q, 311E

G

Gaussian distribution
advantages of, 32
curve area, 147–148, 414–415
diagnostic testing, 49Q, 50E
limitations of, 32
limits of, 340Q, 356E
probabilities, 148–152
empirical vs. theoretical, 152
procedure for, 31–32
properties, 31–32
random sampling, 326Q, 348E
shape, 147
skewedness, 161Q, 163E
standard deviation and, 334Q, 353E
statistical applications, 147
validity, 327Q, 348E
Genetic counseling
binomial theorem and, 83–84
probability assessment, 81–82
recessive traits, 89Q, 90–91E
risk factors, 109
Gold standard
diagnostic testing
new test evaluation, 42
performance evaluation, 34
probability theory and, 18–19, 27Q, 29E
Grand total, in 2 × 2 table, 18
Graphic representation
diagnostic testing, 34–36
sensitivity analysis, 73
threshold analysis, 73

H

Hawthorne effect
randomized controlled clinical trials, 141Q, 143E
uncontrolled clinical trials, 131
Hazard rate. *See* Incidence density (ID)
Health records, disease occurrence and, 100
Health services, disease frequency measurement and, 102
Healthy worker effect, risk assessment, 116
Heritability assessment
Bayes' rule, 84–87
binomial formula, 82–84
conditional and joint probability, 81
Hill's criteria, causality, 120–121
Histogram, 6–7
Historical (nonconcurrent) control group, 133
Historical cohort study
observational study, 142Q, 144E
risk assessment, 112–113
Hypothesis testing
analysis of variance testing
Bonferroni procedure, 303
randomized complete block (RCB) design, 300–301
case–control study, 127Q, 129E
chi-square testing
case–control study design, 221–222
cohort study design, 216–218
cross-sectional study design, 212–213
matched samples, 222–223
one-tailed vs. two-tailed testing, 223–225
clinical research studies
design and interpretation, 198–202
completely random design (CRD), 295–296
inferential statistics, 187
matched sampling
population means testing, 279–281
odds ratio (OR), 241
population means
pooled *t* test, 275–278
procedures in, 192
randomized controlled clinical trials, 132
statistical reasoning, 187–189
clinical example, 189–192
uncertainty quantification, 192–198
effect size, 195–197
false negative errors, 194–195
false positive errors, 192–194
power, 195
test outcome summary, 197–198

I

Incidence
absolute risk ratios, 248Q, 252E
characteristics of, 342Q, 357E
cohort studies, 329Q, 331Q, 350E, 351E
cross-sectional studies, 250Q, 253E
disease occurrence, 100
prediction, 342Q, 357E
risk factors and, 112
disease prevention and control measures, 103
drug therapy and, 327Q, 349E
example of, 107Q, 108E
frequency of disease and, 93–98
attack rate, 100
case-fatality rate, 100
cumulative incidence (CI), 93–95
incidence density (ID), 95–97
morbidity rates, 98
mortality rates, 98–99
prevalence and, 97–98
relationship of CI to ID, 97
hypothesis testing and, 127Q, 129E
prevalence and, 97–98
risk assessment
case–control studies, sampling bias, 117–118
prospective cohort studies, 111
risk quantification, 251Q, 254E
time of onset, 100–101
Incidence density (ID)
calculation formula, 95–96
cumulative incidence (CI) and, 97
defined, 95
example of, 107Q, 108E
Incidence rate. *See* Incidence density
Independent events, multiplication rule of probability, 21, 24
Independent sampling
odds ratio (OR)
calculation formula, 236–237
case–control study, 239
pooled *t* test, 288Q, 291E
population means, 275–278
Independent variables
correlation or regression analysis, 255–256
randomized controlled clinical trials, 132
uncontrolled clinical trials, 132
Inferential statistics, hypothesis testing, 187
Information bias
case–control studies, 117
information bias, 117
Internal validity
randomized controlled clinical trials, 141Q, 143E
risk assessment, 115–118
Interval scale, defined, 2
Interval/ratio data
associations and, 255–259
correlation analysis, 255
data representation and organization, 255–258
confidence interval estimation
single population proportion, 175
statistical methods, 11
t test, 275
scale
decision trees, 209
quantitative variables and, 2
risk assessment, 112
variables, 14Q, 16E
Intra-arterial cannul, biased measurement and, 4

J

Joint probability
calculation formulas for, 21–23
addition rule, 22–23
in terms of conditional probability, 21
summation principle, 23
defined, 19–20
heritability assessment, 81
notation, 19–20, 27Q, 29E

L

Level of confidence, confidence intervals, 168
Level of significance (α)
correlation coefficient (*r*), 423
hypothesis testing
effect size, 196–197
error probability, 189
interpretation guidelines, 201
uncertainty, 198–200
paired *t* test, 283
pooled *t* test, 282
randomized controlled clinical trial, 339Q, 356E
randomized complete block (RCB) design, 336Q, 355E
statistical significance and, 336Q, 354E
t tests
critical values, 289Q, 291E
sample size, 281
validity, 335Q, 354E
Linear data
correlation coefficient (*r*), 257–259
regression analysis, 257–258, 338Q, 355E
Linear relationships, correlation or regression analysis, 257–258
Literature values, decision analysis, 62

M

Management trials, randomized controlled clinical trials, 136
Marginal totals, 2 × 2 table, 18
Markers, risk factors as, 109
Matched (paired) sampling
chi-square testing, 222–223
odds ratio (OR)
calculation formula, 237
case–control study, 239–240

population means, 278–281
randomized complete block (RCB) design, 298–299
Matching cases and controls
risk assessment, 118–119
historical cohort study, 114
control of variables and bias, 118–120
variables
case–control studies, 128Q, 130E
risk assessment, 118–119
McNemar's test, chi-square testing, 222–223
Mean
accuracy in clinical measurements and, 4
calculation formulas, 323Q, 344E
confidence interval estimation, 180Q, 182Q, 183E, 185E
paired studies, 173–174
defined, 7
matched sampling, 279–280
observed values, 14Q, 16E
population of sample mean, 153, 161Q, 163E
probability and, 162Q, 164E
survival rates with, 343Q, 358E
uncontrolled clinical trials, 131–132
See also Population mean
Mean square for error (MSE)
analysis of variance testing
calculation formulas, 305
completely random design (CRD), 297
Mean square for treatment (MST)
analysis of variance testing
calculation formulas, 305
completely random design (CRD), 297
Measurement process, bias and, 4
Measurement scales, 1–2
interval scale (interval data and interval/ratio scale), 2
nominal scale, 1
ordinal scale, 1
ratio scale, 2
Measurement variation (measurement error), 3
Measures of association, 234–241
odds ratio (OR), 236–241
calculation, 236–237
case–control study
independent (unmatched) samples, 239
matched samples, 239–240
confidence intervals, 240–241
definition, 236
hypothesis testing, 241
interpretation, 237
study design, 237–239
relative risk (RR), 234–236, 248Q, 252E
risk factor and disease, 334Q, 354E
Measures of central tendency
calculations and applications, 323Q, 344E
interval/ratio data and, 11
median, 14Q, 16E
numerical summaries and, 7–8
mean, 7
median, 7–8
mode, 8
Measures of frequency
applications of, 102–103
calculation formulas, 93–100
definition, 93–100
evaluation of, 100–102
disease occurrence definition, 100
population definition, 101
recurrence assessment, 100–101
incidence
case-fatality rate, 100
cumulative incidence (CI), 93–95
incidence density (ID), 95–97
morbidity rate, 98
mortality rate, 98–99
prevalence and, 97–98
relationship of CI to ID, 97
specialized measures of, 98–100
prevalence, 93
risk assessment, 233
Measures of potential impact, 241–244
attributable fraction in exposed (AF_E), 242–243
attributable fraction in total population (AF_T), 243–244
attributable risk, 241–242
population attributable risk (PAR), 242
risk assessment, 234
Measures of spread
numerical summaries, 8–10
range, 8
standard deviation, 9
variance, 8–9
Median
advantages of, 14Q, 16E
applications of, 323Q, 344E
calculation formulas, 323Q, 344E
defined, 7–8
ordinal data, 10–11
survival rates with, 343Q, 358E
Method of least squares, regression analysis, 264–265
Migration bias
randomized controlled clinical trials, 134, 136
risk assessment, 118
prospective cohort studies, 116–117
Mode
calculation formulas, 323Q, 344E
defined, 8
descriptive statistics, 10
ordinal data, 10–11
Morbidity rates
defined, 98
double blind studies, 333Q, 353E
incidence density (ID) and, 95
Mortality rate
age-adjusted, 99
age-specific, 99
category-specific, 99
cause-specific, 99
cohort studies and, 128Q, 130E
crude, 99
defined, 98–99
double blind studies, 333Q, 353E
incidence density (ID) and, 95
Multidimensional utility measures, 62
Multiple comparison groups, analysis of variance testing, 293, 302–304
Bonferroni procedure, 302–304
completely random design (CRD), 298
randomized complete block (RCB) design, 301
Multiple comparisons, randomized controlled clinical trials, 137
Multiple control groups, risk assessment, 114
Multiple testing
diagnostic testing
parallel testing, 43–44, 49Q, 51E
serial testing, 43–44, 49Q, 51E
Gaussian distribution with, 32
Multiplication rule of probability, 21–22
formula for, 23–24
independent events, 21
Multivariate regression, risk assessment studies, 120
Mutually exclusive events
addition rule for, 22–23
calculation formula for, 24

N

Natural matching, analysis of variance testing, 300
Negative results
chi-square testing, 225
hypothesis testing, statistical significance, 202
Nodes
decision tree construction, 61–62
diagnostic strategy, 64
Nominal data
description of, 10
qualitative variables, 2
Nominal scales
confidence interval estimation, 174–175
decision trees, 209
defined, 1
examples, 1
Nominal variables, 14Q, 16E
Nonconcurrent cohort study, risk assessment, 112–113
Nonindependent events, multiplication rule of probability formula for, 23
Nonlinear data
correlation coefficient (*r*), 259
regression analysis, 258
Nonresponding subjects, Berkson's bias (Berkson's fallacy), 118
Normal distribution. *See* Gaussian distribution
Normal curve areas, 414–415
Null hypothesis (H_O)
analysis of variance testing
completely random design (CRD), 296–297, 298
randomized complete block (RCB) design, 301
chi-square testing
case–control study design, 221–222
cohort study design, 218–219
cross-sectional study design, 212–213
evidence against, 215
matched (paired) samples, 223
correlation analysis, 261–262
cross-sectional studies, 337Q, 355E
evidence evaluation, 188, 190–191
hypothesis testing, *p*-value, 189
odds ratio (OR), 241
pooled *t* test, 277–278
population means, matched sampling, 281
power characteristics, 195
random sampling and, 228Q, 230E
regression analysis, 266–267, 272Q, 274E
relative frequency distribution, 188
statistical significance, 187, 204–205Q, 207E, 334Q, 354E
test statistics and critical values, 335Q, 354E
t tests
critical values, 289Q, 291E
power of test, 287Q, 291E
Numerical summaries
clinical data, 7–10

measures of central tendency, 7–8
mean, 7
median, 7–8
mode, 8
measures of spread, 8–10
range, 8
standard deviation, 9–10
variance, 8–9

O

Observational studies
analysis of variance (*F* test), 293–294
randomized complete block (RCB) design, 299
risk assessment, 109–110
characteristics of, 126Q, 129E
Observational studies
decision trees, 209
historical cohort study, 142Q, 144E
vs. randomized controlled clinical trials, 141Q, 144E
Observed frequencies, chi-square testing, 213
Odds ratio (OR)
association measurement and, 334Q, 354E
calculation formula, 236–237
independent samples, 236–237
matched (paired) samples, 237
case–control studies, 244–245, 249Q, 253E, 330Q, 350E
cases and controls, 328Q, 349E
confidence intervals, 240–241
cross-sectional studies, 115, 250Q, 253E
definition, 236
formula for, 111
hypothesis testing, 241
incidence, 251Q, 254E
interpretation of results, 237
probability vs. odds, 236
relative risk and, 248Q, 252E
study design and, 237–239
case–control studies, 238–239
One-tailed hypothesis testing
applications, 335Q, 354E
chi-square testing, 223–225
regression analysis, 266–267
Opposing hypotheses, 190
Ordinal data
confidence interval estimation, 175
statistical methods for, 10–11, 140Q, 16E
Ordinal scale
decision trees, 209
defined, 1
quantitative variables and, 2
Outcome variable, risk assessment, 110
Overmatching, risk assessment studies, 119

P

Paired *t* test
applications of, 290Q, 291E
clinical trials, 335Q, 354E
matched sampling, population means, 278–281
sample size, 283–284
Paired studies, confidence interval estimation, 173–174
Parallel testing, diagnostic testing, 43–44, 49Q, 51E
Parameters
confidence interval estimation
single population mean estimation, 171
single population proportion, 175
two population means, 171–172
two population proportions, 176
means estimation, paired studies, 174
statistics and, 145
Parent population
non-normal distribution, 155
normal distribution, 155
sampling distribution, 153
Participant selection, randomized controlled clinical trials, 133
Patient compliance disparities, randomized controlled clinical trials, 138
Patient status, ordinal scale measurement, 1
Pearson's product moment correlation coefficient (Pearson's *r*), 255
Percentile method, positivity criterion and, 32–33
Period referent, cumulative incidence (CI) and, 94
Person-time, defined, 96–97
Pictorial descriptions, for clinical data, 5–7
frequency table, 5–6
histogram, 6–7
Placebo effect
double-blind studies, 141Q, 144E
randomized controlled clinical trials, 133–134
Plausibility, and causality, 121
Point estimate, confidence intervals, 167
computational formula, 168–169
means estimation, paired studies, 174
single population mean estimation, 171
two population proportions, 176
Pooled *t* test
independent sampling, 288Q, 291E
population means, 275–278
sample size, 282
tables, 419–420
Pooled estimates, confidence interval estimation, two population means, 172
Pooled standard deviation, pooled *t* test, 277
Pooled variance
confidence interval estimation, 172
pooled *t* test, 277
Population
defining, disease frequency measurement and, 101–102
parameters and, 145
probability theory and, 17
Population attributable fraction. *See* Attributable fraction in total population (AF_T)
Population attributable risk (PAR), 242
Population attributable risk percent. *See* Attributable fraction in total population (AF_T)
Population correlation coefficient, correlation analysis
defined, 259
power of test, 262
Population mean
confidence interval estimation
formulas, 182Q, 185E
normally distributed variable, 180Q, 183E
single population
known standard deviation, 169–170
unknown standard deviation, 170–171
two population means, unknown standard deviations, 171–172
differences among, 275
gaussian distribution, 147
hypothesis testing, effect size, 197
independent samples, 275–278
difference between two means, 278
hypothesis testing, 275–278
pooled *t* test, 275
matched samples, 278–281
difference between two means, 281
hypothesis testing, 279–281
paired *t* test, 278–279
randomized controlled clinical trial, 339Q, 356E
sampling distribution, 153, 161Q, 163E
sample size, 281–284
paired *t* test, 283–284
pooled *t* test, 282
statistics and, 145
theoretical frequency, 324Q, 345E
Population standard deviation, 145
Positivity criterion
diagnostic testing, 340Q, 356E
hypothesis testing, 193–194
effect size, 197
uncertainty, 198–200
predictive value method, 53–54
quantitative diagnostic testing, 31
receiver operator characteristic (ROC) curve, 55–56
selection criteria for, 31–33
"culturally desirable" method, 33
diagnostic or predictive value method, 33
gaussian distribution, 31–32
percentile method, 32–33
risk factor method, 33
therapeutic method, 33
Post-hoc comparisons, randomized controlled clinical trials, 137
Post-randomization bias, randomized controlled clinical trials, 134, 141Q, 144E
Post-stratification, randomized controlled clinical trials, 137
Posterior probability.
carrier status, 20, 327Q, 348E
See also Predictive value positive (PVP)
Posttest probability. *See* Predictive value positive (PVP)
Power of test concept
double-blind studies, 338Q, 355E
hypothesis testing, 195
effect size, 196–197
interpretation guidelines, 201
null hypothesis (H_O), 205Q, 207E
probability, 206Q, 208E
uncertainty, 198–199
t tests
effect size, 287Q, 291E
sample size, 281
chi-square testing, 225
cross-sectional study design, 216
correlation analysis, 262
effect size, 339Q, 355E
paired *t* test, 283
pooled *t* test, 282

statistical significance and, 338Q, 355E
Precision
defined, 4–5, 15Q, 16E
diagnostic testing evaluation, 42
matched sampling, population means testing, 279
Predictable improvement, uncontrolled clinical trials, 131
Predictive ability, diagnostic testing, 40
Predictive value method
normality defined, 59Q, 60E
positivity criterion, 33
disadvantages, 54
false positive–false negative trade-offs, 53
selection, 53–54
sensitivity and PVN increase, 53
specificity and PVP increase, 53–54
receiver operator characteristic (ROC) curve, 55–56
Predictive value negative (PVN)
Bayes' rule calculations, 41
calculation formula, 38
defined, 38
notation, 38
prevalence and, 40
sensitivity and, positivity criterion selection, 53–54
Predictive value positive (PVP)
Bayes' Rule
calculations, 41
derivation, 412
calculation formulas, 37–38, 325Q, 346E
defined, 37
diagnostic testing, 49Q, 50E, 326Q, 347E
performance evaluation, 340Q, 357E
multiple testing, 43–44
notation, 37
prevalence and, 40
sensitivity
specificity and, 326Q, 347E
positivity criterion selection, 53–54
prevalence and, 41–42
Pretest probability. *See* Prevalence
Prevalence
calculation formulas, 37, 325Q, 346E
frequency of disease, 93
case–control studies, 249Q, 253E
sampling bias, 117–118
characteristics of, 342Q, 357E
cohort studies, 329Q, 331Q, 350E, 351E
cross-sectional studies, 250Q, 253E
defined, 37
diagnostic testing, 40
disease frequency measurement and, 93–94, 106Q, 108E
health care service assessment, 102–103
risk estimation with, 102
disease occurrence, 100
disease prediction, 342Q, 357E
false negative and positive rates, 326Q, 347E
heritability and, 327Q, 348E
hypothesis testing and, 127Q, 129E
incidence and, 97–98
notation, 37
predictive value of diagnostic testing and, 103
revised estimates, 39
risk assessment, 115
case–control studies, sampling bias, 117–118
signs and symptoms testing, 48Q, 50E
Prior probability. *See* Prevalence
Probability
addition rule, 24
assignment
chance node asssignment, 67–68
diagnostic strategy, 64
calculation formulas, 21–24, 325Q, 346E
conditional probability in terms of joint probability, 21
joint probability
in terms of conditional probability, 21–22
using addition rule, 22–23
summation principle
for conditional probability, 23
for joint probability, 23
carrier status, 327Q, 348E
confidence intervals, 167–168
decision analysis, 62
defined, 17
diagnostic testing sensitivity and specificity, 49Q, 50E
example of, 17
frequency distribution and, 324Q, 345E
gaussian distribution, 148–152, 155–159
genetic counseling, 81–82
heritability, 327Q, 348E
Bayes' rule, 84–87
binomial formula, 82–84
conditional and joint probability assessment, 81–82
probability of outcomes, 83
hypothesis testing errors, 188–189
level of significance (α), 336Q, 354E
multiplication rule, 21–22
formula for, 23–24
notation for, gold standard procedure, 27–28Q, 29E
of events, formula for, 23
properties, 17
random sampling, 162Q, 164E
randomized controlled clinical trial, 339Q, 356E
rules of, 19–23
conditional probabilities, 20
joint probabilities, 19–20
sample problems in, 17–18
sensitivity and specificity, 326Q, 347E
signs and symptoms diagnostic testing, 48Q, 50E
standard deviation, 162Q, 165E
statistical tests, null hypothesis, 204Q–205Q, 207E
theoretical frequency, 324Q, 345E
2 × 2 table and, 18–19
value, 17
Probability density function, 146
Proportion
chi-square testing, 212
case-controlled study design, 221
cohort study design, 216–218
confidence interval estimation
formulas, 182Q, 185E
single population estimation, 174–175
two population proportions, 175–176
decision tree construction, 340Q–341Q, 356E
normal and abnormal ranges, 326Q, 348E
Prospective cohort studies
case–control study compared with, 113–114
characteristics of, 127Q, 130E
vs. experimental studies, 140Q, 143E
risk assessment, 110
information bias, 117
selection bias, 116–117
Psychiatric diagnosis, nominal scale measurement and, 1
p-value
analysis of variance testing
completely random design (CRD), 298
randomized complete block (RCB) design, 301
case–control studies, 330Q, 350E
chi-square testing, 228Q, 229Q, 230E, 231E
cohort study design, 218–219
cross-sectional study design, 215
matched (paired) samples, 223
power of test, 225
correlation analysis, 271Q, 274E
hypothesis testing
level of significance (α level), 189
statistical significance, 201–202
statistical tests, 206Q, 208E
null hypothesis (H_O)
correlation analysis, 261–262
regression analysis, 266–267
test statistic, 281
t tests
sample size, 287Q, 291E
test statistics, 288Q, 291E

Q

Qualitative variables, defined, 2
Quality dimension, decision analysis, 62
Quantitative measurements
decision trees, 209
measures of association, 233
Quantitative variables, defined, 2
Quantity dimension, decision analysis, 62
Questionnaires
disease occurrence and, 100–101
statistical significance and, 342Q, 357E

R

Race variables, nominal scale measurement, 1
Random allocation, randomized controlled clinical trials, 134
Random chance, randomized controlled clinical trials, 136
Random effect, biologic variation, 3
Random measurement error, 3
Random number generator, randomized controlled clinical trials, 134
Random number table, randomized controlled clinical trials, 134
Random numbers, risk assessment, 111–112
Random sampling
bias and, 334Q, 353E
confidence interval estimation, 181Q, 182Q, 183–184E, 185E
disease frequency measurement and, 102

gaussian distribution and, 326Q, 348E
heritability assessment, binomial formula, 82–84
probability, 28Q, 30E, 162Q, 164E
risk assessment, case–control study, 112–113
Random selection
case–control studies, 330Q, 350E
probability and, 161Q, 163E
randomized complete block (RCB) design, 309Q, 311E
randomized controlled clinical trials, 133
Random variation
analysis of variance (*F* test), 294
Randomized block design
experimental study, 142Q, 144E
randomized controlled clinical trials, 134
Randomized complete block (RCB) design
analysis of variance testing, 298–302
applications of, 309Q, 311E
artificial matching, 300
calculation formulas, 306
characteristics, 298–299
experimental errors, 301–302
hypothesis testing, 300–301
natural matching, 300
self-matching, 299–300
test statistic calculation, 300–301
Randomized controlled clinical trial
comparison groups, 134, 136
hypothesis formulation, 132–133
participant selection, 133–134
procedures, 132–137
sample size, 339Q, 356E
schematic representation, 135
study vs. target populations, 138
therapy evaluation with
data analysis, 136–137
key features, 131–132
treatment administration and outcome measurement, 136
Randomized controlled clinical trials
characteristics, 332Q, 352E
confidence intervals, 343Q, 358E
hawthorne effect, 141Q, 143E
vs. observational study, 141Q, 144E
Range
calculation formulas, 323Q, 344E
normal and abnormal, 326Q, 348E
standard deviation and, 14Q, 16E
Rate difference. *See* Attributable risk
Ratio scale
defined, 2
quantitative variables and, 2
Recall bias, risk assessment, 117
Receiver operator characteristic (ROC) curve
example of, 59Q, 60E
positivity criterion selection, 55–56
Recurrence measurement, disease frequency measures, 100
Referrent value, quantitative diagnostic testing, 31
Regression analysis, 263–268
calculation formulas, 264–265
coefficient of determination (r^2), 267
adequacy of regression line, 268
defined, 255
dependent and independent variables, 256
interpretation of results, 268
linear relationship, 266, 338Q, 355E
line derivation, 263–264
nonlinear data, 258
regression line variation, 265
Relative frequency
calculation and application of, 324Q, 344E
confidence intervals, 168–169
defined, 6
gaussian distribution, 147–148
probability, 150
hypothesis testing, 188
level of significance (α level), 189
Relative risk (RR)
association measurement and, 334Q, 354E
calculation formula, 234–235, 329Q, 350E
case–control studies, 249Q, 253E, 330Q, 350E
characteristics of, 248Q, 252E
cohort studies, 329Q, 331Q, 350E, 351E
cross-sectional studies, 250Q, 253E
defined, 234
examples of, 235–236, 249Q, 252E
formula for, 111
incidence, 251Q, 254E
odds ratio (OR) and, 248Q, 252E
study designs for, 235
Reliability, in diagnostic testing, 42
Representation, risk assessment, 111–112
Reproducibility. *See* Precision
Research hypothesis, 187
Residual sum of squares, regression analysis, 264–265
Response variable, risk assessment, 110–111
Restriction, risk assessment, 118–120
Retrospective cohort study, risk assessment, 112–114
Risk
characteristics off, 109–120
measures of, 233–234
Risk assessment
advantages, 112
case-control studies, 330Q, 350E
cases and controls, 328Q, 349E
cohort studies, 331Q, 351E, 352E
confounding and bias sources, 115–118
Berkson's bias (Berkson's fallacy), 118
control of, 118–120
information bias, 117
migration bias, 116–117
sampling bias, 117–118
selection bias, 116–117
decision analysis and, 63
direct estimate, 112
disadvantages, 112
disease frequency measurement and, 102
experimental vs. observational studies, 109–110
hypothesis testing, risk of error, 192
outcome evaluation, 111–112
study designs for, 110–115
case–control study, 113–115
cross-sectional study (prevalence study), 115
historical cohort study, 112–113
prospective cohort study, 110–112
Risk difference. *See also* Attributable risk
Risk factors
case–control studies, 332Q, 352E
disease and, 109
frequency data, 212
exposure levels, 109
positivity criterion and, 33
socioeconomic status, 328Q, 349E
types of, 109
Risk quantification
magnitude of associations, 233–234
measures of association (comparative risk), 234–241
odds ratio (OR), 236–241
relative risk (RR), 234–236
measures of potential impact, 241–244
attributable fraction in exposed (AF_E), 242–243
attributable fraction in total population (AF_T), 243–244
attributable risk, 241–242
population attributable risk (PAR), 242
odds ratio (OR) in case–control studies, 244–245
Risk-free populations, disease frequency measurement and, 101

S

Sample mean, statistics and, 145
Sample size
confidence interval estimation, 180Q, 183E
hypothesis testing, uncertainty manipulation, 200
population of sample means, 155
randomized controlled clinical trials, 133, 339Q, 356E
t tests, 281–284
paired *t* test, 283–284
pooled *t* test, 282
Sample standard deviation, statistics and, 145
Sampling bias
case–control studies, 117–118
probability theory and, 17
risk assessment, 117–118
Sampling distribution
central limit theorem, 152–160
population of sample means, 153–155, 155
probability computation, 155–158
Sampling error (sampling variation)
bias and, 4
null hypothesis (H_O), 190
Sampling techniques, disease frequency measurement and, 101–102
Scales. *See also* Measurement scales
Scatter diagram, correlation or regression analysis, 256–258
Selection bias
cohort study, 128Q, 130E
risk assessment, 116–117
confounding variables as, 116
controls for, 118–120
prospective cohort studies, 116–117
random sample cases and controls, 114
Self-matching, analysis of variance testing, 299–300
Self-pairing
matched sampling, population means testing, 278
paired *t* test, 283
Self-selection
analysis of variance (*F* test), 294
chi-square testing, 216

Sensitivity
calculation formulas for, 36, 325Q, 346E
decision analysis, diagnostic strategy, 73
defined, 36, 62
diagnostic strategy and, 78Q, 79E
performance evaluation, 36–38
prevalence and predictive ability, 40
false negative rate (FNR) and, 38
hypothesis testing, power characteristics, 195
notation of, 36
positivity criterion and, 59Q, 60E
predictive value negative (PVN) and, 42
positivity criterion selection, 53
predictive value positive (PVP) and, 42
prevalence and, 326Q, 347E
receiver operator characteristic (ROC) curve, 55–56
revised estimates and, 39
Series testing, 43–44, 49Q, 51E
Sex variables, nominal scale measurement, 1
Sickle-cell anemia (SCA), heritability assessment case study, 81–82
Signs and symptoms diagnostic testing, 48Q, 50E
Simple etiologic fraction. *See* Attributable fraction in exposed (AF_E)
Single population proportions, confidence interval estimation, 174–175
Single-blind studies, randomized controlled clinical trials, 136
Single-dimensional utility measures, 62
Skewed distribution, 14Q, 16E
diagnostic testing, 49Q, 50E
gaussian distribution and, 32
histogram, 7
interval/ratio data and, 11
median values, 8
Socioeconomic status
as risk factor, 109, 328Q, 349E
stratification and, 128Q, 130E
Specificity
calculation formulas, 36, 325Q, 346E
causality, 121
defined, 36
false positive rate (FPR) and, 38–39
notation, 36
predictive value negative (PVN) and, 42
predictive value positive (PVP) and, 42
positivity criterion selection, 53
prevalence and, 326Q, 347E
receiver operator characteristic (ROC) curve, 55–56
Sphygmomanometer
biased measurement and, 4
precision in, 5
Standard deviation
calculation formulas, 9–10, 323Q, 344E
computational formula, 10
confidence interval estimation, 181Q, 182Q, 184E, 185E
formulas, 182Q, 185E
single population mean, 169–170
known deviation, 169–170
unknown deviation, 170–171
two population means, 171–172
normal distribution, 334Q, 353E
paired *t* test, 283
pooled *t* test, 277, 282
population of sample means, 153
precision and, 5
probability and, 162Q, 164E, 165E
range and, 14Q, 16E
standard normal table, 161Q, 163E
t tests, 281
units, 190
Standard error of the estimate
confidence intervals, 168
computational formula, 168–169
interval width, 169
means estimation
paired studies, 174
two populations
means, 172
proportions, 176
matched sampling, 279
regression analysis, 265
Standard error of the mean
applications, 162Q, 165E
characteristics, 162Q, 165E
population of sample means, 153
Standard normal distribution
probability representation, 158–159
z scores, 148–149
table
gaussian distribution, 148
probability, 151
standard deviation, 161Q, 163E
z scores, 149
Standard population, age-adjusted mortality rate and, 99
Standard treatment, randomized controlled clinical trials, 134
Standardization of rates, risk assessment studies, 119–120
Standardized (*z*) scores. *See* *z* scores
Statistical methods
descriptive statistics, 10
hypothesis testing, 187–189
risk assessment studies, 120
Statistical significance
analysis of variance testing, 298
case–control studies, 330Q, 350E
chi-square testing, 229Q, 230E
coefficient of determination (r^2), 263
correlation analysis, 261–262
cross-sectional studies, 337Q, 355E
hypothesis testing, 188, 201–202
levels of significance (α), 336Q, 354E
null hypothesis (H_O) and, 334Q, 354E
of associations, 212
pooled *t* test, 278
power of test, 338Q, 355E
regression analysis, 266–267
risk assessment
dichotomous variables, 111
prospective cohort studies, 111
Statistical test
hypothesis testing, α and β errors, 206Q, 208E
null hypothesis (H_O) , 204Q–205Q, 207E
Statistics
frequency distributions, 145–146
normal (gaussian) distribution, 147–152
empirical (sample) and theoretical (population) probabilities, 152
probability determination, 148
relative frequencies, 147–148, 150
shape, 147
standard normal table, 148–149
z scores, 148–149
parameters, 145
sampling distribution and central limit theorem, 152–160
population of sample means, 153–155
probabilities with sample means, 155–159
Stratification
risk assessment studies, 119
socioeconomic status, 128Q, 130E
Student *t* test. *See* Pooled *t* test
Study designs, randomized controlled clinical trials, 134
Study populations, disease frequency measurement and, 101–102
Subject migration, randomized controlled clinical trials, 136
Subject-to-subject variation
analysis of variance testing
completely random design (CRD), 295–296, 298
coefficient of determination (r^2), 262–263
Sum of probabilities, 17
Sum of squares (SS), analysis of variance testing
calculation formulas, 304–306
completely random design (CRD), 297, 310Q, 311E
Summation notation
applications, 411
defined, 411
Summation principle
calculation formula for, 24
conditional probability, 23
joint probability, 23
Surveillance bias
prospective cohort studies, 117
risk assessment, 117
Symmetric distribution
histogram, 6
interval/ratio data and, 11
median/mean values in, 8
Systematic measurement error, 3
Systematic variation
analysis of variance (*F* test), 294
coefficient of determination (r^2), 262–263
defined, 3

T

Tabulated data
age-adjusted rates, 122
correlation or regression analysis, 256–257
diagnostic testing, 34–36
Target populations, exclusion of disease frequency measurement and, 101–102
randomized controlled clinical trials, 138
t distribution
confidence interval estimation, 170–171
regression analysis, 266–267
table, 416
Temperature, ordinal scale measurement, 2
Temporal relationships, causality, 121
Test statistic
analysis of variance (*F* test), 294
completely random design (CRD), 296–297
randomized complete block (RCB) design, 300–301

chi-square testing
calculated value, 228Q, 230E
case–control study design, 221
cohort study design, 218
correlation analysis, 261–262
critical values and, 335Q, 354E
hypothesis testing
null hypothesis (H_O), 192
positivity criterion, 194
regression analysis, 266–267, 272Q, 274E
population means
frequency distribution, 277
matched sampling, 280–281
pooled *t* test, 275–276
randomized complete block (RCB) design, 309Q, 311E
t tests, 288Q, 291E
critical value, 289Q, 291E
chi-square testing
cross-sectional study design, 213
cross-sectional studies, 337Q, 355E
hypothesis testing, 190
Theoretical frequency distribution, 145–146
characteristics, 162Q, 165E
Theoretical probabilities, gaussian distribution, 152
Therapeutic method for positivity criterion, 33
Therapy evaluation, randomized controlled clinical trials, 136
Threshold analysis
defined, 62
diagnostic strategy and, 73, 78Q, 79E
Threshold effect, causality, 121
Threshold values, diagnostic strategy and, 78Q, 79E
Total etiologic fraction. *See* Attributable fraction in total population (AF_T)
Treatment administration, randomized controlled clinical trials, 136
True negative rate (TNR), diagnostic testing, 36
True negatives, defined, 35
True population mean, confidence intervals, 167–168
True positive rate (TPR), diagnostic testing, 36
True positives, defined, 35
t statistics, analysis of variance testing, 303–304
t tests
analysis of variance testing
completely random design (CRD), 310Q, 311E
multiple comparison groups, 293
population means
independent sampling, 275–278
matched sampling, 278–281
sample size and, 281–284
tables, 417–420
2 × 2 table
decision analysis, 66
diagnostic testing, 35–36
probability theory and, 18–19
risk assessment
case–control study, 114–115
dichotomous variables, 111–112
prospective cohort studies, 111
Two-tailed hypothesis testing
chi-square testing, 223–225
regression analysis, 266–267
Two-way table. *See also* 2 × 2 table

U

Uncertainty, hypothesis testing
clinical research, 198–200
quantification of, 192–198
Unconditional probability, defined, 20
Uncontrolled clinical trials
defined, 131
example of, 140Q, 143E
Utilities
decision analysis
diagnostic strategy, 69–72
probability assignment, 62
See also Expected utilities

V

Validity
case–control studies, 329Q, 350E
chi-square testing, 225
defined, 15Q, 16E
gaussian distribution, 327Q, 348E
level of significance (α), 335Q, 354E
Variables
accuracy, 4
among-patient variation, 3
basis for, 2–3
bias and, 4
chance and, 4
measurement process bias, 4
sampling error, 4
categories, 2
in clinical data, 3–5
accuracy, 4
bias, 4
precision, 4–5
defined, 1
examples of, 1
interval/ratio, 14Q, 16E
interval scale, 2
measurement scales, 1–3
nominal, 1, 14Q, 16E
ordinal, 1, 14Q, 16E
precision, 4
qualitative, 2
quantitative, 2
ratio scale, 2
systematic variation, 3
true biologic variation, 3
within-patient variation, 3
Variance
calculation of, 8–9
completely random design (CRD), 297
defined, 8
pooled, 277
ratio, analysis of variance testing, 297
Volunteer subjects
Berkson's bias (Berkson's fallacy), 118
randomized controlled clinical trials, 138
random selection, 133
uncontrolled clinical trials, 131

W

Weighted average, pooled *t* test, 277
Within-patient variation, 3
Woolf's method, odds ratios confidence intervals, 240–241
"Worst possible" factors, risk assessment studies, 120

Z

z values
gaussian distribution, 148–149
graphic representation, 149
sample mean conversion, 158
table, 414–415
Zero point
interval scale, 2
ratio scale, 2